Springer Series in Optical Sciences Volume 72

Founded by H. K. V. Lotsch

Springer

Berlin
Heidelberg
New York
Barcelona
Hong Kong
London
Milan
Paris
Singapore
Tokyo

Physics and Astronomy

ONLINE LIBRARY

http://www.springer.de/phys/

Springer Series in Optical Sciences

Editorial Board: A. L. Schawlow† A. E. Siegman T. Tamir

Volumes 1-41 are listed at the end of the book

Peter Günter (Ed.)

Nonlinear Optical Effects and Materials

With 174 Figures and 43 Tables

 Springer

Professor Dr. PETER GÜNTER

Nonlinear Optics Laboratory
Institute of Quantum Electronics
ETH Hönggerberg
8093 Zürich, Switzerland
E-mail: nlo@iqe.phys.ethz.ch

Library of Congress Cataloging-in-Publication Data

Nonlinear optical effects and materials / Peter Günter (ed.). p.cm. –
(Springer series in optical sciences, ISSN 0342-4111; v.72)
Includes bibliographical references and index.

1. Nonlinear optics.   2. Optical materials.   I. Günter, P. (Peter), 1944 –   II. Series
QC446.2.N665 2000   621.36'94 – dc21   00-021730

ISSN 0342-4111

ISBN 978-3-642-53694-6   ISBN 978-3-540-49713-4 (eBook)
DOI 10.1007/978-3-540-49713-4

Springer-Verlag is a company in the BertelsmannSpringer publishing group
© Springer-Verlag Berlin Heidelberg 2000

Softcover reprint of the hardcover 1st edition 2000

Typesetting: Data conversion by Steingraeber Satztechnik GmbH, Heidelberg
Cover concept by eStudio Calamar Steinen using a background picture from The Optics Project.
Courtesy of John T. Foley, Professor, Department of Physics and Astronomy, Mississippi State University, USA
Cover production: *design & production* GmbH, Heidelberg

Printed on acid-free paper   SPIN 10693897   56/3144/di -5 4 3 2 1 0

Preface

It is now well established that a unique feature of coherent optical beams is their ability to transmit, process, store and interconnect in parallel a large number of high bandwidth information channels. However, although these techniques possess great potential their development depends critically on the nonlinear optical effects used and on the availability of nonlinear optical materials that work at high speed and low incident optical power. At present, these requirements are stimulating a great deal of research in materials science and are challenging existing technologies, in particular high speed electronics.

This volume devoted to nonlinear optical effects and materials presents a detailed account of selected topics in inorganic and organic materials research. The status of organic crystals and polymers for nonlinear optics is critically compared with their inorganic counterparts. The preparation techniques and a description of the methods used to characterize the nonlinear optical effects relevant for device applications are dealt with, as well as a theoretical description of the nonlinear optical, electro-optical and photorefractive effects observed. The main concepts and potential applications are outlined and developed in the various chapters of this book. This collection of articles provides a broad survey of selected research topics in organic and inorganic nonlinear optics. We hope that it will stimulate further research into the development and optimization of nonlinear optical crystals and polymers, as well as the investigation of new effects and the invention of new optical devices. The development of new materials will certainly play a role in future optical applications such as optical fiber telecommunication, electro-optics, integrated optics, laser technology, and in the processing of large bit arrays and high density data storage.

Finally we would like to thank Prof. A.E. Siegman and the other series editors, as well as Dr. H.K.V. Lotsch, Dr. H.J. Kölsch and Mrs. G. Dimler of Springer-Verlag who carefully revised the text and made a great effort in efficiently producing this book.

Zürich, December 1999   *Peter Günter*

Contents

Contributors

Martin Bösch
Nonlinear Optics Laboratory
Institute of Quantum Electronics
ETH Hönggerberg
CH-8093 Zürich
Switzerland
E-Mail: boesch@iqe.phys.ethz.ch

Christian Bosshard
Nonlinear Optics Laboratory
Institute of Quantum Electronics
ETH Hönggerberg
CH-8093 Zürich
Switzerland
E-Mail: bosshard@iqe.phys.ethz.ch

Markus Duelli
Nonlinear Optics Laboratory
Swiss Federal Institute of Technology
ETH Hönggerberg HPF E18
CH-8093 Zürich
Switzerland

Daniel Fluck
Nonlinear Optics Laboratory
Institute of Quantum Electronics
ETH Hönggerberg
CH-8093 Zürich
Switzerland
E-Mail: fluck@iqe.phys.ethz.ch

Peter Günter
Nonlinear Optics Laboratory
Institute of Quantum Electronics
ETH Hönggerberg
CH-8093 Zürich
Switzerland
E-Mail: nlo@iqe.phys.ethz.ch

Matthias Jäger
Nonlinear Optics Laboratory
Institute of Quantum Electronics
ETH Hönggerberg
CH-8093 Zürich
Switzerland
E-Mail:
matthias-L-Jaeger@agilent.com

Ilias Liakatas
Nonlinear Optics Laboratory
Institute of Quantum Electronics
ETH Hönggerberg
CH-8093 Zürich
Switzerland
E-Mail: liakatas@iqe.phys.ethz.ch

Carolina Medrano
Nonlinear Optics Laboratory
Swiss Federal Institute of Technology
ETH Hönggerberg
CH-8093 Zürich
Switzerland
E-Mail: medrano@iqe.phys.ethz.ch

Germano Montemezzani
Nonlinear Optics Laboratory
Institute of Quantum Electronics
ETH Hönggerberg
CH-8093 Zürich
Switzerland
E-Mail:
montemezzani@iqe.phys.ethz.ch

Marko Zgonik
Department of Physics
University of Ljubljana
Jadranska 19
SI-1000 Ljubljana
Slovenia
E-Mail: zgonik@samson.fiz.uni-lj.si

Tomas Pliska
Nonlinear Optics Laboratory
Institute of Quantum Electronics
ETH Hönggerberg
CH-8093 Zürich
Switzerland
E-Mail: tomas.pliska@ch.jdsunph.com

1   Introduction

P. Günter

Much progress has been made in the last decade in using light as an information carrier. Optical telecommunication using optical fibers and optical displays using liquid crystalline or polymeric light emitting materials are the best-known examples of these fast growing areas. It is expected that optical information technology will expand rapidly at a rate that roughly doubles transport, processing and storage capacity every three years. Within a decade, information transport networks of terabits per second, storage of terabytes and ultrafast processing systems with teraoperations per second should be realized [1]. It is believed that a large fraction of the elements for these technologies will use light as the information carrier. For this reason the search for new materials and the investigation of older materials are very important tasks.

Nonlinear optical materials, defined as materials in which light waves can interact with each other, are the key materials for the fast processing of information and for dynamic or permanent optical storage applications. The field of research and development of nonlinear optical materials has progressed impressively since the invention of the laser in the late 1950s. Semiconducting materials, mainly GaAs and InP, but also inorganic and organic insulating materials, have been proposed for a series of nonlinear optical devices. Semiconductor waveguide devices are already used for modulators, switches and other electro-optical applications needed for fiber-optic telecommunication. For nonlinear optical frequency conversion and electro-optical modulation and switching, a series of polar, ferroelectric or piezoelectric oxides have been used for applications using different pump wavelengths and pump powers and for generating new optical waves in wavelength ranges extending from the ultraviolet to the far infrared. Figure 1.1 shows the figure of merit for optical frequency conversion for the most important materials already used and for those recently investigated. It can be seen from the figure that a one-million-fold increase in the figure of merit, and thus in the efficiency of the wavelength conversion, has been achieved since the first second-harmonic generation experiment was reported in quartz [2]. Subsequent experiments using KH_2PO_4 and $(NH_4)H_2PO_4$ showed an increase in efficiency of almost an order of magnitude, and the materials most commonly used nowadays ($LiNbO_3$, $KTiOPO_4$) show a 100-fold increase in the figure of merit. The more densely packed highly polarizable potassium niobate ($KNbO_3$) shows a figure of merit three orders of magnitude larger than that measured in quartz [3]. In the last few years the technology for producing controlled inhomogeneous materials with antiparallel ferroelectric domains following each other has allowed the use of the component d_{333} of the first-order nonlinear optical susceptibility tensor, which can be used for efficient frequency conversion only in these so-called periodically poled structures [4, 5]. By using this coefficient, the figure of merit in periodically poled $LiNbO_3$, $LiTaO_3$ and $KTiOPO_4$ is also increased by about an order of magnitude over the efficiency obtained by using the birefringence phase-matchable nonlinear optical susceptibility coefficients.

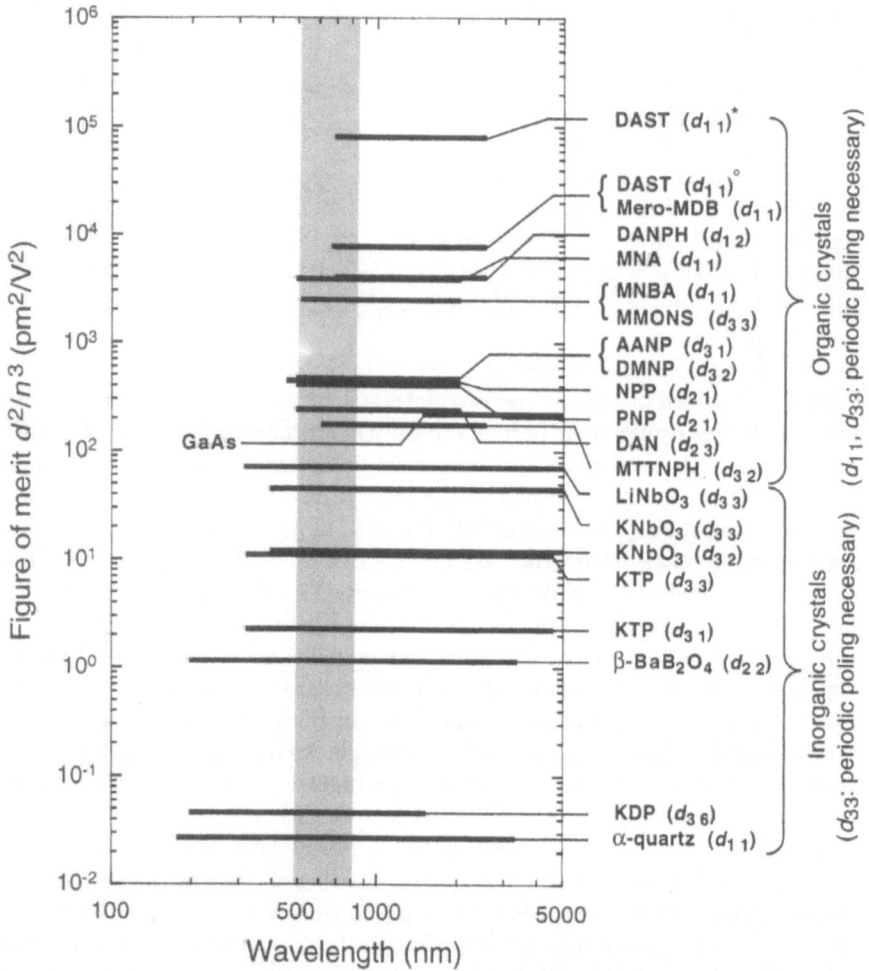

Fig. 1.1. Figure of merit d^2/n^3 for optical frequency conversion vs. transmission range of inorganic and organic nonlinear optical crystals (d: effective nonlinear optical susceptibility; n: refractive index; for the full names of organic compounds see Chap. 3; $*$: $\lambda = 1.3\,\mu\text{m}$; $\circ$: $\lambda = 1.9\,\mu\text{m}$)

It is shown in Fig. 1.1 that the figure of merit for optical frequency conversion can be further increased by almost three orders of magnitude by using organic nonlinear optical materials. The development of new organic crystals is a very challenging field of research. A multidisciplinary effort by synthetic organic chemists, crystal growers, physicists and engineers has led to materials with much improved properties. New effects with these highly nonlinear optical materials are presently being investigated and are still to be discovered in the future. The large nonlinear susceptibilities of the newest materials allow the preparation and investigation of nonlinear optical elements requiring only modest power, by using cw laser diodes as the pump sources, for example.

The nonlinear optical figure of merit, however, is only one of the criteria necessary for the application of a new material in an optical device. Ease of production or processing, the optical damage threshold, absorption losses etc., must also be considered. Some organic materials still need further improvements; purification and crystal growth still need to be developed further in many of the high optical nonlinearity materials. Therefore the field of organic nonlinear optics is a rapidly progressing and challenging topic in materials research and development.

This volume presents an overview of the development of insulating inorganic and organic solid materials. Most of the work presented here has been performed in the last few years at the Nonlinear Optics Laboratory of the Institute of Quantum Electronics at ETH. The chapters present reviews of selected topics of the present worldwide activities in the field of materials development for nonlinear optics. Figure 1.2 shows the different areas of investigation in first- and second-order nonlinear optical materials. In these materials the induced nonlinear polarization P^{NL} is proportional to either

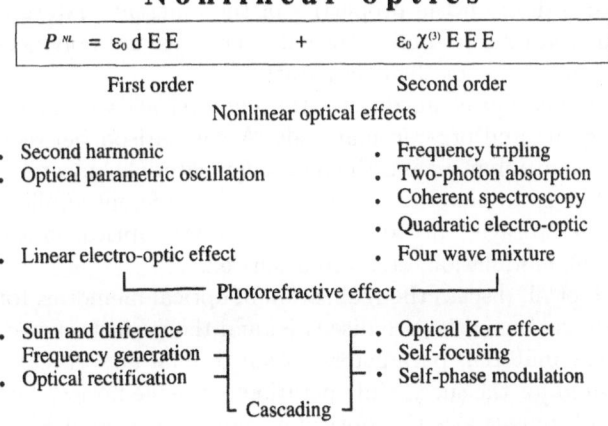

Fig. 1.2. First- and second-order nonlinear optical effects

the square or the third power of the electric field at optical or lower frequencies. Pure first-order nonlinear optical effects are listed in the first column, pure second-order effects in the last one. Inbetween are listed some effects related to both types of nonlinearity: the photorefractive effect, which is an electro-optical effect driven by photoinduced space charges, and the cascaded optical Kerr effect, where two cascaded second-order wave-mixing processes produce an optical Kerr effect in a first-order nonlinear optical material. In this volume, work on both first- and second-order nonlinear optical effects is presented. In addition, the photorefractive effect and the cascading of first-order nonlinear optical effects for applications in second-order nonlinear optics as effects connecting both types of nonlinearity are discussed.

In Chaps. 2 and 3 the preparation and investigation of organic nonlinear optical materials (molecules, crystals and polymers) for first- and second-order nonlinear optics are summarized. Chapter 2 also gives an introduction to the definitions of nonlinear optical susceptibilities commonly used for describing nonlinear optical effects in molecules, crystals and polymers and how the differently defined tensor coefficients used in the literature are interrelated. The effects are described theoretically and measurements on a wide variety of organic molecules and crystals are presented.

The second chapter also describes the main nonlinear optical techniques for measuring the properties of molecular or bulk crystals. In addition, experiments on the cascaded optical Kerr effect are discussed. In cascading, the second-order nonlinearities are used to optically induce refractive index changes by a cascading of a sum and difference frequency generation ($\omega + \omega = 2\omega$ and $2\omega - \omega = \omega$) or optical rectification and sum frequency generation (electro-optic effect; $\omega - \omega = 0$, and $0 + \omega = \omega$). It is shown that this indirect process can yield much larger nonlinearities than the optical Kerr effect does in a second-order nonlinear optical $\chi^{(3)}$ material.

Chapter 3 reviews progress in first-order nonlinear optical organic materials. It presents extensive data on the properties of the different materials synthesized and investigated to date. Polar molecules, crystals and polymers are discussed, as are the methods for their preparation.

In Chap. 4, G. Montemezzani et al. discuss the photorefractive and related properties of inorganic and organic materials. A comparison between the newer organic crystals or polymers and the inorganic materials known for a longer time is made. The main materials requirements for different applications, such as holographic storage, optical phase conjugation, optical image amplification, spatial light modulation, etc., are discussed.

In Chap. 5, M. Duelli et al. discuss the realization of optical memories for real-time processing. Several memories are discussed and the performance of the devices described. The materials parameters obtained so far are critically related to the ones required for the successful operation of the memories. The function and properties of some associative optical memories realized at ETH are discussed.

The final chapter describes the production of optical waveguides in ferroelectric materials to be used for optical frequency conversion using ion exchange, in- and outdiffusion of ions as well as ion implantation. High-energy ($> 1\,\text{MeV}$) ion implantation in $KNbO_3$ is described in detail. This process is currently the only one that allows high-quality optical waveguides to be produced for optical frequency doubling in this high optical nonlinearity material. The waveguide parameters achieved are discussed and compared with the parameters obtained in other waveguiding materials. Optical second-harmonic generation experiments in $KNbO_3$ waveguides are reviewed. This review demonstrates that direct diode laser frequency doubling with high efficiency can be achieved in $KNbO_3$ waveguides, making this material the prime candidate for the realization of a direct doubled blue light source [6].

This book constitutes a report on the present status of a selection of topics in nonlinear optical materials research. The selection is rather personal since mainly research topics within the activities of the editor's laboratory have been covered. All the areas treated, however, are areas in which active research is on-going in a series of research laboratories worldwide.

Even though some materials are already being used in nonlinear photonic devices we believe that several applications require nonlinear optical materials with even better properties. Particularly, but not exclusively, the field of organic nonlinear optical materials still has a bright future, since the potential for almost unlimited design possibilities of organic molecules represents a very interesting field of research.

I hope that this book, with its summary of the present situation in some selected topics, will encourage future materials and device research which then facilitates the realization of new nonlinear photonic systems. I hope also that this book will be useful to scientists and engineers in the different areas that represent the multidisciplinary field of nonlinear optical materials research and in the application of such materials in future devices.

References

1.1 National Research Council, *Harnessing Light: Optical Science and Engineering for the 21st Century* (National Academy Press, Washington, DC 1998).
1.2 P. Franken, A. Hall, C. Peters, G. Weinreich, Phys. Rev. Lett. **7**, 118 (1961).
1.3 P. Günter, Appl. Phys. Lett. **34**, 650 (1979).
1.4 J. A. Armstrong, N. Bloembergen, J. Ducuing, P. S. Pershan, Phys. Rev. **127**, 1918 (1962).
1.5 M. M. Fejer, G. A. Magel, D. H. Jundt, R. L. Byer, IEEE J. Quantum Electron. **28**, 2631–2654 (1992).
1.6 D. Fluck, P. Günter, Optics Commun. **136**, 257–260 (1997).

2 Third-Order Nonlinear Optics in Polar Materials

Ch. Bosshard

2.1 Introduction

2.1.1 Motivation

Nonlinear optics refers to any light-induced change in the optical properties of a material. The most commonly used nonlinear optical effects are of second order: frequency doubling and linear electro-optics. For these second-order nonlinear optical effects, which also include optical sum and difference frequency generation, optical parametric amplification, and in most cases also photorefractive effects, noncentrosymmetry is required. Otherwise they are absent.

The use of third-order optical nonlinearities for all-optical signal processing has been a goal for many years [2.1]. All-optical processes allow the manipulation of an output beam in a nonlinear material through the input signal or a separate control beam, that is, by purely optical means [2.2]. As an example, Fig. 2.1 shows a schematic of the principle of all-optical switching. All the relevant parameters for this type of operation are well established and we know that the most important factor is the light-induced nonlinear phase shift. However, it has proved difficult to find materials with large nonlinearities combined with low enough linear and nonlinear losses. There are no criteria for the design of appropriate materials with large third-order nonlinearities. Although very promising large third-order effects were reported, for example in poly[2,4-hexadiyne-1,6-diol-bis-p-toluene-sulfonate] (PTS), there still remain many problems with the optical quality [2.3–4]. Therefore it is still challenging to make high throughput all-optical switching devices responding on picosecond time-scales.

Third-order nonlinear optics imposes no symmetry requirements for the effects to occur. In this chapter the extent to which noncentrosymmetry can play a role in third-order nonlinear optics is investigated. On the one hand, the direct third-order nonlinear optical response was determined in organic molecules as well as in macroscopic crystals and a comparison with corresponding centrosymmetric materials was carried out. On the other hand, the influence of *cascaded* second-order effects [2.5–7] on third-order nonlinearities was investigated. These contributions consist of a combination of two second-order nonlinear optical processes and only occur in noncentrosymmetric media. In addition to the third-order nonlinear susceptibility $\chi^{(3)}$ symme-

(a)

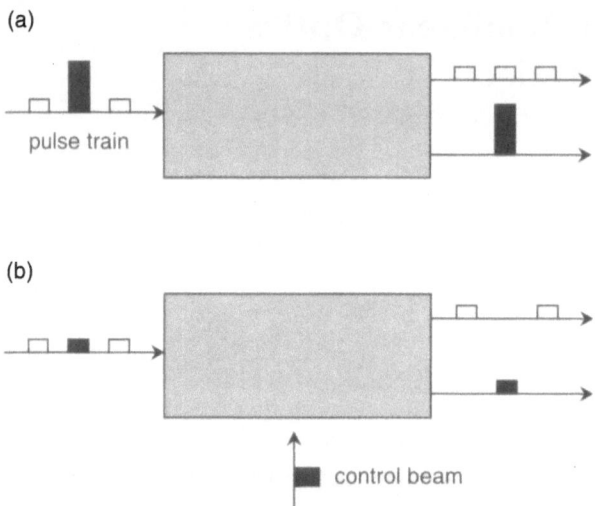

(b)

control beam

Fig. 2.1. Schematic of an all-optical switch for (**a**) self-switching and (**b**) switching induced by a control pulse. Devices with several input and output ports are possible

try allowed in all materials, they can be very important in all-optical signal processing applications. One of these cascaded effects, which has been known for a long time, leads to a nonlinear phase shift of the interacting waves in nearly phase-matched second-harmonic generation and other parametric processes [2.8–11]. Such effects were first observed in high-power pulsed second-harmonic generation experiments [2.10–11]. Another interesting process contributing to cubic nonlinear effects comes from cascaded optical rectification and the linear electro-optic effect [2.12–14].

Some years ago the possibility of increasing the magnitude of a third-order nonlinear effect by a combination of cascaded second-order effects was discovered and demonstrated by using a nonlinear transmission line resonator [2.15]. Recently, the potential for all-optical switching was realized [2.16–17] and large nonlinear phase shifts due to cascading were observed for the first time in potassium titanyl phosphate (KTP) waveguides [2.16–17]. Unexpectedly large nonlinear refractive index changes observed in fibers of 4-(N, N-dimethylamino)-3-acetamidonitrobenzene (DAN) could also be explained by cascaded nonlinearities [2.18].

For second-order optical nonlinearities the criteria for obtaining large effects are by now reasonably well understood and appropriate materials can be designed. It may therefore be that the development of materials with large second-order nonlinear optical susceptibilities $\chi^{(2)}$ will now pay off for cubic nonlinearities through the cascaded processes discussed in this chapter.

This chapter is structured as follows. First the basic concepts of nonlinear optics relevant for this work are discussed and the important material requirements are presented. Most experiments have concentrated on organic

materials which can have extremely large nonlinear optical effects. Therefore the basic molecular units for second- and third-order nonlinear optics
will be presented along with the different energy levels needed to describe
the wavelength dispersion of the effects. Based on this discussion, the maximum macroscopic nonlinear optical coefficients will be approximated. In the
following the concept of cascaded $\chi^{(2)} \cdot \chi^{(2)}$ is presented in detail with special emphasis on cascaded optical rectification and the linear electro-optic
effect, an important contribution to the nonlinear refractive index that has
been overlooked prior to this work. Then experimental results are presented
showing (i) that noncentrosymmetric molecules show superior third-order
nonlinear optical effects compared with the centrosymmetric analogues for
the molecular classes presented here,[1] (ii) that there exist two contributions
to the nonlinear phase shifts due to cascaded $\chi^{(2)} \cdot \chi^{(2)}$, namely cascaded
second-harmonic generation and difference-frequency mixing, and cascaded
optical rectification and the linear electro-optic effect, and (iii) that cascaded
second-order nonlinearities can be used to calibrate third-order nonlinearities, therefore yielding absolute values of $\chi^{(3)}$. In addition, the importance of
cascading through the reaction field and the local field will be evaluated. Applications of cascaded second-order nonlinearities, such as optical switching,
spatial solitons and beam clean-up will be discussed in the following. Finally,
the nonlinear optical properties and effects are confronted with theoretical
predictions and requirements for real applications.

2.1.2 Basic Concept of Cascading

Cascading is a process where lower-order effects are combined to contribute
to a higher-order nonlinear process. An illustrative example can be given for
the case of third-harmonic generation. Third-harmonic generation can occur in any material, even air (see Sect. 2.4). In this process a fundamental
wave at frequency ω produces a wave at frequency $3\omega(\omega + \omega + \omega = 3\omega)$.
In noncentrosymmetric materials sum frequency mixing $(\omega_1 + \omega_2 = \omega_3)$
and second-harmonic generation $(\omega + \omega = 2\omega)$ are also allowed. We can
use the two latter processes to obtain a wave at frequency 3ω as well: we
first generate an intermediate field at frequency 2ω through second-harmonic
generation. This field can interact with the fundamental wave through sum-
frequency mixing $(2\omega + \omega = 3\omega)$ to generate a field at frequency 3ω itself
[2.7]. This combined process is called cascading. To summarize: if we illuminate a noncentrosymmetric material with a wave at frequency ω and detect
the third-harmonic wave (3ω) at the output, this output can contain two contributions: one from direct third-harmonic generation $(\propto \chi^{(3)}(-3\omega, \omega, \omega, \omega))$
and one from cascaded second-harmonic generation and sum frequency mix-
ing $(\propto \chi^{(2)}(-3\omega, 2\omega, \omega) \cdot \chi^{(2)}(-2\omega, \omega, \omega))$. Figure 2.2 illustrates this process
schematically.

---

[1] Theoretical considerations have shown that centrosymmetric systems potentially
offer the largest third-order nonlinear optical susceptibilities [2.19].

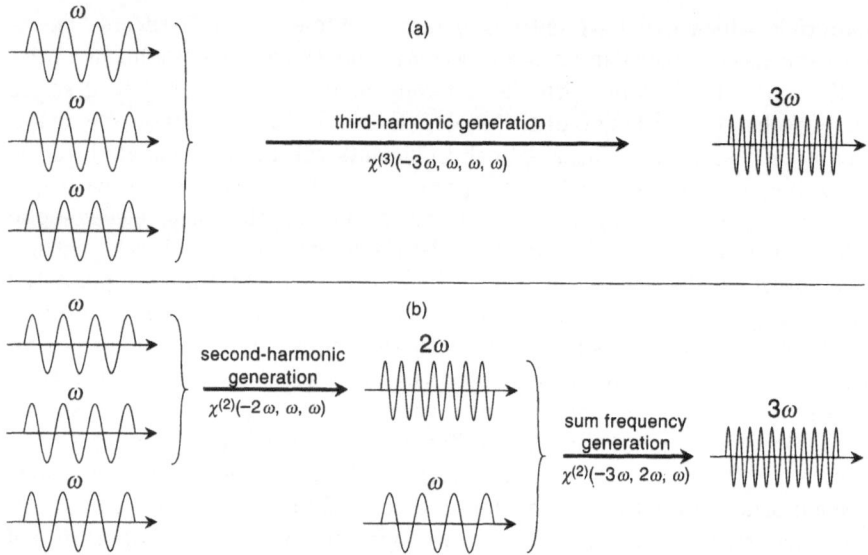

Fig. 2.2. Schematic of the generation of a wave at frequency 3ω through (**a**) direct third-harmonic generation ($\propto \chi^{(3)}(-3\omega, \omega, \omega, \omega)$) and (**b**) cascaded second-harmonic and sum frequency generation ($\propto \chi^{(2)}(-3\omega, 2\omega, \omega) \cdot \chi^{(2)}(-2\omega, \omega, \omega)$)

2.1.3   Definition of Nonlinear Optical Coefficients

For the definition of the nonlinear optical susceptibilities given below in (2.1) and (2.2), alternative conventions are frequently used. This has led to some confusion in the literature concerning the comparison of experimentally determined values obtained with different techniques. In particular, for the comparison of calculated and experimental values the hyperpolarizabilities must be defined using consistent conventions. Very often, however, experimental values are directly compared to theoretical values that use a completely different convention. The fact that most often the precise definitions in use are not clearly stated further complicates the comparison of nonlinear optical hyperpolarizabilities.

A   Definition of the Molecular Hyperpolarizabilities

The basic equation describing nonlinear optical effects expresses the macroscopic polarization $\boldsymbol{P}$ as a power series in the strength of the applied electric field $\boldsymbol{E}$ as

$$P_i = P_{0,i} + \varepsilon_0(\chi_{ij}^{(1)} E_j + \chi_{ijk}^{(2)} E_j E_k + \chi_{ijkl}^{(3)} E_j E_k E_l + \ldots \,, \tag{2.1}$$

where the Einstein summation convention has to be applied. $\boldsymbol{P}_0$ is the spontaneous polarization and $\chi^{(n)}$ is the nth order susceptibility tensor. ε_0 is the vacuum permittivity.

In organic materials the microscopic optical properties are determined by the molecular polarizabilities. These are defined by an expansion analogous to (2.1) expressing the molecular dipole moment p as

$$p_i = \mu_{\mathrm{g},i} + \varepsilon_0(\alpha_{ij}E_j + b_{ijk}E_jE_k + g_{ijkl}E_jE_kE_l \ldots) , \qquad (2.2)$$

where μ_{g} is the ground state dipole moment, α_{ij} is the linear polarizability, β_{ijk} is the second-order polarizability or first-order hyperpolarizability, and γ_{ijkl} is the third-order polarizability or second-order hyperpolarizability. The electric field to be inserted in (2.2) is the electric field at the location of the molecule. If the molecule is embedded in an optically dense environment, appropriate local field factors have to be used for the calculation of this electric field (see Sects. 2.2 and 2.3).

The fundamental quantities describing second-order nonlinear optical and electro-optic effects are the tensors $\chi_{ijk}^{(2)}$ and β_{ijk}. These are third rank tensors which, in the electric dipole approximation, contain nonvanishing elements only for noncentrosymmetric molecular and crystalline structures [2.20], respectively. Third-order nonlinear optical effects are described by the tensors $\chi_{ijkl}^{(3)}$ and γ_{ijkl}. No symmetry requirements are imposed on these effects to occur.

Systematic studies have shown that conjugated organic molecules with large delocalized π electron systems show very large nonlinear optical effects. The attachment of functional groups with electron-accepting and donating character at opposite ends of the conjugation bridge leads to an essentially one-dimensional *charge transfer*, enhancing especially the second-order nonlinearity. The one-dimensional character of the charge transfer usually leads to a single dominating element $\beta = \beta_{zzz}$ along the charge transfer direction usually referred to as the z direction. In this case, all other tensor elements can be neglected.

B   Comparison of Nonlinear Optical Susceptibilities Based on Different Definitions

We compare here some of the nonlinear optical susceptibilities important to this work. A comprehensive list containing all degeneracy factors can be found in Appendix A.1.

As discussed before, care must be taken in defining the appropriate nonlinear optical coefficients in order not to create confusion. One example illustrating these uncertainties is the different values that have been reported in the literature for the molecular hyperpolarizability of para-nitroaniline (p-NA): Oudar gave a value of 34.5×10^{-30} esu [2.21], whereas Teng and Garito reported 16.7×10^{-30} esu [2.22–23]. This disagreement is often attributed to the use of different solvents.[2] Some authors, however, believe that these differences are caused by different definitions of the hyperpolarizability [2.24].

---

[2] Oudar's value was measured in methanol whereas Teng and Garito used dioxane as a solvent.

We therefore dedicate this section to a discussion of the most widely used conventions and their mutual relations, although some aspects have already been treated in the literature [2.25–27].

A frequent alternative formulation for the microscopic expression (2.2) writes the induced dipole moment as a Taylor series [2.28]

$$p_i = \mu_{g,i} + \varepsilon_0 \left(\alpha_{ij}^{\mathrm{T}} E_j + \frac{1}{2!} \beta_{ijk}^{\mathrm{T}} E_j E_k + \frac{1}{3!} \gamma_{ijkl}^{\mathrm{T}} E_j E_k E_l + \dots \right). \quad (2.3)$$

Note that this Taylor series expansion is usually applied in theoretical studies where the cgs system of units or atomic units are used, with the consequence that the factor ε_0 does not appear. Since we are using SI units consistently in this work the factor ε_0 is also included in (2.3). Conversion factors between the different systems of units are given in Sect. A.2 for the most important quantities related with nonlinear optics.

The hyperpolarizabilities defined in (2.2) and (2.3) are thus related by

$$\beta_{ijk}^{\mathrm{T}} = 2 \cdot \beta_{ijk} \quad \text{and} \quad \gamma_{ijkl}^{\mathrm{T}} = 6 \cdot \gamma_{ijkl} . \quad (2.4)$$

Apart from the difference of the factor of 2! and 3! in the definitions of the hyperpolarizabilities in (2.2) and (2.3), additional conventions have been used to define the hyperpolarizabilities.

For the definition of the electric field amplitudes different definitions are also in use. The most widely used convention is to define the electric field by

$$\boldsymbol{E}(\boldsymbol{r},t) = \frac{1}{2} \sum_{\omega;k} \left[\boldsymbol{E}(\omega, \boldsymbol{k}) \mathrm{e}^{\mathrm{i}(\boldsymbol{kr}-\omega t)} + \text{c.c.} \right]. \quad (2.5)$$

For the present discussion all quantities can be assumed to be real. Considering only the time dependence, (2.5) then reduces to

$$\boldsymbol{E}(t) = \boldsymbol{E}^\omega \cos(\omega t) \quad (2.6)$$

for monochromatic waves.

The most widely used technique for the determination of first-order molecular hyperpolarizabilities is electric field-induced second-harmonic generation (EFISH; see Sect. 2.4.2). In this case the total electric field to be considered is a superposition of the applied dc field orienting the molecular dipoles and the oscillating field of the light wave:

$$\boldsymbol{E} = \boldsymbol{E}^0 + \boldsymbol{E}^\omega \cos(\omega t) . \quad (2.7)$$

Inserting this field into (2.2) and adopting convention (2.5) for the definition of the polarization,

$$\boldsymbol{p} = \boldsymbol{p}^0 + \boldsymbol{p}^\omega \cos(\omega t) + \boldsymbol{p}^{2\omega} \cos(2\omega t) + \boldsymbol{p}^{3\omega} \cos(3\omega t) , \quad (2.8)$$

we obtain for the Fourier components at the second- and third-harmonic frequency

$$p_i^{2\omega} = \varepsilon_0 \left(\frac{1}{2} \beta_{ijk}(-2\omega, \omega, \omega) E_j^\omega E_k^\omega + \frac{3}{2} \gamma_{ijkl}(-2\omega, \omega, \omega, 0) E_j^\omega E_k^\omega E_l^0 \right) \quad (2.9)$$

$$p_i^{3\omega} = \varepsilon_0 \frac{1}{4} \gamma_{ijkl}(-3\omega, \omega, \omega, \omega) E_j^\omega E_k^\omega E_l^\omega. \quad (2.10)$$

The factors $1/2^{(n-1)}$ at the frequency $n\omega$ are due to the definition of the field amplitudes (see (2.5)). In order to prevent these factors some authors dropped the factor $1/2$ in the definition of (2.5). The disadvantage of this convention is the unusual convergence behaviour of the electric field as ω approaches zero. This different field definition of course additionally complicates the comparison of different hyperpolarizability values. The factor 3 in front of $\gamma_{ijkl}(-2\omega,\omega,\omega,0)$ accounts for the three different ways of permuting the input frequencies.[3]

For the macroscopic polarization analogous expressions can be obtained by replacing the tensors β and γ by the corresponding macroscopic susceptibilities $\chi^{(2)}$ and $\chi^{(3)}$, respectively. Considering frequency doubling in a non-centrosymmetric material we get for the second-harmonic polarization

$$P_i^{2\omega} = \frac{1}{2}\varepsilon_0\chi_{ijk}(-2\omega,\omega,\omega)E_j^\omega E_k^\omega = \varepsilon_0 d_{ijk}(-2\omega,\omega,\omega)E_j^\omega E_k^\omega. \qquad (2.11)$$

This defines the well-known d-coefficient, which is usually used for the nonlinear optical characterization of macroscopic samples.

With the convention based on the Taylor series (see (2.3)) the equations analogous to (2.9) and (2.10) become

$$P^{2\omega} = \frac{1}{4}\varepsilon_0(\beta^{\mathrm{T}}(-2\omega,\omega,\omega) + \gamma^{\mathrm{T}}(-2\omega,\omega,\omega,0)E^0)(E^\omega)^2 \qquad (2.12)$$

$$p^{3\omega} = \frac{1}{24}\varepsilon_0\gamma^{\mathrm{T}}(-3\omega,\omega,\omega,\omega)(E^\omega)^3, \qquad (2.13)$$

where the tensor notation has been dropped for simplicity.

Both definitions of the hyperpolarizability tensors introduced up to now have the important property of converging towards their corresponding static values as the frequency of the applied electric field approaches zero, i.e.

$$\lim_{\omega\to 0}\beta(-2\omega,\omega,\omega) = \beta_0 \qquad (2.14)$$

$$\lim_{\omega\to 0}\gamma(-2\omega,\omega,\omega,0) = \lim_{\omega\to 0}\gamma(-3\omega,\omega,\omega,\omega) = \gamma_0 \qquad (2.15)$$

and accordingly for β^{T} and γ^{T}. Calculated static values can thus directly be compared with experimental values.

An additional definition for the hyperpolarizabilities, mainly used by experimentalists, directly defines the Fourier components of the polarization as the quantities of interest:

$$p^{2\omega} = \varepsilon_0\left(\beta^{\mathrm{exp.}}(-2\omega,\omega,\omega) + \gamma^{\mathrm{dc-SHG}}(-2\omega,\omega,\omega,0)E^0\right)(E^\omega)^2 \qquad (2.16)$$

$$p^{3\omega} = \varepsilon_0\gamma^{\mathrm{THG}}(-3\omega,\omega,\omega,\omega)(E^\omega)^3. \qquad (2.17)$$

---

[3] More precisely, the expression for the static field-induced second-harmonic component should be written as $\gamma_{ijkl}(-2\omega,\omega,\omega,0)E_j^\omega E_k^\omega E_l^0 + \gamma_{ijkl}(-2\omega,0,\omega,\omega)E_j^0 E_k^\omega E_l^\omega + \gamma_{ijkl}(-2\omega,\omega,0,\omega)E_j^\omega E_k^0 E_l^\omega$. Equation (2.9) is thus obtained using the permutation symmetry relation $\gamma_{ijkl}(-2\omega,\omega,\omega,0) = \gamma_{ijlk}(-2\omega,\omega,0,\omega) = \gamma_{ilkj}(-2\omega,0,\omega,\omega)$.

Table 2.1. Frequency components of the polarization for the different definitions. For simplicity, the superscripts pointing out on which definition the hyperpolarizabilities are based have been dropped

Definition	$\dfrac{p^{2\omega}}{\varepsilon_0}$	$\dfrac{p^{3\omega}}{\varepsilon_0}$
Standard (2.2), this work	$\frac{1}{2}\beta + \frac{3}{2}\gamma$	$\frac{1}{4}\gamma$
Taylor series (2.3)	$\frac{1}{4}\beta + \frac{1}{4}\gamma$	$\frac{1}{24}\gamma$
Experimental (2.16), (2.17)	$\beta + \gamma$	γ

In comparison with (2.9) and (2.10), the factors 1/2, 3/2, and 1/4 are absorbed in the hyperpolarizabilities. This definition of the first-order hyperpolarizability $\beta^{\text{exp.}}$ is thus equivalent to the introduction of the d-tensor (2.11) for macroscopic samples. The different superscripts for the second-order hyperpolarizability in (2.16) and (2.17) are necessary, since they are different in that they do not converge towards the same values in the zero frequency limit. The static hyperpolarizabilities are related to the corresponding quantities defined in (2.16) and (2.17) by

$$\beta^{\text{exp.}}(0,0,0) = \frac{1}{2}\beta_0 \tag{2.18}$$

$$4\gamma^{\text{THG}}(0,0,0,0) = \frac{2}{3}\gamma^{\text{dc-SHG}}(0,0,0,0) = \gamma_0 . \tag{2.19}$$

Experimental values for γ obtained with third-harmonic generation and using (2.17) must thus not be compared directly with γ-values resulting from field-induced second-harmonic generation experiments evaluated using (2.16). In Table 2.1 the different definitions are summarized.

In this work the correct definitions for β and γ (based on (2.2), (2.9) and (2.10)) are used.

Let us now consider the macroscopic polarization generated in an EFISH experiment. From a macroscopic point of view the main effect of the static electric field is the orientation of the dissolved molecules. The macroscopic polarization thus has to be calculated as the thermodynamical expectation value of the molecular polarization multiplied with the molecular number density:[4]

$$P_{\mathrm{I}}^{2\omega} = N \langle p_{\mathrm{I}}^{2\omega} \rangle . \tag{2.20}$$

Writing the microscopic polarization in a general way with coefficients A and B representing the different factors due to the three definitions (2.9), (2.12) and (2.16) as

$$p_i^{2\omega} = \varepsilon_0 \left(A\beta_{ijk}E_j^\omega E_k^\omega + B\gamma_{ijkl}E_j^0 E_k^\omega E_l^\omega \right) \tag{2.21}$$

---

[4] For the present discussion local field factors are neglected.

Table 2.2. Effective d-coefficients for EFISH experiments for different definitions of the hyperpolarizabilities

Definition	A	B	$\frac{d(E^0)}{NE^0}$
Standard, this work	$\frac{1}{2}$	$\frac{3}{2}$	$\frac{3}{2}\bar\gamma + \frac{1}{2}\frac{\mu\beta_z}{5kT}$
Taylor	$\frac{1}{4}$	$\frac{1}{4}$	$\frac{1}{4}\bar\gamma + \frac{1}{4}\frac{\mu\beta_z}{5kT}$
Experimental	1	1	$\bar\gamma + \frac{\mu\beta_z}{5kT}$

and performing the thermodynamical averaging with the assumption of a weak electric field, i.e. $\mu E^0 \ll kT$, the macroscopic polarization can be written as (see, for example, [2.29])

$$P^{2\omega} = \varepsilon_0 N \left(B\bar\gamma + A\frac{\mu\beta_z}{5kT} \right) (E^\omega)^2 E^0, \tag{2.22}$$

where β_z is the vector part of the hyperpolarizability tensor along the molecular dipole moment and $\bar\gamma$ is the scalar part of the second-order hyperpolarizability:

$$\beta_z = \beta_{zzz} + \frac{1}{3}(\beta_{zxx} + 2\beta_{xxz} + \beta_{zyy} + 2\beta_{yyz}) \tag{2.23}$$

$$\bar\gamma = \frac{1}{15}\sum_{i,j}(2\gamma_{iijj} + \gamma_{ijji}) . \tag{2.24}$$

These equations apply for any of the different conventions.

Comparing (2.22) with (2.11) we can define a field induced d-coefficient as (Table 2.2)

$$d(E^0) = N \left(B\bar\gamma + A\frac{\mu\beta_z}{5kT} \right) E^0 , \tag{2.25}$$

which is the quantity that has to be compared with a known d-coefficient of a nonlinear optical crystal in an EFISH experiment.

A fourth convention which is closely related to the standard definition has been introduced more recently for the evaluation of EFISH experiments [2.26, 2.30]. With this definition the effective d-coefficient is written as

$$d(E^0) = \frac{3}{2}N \left(\bar\gamma + \frac{\mu\beta_z'}{5kT} \right) E^0 . \tag{2.26}$$

Thus β' must be multiplied by $1/3$ in order to compare with EFISH values obtained with the standard definition.[5]

---

[5] This definition arises from using a different convention on the microscopic than on the macroscopic level. Using the experimental EFISH definition (2.16), the macroscopic third-order susceptibility is written as $\chi^{(3)} = N\left(\bar\gamma + \mu\beta_z'/5kT\right)$. This quantity is then related to the crystalline d-coefficient using the standard definition (2.9) via $d = 3/2\chi^{(3)}E^0$.

Theoretical calculations performed with the finite-field method which compute the hyperpolarizability in the presence of a static electric field are usually based on the Taylor series definition (2.3). This is often done in the semiempirical quantum mechanical program MOPAC. The hyperpolarizability calculations of MOPAC are based on a finite-field calculation and give as the results for the first-order hyperpolarizability of the single elements of the β^{T} tensor in atomic units (a.u.) as well as the vector part of β^{T} along the molecular dipole moment in both atomic and esu units. The definition of this vector part, however, is different from the one given in (2.23):

$$\beta^{\mathrm{MOPAC}} = \beta_{||}^{\mathrm{T}} = \frac{3}{5}(\beta_{zzz}^{\mathrm{T}} + \beta_{zxx}^{\mathrm{T}} + \beta_{zyy}^{\mathrm{T}}) \ . \tag{2.27}$$

If Kleinman symmetry is assumed in (2.23) we thus have

$$\beta_{||}^{\mathrm{T}} = \frac{3}{5}\beta_z^{\mathrm{T}} \ . \tag{2.28}$$

In order to compare the experimental results of an EFISH experiment with the theoretical results obtained with MOPAC all differences regarding the hyperpolarizability definition have to be taken into account. With (2.4), (2.18) and (2.27) we thus get

$$\beta^{\mathrm{MOPAC}} = \frac{6}{5} \cdot \beta_0 = \frac{12}{5} \cdot \beta_0^{\mathrm{exp}} \ , \tag{2.29}$$

where β_0^{exp} is the EFISH result extrapolated to zero optical frequency, which is usually calculated from the measured values with a theoretical two-level dispersion formula (2.39).

Experimental values of the hyperpolarizability are often compared with theoretical calculations based on quantum-mechanical sum over states expressions or with solvatochromic measurements [2.31–32] which are based on a two-level reduction of the complete sum-over-states expression (see Sect. 2.2.1.A). Bloembergen and Shen [2.33] derived this quantum mechanical expression for the hyperpolarizability $\beta_{ijk}(-\omega_3, \omega_1, \omega_2)$ with the perturbing Hamiltonian defined as

$$H' = \frac{1}{2}E_0^{\omega_1}e^{-i\omega_1 t} + \frac{1}{2}(E^{\omega_1})^*e^{i\omega_1 t} + \frac{1}{2}E^{\omega_2}e^{-i\omega_2 t} + \frac{1}{2}(E^{\omega_2})^*e^{i\omega_2 t} \ , \tag{2.30}$$

which is consistent with our definition of the electric field amplitudes (2.5).

The third-order optical nonlinearity of molecules is often studied using third-harmonic generation (THG) in solution. In an isotropic liquid, one has to consider only the rotational average of γ_{ijkl}, for which one gets:

$$\begin{aligned}
\gamma &\equiv \langle\gamma\rangle_{3333} \\
&= \frac{1}{5}(\gamma_{1111} + \gamma_{2222} + \gamma_{3333} + \gamma_{1122} + \gamma_{1133} + \gamma_{2211} + \gamma_{2233} + \gamma_{3311} + \gamma_{3322}) \\
&= \frac{1}{5}\sum_{i,j}\gamma_{iijj} \ .
\end{aligned} \tag{2.31}$$

The detailed experimental set-ups for THG and EFISH are described in Sects. 2.4.2 and 2.4.1.

2.1.4   Materials Requirements for All-Optical Signal Processing

For all-optical signal processing (based on pure $\chi^{(3)}$ nonlinearities) the important quantity is the nonlinear refractive index n_2 defined through (see also Sect. 2.2.3.B)

$$n = n_0 + n_2 I ,\tag{2.32}$$

where n_0 is the normal linear refractive index at low powers and I is the intensity of the incident light beam. Generally, the most important factor for device operation based on third-order effects is the maximum nonlinear phase shift $\Delta\phi^{\mathrm{NL}}$ that can be produced at relatively low intensity levels. This phase shift depends on the effective interaction length L_{eff}. A fundamental limitation to L_{eff} is the absorption, with both a linear (α_0) and a nonlinear (α_2, α_3) contribution. For device purposes these contributions must be small [2.2]. Unfortunately there is often also an intensity-dependent absorption α_2 associated with large values of n_2. Sometimes even three-photon absorption, represented through α_3, can be considerable. α_2 and α_3 are defined through

$$\alpha = \alpha_0 + \alpha_2 I + \alpha_3 I^2 ,\tag{2.33}$$

where α_0 is the linear absorption coefficient at low powers. For pure Kerr nonlinearities the requirements of a sufficient nonlinear phase shift and a large signal throughput can be expressed with three figures of merit which can be obtained from the maximum possible nonlinear phase shift in a sample of length L

$$\Delta\phi^{\mathrm{NL}} = \int_0^L \frac{2\pi}{\lambda} n_2 I(z) \mathrm{d}z .\tag{2.34}$$

First, if the linear absorption dominates such that $\alpha_0 \gg \alpha_2 I + \alpha_3 I^2$ we have $I(z) = I_0 \exp(-\alpha_0 z)$. This leads to the effective length $L_{\mathrm{eff}} = (1 - \exp(-\alpha_0 L))/\alpha_0$. From $\alpha_0^{-1} > L_{\mathrm{eff}}$ we obtain $n_2 I(0) 2\pi/\lambda)/\alpha_0 > \Delta\phi^{\mathrm{NL}}$. From the requirement of a minimum phase shift of 2π (often needed, for example in the nonlinear directional coupler) we get

$$W = \frac{\Delta n}{\alpha_0 \lambda} = \frac{\Delta\phi^{\mathrm{NL}}}{2\pi} > 1 .\tag{2.35}$$

That is, $W > 1$ is needed for the directional coupler with high throughput. Δn can either be $n_2 I$ or, for example, a saturating nonlinearity. Note that for saturating nonlinearities the required value for W is more or less doubled [2.2]. Second, for the two-photon absorption figure of merit for optical switching (for $\alpha_2 I \gg \alpha_0 + \alpha_3 I^2$ and $1/\alpha_2 I > L_{\mathrm{eff}}$), we have (defining $T = 2\pi/\Delta\phi^{\mathrm{NL}}$)

$$T^{-1} = \frac{n_2}{\alpha_2 \lambda} > 1 .\tag{2.36}$$

T^{-1} quantifies the nonlinear phase shift that can be achieved over typical absorption lengths associated with two-photon absorption. If three-photon absorption dominates ($\alpha_3 I^2 \gg \alpha_0 + \alpha_2 I$) and for $1/\alpha_3 I^2 > L_{\text{eff}}$ the corresponding requirement is (defining $V = 2\pi/\Delta\phi^{\text{NL}}$)

$$V^{-1} = \frac{n_2}{\alpha_3 I_{\text{sw}} \lambda} > 1 . \tag{2.37}$$

Note that V is proportional to the switching intensity I_{sw} of the device of interest [2.34]. Therefore, since we usually have $I_{\text{sw}} \propto 1/L$, one can reduce V by increasing the device length. Most often V can be ignored since α_3 is small. The requirements for the three figures of merit W, T and V may vary depending on the device under consideration. They can be combined to give

$$\frac{\Delta\phi^{\text{NL}}}{2\pi} < \frac{W}{1 + WT + WV} . \tag{2.38}$$

This condition depends on the nonlinear phase shift required [2.34].

2.2   Optical Nonlinearities

Different effects can give rise to a material's nonlinear optical response. The polarizability of a crystal is generally composed of contributions from, first, the lattice components (atoms, molecules) and, second, the interaction between these components. Whereas the second effect is dominant in inorganic materials with their weakly polarizable atoms and complexes, the first contribution is dominant in organic materials because of the weak intermolecular bonding (van der Waals, dipole–dipole interactions, hydrogen bonds). Nonlinear optical effects that mainly depend on the change of polarizability of the electrons in the π bonding orbitals are much faster than effects due to lattice vibrations (occurring in a frequency range of typically 1 MHz up to 100 MHz). Therefore organic materials are well suited for high-speed applications, e.g. nonlinear optics with ultrashort pulses or high data rate electro-optic modulation.

2.2.1   Organic Nonlinear Optical Materials

Here we discuss the basic molecular units and structures that are essential for nonlinear optics in organic materials. We briefly introduce the concepts necessary to obtain an efficient electro-optic and nonlinear optical response.

   The exact definition of molecular polarizabilities is given by quantum mechanics through time-dependent perturbation theory [2.35]. These expressions are usually given in terms of sums of transition matrix elements over energy denominators involving the full electronic structure of the molecule. Although accurate, these expressions offer little physical insight into the synthetic methods that are suitable for optimizing the molecular nonlinear optical response.

The wide diversity of the properties of organic compounds is primarily due to the unsurpassed ability of the carbon atom to form a variety of stable bonds with itself and with many other elements. This bonding is primarily of two types, which differ considerably in the localization of electron charge density [2.36]. A two-electron covalent σ C–C bond is spatially confined along the internuclear axis of the C–C bond. In contrast to σ bonds, π bonds are regions of delocalized electronic charge distributions above and below the interatomic axis. The electron density of π bonds is much more mobile than that of the σ bonds. This electron distribution can also be skewed by substituents; the extent of redistribution is measured by the dipole moment, and the ease of redistribution in response to an externally applied electric field is measured by the (hyper)polarizability. If the perturbation to the molecular electronic distribution caused by an intense optical field is asymmetric a quadratic nonlinearity results [2.37]. Virtually all significantly interesting nonlinear optical organic molecules exhibit π bond formation between various nuclei.

A   Molecular Second-Order Polarizabilities

Basic Molecular Units for Second-Order Nonlinear Optical Effects. For the case of the second-order polarizability, β, the understanding was considerably clarified by the recognition that β is primarily determined by strong low-energy charge transfer electronic excitations. When this charge transfer (CT) state dominates the perturbation expression for β, one arrives at a two-level approximation between the highest occupied (HOMO) and lowest unoccupied (LUMO) molecular orbitals (see Sect. 2.2.1.A) [2.38–39]. The resulting β_{CT} is a function of the energy gap between the two states, the oscillator strength of the CT transition, the dipole moment associated with that transition, and the fundamental laser photon energy. The spectroscopic energy gap is related to the frequency (or wavelength) of the UV-VIS absorption spectra of the molecule, and the oscillator strength is related to the extinction coefficient of the absorption [2.40].

A further simplified model (the Equivalent Internal Field (EIF) model) of a free electron gas corresponding to the delocalized π electron density of a conjugated system of length L has been derived [2.41–42]. In this approximation $\beta \propto L^3$, which shows the strong nonlinear dependence of the hyperpolarizability on the length of the conjugated π system. These two models, although somewhat simplistic, have provided the foundation of a rational strategy for designing highly polarizable nonlinear optical chromophores.

An optimized nonlinear optical chromophore can then be expected to have an extended conjugated system (large L), a low-energy transition (long wavelength absorption) with a high extinction coefficient, and a large dipole moment between the excited and ground state electronic configurations (charge asymmetry). Charge asymmetry is introduced when different functional groups are substituted to the molecule.

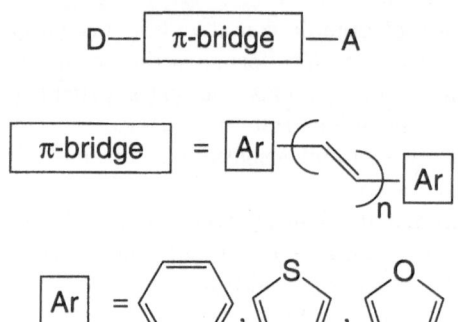

Fig. 2.3. Typical organic molecules for second-order nonlinear optics. These are donor (D)–acceptor (A) disubstituted π-conjugated systems

The second-order optical nonlinearity of organic molecules can be increased either by adding conjugated bonds (increasing L) or by substituting electron donors and electron acceptors (Fig. 2.3). The addition of the appropriate functionality at the ends of the π-conjugated system can enhance the asymmetric electronic distribution in either or both the ground state and excited state configurations. Functional groups are divided into two categories based on their ability to accept (acceptor) or donate (donor) electrons into the π-conjugated system.

A simple physical picture for the effectiveness of the attachment of functional groups with electron accepting and donating character at opposite ends of the conjugation bridge leading to highly polarizable charge transfer molecules can be given in terms of Mulliken resonance structures of the ground and excited states (see Fig. 2.4 [2.43]). When an oscillating electric field is applied to the π electrons of a disubstituted donor–acceptor molecule, charge flow is enhanced towards the acceptor, whereas the motion in the other direction (towards the donor) is highly unfavorable: the charge transfer molecules have a very asymmetric response to an applied electric field.

However, the presence of conjugation and donor and acceptor groups generally introduces an undesirable effect, the so-called transparency–efficiency trade-off. The increase of conjugation by linking double bonds and donor and acceptor substituents leads to a shift of the absorption edge towards longer wavelengths. Therefore it can generally be said that the higher the nonlinearity of such materials the more the absorption edge is shifted towards the red.

Recently, multiple donor-and-acceptor chromophores have gained increased attention [2.44–45]. These chromophores refer to those molecules that contain more than one donor–acceptor pair attached to a π-conjugated core. The octupolar molecule, 1,3,5-triamino-2,4,6-trinitrobenzene is one of the prototype examples (Fig. 2.5) [2.45]. Unlike the elongated, dipolar chromophores, the octupolar molecules do not possess a permanent dipole moment and the vectorial components of the molecular hyperpolarizability, β_μ,

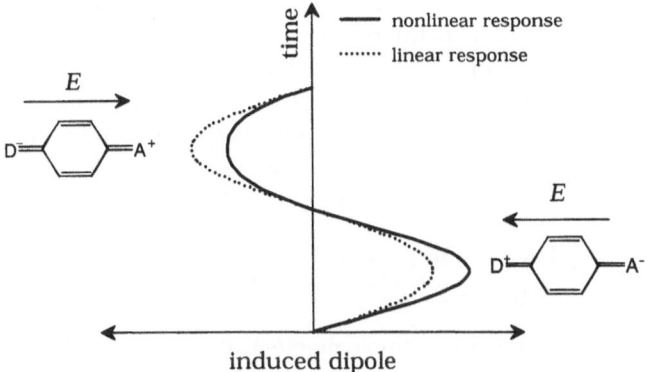

higher excited state |2>   ground state |0>   first excited state |1>

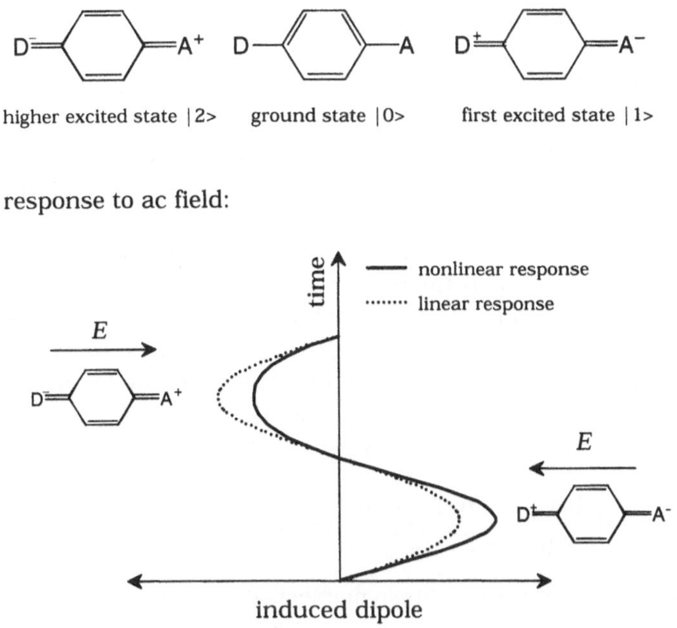

response to ac field:

Fig. 2.4. Simple picture of the physical mechanisms of the nonlinearity of conjugated molecules. $|0\rangle$: ground state. $|1\rangle$: situation where one electron is transferred from the donor to the acceptor; maximum charge transfer state. $|2\rangle$: situation where one electron is transferred from the acceptor to the donor; very unlikely. We get an asymmetric electronic response of the polarization on application of an optical field for molecules with donor and/or acceptor groups in contrast to the symmetric electronic response for centrosymmetric molecules (adapted from [2.43])

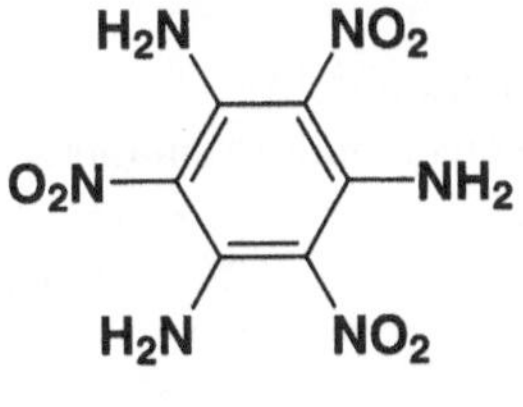

Fig. 2.5. Prototype example of an octupolar molecule. This molecule is noncentrosymmetric, although it does not have a polar axis

are zero. In this work we concentrate on the classical type of molecules shown in Fig. 2.3.

The Two-Level Model. The nonlinear optical and electro-optic responses in organic materials arise from the nonlinearity of the basic molecular units. Many properties of the solid can therefore be explained by the properties of the molecules.

Molecular hyperpolarizabilities are usually measured at a fixed laser wavelength. In order to estimate the size of the nonlinear optical effects at different wavelengths, the knowledge of the frequency dependence of the hyperpolarizability is of importance.

Bloembergen and Shen [2.33] derived a quantum mechanical expression for the first-order hyperpolarizability, based on perturbation theory.

For most molecules used for second-order nonlinear optical applications the expression can be substantially simplified. These donor–acceptor disubstituted systems are characterized by a single low-lying electronic charge transfer excitation that is mainly responsible for the linear as well as the nonlinear optical properties. Furthermore one assumes that this so-called charge transfer transition extends along one spatial direction. The hyperpolarizability tensor is thus dominated by one single diagonal element for which the quantum mechanical expression reduces to (for the case of sum-frequency generation) [2.21, 2.38]

$$\beta_{zzz}(-\omega_3, \omega_1, \omega_2)$$
$$= \frac{1}{2\varepsilon_0 \hbar^2} \frac{\omega_{eg}^2 \left(3\omega_{eg}^2 + \omega_1\omega_2 - \omega_3^2\right)}{\left(\omega_{eg}^2 - \omega_1^2\right)\left(\omega_{eg}^2 - \omega_2^2\right)\left(\omega_{eg}^2 - \omega_3^2\right)} \Delta\mu\mu_{eg}^2 , \qquad (2.39)$$

where $\mu_{eg} = \langle e|\mu_z|g\rangle$ is the transition dipole moment of the dominant charge transfer transition, $\Delta\mu = \mu_e - \mu_g = \langle e|\mu_z|e\rangle - \langle g|\mu_z|g\rangle$ is the difference between the dipole moments of the ground and the excited state, and where ω_{eg} denotes the resonance frequency of the transition. Note that ω_{eg} of the nonlinear optical molecules is red-shifted in a dielectric medium with $\varepsilon > 1$ due to local field effects. The approximation given by (2.39) is usually referred to as the two-level model.

We can derive the appropriate hyperpolarizability describing the electro-optic effect by setting $\omega_3 = \omega$, $\omega_1 = \omega$ and $\omega_2 = 0$ and frequency doubling by setting $\omega_3 = 2\omega$, $\omega_1 = \omega$ and $\omega_2 = \omega$.

We can now introduce the dispersion-free hyperpolarizability β_0 extrapolated to infinite optical wavelengths away from the electronic resonances

$$\beta_0 = \frac{3}{2\varepsilon_0 \hbar^2} \frac{\Delta\mu\, \mu_{eg}^2}{\omega_{eg}^2} . \qquad (2.40)$$

From this equation we obtain for the linear electro-optic effect

$$\beta_{zzz}(-\omega, \omega, 0) = \frac{\omega_{eg}^2 \left(3\omega_{eg}^2 - \omega^2\right)}{3 \left(\omega_{eg}^2 - \omega^2\right)^2} \beta_0 \qquad (2.41)$$

and for frequency doubling

$$\beta_{zzz}(-2\omega, \omega, \omega) = \frac{\omega_{eg}^4}{\left(\omega_{eg}^2 - 4\omega^2\right)\left(\omega_{eg}^2 - \omega^2\right)} \beta_0 . \qquad (2.42)$$

The two-level model also illustrates that β is not optimized when the ground and first excited states are either fully localized or delocalized but at some intermediate point where there exists a non-negligible overlap between

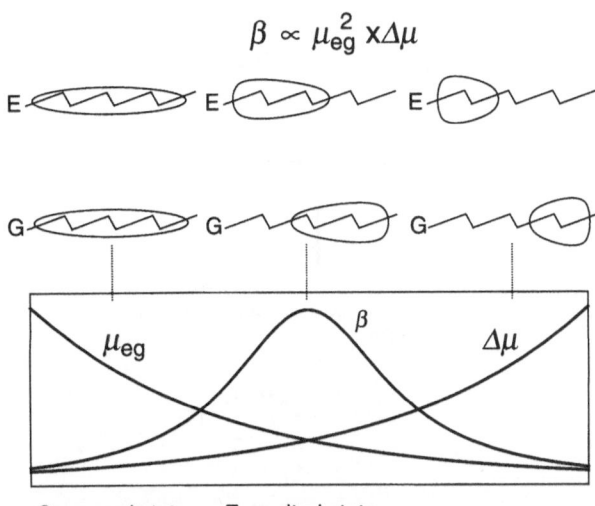

$$\beta \propto \mu_{eg}^2 \times \Delta\mu$$

G: ground state   E: excited state

Fig. 2.6. A phenomenological diagram describing the charge separation necessary for obtaining optimal second-order nonlinear optical responses (adapted from [2.46])

the two states. Complete charge separation (i.e. wave function localization) therefore does not lead to optimal responses [2.46]. Figure 2.6 shows this behaviour schematically.

An additional important feature of the two-level model is that it predicts the sign of β through the term $\Delta\mu = \mu_e - \mu_g$. For a typical molecule such as 4-dimethylamino-4'-nitrostilbene (DANS) with a positive value of β, the excited state dipole moment is greater than and aligned in the same direction as the ground state dipole moment (Fig. 2.7). The electron density in the ground state is weakly biased towards the π-acceptor but strongly localized in the excited state resulting in a positive $\Delta\mu$. For certain molecules such as the merocyanine dye shown in Fig. 2.7 the situation is different. The zwitterionic resonance structure accurately describes the molecular ground state in highly polar solvents, since the aromatic resonance structure is stabilized [2.47]. It is obvious that in such molecules the ground state dipole moment is larger than that in the excited state leading to a negative value of $\Delta\mu$ and therefore β.

B   Molecular Third-Order Polarizabilities

Basic Molecular Units for Third-Order Nonlinear Optical Effects. The development of highly active third-order nonlinear optical materials is important for all-optical signal processing [2.2]. In contrast to second-order nonlinear optical molecular systems, there are few rational strategies for optimizing the third-order nonlinear optical response of molecular materials. Unlike second-order materials, there exist no molecular symmetry restrictions for the observation of a third-order nonlinear optical response. It is the instantaneous

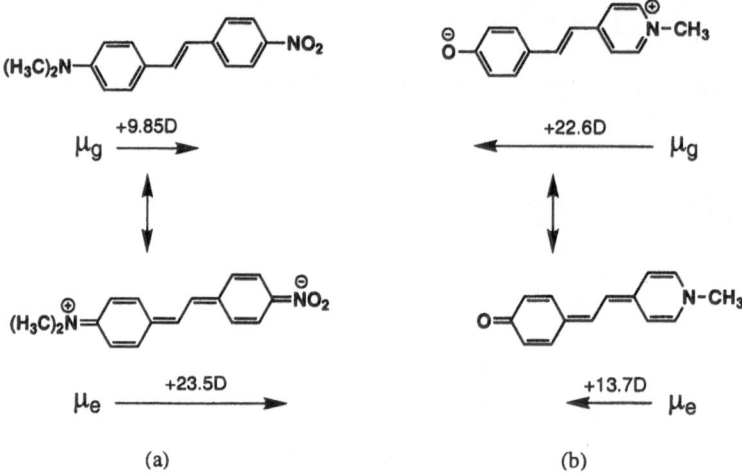

Fig. 2.7. Schematic diagram of two prototypical chromophores with (**a**) positive value of β and (**b**) a negative value of β (adapted from [2.47]). The dipole moments were calculated using ZINDO (formalism based on intermediate neglect of differential overlap/spectroscopy) [2.47]. The sign of β for these two chromophores has also been verified experimentally

shift in π electron densities over a molecule that occur on excitation which explains the large and fast polarizabilities of π electron networks.

The two-level model that has proved so useful for analyzing second-order materials is not sufficient for describing the third-order nonlinear optical response [2.48–49]. A minimum of three states (ground state and two excited states) are necessary to adequately characterize the third-order response (see below). Due to the inherent difficulties of characterizing excited states of molecular systems, very few molecules have been fully analyzed both experimentally and theoretically.

Most of the experimental basis for optimizing the third-order optical response of molecular materials can be understood and interpreted in terms of an early simple model due to Rustagi et al. [2.50]. For extended linear chains of conjugated molecules, they showed that the π electrons are reasonably well modeled in a first approximation by a free electron gas. In this model, the third-order molecular polarizability, γ, scales as the fifth power of the conjugation length L, i.e. $\gamma \propto L^5$. Concurrent experimental work on the characterization of some alkanes and conjugated molecules (polyenes and cyanines) by third harmonic generation showed the enhanced nonlinearity of the delocalized π electron systems [2.51].

Well-defined structure–property relationships for third-order nonlinear optical chromophores are still absent, although bond-length alternation has recently been identified as one useful structural parameter to predict trends for the second-order hyperpolarizability γ (see below) [2.52].

The Three-Level Model. A summation over electronic states (SOS, sum over states) can be used to calculate the second-order hyperpolarizability γ [2.53]

$$\gamma_{ijkl}\left(-\omega_4, \omega_1, \omega_2, \omega_3\right) =$$

$$K \sum_{\text{perm}} \left[\left\{ \sum_{m,n,p(\neq r)} \frac{\langle r|\mu_i|m\rangle \left\langle \overline{m|\mu_j|n} \right\rangle \left\langle \overline{n|\mu_k|p} \right\rangle \langle p|\mu_l|r\rangle}{(\omega_{mr}-\omega_4-i\Gamma_{mr})(\omega_{nr}-\omega_2-\omega_3-i\Gamma_{nr})(\omega_{pr}-\omega_3-i\Gamma_{pr})} \right\} \right.$$
$$\left. - \left\{ \sum_{m,n(\neq r)} \frac{\langle r|\mu_i|m\rangle \langle m|\mu_j|r\rangle \langle r|\mu_k|n\rangle \langle n|\mu_l|r\rangle}{(\omega_{mr}-\omega_4-i\Gamma_{mr})(\omega_{nr}-\omega_3-i\Gamma_{nr})(\omega_{nr}+\omega_2-i\Gamma_{nr})} \right\} \right],$$

$$(2.43)$$

where K is a constant. $\langle r|\mu_i|m\rangle$ is the electronic transition moment between the reference state (usually but not necessarily the ground state) described by the wave function $|r\rangle$ and the excited state $|m\rangle$ along the ith Cartesian axis of the molecular coordinate system. $\left\langle \overline{m|\mu_j|n} \right\rangle$ is the dipole difference operator given by $\langle m|\mu_j|n\rangle - \langle r|\mu_j|r\rangle \delta_{mn}$. ω_{mr} (times $\hbar$) is the energy difference between the states m and r. $\omega_1, \omega_2, \omega_3$ are the frequencies of the incoming light fields. $\omega_4 = \omega_1 + \omega_2 + \omega_3$ is the frequency of the polarization response. Γ_{mr} is the damping associated with the excited state m. $\sum_{\text{perm}}$ represents the summation over the 24 terms obtained by permuting the frequencies. We can see that the SOS technique allows an easy implementation of the frequency dependence of the third-order response. For each term in the summations we have a product of four transition or state dipoles (state dipoles exist only in noncentrosymmetric materials) in the numerator. The denominator contains a product of three energy differences between excited states and the reference state. The first set of summations (positive sign) has terms with three excited states (Fig. 2.8a), for example an excitation where the system evolves from state $|r\rangle$ to state $|m\rangle$ to state $|n\rangle$ to state $|p\rangle$ and back to state $|r\rangle$.

The second set of summations (negative sign) contains only two excited states since state $|r\rangle$ also corresponds to an intermediate state (Fig. 2.8b). Equation (2.43) leads to the following facts:

1. The selection rules for excitations apply. This means that in highly symmetric systems only a few excitation paths exist.
2. The higher the state that is involved, the lower its contribution (due to the energy denominator).
3. Large transition moments are important for large effects.
4. Resonance enhancements can occur if individual or sum of frequencies are close to one of the excitation energies.

A significant simplification occurs if only three states are considered: the ground state, a first low-lying (one-photon allowed) excited state and a second excited state that is accessible from the first excited state [2.19, 2.55–56]. Equation (2.43) is then considerably reduced: there remain one positive and

energy

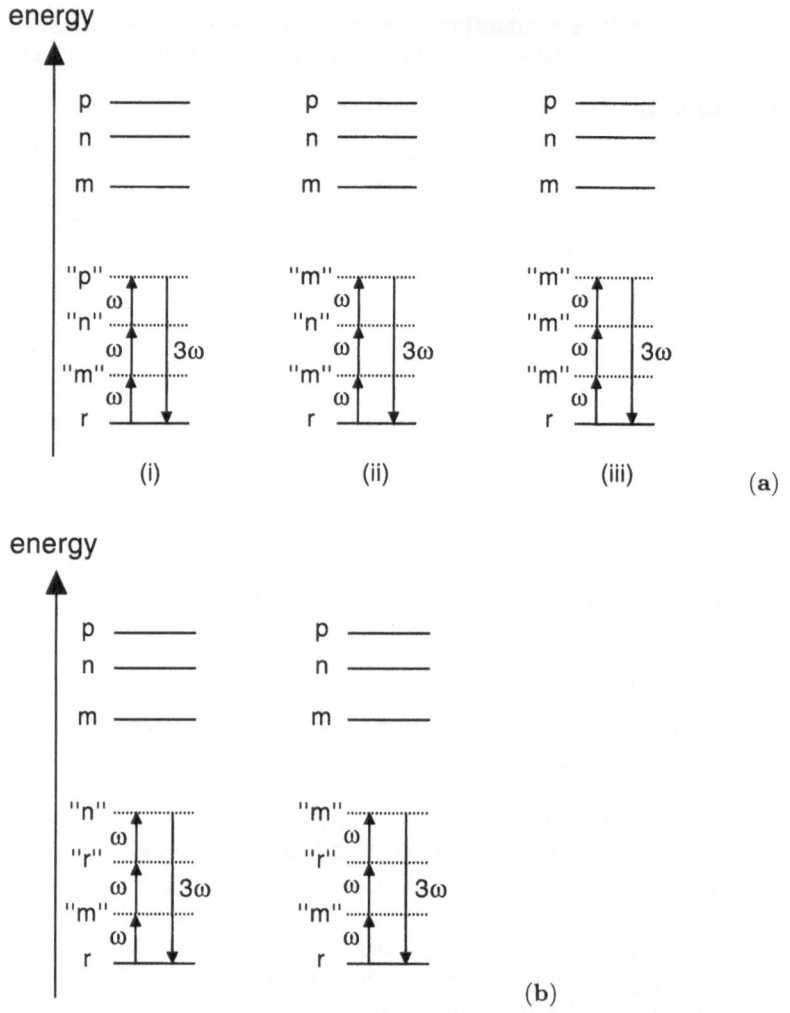

Fig. 2.8a,b. Schematic drawing of the contributions from the different excitations on the off-resonance second-order hyperpolarizability γ for third-harmonic generation (adapted from [2.54]). (**a**) 'positive' contributions; (**b**) 'negative' contributions. The states in quotes refer to the primary stationary state which contributes to the virtual state. Note that part (iii) describes excitations that only exist in noncentrosymmetric materials (states of the same parity are not one-photon allowed)

one negative contribution for centrosymmetric molecules and a third contribution (always positive!) that only exists in noncentrosymmetric systems (Fig. 2.9). The dispersion of γ is then described by

$$
\gamma_{iiii} = K' \left[\underbrace{\frac{\langle r \,|\mu_i|\, m\rangle^2 \,\Delta\mu_{rm}^2}{E_{rm}^3}}_{\substack{\text{D} \\ \text{(applies to noncentro-} \\ \text{symmetric systems)}}} - \underbrace{\frac{\langle r \,|\mu_i|\, m\rangle^4}{E_{rm}^3}}_{\text{N}} + \underbrace{\frac{\langle r \,|\mu_i|\, m\rangle^2 \,\langle m \,|\mu_i|\, n\rangle^2}{E_{rm}^2 E_{mn}}}_{\text{T}} \right],
$$

(2.44)

where K' is a constant and E_{rm} is the electronic transition energy. The term D (dipolar) only appears in noncentrosymmetric materials since it depends on $\Delta\mu$ and it gives a positive contribution. The second term N (negative) gives a negative contribution and depends only on $\langle r \,|\mu_i|\, m\rangle$ and E_{rm} (just as the linear polarizability α). These two terms form the two-state expression for γ with only the ground state $|r\rangle$ and one excited state $|m\rangle$. The third term T (two-photon) includes one more excited state $|n\rangle$ within two-photon like processes. Often these three states are enough to describe the dispersion of γ to a first approximation.

From the above equation we can see that, especially in the case of a small contribution from N, the value of γ can be increased in noncentrosymmetric molecules.

energy

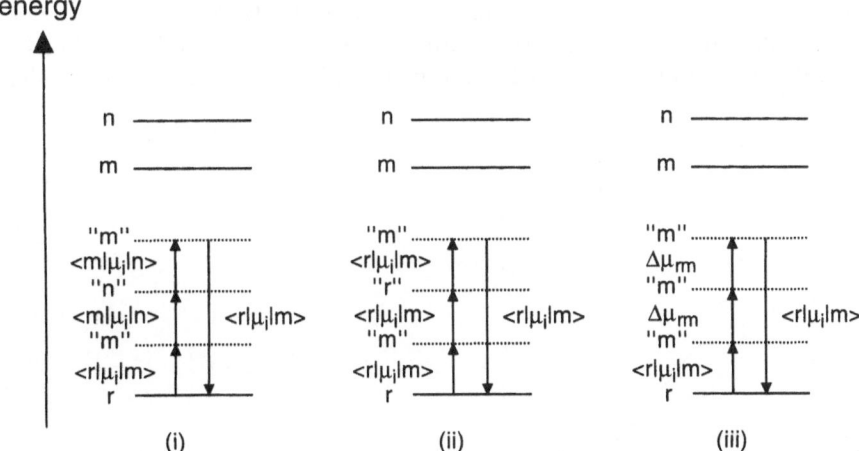

Fig. 2.9. Schematic drawing of the contributions from the different excitations on the off-resonance second-order hyperpolarizability γ for third-harmonic generation in the three-level model (adapted from [2.54]). The states in quotes refer to the primary stationary state which contributes to the virtual state. Note that part (iii) describes excitations that only exist in noncentrosymmetric materials

Equation (2.44) can be written in a different way by taking the factor $\langle r\,|\mu_i|\,m\rangle^2/E_{rm}$ outside of the parentheses. This ratio is proportional to the linear polarizability α_π coming from the π electrons. We obtain

$$\gamma_{iiii} \propto \alpha_\pi \left[\frac{\Delta\mu_{rm}^2 - \langle r\,|\mu_i|\,m\rangle^2}{E_{rm}^2} + \frac{\langle m\,|\mu_i|\,n\rangle^2}{E_{rm}E_{mn}} \right]. \tag{2.45}$$

Bond-Length Alternation. Bond-length alternation (BLA) has recently been identified as one useful structural parameter to predict trends for the second-order hyperpolarizability γ [2.52, 2.57–58]. Bond-length alternation is defined as the difference between the average lengths of carbon–carbon single and double bonds in a molecule. The dependence of α, β and γ on the bond length alternation is schematically shown in Fig. 2.10 [2.57]. The whole curve could be also experimentally confirmed using different solvents and several different molecules for the first-order hyperpolarizability β and in part for the second-order hyperpolarizability γ [2.58]. However, BLA has so far only been able to describe the properties of elongated molecules (containing typically four double bonds and more).

Schematically one has the following behaviour: α is maximum for a BLA of 0. β first increases up to a positive maximum, then passes through zero to a negative extremum, and finally decreases again. γ first increases to a small positive maximum, then passes through zero (=0 when β is maximized), then becomes negative with a maximum negative value for BLA=0 where β is zero. The evolution of γ is symmetrical around the point BLA=0. Note that the evolutions of β and γ with BLA are first- and second-order derivative-like

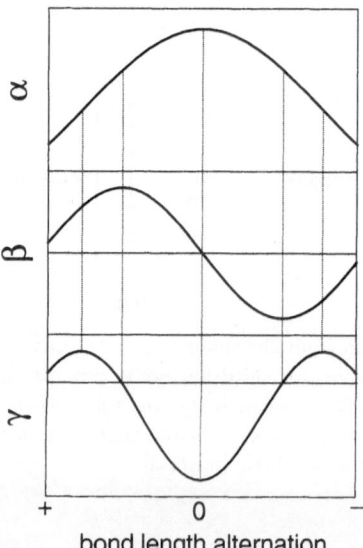

bond length alternation

Fig. 2.10. Schematic illustration of the dependence of α, β and γ on the bond-length alternation. The dashed lines indicate maximum values of β and γ

with respect to the evolution of α [2.58]. In principle, the largest values of γ can be obtained for a BLA of 0. Such materials have not yet been identified except perhaps for squaraine dyes [2.59–60]. However, there are regions in which an increase of β is associated with an increase of γ. It will be shown that these values of γ are already of great interest and can be improved even further.

An analytical structure–(hyper)polarizability relationship based on a two-state description has also been derived. In this model a parameter MIX is introduced that describes the mixture between the neutral and charge-separated resonance forms of donor–acceptor substituted conjugated molecules. This parameter can be directly related to BLA and can explain solvent effects on the molecular hyperpolarizabilities. NMR studies in solution (e.g. in $CDCl_3$) can give an estimate of the BLA and therefore allow a direct correlation with the nonlinear optical experiments.

2.2.2   Macroscopic Second-Order Nonlinear Optical Effects

A   Frequency Mixing

Frequency mixing describes the interaction of three fields which, in general, have frequencies ω_1, ω_2 and ω_3. If the generated macroscopic polarization is $P_i^{\omega_3}$ we have

$$P_i^{\omega_3}(\boldsymbol{r}) = \frac{1}{2}\varepsilon_0 \left[\chi_{ijk}(-\omega_3,\omega_1,\omega_2)E_j^{\omega_1}(\boldsymbol{r})E_k^{\omega_2}(\boldsymbol{r}) \right.$$
$$\left. +\chi_{ijk}(-\omega_3,\omega_2,\omega_1)E_j^{\omega_2}(\boldsymbol{r})E_k^{\omega_1}(\boldsymbol{r}) \right] . \tag{2.46}$$

The negative sign of ω_3 indicates an outcoming photon. In the case of frequency doubling we have (see also (2.11))

$$P_i^{2\omega} = \frac{1}{2}\varepsilon_0\chi_{ijk}(-2\omega,\omega,\omega)E_j^{\omega}E_k^{\omega} \tag{2.47}$$

and

$$P_i^{2\omega} = \varepsilon_0 d_{ijk}(-2\omega,\omega,\omega)E_j^{\omega}E_k^{\omega} . \tag{2.48}$$

The nonlinear optical coefficient d_{ijk} is symmetric in j and k and can be written in a contracted notation

$$d_{ijk} = d_{ikj} = d_{im} \quad \text{where} \quad m = \begin{cases} j & \text{for } j = k \\ 9-(j+k) & \text{otherwise} . \end{cases} \tag{2.49}$$

In the case of non-absorbing, dispersionless materials the Kleinman relation [2.61] yields

$$d_{ijk} = d_{ikj} = d_{jki} = d_{kij} = \dots . \tag{2.50}$$

B   The Linear Electro-Optic Effect

Changes (deformation, rotation) of the optical indicatrix in noncentrosymmetric substances can be induced by application of an electric field [2.62]. These electro-optic effects are most commonly described by the tensors r_{ijk} and R_{ijkl} for the linear (Pockels electro-optic coefficient) and quadratic case:

$$\Delta \left(\frac{1}{n^2}\right)_{ij} = \Delta \left(\frac{1}{\varepsilon}\right)_{ij} = r_{ijk} E_k + R_{ijkl} E_k E_l \ . \tag{2.51}$$

The change in the refractive index Δn resulting from the linear effect can be approximated as follows (for $r \cdot E \ll 1/n^2$) [2.63]

$$\Delta n_i = -\frac{n_i^3 r_{ijk} E_k}{2} \quad \text{(for } i = j) \ . \tag{2.52}$$

The relation between the linear electro-optic coefficient r_{ijk} and the nonlinear optical susceptibility $\chi_{ijk}(-\omega, \omega, 0)$ is given by

$$r_{ijk} = -\frac{2}{n_i^2 n_j^2} \chi_{ijk}(-\omega, \omega, 0) \ . \tag{2.53}$$

An important parameter for the characterization of electro-optic materials for modulator applications is the half-wave voltage, V_π, which is the voltage required to shift the phase of the transmitted beam by π in an appropriate modulator configuration [2.64]

$$V_\pi = \frac{\lambda}{(n^3 r)_{\text{eff}}} \times \frac{d}{L} = v_\pi \frac{d}{L} \ , \tag{2.54}$$

where $(n^3 r)_{\text{eff}}$ is the effective electro-optic coefficient. Low modulation voltages require a high value of $(n^3 r)_{\text{eff}}$ as well as large geometry factors L/d (L: sample thickness, d: electrode distance) achieved using the transverse modulator configuration where the light propagation direction is perpendicular to the electric field. The reduced half-wave voltage $v_\pi = \lambda/(n^3 r)_{\text{eff}}$ is often used for characterizing an electro-optic material as a sample independent quantity.

An appropriate figure of merit (FM_{EO}) for linear electro-optic effects to be used for amplitude modulators is [2.65]

$$FM_{\text{EO}} = n^3 r_{ijk} \quad \text{with} \quad n = n_i \qquad\qquad \text{for } i = j$$
$$n = \sqrt{2} \frac{n_i n_j}{\sqrt{n_i^2 + n_j^2}} \quad \text{for } i \neq j \ . \tag{2.55}$$

Using the field-induced polarization P_k the electro-optic effect is frequently expressed as [2.62]

$$\Delta \left(\frac{1}{n^2}\right)_{ij} = f_{ijk} P_k + g_{ijkl} P_k P_l \ , \tag{2.56}$$

where f_{ijk} and g_{ijkl} are the linear and quadratic polarization optic coefficients. The connection between these coefficients and the electro-optic coefficients is given by

$$r_{ijk} = \varepsilon_0 \, f_{ijm} \left(\varepsilon_{mk} - \delta_{mk} \right)$$
$$R_{ijkl} = \varepsilon_0^2 \, g_{ijmn} \left(\varepsilon_{mk} - \delta_{mk} \right) \left(\varepsilon_{nl} - \delta_{nl} \right) \tag{2.57}$$
$$\delta_{mk} = \begin{cases} 1, & \text{for } m = k \\ 0, & \text{otherwise}, \end{cases}$$

where ε_{kl} is the dielectric constant at the frequency of the modulating field. Polarization-optic coefficients are of special importance in the characterization of oxygen-octahedra ferroelectric materials since they can be used to describe the linear electro-optic effect as well as the birefringence in these materials.

C   Phase-Matching Considerations

For phase-matched parametric interactions among three waves, where $\omega_3 = \omega_1 + \omega_2$, the interacting waves must satisfy $\boldsymbol{k}_3 = \boldsymbol{k}_1 + \boldsymbol{k}_2$. That is, in the case of collinear interaction if all the $\boldsymbol{k}_i$ are parallel to one another, we have $n(\omega_3)\omega_3 = n(\omega_1)\omega_1 + n(\omega_2)\omega_2$, where $n(\omega_i)$ are the refractive indices for the waves at frequencies ω_i. For general directions of the wave vectors and polarizations in the crystal the induced nonlinear polarization at ω_3 can be written as

$$|\boldsymbol{P}^{\omega_3}| = 2\varepsilon_0 d_{\text{eff}} |\boldsymbol{E}^{\omega_1}| \, |\boldsymbol{E}^{\omega_2}| \quad \text{(type II)}$$
$$\text{and } |\boldsymbol{P}^{2\omega}| = \varepsilon_0 d_{\text{eff}} |\boldsymbol{E}^{\omega}|^2 \quad \text{(type I)} \tag{2.58}$$

for second-harmonic generation with the effective nonlinear optical coefficient

$$d_{\text{eff}} = \sum_{ijk} d_{ijk}(-\omega_3, \omega_1, \omega_2) \cos(\alpha_i^{\omega_3}) \cos(\alpha_j^{\omega_1}) \cos(\alpha_k^{\omega_2}) \,, \tag{2.59}$$

where α_i^{ω} is the angle between the electric-field vector of the wave ω and the main axis i of the indicatrix [2.66–67]. Note that the walk-off angle should be taken into account in order to calculate the angles α_i^{ω} correctly.

In the case of electro-optic modulation, the situation is different compared with frequency conversion experiments. As long as the frequency of the modulating electric field is much smaller than that of the light field, the phase-matching condition is always fulfilled. Phase differences between the modulating and light fields may become important for modulation frequencies above 1 GHz.

D   Combination of Piezoelectricity and the Elasto-Optic Effect: an Example of Cascading

In strongly polar inorganic materials such as $KNbO_3$ [2.65] lattice vibrational effects due to acoustic and optic phonons can significantly contribute to the linear electro-optic effect and are often similar in size to the electronic contributions. In organic materials electro-optic effects are assumed to be of purely electronic origin in the visible spectral range.

The free electro-optic coefficient r^{T} contains three contributions: the electronic part (r^{e}) and the effects from optic (r^{o}) and acoustic (r^{a}) phonons [2.68]

$$r^{\mathrm{T}} = r^{\mathrm{e}} + r^{\mathrm{o}} + r^{\mathrm{a}} = r^{\mathrm{s}} + r^{\mathrm{a}} . \tag{2.60}$$

r^{s} is called the clamped electro-optic coefficient and is defined through (2.60). The various physical processes contributing to the electro-optic effect have been explained by considering field-induced changes of the optical dielectric constant with contributions from optic (Raman scattering processes) and acoustic vibrational modes (Brillouin scattering processes) [2.69–71].

We first have a contribution (r^{e}) that is purely electronic in origin. The second contribution (r^{o}) is due to the optic modes and is nonzero only for vibrational modes being both Raman and infrared active. The third contribution (r^{a}) arises from the photoelastic effect, i.e. from lattice strains induced by the inverse piezoelectric effect driven by the applied electric field. It should be mentioned that all three contributions can have either sign.

The acoustic mode contribution can be described by considering elasto-optic effects due to strains $s_{jk} = a_{ijk}E_i$ induced by the inverse piezoelectric effect for electric fields E_i [2.20]. In measurements of electro-optic coefficients the applied electric field first induces a change of the refractive index via the electro-optic coefficient r^{s} (clamped electro-optic effect). In addition the applied field can also induce a strain (inverse piezoelectric effect) which leads to a change in the refractive index via the elasto-optic coefficients p_{ijmn} defined by

$$\Delta \left(\frac{1}{n^2} \right)_{ij} = p_{ijmn}\, s_{mn} . \tag{2.61}$$

This leads to

$$\Delta \left(\frac{1}{n^2} \right)_{ij} = \left(r^{\mathrm{s}}_{ijk} + p_{ijmn}\, a_{kmn} \right) E_k = r^{\mathrm{T}}_{ijk} E_k \tag{2.62}$$

with

$$r^{\mathrm{a}}_{ijk} = p_{ijmn}\, a_{kmn} . \tag{2.63}$$

Obviously the combination of piezoelectricity and the elasto-optic effect can yield an important contribution to the linear electro-optic effect and is a very illustrative example of cascading.

At low frequencies acoustic and optic phonons can be excited (already below 1 MHz). Above the fundamental acoustic resonances the acoustic modes are clamped and above the 'Reststrahlen' region (around 10^{12} Hz) also the optic mode contributions disappear. The electro-optic coefficients thus often

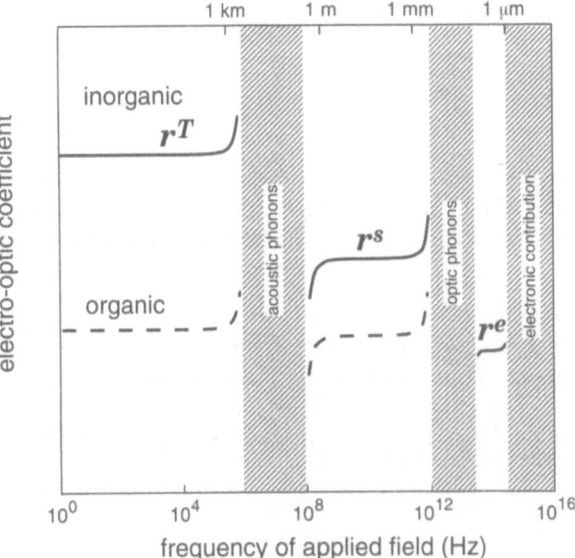

Fig. 2.11. Sketch of the dependence of the electro-optic coefficient on the frequency of the applied electric field. A different behavior is expected as well as experimentally confirmed for inorganic and organic materials

show a significant dependence on the frequency of the applied electric field (Fig. 2.11). Since the phonon contributions are small in organic materials this frequency dependence is expected to be of minor importance in these materials.

E   The Local Field Problem

If an electric field is applied to a solid, the molecules or atoms in the solid experience a different field. This field which is very difficult to calculate in most cases is called the *local field*. Possible local field corrections are discussed in the following.

The Reaction Field. The reaction field R is the electric field felt by the solute due to the orientation and/or electronic polarization of the solvent (reaction) induced by the solute dipole [2.72–75]. The field of the dipole in a cavity polarizes the surrounding matter and the resulting inhomogeneous polarization of the environment will give rise to a field at the dipole, R. Its magnitude is proportional to the solute dipole moment μ

$$R = r\mu ,\tag{2.64}$$

where r is the reaction field factor. We here assume a spherical cavity and neglect possible anisotropies. For a non-polarizable point dipole the reaction field factor r is given by [2.73]

$$r = \frac{1}{4\pi\varepsilon_0 a^3} \frac{2(\varepsilon - 1)}{2\varepsilon + 1} \ , \tag{2.65}$$

where ε is the static dielectric constant and a is the radius of the cavity occupied by a solute molecule. This radius will be treated below. If the dipole is polarizable[6] (with an average polarizability α[7]) the reaction field is given by

$$R = r\left(\mu_0 + \varepsilon_0 \alpha R\right) \ , \tag{2.66}$$

where μ_0 is the dipole moment. This yields the following equation for R

$$R = \frac{1}{4\pi\varepsilon_0 a^3} \frac{1}{[2(\varepsilon + 1)/(2\varepsilon - 1)] - (\alpha/4\pi a^3)} \mu_0 \tag{2.67}$$

and leads to

$$R = \frac{1}{4\pi\varepsilon_0 a^3} \frac{2(\varepsilon - 1)}{\varepsilon + 2} \mu_0 \tag{2.68}$$

with the approximation $\alpha = 2\pi a^3$ [2.74] and to

$$R = \frac{1}{4\pi\varepsilon_0 a^3} \frac{2(\varepsilon - 1)(\varepsilon + 2)}{9\varepsilon} \mu_0 \tag{2.69}$$

with the Clausius–Mossotti equation

$$\alpha = \frac{\varepsilon - 1}{\varepsilon + 2} 4\pi a^3 = \frac{\varepsilon - 1}{\varepsilon + 2} \frac{3}{N} \ . \tag{2.70}$$

For optical frequencies (2.68) and (2.69) are [2.73–74]

$$R_{\text{opt}} = \frac{1}{4\pi\varepsilon_0 a^3} \frac{2\left(n^2 - 1\right)}{n^2 + 2} \mu_{\text{opt}} \tag{2.71}$$

$$R_{\text{opt}} = \frac{1}{4\pi\varepsilon_0 a^3} \frac{2\left(n^2 - 1\right)\left(n^2 + 2\right)}{9n^2} \mu_{\text{opt}} \ . \tag{2.72}$$

Meredith [2.76] has defined (2.72) as

$$R_{i,\text{opt}} = \frac{1}{4\pi\varepsilon_0 a^3} \frac{2\left(n^2 - 1\right)\left(n_i^2 + 2\right)}{3\left(2n^2 + n_i^2\right)} \mu_{\text{opt}} \tag{2.73}$$

---

[6] All real molecules have a positive polarizability α.

[7] Throughout this work we assume the polarizability tensors of the solvent and solute molecules to be isotropic. The basic theory is still quite crude and refinements of the polarizability anisotropies would not necessarily yield more information and would considerably complicate the present discussion.

by taking into account that if species i is more polarizable its reaction field must be stronger. Equation (2.73) reduces to (2.72) if we take $n = n_i$. We can illustrate the magnitude of the reaction field with an example by assuming the following parameters: a cavity radius of $a = 5\,\text{Å}$, a refractive index of 1.5, and a dipole moment μ_{opt} of $3.35 \times 10^{-29}\,\text{Cm}$ (10 Debye). This gives values for the reaction field of $1.4 \times 10^7\,\text{V/cm}$ and $1 \times 10^7\,\text{V/cm}$ ((2.71) and (2.72), respectively) of considerable strength. The reaction field will be used in Sects. 2.3 and 2.6 to calculate the contribution of cascading through the reaction field to third-harmonic generation in solution.

The Cavity Radius of a Molecule. The calculation of the radius of the cavity occupied by a solute molecule is a difficult task. The simplest approximation was given by Onsager [2.73]

$$\frac{4\pi N a^3}{3} = 1 \tag{2.74}$$

with N the number density. Amos proposed using [2.74]

$$a = r_{solute} + \frac{1}{2}r_{solvent} \tag{2.75}$$

with

$$r_{solute} = \left(\frac{3M_{solute}}{4\pi N_A \rho_{solute}}\right)^{1/3} , \quad r_{solvent} = \left(\frac{3M_{solvent}}{4\pi N_A \rho_{solvent}}\right)^{1/3} . \tag{2.76}$$

r_{solute} and r_{sovent} are the spherical radii of the solute and solvent molecules, respectively. M is the molecular mass, N_A is Avogadro's number, and ρ is the molecular density. A different estimation was used by Wong et al. [2.77]. Based on theoretical considerations they obtained

$$a = r_{solute} + 0.5\,\text{Å} . \tag{2.77}$$

The addition of 0.5Å accounts for the van der Waals radii of the surrounding solvent molecules. The approximation of Wong yields values that lie between those of Onsager and Amos. As an example, we obtain for the stilbene derivative 4-(N,N-dimethylamino)-4'-nitrostilbene (DANS, $M = 268.32\,\text{g}$, $\rho \approx 1.3\,\text{g/cm}^3$)) dissolved in 1,4-dioxane ($M = 84.07\,\text{g}$, $\rho \approx 1.3\,\text{g/cm}^3$)

$$a = 4.34\,\text{Å (Onsager)} \quad a = 5.81\,\text{Å (Amos)} \quad a = 4.84\,\text{Å (Wong)}.$$

The Local Field. Let us start with a sphere in a dielectric with a uniform field E_{ext}. The corresponding cavity field E_{cav} inside the sphere is given by [2.73]

$$E_{cav} = \frac{3}{2\varepsilon + 1}E_{ext} . \tag{2.78}$$

In the Onsager theory the local field can be described as the sum of E_{cav} and the reaction field. For a two-component dipolar liquid (e.g. solvent and solute) considered as a non-polar dielectric the local field (for optical frequencies) can be written as

$$E_{L,i} = E_{\text{cav}} + R = \frac{n^2\left(n_i^2 + 2\right)}{2n^2 + n_i^2} E \tag{2.79}$$

with the use of the Onsager approximation for the cavity radius a, the Clausius–Mossotti equation and (2.65). n is the refractive index of the solution (concentration dependent) and n_i is the refractive index of species i. Equation (2.79) defines the Onsager local field.

Lorentz [2.73] developed a different well-known model that works well for non-polar dielectrics to account for the local field:

$$E_L = \frac{\left(n^2 + 2\right)}{3} E = fE \ . \tag{2.80}$$

The Lorentz local field is completely accurate in the case of a cubic or isotropic arrangement of identical particles, such as solutions (assuming no interaction between molecules). It is interesting to note that the Onsager local field yields the Lorentz local field if we take $n = n_i$.

The local field can also be considered in a different way: The microscopic field E_L^ω acting on e.g. a molecule due to a macroscopic field E^ω differs due to the polarization field P^ω:

$$E_L^\omega = E^\omega + L^\omega P^\omega = f^\omega E^\omega \ . \tag{2.81}$$

One finds

$$f^\omega = 1 + \varepsilon_0 \chi^{(1)} L^\omega \ . \tag{2.82}$$

In the Lorentz approximation L^ω is equal to $1/(3\varepsilon_0)$.

The difference between the reaction field and the local field is that, in the first case, only the response of the surrounding molecules polarized by that *one* molecule is considered, whereas in the second case the total influence of all other dipoles is taken into account.

F   Relation Between Microscopic and Macroscopic Nonlinear Optical Coefficients

Based on bond hyperpolarizabilities in inorganic materials, microscopic and macroscopic nonlinear optical properties were first related by Bergman and Crane [2.78]. Zyss and Oudar later applied their model to organic crystals [2.79]. Their so-called oriented gas model neglects all intermolecular interactions contributing to the optical nonlinearity except for accounts for field corrections. Therefore only intramolecular interactions are taken into account. This leads to a simple dependence of the nonlinear optical susceptibilities on structural parameters and molecular hyperpolarizabilities

$$\chi_{ijk}^{(2)} = N \frac{1}{n(g)} f_i^{2\omega} f_j^\omega f_k^\omega \sum_s^{n(g)} \sum_{mnp}^{3} \cos\left(\theta_{im}^s\right) \cos\left(\theta_{jn}^s\right) \cos\left(\theta_{kp}^s\right) \beta_{mnp} \ . \tag{2.83}$$

θ^{s}_{im} are the angles between dielectric and molecular axes i and m, N is the number of molecules per unit volume, $n(g)$ is the number of symmetry equivalent positions in the unit cell, s denotes a site in the unit cell, f^{ω}_i are local field corrections at the different frequencies and β_{mnp} is the molecular hyperpolarizability.

For frequency doubling (2.83) yields

$$
\begin{aligned}
d_{ijk}\left(-2\omega,\omega,\omega\right) &= \frac{1}{2}\chi^{(2)}_{ijk}\left(-2\omega,\omega,\omega\right) \\
&= \frac{1}{2}N\frac{1}{n(g)}f^{2\omega}_i f^{\omega}_j f^{\omega}_k \sum_{\mathrm{s}}^{n(g)}\sum_{mnp}^{3}\cos\left(\theta^{\mathrm{s}}_{im}\right) \\
&\quad \cos\left(\theta^{\mathrm{s}}_{jn}\right)\cos\left(\theta^{\mathrm{s}}_{kp}\right)\beta_{mnp}\left(-2\omega,\omega,\omega\right) \ .
\end{aligned}
\tag{2.84}
$$

A similar relation can be obtained for the linear electro-optic effect:

$$
\begin{aligned}
r_{ijk}\left(-\omega,\omega,0\right) &= -\frac{2}{n^2_i(\omega)n^2_j(\omega)}\chi^{(2)}_{ijk}\left(-\omega,\omega,0\right) \\
&= -N\frac{1}{n(g)}\frac{2}{n^2_i(\omega)n^2_j(\omega)}f^{(\omega)}_i f^{(\omega)}_j f^{(0)}_k \sum_{\mathrm{s}}^{n(g)}\sum_{mnp}^{3}\cos\left(\theta^{\mathrm{s}}_{im}\right) \\
&\quad \cos\left(\theta^{\mathrm{s}}_{jn}\right)\cos\left(\theta^{\mathrm{s}}_{kp}\right)\beta_{mnp}\left(-\omega,\omega,0\right) \ .
\end{aligned}
\tag{2.85}
$$

As mentioned above it is appropriate for molecules with strong nonlinearities along a single charge transfer axis to assume a one-dimensional hyperpolarizability β_{zzz} along such an axis. Using the oriented gas model measured values of β_{zzz} could be compared reasonably well with the measured d_{ijk} and r_{ijk} [2.80–81].

When the intermolecular interactions do not contribute significantly to the optical nonlinearity, one generally only has to take into account a red shift of the resonance frequency when describing macroscopic electro-optic or nonlinear optical coefficients with microscopic hyperpolarizabilities. Such red shifts can be deduced from, for example, a comparison of Sellmeier coefficients of bulk crystals and absorption measurements of molecules in a solution (see below). The local field correction factors f^{ω} are most often calculated within the simple approximation of the Lorentz model [2.82] (see above).

The electronic contributions to the electro-optic effect r^{e}_{ijk} can be calculated directly from the nonlinear optical coefficients d^{EO}_{ijk} using (see also (2.53)) [2.25]

$$
r^{\mathrm{e}}_{ijk} = -\frac{4}{n^2_i(\omega)n^2_j(\omega)}d^{\mathrm{EO}}_{ijk} \ .
\tag{2.86}
$$

Again, assuming a one-dimensional two-level model for the charge-transfer, d^{EO}_{ijk} is given by

$$
d^{\mathrm{EO}}_{ijk} = d^{(-\omega,\omega,0)}_{ijk}
$$

$$= \frac{f_i^\omega f_j^\omega f_k^0}{f_k^{2\omega'} f_i^{\omega'} f_j^{\omega'}} \frac{\left(3\omega_{eg}^2 - \omega^2\right)\left(\omega_{eg}^2 - \omega'^2\right)\left(\omega_{eg}^2 - 4\omega'^2\right)}{3\left(\omega_{eg}^2 - \omega^2\right)^2 \omega_{eg}^2} d_{kij}^{(-2\omega',\omega',\omega')} .$$

$$(2.87)$$

With the above two equations nonlinear optical coefficients measured at frequency ω' can be compared to electro-optic coefficients measured at frequency (of light) ω.

G Limits of the Electro-Optic and Nonlinear Optical Response in Molecular Crystals

Equations (2.39), (2.84) and (2.85) describe the nonlinear optical and electro-optic coefficients d_{ijk} and r_{ijk} in terms of molecular and crystallographic parameters and refractive indices. They also describe their dispersion by the frequency dependence of the second-order polarizability β and the local field factors. In order to estimate the maximum nonlinear optical and electro-optic coefficients of a material based on charge transfer molecules, the limits of the following parameters must be known [2.71]:

- The molecular second-order polarizability β (2.39). This is the parameter that will have the strongest influence on the coefficients. It primarily depends on the shape and size of the molecule and the nature of its donor and acceptor substituents. It also depends strongly on the frequency of the optical fields and shows resonance enhancement near the charge transfer transition.
- The structural parameters N (number of molecules per unit volume) and θ_{ii}^s (angle between molecular charge transfer axis and reference system) or θ_p (angle between molecular charge transfer axis and polar crystal axis).
- The refractive indices n_i, which are used in (2.84) and (2.85) and which give an estimate of the local field factors f_i. The refractive indices are estimated to about 2.2–2.4 for stilbene and extended thiophene derivatives and around 2.0 for benzene derivatives in spectral regions of moderate absorption.

The first two parameters in this list, the hyperpolarizability and the molecular density, are discussed in more detail in the following two sections.

Second-Order Polarizabilities. Maximum second-order polarizabilities can be estimated from measured values. For a semiempirical estimation of β we used experimental results published by L.-T. Cheng et al. and Rao et al. [2.83–86]. The measurements were performed at $\lambda = 1.907\,\mu$m to be truly nonresonant. These authors have measured the hyperpolarizabilities of a large number of benzene, stilbene and thiophene derivatives by electric field-induced second harmonic generation (EFISH). Also given are values of hydrazone derivatives determined in this work (see Sect. 2.4). Figure 2.12 shows a plot of the hyperpolarizability extrapolated to infinite wavelengths, β_0 (see (2.42)) as a function of the maximum absorption wavelength λ_{eg}. Donor and acceptor influence both ω_{eg} and β so that a direct relationship between β and ω_{eg} can be deduced. It is also important to notice that in the low wavelength region the transparency–efficiency trade-off is much more favourable for the hydrazone derivatives when compared with the stilbene and extended thiophene derivatives.

From the two parameters λ_{eg} and β_0 it is possible to calculate the hyperpolarizabilities $\beta_{zzz}(-\omega,\omega,0)$ and $\beta_{zzz}(-2\omega,\omega,\omega)$ at the desired wavelengths using the relations (2.41) and (2.42) [2.71].

Note that when transferring the data in Fig. 2.12 to solids a shift $\Delta\lambda$ of λ_{eg} ($= 2\pi c/\omega_{eg}$) towards the red is observed due to higher dipole-dipole interactions: $\lambda_{eg}^{solid} = \lambda_{eg}^{solution} + \Delta\lambda$. For benzenes we have $\Delta\lambda \approx 40\,nm$, for stilbenes $\Delta\lambda \approx 60\,nm$ (for solutions in 1,4-dioxane) as deduced from a comparison of Sellmeier coefficients and absorption measurements in solutions of various compounds in nonpolar solvents [2.71]. To take full advantage of the resonance enhancement, the operating wavelength should lie close to the absorption edge. As a crude rule the absorption edge for the molecules considered here lies 100 nm from its solid state peak absorption wavelength λ_{eg}^{solid}, i.e. $\lambda_{eg}^{solid} = \lambda_c - 100\,nm$. Therefore we will calculate the electro-optic coefficients at $\lambda_{eg}^{solution}+100\,nm$ but plot them vs. $\lambda_c = \lambda_{eg}^{solution}+100\,nm+\Delta\lambda$ (see below).

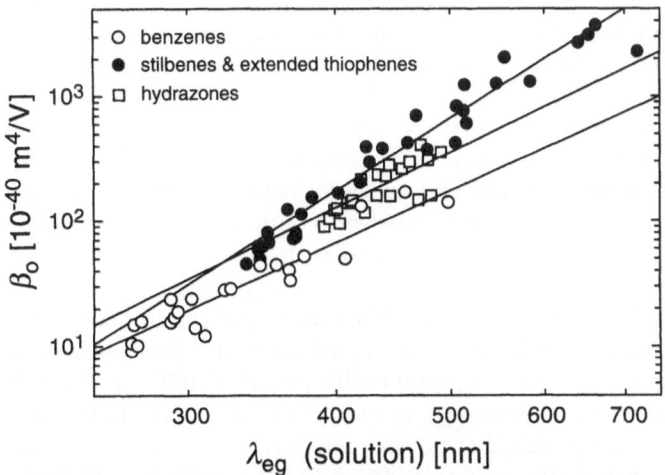

Fig. 2.12. Second-order polarizability β_0 of benzene, stilbene, thiophene and hydrazone derivatives extrapolated to infinite wavelengths vs. wavelength of maximum absorption. For the extended donor–acceptor substituted thiophene derivatives benzene rings were replaced by thiophene rings. Data points originate from Cheng et al. [2.83], Jen et al. [2.84], Rao et al. [2.85–86], and Wong et al. [2.87]. All data points were adjusted for a reference value of quartz of $d_{11} = 0.28\,pm/V$

Limitations Due to Crystal Structure. The relations between the microscopic and macroscopic nonlinear optical and electro-optic response ((2.84) and (2.85)) depend on the structural parameters N and θ_p. N is the number of molecules per unit volume and θ_p the angle between the molecular charge transfer axis and the polar crystalline axis. In order to maximize a coefficient r_{ijk} (or d_{ijk}), it is easy to see from (2.85) (or (2.84)) that θ_{im}^s (or θ_p) should be close to zero. Such a configuration optimizes the diagonal coefficient r_{iii} (or d_{iii}). In other words, the charge transfer axes of all molecules should be parallel.

For nonlinear optical applications in bulk crystals, such as frequency doubling, phase matching considerations come into play. In this case theoretical considerations lead to an optimum angle $\theta_p^{PM} = 54.7°$ (or 125.3°) between the charge transfer axis (assumed one-dimensional) and the polar axis of the crystal (for the most favourable point groups) for phase-matchable nonlinear optical coefficients [2.79]. In addition, highly efficient nonlinear optical interactions are only possible for the case of noncritical phase matching.

For quasi phase-matched frequency conversion in bulk crystals and waveguides or phase matching using modal conversion in waveguides, the highest conversion efficiencies are obtained for the same orientation of the molecules as for electro-optics.

The molecular density N can be estimated from typical densities and molecular weights for benzene and stilbene derivatives. A typical benzene derivative with a nearly optimized structure is 2-methyl-4-nitroaniline (MNA) [2.88]; typical stilbene derivatives are 3-methyl-4-methoxy-4'-nitrostilbene (MMONS) [2.89] and 4'-nitrobenzylidene-3-acetamino-4-methoxyaniline (MNBA) [2.90].

The molecular density for MNA is $N \approx 6 \times 10^{27}\,\mathrm{m}^{-3}$; for MMONS and MNBA it is $N \approx 3 \times 10^{27}\,\mathrm{m}^{-3}$. Since the packing in most molecular crystals is very tight, these numbers do not vary much for other molecules. For comparison, typical chromophore densities of electro-optic polymers are around $N \approx 1.5 \times 10^{27}\,\mathrm{m}^{-3}$.

Electro-Optic and Nonlinear Optical Coefficients. Using the optimized hyperpolarizabilities β_{max} as explained above we can now find the upper limits of the nonlinear optical and electro-optic coefficients d_{ijk} and r_{ijk} [2.71]. If all intermolecular contributions (except for local field corrections) to the nonlinearity are neglected, optimized structures for electro-optic and nonlinear optical applications can easily be obtained (see (2.84) and (2.85)). For a series of compounds such as para-disubstituted benzene derivatives we chose a small molecule to use the corresponding large number of molecules per unit volume and refractive index in order to get an approximation for the upper limit for electro-optic and nonlinear optical effects. These values were then used for all compounds in this series. Since MNA and MMONS are well characterized, we chose the values from MNA [2.88, 2.91] ($N \approx 6 \times 10^{27}\,\mathrm{m}^{-3}$, $n^{2\omega} = 2.2$, $n^\omega = 2.2 - 0.3$) for the benzene derivatives and MMONS [2.89]

$(N \approx 3 \times 10^{27}\,\mathrm{m}^{-3},\ n^{2\omega} = 2.4,\ n^{\omega} = 2.4 - 0.3)$ for the stilbene and thiophene derivatives as was already mentioned above.

Taking these conditions into account and using (2.84) and (2.85) one obtains the upper limits for the nonlinear optical coefficients d_{max} in Fig. 2.13a and for the electro-optic coefficients r_{max} in Fig. 2.13b. Note that the values are calculated at $2 \times (\lambda^{\mathrm{solution}}_{\mathrm{eg}} + 100\,\mathrm{nm})$ but plotted vs. $\lambda_{\mathrm{c}} = 2 \times (\lambda^{\mathrm{solution}}_{\mathrm{eg}} + 100\,\mathrm{nm} + \Delta\lambda)$ ($\Delta\lambda = 40\,\mathrm{nm}$ for benzene derivatives and $\Delta\lambda = 60\,\mathrm{nm}$ for the others) for d_{max} to account for the red shift as explained above. By analogy, r_{max} is calculated at $\lambda^{\mathrm{solution}}_{\mathrm{eg}} + 100\,\mathrm{nm}$ but plotted vs $\lambda_{\mathrm{c}} = \lambda^{\mathrm{solution}}_{\mathrm{eg}} + 100\,\mathrm{nm} + \Delta\lambda$. The Lorentz approximations for the local field factors (2.80) were used in all cases.

It can be seen from Figs. 2.13a and 2.13b that shifting the absorption edge to longer wavelengths by adding stronger donors and acceptors will consequently increase the maximum value of the electro-optic coefficient. However, it should be noted that the absorption edge can only be shifted up to values around 700–800 nm for very strong donors and acceptors. Also, the addition of strong donors and acceptors increases the ground state dipole moment of a molecule, which often leads to centric crystal structures.

For poled electro-optic polymers the situation is slightly different. The refractive indices of common electro-optic polymers are smaller than the ones of molecular crystals (typically around 1.6–1.9). In comparison with stilbene-type crystals the chromophore density is about half as big (typically around $N \approx 1.5 \times 10^{27}\,\mathrm{m}^{-3}$). Taking these considerations into account we obtain upper limits for electro-optic coefficients r_{max} for poled polymers that are about half as large as the ones for the corresponding stilbene and thiophene-type molecular crystals.

Figures 2.13a and 2.13b also display measured values of nonlinear optical and electro-optic coefficients of different organic single crystals. We see that most materials known up to now are far from being optimized. Our estimation indicates that several 1000 pm/V might be possible for both d and r. The major problem is not the molecular but the crystal engineering. There are already many new molecules with exceptionally large nonlinearities. Few of them could, however, be used to grow optimized noncentrosymmetric crystals. Therefore a great potential in the development of better nonlinear optical and electro-optic molecular crystals remains to be exploited.

2.2.3   Macroscopic Third-Order Nonlinear Optical Effects

A   Frequency Mixing

Frequency mixing phenomena in third-order nonlinear optics include general four-wave mixing ($\chi^{(3)}(-\omega_4, \omega_1, \omega_2, \omega_3)$), third-order sum and difference frequency generation ($\chi^{(3)}(-\omega_3, \pm\omega_1, \omega_2, \omega_2)$), dc-induced second-harmonic generation ($\chi^{(3)}(-2\omega, 0, \omega, \omega)$), third-harmonic generation ($\chi^{(3)}(-3\omega, \omega, \omega, \omega)$),

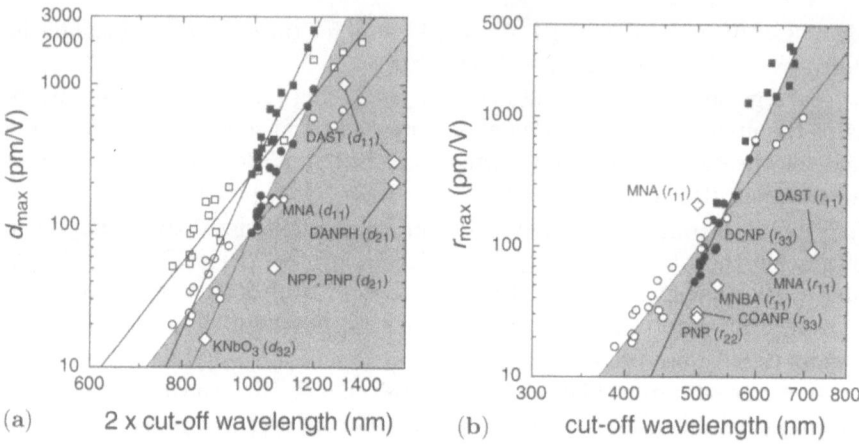

Fig. 2.13. (a) Upper limits of nonlinear optical coefficients. The graph shows maximum values of d_{ij} vs. $2\times$ cut-off wavelength calculated from measured values of hyperpolarizabilities assuming an optimum parallel alignment of all molecules for either phase-matched interactions in the bulk (dashed line) or in quasi-phase-matched (QPM) structures. ∘: donor–acceptor disubstituted benzene derivatives (bulk); □: donor–acceptor disubstituted benzene derivatives (QPM), •: donor–acceptor disubstituted stilbene derivatives (bulk); ■: donor–acceptor disubstituted stilbene derivatives (QPM). The shaded area represents values of d_{ij} that should be reachable by optimized crystalline structures for phase-matched interactions in the bulk. Measured values of nonlinear optical coefficients of some molecular crystals are also plotted. (b) Upper limits of electro-optic coefficients. The graph shows maximum values of r_{ij} vs. cut-off wavelength calculated from measured values of hyperpolarizabilities assuming an optimum parallel alignment of all molecules. ∘: donor–acceptor disubstituted benzene derivatives; •: donor–acceptor disubstituted stilbene derivatives; ■: extended donor–acceptor disubstituted thiophene derivatives. The solid line is a fit for the stilbene derivatives and the dashed line a fit for benzene derivatives. The shaded area represents values of r_{ij} that should be reachable by optimized crystalline structures. Measured values of electro-optic coefficients of some molecular crystals are also plotted. All data points were adjusted for a reference value of quartz of $d_{11} = 0.28\,\mathrm{pm/V}$

and others [2.92]. As in second-order nonlinear optical frequency processes, phase-matching can strongly enhance the observed effects.

Third-harmonic generation is the most widely used technique to measure elements of $\chi^{(3)}_{ijkl}$. The biggest advantage is that only electronic contributions to the third-order nonlinearity are measured since only these effects can follow the frequency of the light field (see Appendix A.1 for definition).

B   The Optical Kerr Effect

The refractive index of many optical materials depends on the intensity of the light propagating through the material. This phenomena is the optical Kerr

Table 2.3. Mechanism leading to a nonlinear refractive index. Typical values for the magnitude and the response time of the effects are given

Mechanism	n_2 (cm^2/W)	$\chi^{(3)}$ (m^2/V^2)	Response time (s)
Electronic polarization	10^{-16}	10^{-18}	10^{-15}
Molecular orientation	10^{-14}	10^{-14}	10^{-12}
Electrostriction	10^{-14}	10^{-14}	10^{-9}
Saturated atomic absorption	10^{-10}	10^{-12}	10^{-8}
Thermal effects	10^{-6}	10^{-8}	10^{-3}
Photorefractive effects	Large	Large	Intensity-dependent
Example of KNbO$_3$: $I \approx 1\,\mathrm{mW/cm^2}$			
$-1\,\mathrm{W/cm^2}$	$10^{-5} - 10^{-1}$	$1 \times 10^{-12} - 5 \times 10^{-8}$	$10^4 - 10^{-4}$

effect. In centrosymmetric media the third-order susceptibility $\chi^{(3)}$ is the dominant contribution to the nonlinear refractive index n_2 which is defined as (see also (2.32))

$$n = n_0 + \overline{n_2} \langle E(t)^2 \rangle = n_0 + \frac{\overline{n_2}}{2} |E(\omega)|^2 = n_0 + n_2 I \,, \qquad (2.88)$$

where n_0 is the linear refractive index and I is the intensity of the incoming beam. The relation between the optical Kerr susceptibility $\chi^{(3)} = \chi^{(3)}$ $(-\omega, \omega, -\omega, \omega)$ (assumed to be real) and n_2 can be deduced from (2.1) and (2.88) for the case of negligible absorption and is given by [2.93]

$$n_2 = \frac{3\chi^{(3)}}{4c\varepsilon_0 n_0^2} \,, \qquad (2.89)$$

where c is the speed of light in vacuum. Note that in the case of significant linear and nonlinear absorption n_2 may depend on both the real and imaginary parts of $\chi^{(3)}$.

There are many different mechanisms that produce a nonlinear change in the refractive index, which differ in magnitude and response time. Table 2.3 lists several typical processes [2.94].

Changes in the refractive index through the cascading of second-order nonlinearities will be discussed in Sect. 2.3.

There are many different processes that result from the nonlinear refractive index. Among them are optical phase conjugation [2.95], the self-focusing and self-trapping of light [2.96–99], optical bistability [2.100], and optical solitons in time [2.101–102] and space [2.103–104].

C   Relation Between Microscopic and Macroscopic Nonlinear Optical Coefficients

By analogy with the second-order nonlinearities a relation between the microscopic second-order hyperpolarizability γ and the macroscopic susceptibility

$\chi^{(3)}$ can be found. For the case of third-harmonic generation in single crystals this relation is as follows:

$$\chi_{ijkl}^{(3)} = N \frac{1}{n(g)} f_i^{3\omega} f_j^{\omega} f_k^{\omega} f_l^{\omega}$$

$$\times \sum_{s}^{n(g)} \sum_{mnpq}^{3} \cos\left(\theta_{im}^{s}\right) \cos\left(\theta_{jn}^{s}\right) \cos\left(\theta_{kp}^{s}\right) \cos\left(\theta_{lq}^{s}\right)$$

$$\times \gamma_{mnpq}\left(-3\omega, \omega, \omega, \omega\right) , \qquad (2.90)$$

whereas it is given by

$$\chi^{(3)} = N f_i^{3\omega} f_j^{\omega} f_k^{\omega} f_l^{\omega} \gamma\left(-3\omega, \omega, \omega, \omega\right) \qquad (2.91)$$

for the case of a random polymer ($\chi^{(3)}$ and γ are rotational averages (see Sect. 2.1)). As can be seen from (2.90) the third-order nonlinear optical susceptibility is largest if all molecules are completely parallel to each other (that is for $\theta_{im} = 0$). In contrast to second-order nonlinear optical effects it does not matter if the molecules are parallel or antiparallel to each other (except for different molecular interactions).

2.3   Cascaded Second-Order Nonlinearities $\chi^{(2)} : \chi^{(2)}$

Cascading is a process where lower-order effects are combined to contribute to a higher-order nonlinear process. The basic concept of cascading has already been introduced in Sect. 2.1.2. Here we present different possible mechanisms of cascading and show their relevance for future applications. All definitions are based on the equations in Sect. 2.1.3.

2.3.1   Second-Harmonic Generation
and Sum-Frequency Generation

Third-harmonic generation can occur in any material. In this process a fundamental wave at frequency ω produces a wave at frequency 3ω ($\propto \chi^{(3)}$ $(-3\omega, \omega, \omega, \omega)$). As was described in Sect. 2.1.2 in noncentrosymmetric materials sum frequency mixing ($\omega_1 + \omega_2 = \omega_3$) and second-harmonic generation ($\omega + \omega = 2\omega$) are also allowed: an intermediate field at frequency 2ω is generated through second-harmonic generation. This field can interact with the fundamental wave through sum frequency mixing ($2\omega + \omega = 3\omega$) to generate a field at frequency 3ω itself ($\propto \chi^{(2)} (-3\omega, 2\omega, \omega) \times \chi^{(2)} (-2\omega, \omega, \omega)$) [2.7]. Figure 2.2 illustrates this process schematically. In this work third-harmonic generation experiments were carried out with the Maker fringe technique. The detailed experimental setup as well as the theoretical description of the experimental curves will be given in Sects. 2.5 and A.3.

It is important to note that these intermediate fields do not need to be able to propagate. It is possible to generate longitudinal fields at frequency

2ω that can subsequently interact with photons at the fundamental frequency ω.

A special feature of cascading is the fact that it often is a self-calibrating process. In the case of cascading in third-harmonic generation this means the following: from a theoretical fit of an experimental curve that shows cascaded processes one can obtain the ratio

$$\frac{\chi^{(3)}\left(-3\omega,\omega,\omega,\omega\right)}{\chi^{(2)}\left(-3\omega,2\omega,\omega\right)\times\chi^{(2)}\left(-2\omega,\omega,\omega\right)} \,. \tag{2.92}$$

Therefore, if the values of $\chi^{(2)}$ are known, the value of $\chi^{(3)}$ can be determined. This principle has been applied by Meredith [2.7] to find the values of $\chi^{(3)}$ of crystalline quartz (α-quartz) and fused silica, based on known values of $\chi^{(2)}$ of α-quartz. An advantage of cascaded second-harmonic and sum frequency generation is the fact that only electronic effects are measured; this in contrast to, for example, the z-scan technique, in which all effects that are fast enough (and which can react within the duration of the pulses used in the experiment) can contribute (especially optical phonons).

2.3.2   Cascading Through the Reaction Field in Centrosymmetric Media

It is possible to get cascaded processes through the reaction field (Sect. 2.2.2.E) [2.76, 2.105]. We will discuss this for the case of third-harmonic generation in solutions of noncentrosymmetric molecules. All tensor notations will be dropped for the moment. A solution of noncentrosymmetric molecules is centrosymmetric and, therefore, no second-harmonic light is coherently generated on illumination with a beam at frequency ω. However, each molecule generates a microscopic polarization

$$p^{2\omega} = K\varepsilon_0\beta\left(-2\omega,\omega,\omega\right)E^\omega E^\omega \,. \tag{2.93}$$

K and K' (see below) account for degeneracy factors due to the different nonlinear optical processes involved. An average over all polarizations will yield zero. $p^{2\omega}$ will, however, polarize its environment. The associated reaction field at the position of the molecule in discussion (see Sect. 2.2.2.E)

$$R^{2\omega} = r^{2\omega}p^{2\omega} \tag{2.94}$$

can then interact with the fundamental wave to generate a field at frequency 3ω through

$$p^{3\omega} = K'\varepsilon_0\beta\left(-3\omega,2\omega,\omega\right)R^{2\omega}E^\omega \,. \tag{2.95}$$

The total polarization generated at the third-harmonic is then given by (now taking all degeneracy factors and tensor elements into account)

$$p_3^{3\omega} = \varepsilon_0\frac{1}{4}\gamma_{\text{eff}}E_3^\omega E_3^\omega E_3^\omega \tag{2.96}$$

$$\gamma_{\text{eff}} = \langle \gamma \left(-3\omega, \omega, \omega, \omega, \right) \rangle_{3333}$$

$$+ 2r^{2\omega} \langle \beta \left(-3\omega, 2\omega, \omega\right) \cdot \beta \left(-2\omega, \omega, \omega\right) \rangle_{3333}$$

$$= \langle \gamma \left(-3\omega, \omega, \omega, \omega, \right) \rangle_{3333}$$

$$+ 2r^{2\omega} \left\langle \sum_n \beta'_{ijn} \left(-3\omega, 2\omega, \omega\right) \cdot \beta_{nkl} \left(-2\omega, \omega, \omega\right) \right\rangle_{3333} . \tag{2.97}$$

$\langle \beta' \cdot \beta \rangle_{3333}$ represents the rotational average with the polarization of the fundamental and the third-harmonic waves along the macroscopic 3-direction. As a first approximation, (2.72) can be used for $r^{2\omega}$. The rotational average $\langle \gamma \rangle_{3333}$ is given by (2.31):

$$\gamma \equiv \langle \gamma \rangle_{3333}$$

$$= \frac{1}{5} \left(\gamma_{1111} + \gamma_{2222} + \gamma_{3333} + \gamma_{1122} + \gamma_{1133} + \gamma_{2211} \right.$$

$$\left. + \gamma_{2233} + \gamma_{3311} + \gamma_{3322} \right) \tag{2.31}$$

The rotational average $\langle \beta' \cdot \beta \rangle_{3333}$ depends on the symmetry of the molecule [2.105]. For the two molecular symmetry point groups 2 and $mm2$ that are important here the rotational averages are given by

point group 2

$$\langle \beta' \cdot \beta \rangle_{3333} = \frac{1}{15}$$

$$\times \left\{ \begin{array}{l} \beta'_{xzx} \left(\beta_{xxz} + \beta_{xzx}\right) + \beta'_{xzy}\beta_{yxz} + \beta'_{yzx}\beta_{xyz} + \beta'_{yzx}\beta_{xzy} + \beta'_{xzy}\beta_{yzx} \\ + \beta'_{yzy} \left(\beta_{yyz} + \beta_{yzy}\right) + \beta'_{zxx} \left(\beta_{xxz} + \beta_{xzx}\right) + \beta_{zxx} \left(3\beta'_{xxz} + \beta'_{yyz}\right) \\ + \beta'_{zxy} \left(\beta_{yxz} + \beta_{yzx}\right) + \beta_{zxy} \left(\beta'_{xyz} + \beta'_{yxz}\right) + \beta'_{zyx} \left(\beta_{xyz} + \beta_{xzy}\right) \\ + \beta_{zyx} \left(\beta'_{xyz} + \beta'_{yxz}\right) + \beta'_{zyy} \left(\beta_{yyz} + \beta_{yzy}\right) + \beta_{zyy} \left(\beta'_{xxz} + 3\beta'_{yyz}\right) \\ + \beta'_{zzz} \left(\beta_{zxx} + \beta_{zyy}\right) + \beta_{zzz} \left(\beta'_{xxz} + \beta'_{yyz}\right) + 3\beta'_{zzz}\beta_{zzz} \end{array} \right\} \tag{2.98}$$

point group mm2

$$\langle \beta' \cdot \beta \rangle_{3333} = \frac{1}{15}$$

$$\times \left\{ \begin{array}{l} \beta'_{xzx} \left(\beta_{xxz} + \beta_{xzx}\right) + \beta'_{yzy} \left(\beta_{yyz} + \beta_{yzy}\right) + \beta'_{zxx} \left(\beta_{xxz} + \beta_{xzx}\right) \\ + \beta_{zxx} \left(3\beta'_{xxz} + \beta'_{yyz}\right) + \beta'_{zyy} \left(\beta_{yyz} + \beta_{yzy}\right) + \beta_{zyy} \left(\beta'_{xxz} + 3\beta'_{yyz}\right) \\ + \beta'_{zzz} \left(\beta_{zxx} + \beta_{zyy}\right) + \beta_{zzz} \left(\beta'_{xxz} + \beta'_{yyz}\right) + 3\beta'_{zzz}\beta_{zzz} \end{array} \right\} . \tag{2.99}$$

If one tensor element β_{zzz} dominates, the expressions reduce to

$$\langle \beta' \cdot \beta \rangle_{3333} = \frac{1}{5} \left(\beta'_{zzz} \cdot \beta_{zzz} \right) , \tag{2.100}$$

which gives

$$\gamma_{\text{eff}} = \langle \gamma \rangle_{3333} + \frac{2}{5} r^{2\omega} \beta'_{zzz} \cdot \beta_{zzz}. \tag{2.101}$$

Meredith [2.76] has demonstrated that these contributions are important in the case of para-nitroaniline (p-NA). In Sect. 2.6 we will give calculated contributions of cascading through the reaction field for a few examples of extended molecules based on our own experimental data.

A few remarks must be made here. First, intermolecular interactions have been neglected completely. Second, the exact form of the reaction field can only be approximated. Nevertheless, the estimations presented here and in Sect. 2.6 give the magnitude of these effects. Generally, it will always be difficult to determine the contribution of the cascading through the reaction field with absolute certainty. One possible solution might be dispersion measurements of third-harmonic generation, since the dispersion of β is known quite well (two-level model). From the dispersion of measured values of γ one might be able to draw further conclusions.

2.3.3   Cascading Through the Local Field

In Sect. 2.3.2 we have considered the influence of the reaction field on third-harmonic generation. Now we go on and investigate the influence of the local field in cascaded second-harmonic generation and sum frequency generation on third-harmonic generation in noncentrosymmetric media. This topic has been treated by various authors [2.5, 2.7, 2.106]. The following discussion can be generalized to other cascaded processes as well.

In the case of second-harmonic generation in noncentrosymmetric materials we have a source polarization $P_{NLS}^{2\omega}$ and a bound second-harmonic wave $E^{2\omega}$, to be inserted in the wave equation, that are related by (from the wave equation, neglecting wave vector dependencies)

$$E^{2\omega} = \frac{P_{NLS}^{2\omega}}{\varepsilon_0 \left(\varepsilon^{\omega} - \varepsilon^{2\omega} \right)} \qquad \text{transverse case,} \qquad (2.102)$$

$$E^{2\omega} = -\frac{P_{NLS}^{2\omega}}{\varepsilon_0 \varepsilon^{2\omega}}, \qquad \text{longitudinal case.} \qquad (2.103)$$

Each molecule experiences a local field described by [2.5]

$$E_L^{2\omega} = E^{2\omega} + L^{2\omega} P^{2\omega} + L^{2\omega} P_{LS}^{2\omega} = f^{2\omega} E^{2\omega} + L^{2\omega} P_{NLS}^{2\omega} . \qquad (2.104)$$

Note that this equation differs from the linear case (2.81). Equation (2.104) can be further changed to

$$E_L^{2\omega} = f^{2\omega} \left\{ E^{2\omega} + \frac{L^{2\omega}}{f^{2\omega}} P_{NLS}^{2\omega} \right\} \equiv f^{2\omega} E_{\text{eff}}^{2\omega} , \qquad (2.105)$$

where $E_{\text{eff}}^{2\omega}$ is the effective field that has to be applied when deriving $P_{NLS}^{2\omega}$ from macroscopic fields. $E_{\text{eff}}^{2\omega}$ should only be applied for the generation of $P^{3\omega}$ (the bound wave for second-harmonic generation remains unchanged).

In the Lorentz approximation, $E_{\text{eff}}^{2\omega}$ is given by

$$E_{\text{eff}}^{2\omega} = \left\{ \frac{1}{\varepsilon^\omega - \varepsilon^{2\omega}} + \frac{1}{\varepsilon^{2\omega} + 2} \right\} \frac{P_{\text{NLS}}^{2\omega}}{\varepsilon_0} \qquad \text{transverse case} \qquad (2.106)$$

$$E_{\text{eff}}^{2\omega} = \left\{ -\frac{1}{\varepsilon^{2\omega}} + \frac{1}{\varepsilon^{2\omega} + 2} \right\} \frac{P_{\text{NLS}}^{2\omega}}{\varepsilon_0} \qquad \text{longitudinal case} . \qquad (2.107)$$

The additional term in $E_{\text{eff}}^{2\omega}$ leads to a modification of the effective third-order nonlinear optical susceptibility (in addition to the 'usual' cascading). This effect is called *cascading through the local field*. In the case of transverse fields the modified third-order nonlinear optical susceptibility is

$$\chi_{\text{eff}}^{3\omega} = \chi^{3\omega} + 2\chi_{\text{SHG}}^{(2)} \cdot \chi_{\text{SFG}}^{(2)} \underbrace{\left\{ \frac{1}{\varepsilon^\omega - \varepsilon^{2\omega}} + \frac{1}{\varepsilon^{2\omega} + 2} \right\}}_{\text{modification}} . \qquad (2.108)$$

Note that in the case of phase-matched interactions ($\varepsilon^\omega \approx \varepsilon^{2\omega}$) the term

$$\frac{L^{2\omega}}{f^{2\omega}} P_{\text{NLS}}^{2\omega}$$

can be safely neglected.

2.3.4   Second-Harmonic Generation and Difference Frequency Mixing

The remainder of this section is devoted to cascaded processes that are relevant for large, nonresonant optical Kerr-like nonlinearities, which are a basic requirement for all-optical switching and related applications [2.2].

As already mentioned above, in centrosymmetric media the third-order susceptibility $\chi^{(3)}$ is the dominant contribution to the nonlinear refractive index n_2, which is defined as $n = n_0 + n_2 I$, where n_0 is the linear refractive index and I is the intensity of the incoming beam. The induced nonlinear phase shift in a noncentrosymmetric material is in general given by

$$\Delta\phi = \Delta\phi^{\text{direct}} + \Delta\phi^{\text{SHG}} + \Delta\phi^{\text{OR}}, \qquad (2.109)$$

where $\Delta\phi^{\text{direct}} \propto n_2^{\text{direct}} I$ contains the contribution from direct $\chi^{(3)}$ processes in all materials, both polar and nonpolar. As just mentioned, in centrosymmetric media the third-order susceptibility $\chi^{(3)}$ is the dominant contribution to the nonlinear refractive index n_2.

In noncentrosymmetric materials cascaded second-order processes can lead to an effective nonlinear phase shift ($\Delta\phi^{\text{SHG}}$, $\Delta\phi^{\text{OR}}$). One possibility is due to the combined processes of second-harmonic generation of the incident wave (SHG) and difference frequency mixing of the generated second-harmonic wave with the incident wave (Fig. 2.14a). The origin of the induced

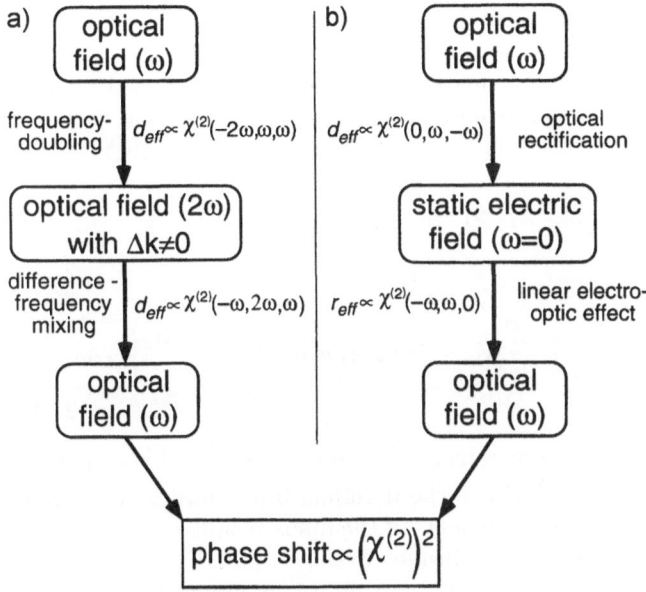

Fig. 2.14. Principle of (**a**) cascaded second-harmonic generation and difference frequency mixing and (**b**) optical rectification and the linear electro-optic effect

nonlinear phase shift can be sketched as follows: the total process involves up-conversion from ω to 2ω and down-conversion from 2ω to ω with the requirement of a non zero wave vector mismatch between the fundamental and the second-harmonic wave. Since the phase fronts of the waves with frequency ω and 2ω travel at different velocities the relative phase accumulated by the phase fronts during the down-conversion process leads to a net nonlinear phase shift of the fundamental wave.

One can immediately see that the sign of the phase change depends on which wave travels faster (whether $n^\omega > n^{2\omega}$ or $n^{2\omega} > n^\omega$). For low conversion rates (undepleted wave approximation) this phase shift is proportional to the input intensity and one can write for the contribution to n_2 from this type of interaction [2.107]

$$n_2^{\mathrm{SHG}} = -\frac{4\pi}{c\varepsilon_0}\left(\frac{L}{\lambda_0}\right)\frac{d_{\mathrm{eff}}^2}{n^{2\omega}(n^\omega)^2}\frac{1}{\Delta k\,L}\,, \qquad (2.110)$$

where λ_0 is the wavelength of the fundamental in vacuum, d_{eff} is the effective nonlinear optical coefficient ($= \chi_{\mathrm{eff}}/2$), $\Delta k = k^{2\omega} - 2k^\omega$ is the phase mismatch, with $k^\omega = 2\pi n^\omega/\lambda_0$ and $k^{2\omega} = 4\pi n^{2\omega}/\lambda_0$ being the wave vectors in the medium. The sign of n_2^{SHG} depends on the sign of the phase mismatch Δk, with n_2^{SHG} being negative for normal dispersion ($n^{2\omega} > n^\omega$). Far from phase matching, n_2^{SHG} is small but always present if $d_{\mathrm{eff}} \neq 0$. The phase shift is proportional to the figure of merit d_{eff}^2/n^3 for frequency doubling [2.80].

Therefore, good materials for second-harmonic generation are also good candidates for cascading.

Generally the process of cascaded nonlinearities can be modeled with appropriate coupled time domain equations governing pulsed second-harmonic generation in a dispersive medium [2.108–109]

$$\frac{\partial}{\partial \xi} A_1 - \frac{1}{2} \frac{\partial}{\partial \tau} A_1 - i \frac{k_1'' L_w}{2T_0^2} \frac{\partial^2}{\partial \tau^2} A_1 = i \kappa L_w A_2 A_1^* e^{i \Delta k \xi L_w} - \frac{\alpha_1 L_w}{2} A_1$$

$$(2.111)$$

$$\frac{\partial}{\partial \xi} A_2 + \frac{1}{2} \frac{\partial}{\partial \tau} A_2 - i \frac{k_2'' L_w}{2T_0^2} \frac{\partial^2}{\partial \tau^2} A_2 = i \kappa L_w A_1 A_1 e^{-i \Delta k \xi L_w} - \frac{\alpha_2 L_w}{2} A_2 \, ,$$

$$(2.112)$$

where κ is proportional to the nonlinear optical coefficient and is defined by $\kappa = 2\omega d_{\text{eff}} / \left(2n_{2\omega}^2 n_{2\omega} \varepsilon_0 c^3\right)^{1/2}$, Δk is the detuning from quasi-phase matching, the α_i are the absorption coefficients of the fundamental and second harmonic, and the A_i are the field amplitude envelopes. A reduced time which averages the group velocities of the fundamental and second harmonic can be used, and the time and space coordinates can be normalized to the pulse width and walk-off length, respectively. The walk-off length is defined by $L_w = T_0/(k_2' - k_1')$, where T_0 is the intensity FWHM pulse width and $k_{1,2}'$ are the first frequency derivatives of the fundamental and second-harmonic wave vectors. Likewise, the double-primed wave vectors are the second derivatives which represent group velocity dispersion. The above equations are most conveniently solved using numerical methods. The results of such numerical simulations reveal many of the features of cascaded second-order processes [2.16, 2.110–111]: the optimum nonlinear effect depends strongly on the detuning from phase matching, the sample length, the intensity of the fundamental beam, and the sample nonlinearity. As an example, Fig. 2.15 shows the dependence of the nonlinear phase as a function of detuning $\Delta k L$ for several values of the parameter $\kappa |A_1(0)| L$ (L: sample length (for details see [2.111])).

As expected, the sign of the phase shift changes as the wave vector mismatch changes sign. Note that such an easy change of $\Delta \phi$ is not possible with nonlinearities using direct $\chi^{(3)}$. In addition, there is no phase shift for zero phase-mismatch ((2.110) is not valid). With additional seeding of the second harmonic this situation can easily be changed [2.110]. Numerical simulations also show that the largest phase shifts are obtained for a phase mismatch of $\Delta k L \approx 3$ [2.107].

Results from numerical simulations of the intensity dependence of the nonlinear phase shift (important for applications) and the fundamental throughput for different detunings and for fixed nonlinear optical coefficients and sample lengths are shown in Fig. 2.16. There are strong dependencies on the detuning from phase matching, the sample length, and the intensity of the fundamental beam. The relations between the different parameters are quite

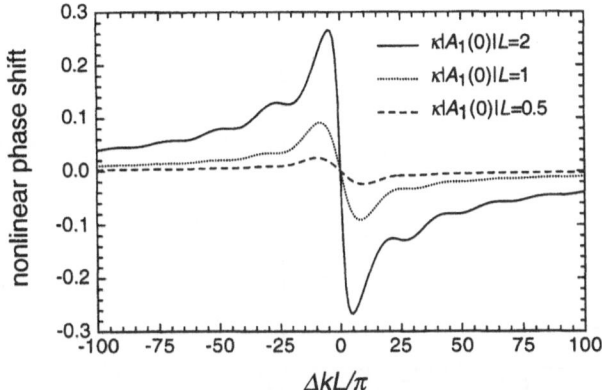

Fig. 2.15. Nonlinear phase shift induced by cascading as a function of the phase-mismatch Δk. Larger second-order nonlinearities and/or intensities lead to larger phase shifts. As the wave vector mismatch decreases the phase shift first increases and then changes sign, as expected (with permission from M. Sundheimer)

complex. The nonlinear phase is only linear with intensity for large detunings or low intensities. Only in these cases can we describe the process by an effective n_2 (Kerr nonlinearity; see (2.110)). In general, this limitation presents no problem, since for all-optical devices the large nonlinear phase shift is the only requirement.

2.3.5   Optical Rectification and Linear Electro-Optic Effect

There is an additional contribution to n_2 that has often been overlooked previously [2.12–13]. We have demonstrated, for the first time to our knowledge, that the combined processes of optical rectification and the linear electro-optic effect give rise to a large effective nonlinear refractive index n_2. This combined effect had already been observed in four-wave mixing experiments with ps lasers in $KNbO_3$ [2.112], but its contribution to nonlinear light-induced phase shifts has previously been overlooked. In Sect. 2.5 we will present the first experimental evidence of this new contribution to n_2 in $KNbO_3$ crystals. In order to analyze the results obtained in polar $KNbO_3$ we also performed similar experiments in cubic $KTaO_3$, where only the direct third-order processes are allowed by symmetry.

We will use a continuous wave approach to describe the combined processes of optical rectification and the linear electro-optic effect. In noncentrosymmetric materials an input beam at frequency ω can also generate a quasi-static electric field at frequency zero via optical rectification (and therefore induce a polarization)

$$P_i^0 = \frac{1}{2}\varepsilon_0\chi_{ijk}^{(2)}(0,\omega,-\omega)E_j^\omega E_k^{\omega^*} , \qquad (2.113)$$

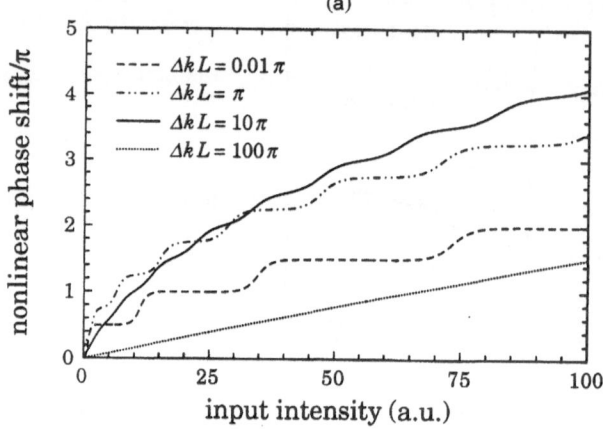

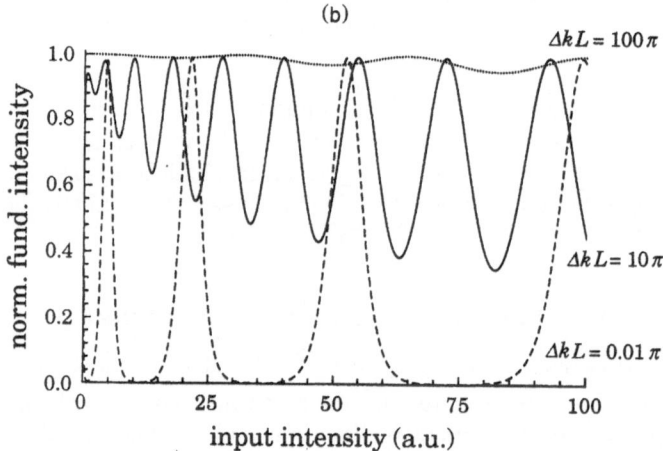

Fig. 2.16. Intensity dependence of (**a**) the nonlinear phase shift and (**b**) the fundamental throughput. In all cases we have $\kappa L = 4$. (**a**) For large $\Delta k \times L$ and/or low intensities the phase shift is linear in intensity and behaves like a true n_2. If we decrease ΔkL the phase change increases but the dependence on intensity becomes step-like. If we get too close to phase matching the phase shift decreases again and exactly on phase matching the phase shift is zero. At high intensities the phase shift starts to saturate. (**b**) For low depletion we have the well-known behavior of the fundamental wave with intensity. As the intensity is increased the coherence length changes due to the interplay of second-harmonic generation and down-conversion. The steps in the phase shift correspond exactly to the points of total dephasing (with permission from M. Sundheimer)

where $\chi_{ijk}^{(2)}(0,\omega,-\omega)$ is the nonlinear optical susceptibility describing optical rectification [2.113]. This generated field can induce a refractive index change via the linear electro-optic effect (2.53). Figure 2.14b shows a schematic drawing of the principle.

We can illustrate the combined processes of optical rectification and the linear electro-optic effect with the following example: we illuminate a plate of a polar crystal with light polarized parallel to the polar axis. The effect of optical rectification will produce a quasi-static field along the polar direction. This field can then act on the beam through a refractive index change produced by the linear electro-optic effect. Obviously, a contribution from second-harmonic generation will also occur at the same time. If we illuminate a plate of a polar crystal with light polarized perpendicularly to the polar axis, cascaded optical rectification and the linear electro-optic effect will also appear (e.g. for light propagation along the polar axis), in which case the longitudinal electro-optic effect becomes active. This is in contrast to second-harmonic generation, for which the appearance of n_2^{SHG} depends on the propagation direction. For light polarized along a main axis of the optical indicatrix we obtain the contribution n_2^{OR} to the nonlinear refractive index n_2 along the same direction as

$$(n_2)_i^{\mathrm{OR}} = \frac{1}{c\varepsilon_0 n_i^2} \frac{\chi_{iik}^{(2)}(-\omega,\omega,0)\,\chi_{kii}^{(2)}(0,\omega,-\omega)}{(\varepsilon_{kk}-1)}, \tag{2.114}$$

where ε_{kk} is the static dielectric constant. By permutation of indices and optical frequencies in $\chi_{ijk}^{(2)}(0,\omega,-\omega)$ we obtain $\chi_{ijk}^{(2)}(0,\omega,-\omega) = \chi_{kji}^{(2)}(-\omega,\omega,0)$ and

$$(n_2)_i^{\mathrm{OR}} = \frac{1}{c\varepsilon_0 n_i^2} \frac{\left(\chi_{iik}(-\omega,\omega,0)\right)^2}{(\varepsilon_{kk}-1)} = \frac{n_i^6}{4c\varepsilon_0} \frac{(r_{iik})^2}{(\varepsilon_{kk}-1)}. \tag{2.115}$$

The derivation of (2.115) is based on the linear polarization optic coefficient (2.56). Taking a different approach [2.5], n_2^{OR} is given by

$$(n_2)_i^{\mathrm{OR}} = \frac{1}{c\varepsilon_0 n_i^2} \frac{\left(\chi_{iik}(-\omega,\omega,0)\right)^2}{(\varepsilon_{kk}+2)} = \frac{n_i^6}{4c\varepsilon_0} \frac{(r_{iik})^2}{(\varepsilon_{kk}+2)}. \tag{2.116}$$

The origin of this difference is very subtle and it is not clear at the moment which approach more accurately describes the experimental situation. For oxides with large dielectric constants the difference in n_2^{OR} is negligible. For organic materials with $n^2 \approx \varepsilon$ it becomes very important.

Note that the contribution n_2^{OR} to the nonlinear refractive index is always phase-matched since all interacting waves have either the same optical frequency or have frequency zero. In addition, n_2^{OR} is always positive. The phase shift is proportional to the square of the electro-optic figure of merit $\left(n^3 r_{\mathrm{eff}}\right)^2$. Therefore, good materials for electro-optics are also good candidates for this type of cascading.

An important effect has to be included in (2.115) and (2.116): the effect of depolarization. If we apply an external field $\boldsymbol{E}_{\text{ext}}$ to a dielectric material a polarization $\boldsymbol{P}$ is generated inside the material. This polarization will decrease the electric field $\boldsymbol{E}_{\text{int}}$ inside the dielectric: $\boldsymbol{E}_{\text{int}} = \boldsymbol{E}_{\text{ext}} - \boldsymbol{L} \times \boldsymbol{P}$, where $\boldsymbol{L}$ is the depolarizing tensor [2.73]. For our case of a single beam that generates a static polarization through optical rectification (see the z-scan experiments in Sect. 2.5) we can approximate our beam by a long cylinder. For the case of an isotropic medium (not the case here!) $\boldsymbol{L}$ is equal to 0 for a longitudinal field and equal to 0.5 for a transverse field [2.114]:

$$E_{\text{transverse}}^{\text{depol}} = -\frac{1}{2}\frac{P^0}{\varepsilon_0}. \tag{2.117}$$

This leads to a decrease of the overall static polarization which becomes

$$P_{\text{total},i}^0 = \frac{2P_i^0}{1 + \varepsilon_i} \tag{2.118}$$

and therefore to a reduction of n_2^{OR} by $2/(1 + \varepsilon)$ (for (2.115)). For the longitudinal fields no reduction appears since the depolarizing factor is zero for our geometry. In organic crystals with ε between 3 and 6 this reduction is not dramatic. For inorganic crystals, however, with values of ε of several tens to hundreds, the depolarization can completely cancel P^0. For special experimental geometries even such large values of ε can be overcome to achieve large phase shifts (Sect. 2.5). However, since most nonlinear optical materials of interest are strongly anisotropic, this simple model for the depolarization does not hold and the effect of depolarization is therefore not considered just yet. In addition, the contribution from the propagating electric field generated by the process of optical rectification merits further investigation, especially for the case of organic materials [2.13].

As will be discussed in Sect. 2.6, the combined effect of optical rectification and the linear electro-optic effect was also observed and theoretically explained in optical four-wave mixing [2.14, 2.112, 2.115].

2.3.6 Limits of the Cascaded Response in Molecular Crystals

Owing to their large nonlinearities, molecular crystals have a significant potential for use in cascaded second-order processes. Based on the above estimation of the limits of the electro-optic and nonlinear optical response in molecular crystals we have estimated the upper limits of the cascaded response.

A   Second-Harmonic Generation and Difference Frequency Mixing

The nonlinear phase shift is proportional to the figure of merit d_{eff}^2/n^3 for frequency doubling. Only in the limit of large detunings or low intensities is

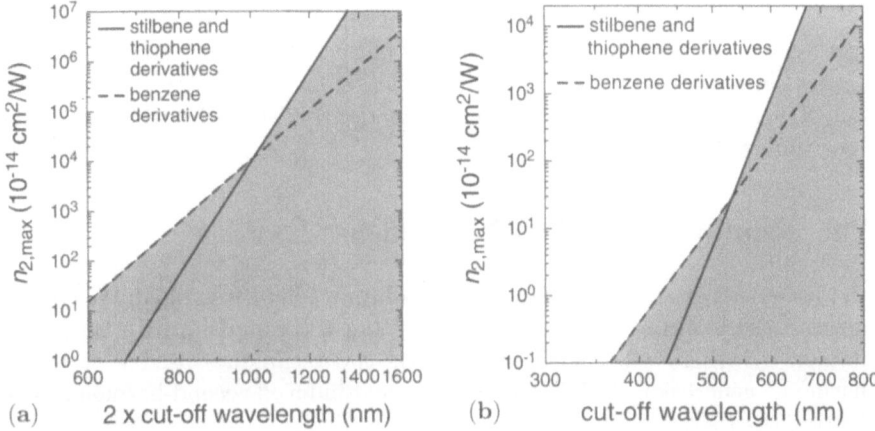

Fig. 2.17a,b. Upper limits of the nonlinear refractive index based on cascaded second-order processes vs. the cut-off wavelength. (**a**) Cascaded frequency doubling and sum frequency generation. The graph shows maximum values of n_2 vs. 2× cut-off wavelength calculated from measured values of hyperpolarizabilities assuming an optimum parallel alignment of all molecules for phase-matched interactions in a waveguide. (**b**) Cascaded optical rectification and the linear electro-optic effect. The graph shows maximum values of n_2 vs. cut-off wavelength calculated from measured values of hyperpolarizabilities assuming an optimum parallel alignment of all molecules. The shaded areas represent values that should be achievable. We can see that extremely large values might be obtained under optimized conditions

this phase shift linear in intensity and can thus be described by (2.110). With the approximation of $\Delta kL \approx 3$ (maximum nonlinear phase shift), $L = 1\,\mathrm{mm}$, $n^{2\omega} = 2.4$, $n^{\omega} = 2.1$ (for stilbene derivatives), $n^{2\omega} = 2.2$, $n^{\omega} = 1.9$ (for benzene derivatives) and the results of Sect. 2.2.2.G we can get an estimate of maximum obtainable values of n_2. Figure 2.17a shows the results based on Fig. 2.13a. A summary of these estimates together with actual experimental data will be given in Sect. 2.6.

B   Optical Rectification and Linear Electro-Optic Effect

We have seen above that the nonlinear phase shift is proportional to the square of the electro-optic figure of merit $\left(n^3 r_{\mathrm{eff}}\right)^2$. As in the above case we can get an estimate of maximum obtainable values of n_2 (using (2.115)). We only consider propagation along a main axis of the index ellipsoid. For a transversely generated static polarization the largest values are obtained for diagonal elements of $\chi^{(2)}$. For the longitudinal case the effect is optimized for non-diagonal elements of $\chi^{(2)}$, since a longitudinal static field has to be generated. As in the case of frequency doubling in the bulk the effective second-order susceptibility is given by $\chi^{(2)}_{\mathrm{eff}} = \chi^{(2)}_{ijj} \sin^2(\theta) \cos(\theta)$ which is maximized for $\theta = 54.7°$ in the oriented gas model. With the approximation of $\varepsilon = 5$,

$n^\omega = 2.1$ (for stilbene and extended thiophene derivatives), $\varepsilon = 4$, $n^\omega = 1.9$ (for benzene derivatives) and the results of Sect. 2.2.2.G we can get the estimate shown in Fig. 2.17b. Since the reduction factors $\sin^2(54.7°)\cos(54.7°)$ and $2/(\varepsilon + 1)$ are almost equal no distinction between the transverse and longitudinal cases was made.

2.4 Nonlinear Optical Molecules

In this section experiments performed in solution will be described. We mainly discuss results obtained in our laboratory, but a comparison with work performed by others will also be made. The two techniques used were third-harmonic generation (THG) and electric field-induced second-harmonic generation (EFISH).

2.4.1 Third-Harmonic Generation

Third-harmonic generation can occur in any material. In this process a fundamental wave at frequency ω produces a wave at frequency 3ω. It differs from other techniques that also measure third-order nonlinear optical susceptibilities such as z-scan (described in Sect. 2.5) and degenerate four-wave mixing in that only the electronic third-order nonlinear optical response is measured.

A   Theory

In an isotropic liquid, one need consider only the rotational average of γ_{ijkl}, for which one gets

$$
\begin{aligned}
\gamma &\equiv \langle \gamma \rangle_{3333} \\
&= \frac{1}{5}\left(\gamma_{1111} + \gamma_{2222} + \gamma_{3333} + \gamma_{1122} + \gamma_{1133} + \gamma_{2211} \right. \\
&\quad \left. + \gamma_{2233} + \gamma_{3311} + \gamma_{3322} \right) .
\end{aligned}
\tag{2.31}
$$

In third-harmonic generation experiments the macroscopic third-order nonlinear optical susceptibility $\chi^{(3)}(-3\omega,\omega,\omega,\omega)$ is measured. It is related to the molecular averaged γ by [2.30]

$$
\chi^{(3)}(C) = N_A \frac{d(C)\,f(C)}{1 + C}\left[\frac{\gamma_s}{M_s} + C\frac{\gamma_m}{M_m} \right],
\tag{2.119}
$$

where C is the concentration, expressed as the ratio between the total weight of the solute molecules to the weight of the used solvent. N_A is Avogadro's number, $d(C)$ is the density of the solution, and $f(C)$ is the local field correction factor in the Lorentz approximation (using the refractive indices of the solvent). M_s and M_m are the molecular masses of the solvent and the solute molecules, respectively. The solution densities were assumed to be constant with concentration.

The description of the theoretical curves used to evaluate the experimental results presented below is given in Sect. A.3.

B   Experimental Description

Solutions with concentrations between 0.25 and 5 wt.% were investigated in 1 mm thick fused silica cuvettes. The laser source was a pulsed Nd:YAG Laser ($\lambda = 1.064\,\mu m$, 10 Hz repetition rate, pulse duration of 5 ns) which was used to pump an H_2 gas Raman cell yielding a frequency shifted wavelength of $1.907\,\mu m$. The s-polarized beam was then focused onto the sample with am $f = 500\,mm$ lens. Third-harmonic generation measurements were performed by rotating the samples around an axis parallel to the polarization to generate the well-known Maker–Fringe interference patterns. Figure 2.18 shows the experimental set-up used.

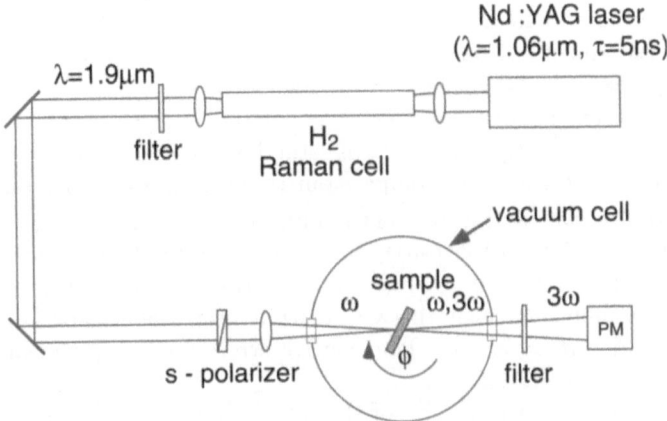

Fig. 2.18. Experimental setup for third-harmonic generation

It is important to realize that reliable measurements have to be carried out in vacuum (0.05–0.1 bar) because air also gives an important contribution to the third-harmonic signal. This can be explained as follows: the signal at the third harmonic is proportional to the product $\left[\chi^{(3)}\right]^2 l_c^2$, where l_c is the coherence length given by $l_c = \lambda/(6(n^{3\omega} - n^{\omega}))$. The materials of interest have a large value of $\chi^{(3)}$ but a small coherence length of the order of 10 to $100\,\mu m$. Air, on the other hand, has a very small value of $\chi^{(3)}$ but almost no dispersion of the refractive index with wavelength and, therefore, a huge coherence length. This leads to a significant contribution of air to the third-harmonic signal, as shown in Fig. 2.19.

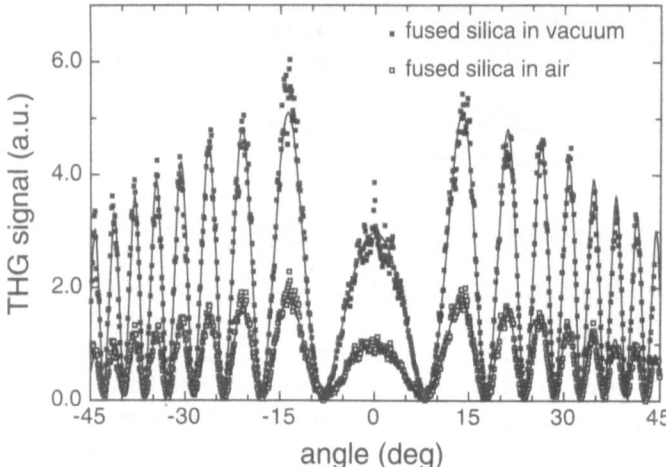

Fig. 2.19. Experimental third-harmonic generation curves of fused silica showing the influence of air

A comparison of measurements of fused silica in vacuum and air allowed all subsequent measurements of our solutions to be performed in air. The Maker–Fringe patterns were analyzed as described in Sect. A.3. All measurements were calibrated against fused silica ($\chi_{\text{fs}}^{(3)}(-3\omega, \omega, \omega, \omega) = 2.62 \times 10^{-22}\,\text{m}^2/\text{V}^2$ (2.16×10^{-14} esu)) a new reliable standard value that was derived in this work (see Sect. 2.5.1). To compare our results with experimental results by other groups that used a different standard ($\chi_{\text{fs}}^{(3)}(-3\omega, \omega, \omega, \omega) = 3.89 \times 10^{-22}\,\text{m}^2/\text{V}^2$ ($= 2.79 \times 10^{-14}$ esu)), our values of γ (and $\chi^{(3)}$) must be multiplied by a factor of 2.40. If necessary, a correction for minor absorptions at the third harmonic wavelength ($\lambda = 636$ nm) was made using the complex refractive index $n^* = n + \mathrm{i}k$. The relative errors for the nonlinear susceptibilities are about 10%.

For each substance we measured $\chi^{(3)}$ of the pure solvent and for two to five different concentrations. γ was obtained from a fit to the concentration dependence of $\chi^{(3)}$ (see (2.119)).

Tetraethynylethenes. Recently, synthetic methods for fully cross-conjugated tetraethynylethene (3,4-diethynyl-hex-3-ene-1,5-diyne) (TEE) and its derivatives have been developed [2.116–117]. These compounds are versatile precursors to multinanometer-sized molecular rods with all-carbon backbones and provide a 'molecular construction kit' for the development of linearly conjugated polymers with the novel polytriacetylene (PTA) backbone [2.118]. The unique TEE framework facilitates π-conjugation with the pendant aromatic substituents by allowing coplanar orientation throughout the molecular core [2.119], in contrast to such molecules as *cis*-stilbenes or tetraphenylethenes where steric interactions prevent coplanarity [2.120]. The core also allows an almost unlimited flexibility in the design of molecules with specific properties and functionalities.

In the following we describe structure–property relationships in this novel class of materials that provide further insight into routes leading to the desired optical nonlinearities and show their importance for third-order nonlinear optics [2.121]. The synthesis as well as chemical properties of the novel functionalized TEEs, **T1–T16**, shown in Table 2.4, are given in [2.122–123].

The monomers described here allow six potential conjugation paths, as illustrated in Fig. 2.20 [2.121]. The substituents R_1–R_4 can be found in Table 2.4. We will show below that the measured nonlinear optical effects depend strongly on these conjugation paths and the substitution patterns.

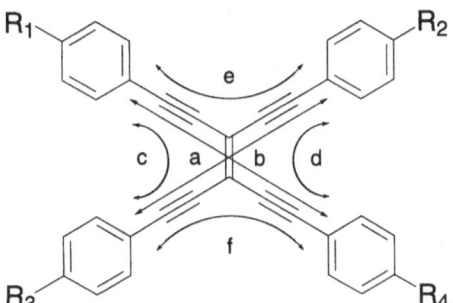

Fig. 2.20. Basic structure of TEE monomers with the four possible substitutions R_1–R_4. The six potential conjugation paths are indicated with arrows, where a–d are linear- and e–f are cross-conjugated paths

For the TEE compounds all measurements were carried out in $CHCl_3$. Figure 2.21 shows an example of the change of the Maker–Fringe pattern with a change in concentration as well as the concentration dependence of $\chi^{(3)}$ for the molecule **T19**. The results of these experiments are summarized in Table 2.4. In addition to γ, $\chi^{(3)}$ values extrapolated from measured values of γ (assuming a linear dependence on concentration of our molecules) are also provided in Table 2.4 as an indication of the maximum macroscopic nonlinearity per unit volume. It should be noted that an isotropic arrangement of the molecules was assumed for these values, and an optimized alignment of all molecules in linear chains or stacks would lead to even larger values of $\chi^{(3)}$. Furthermore, our estimates present a lower limit, since the small (with respect to the pure compounds) refractive index of the solvent was used in the calculation of $\chi^{(3)}$. If we assume a realistic refractive index of 1.7 for the solid state we get values of $\chi^{(3)}$ $100\%/\chi_{fs}^{(3)}$ that are typically a factor of two larger. As we will see in the case of the hydrazone and stilbene derivatives, another factor of 5 (due to optimized molecular orientation) to 10 (due to the combination of optimized molecular orientation and large refractive indices $n \approx 2.0$–2.4) can be gained if crystals of the materials of interest can be grown.

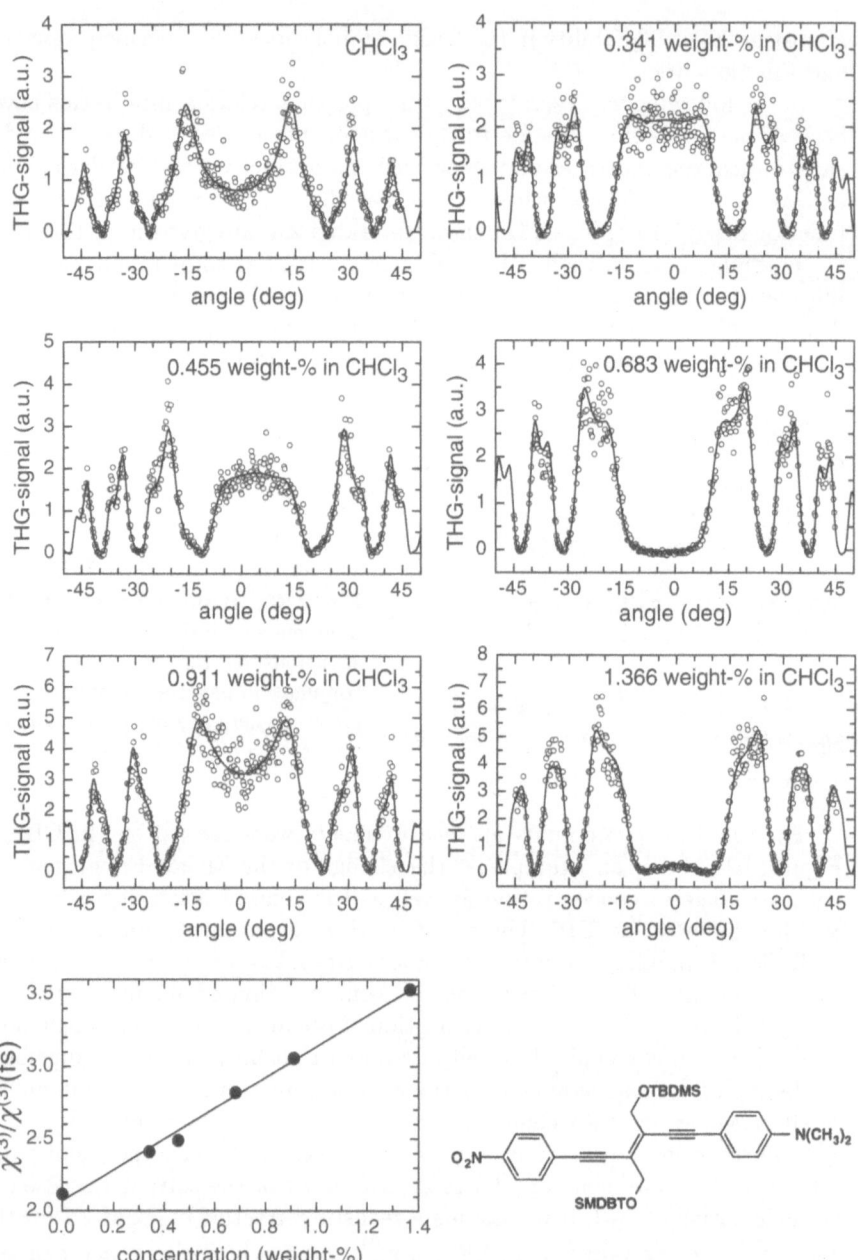

Fig. 2.21. Maker–Fringe pattern and resulting values of $\chi^{(3)}$ as a function of concentration for molecule **T19**. From a fit to the values of $\chi^{(3)}$ vs. concentration, γ can be obtained using (2.119)

Table 2.4. Results of the third-harmonic generation experiments (γ (esu: 10^{-36}) and (SI: $10^{-48}\,\mathrm{m}^5/\mathrm{V}^2$)). $\chi^{(3)}$ was measured relative to $\chi^{(3)}$ of fused silica (fs), for which a value of $\chi^{(3)} = 2.62 \times 10^{-22}\,\mathrm{m}^2/\mathrm{V}^2$ (2.16×10^{-14} esu) was used. To compare our results with experimental results by other groups that used a different standard ($\chi_{\mathrm{fs}}^{(3)}(-3\omega,\omega,\omega,\omega) = 3.89 \times 10^{-22}\,\mathrm{m}^2/\mathrm{V}^2$ ($= 2.79 \times 10^{-14}$ esu)), our values of γ (and $\chi^{(3)}$) must be multiplied by a factor of 2.40. All measurements were carried out at $\lambda = 1907\,\mathrm{nm}$ in CHCl_3, except for compounds **T10**, **T11**, **T20** and **T21** where measurements were also performed at $\lambda = 2100\,\mathrm{nm}$ (indicated by $*$) to get a reliable comparison without possible resonance contributions. The relative errors are estimated to be 15%. λ_{eg} (nm) is the longest wavelength electronic absorption maximum, and ε ($\mathrm{dm}^3/(\mathrm{cm\,mol})$) is the molar extinction coefficient defined by $\varepsilon = \alpha/C\,\ln 10$ (α: absorption coefficient in cm^{-1}, C: concentration in $\mathrm{mol/dm}^3$) given at the third-harmonic wavelength (636 nm or 700 nm)

Compound	λ_{eg} (ε)	ε	γ (esu)	γ (SI)	$\chi^{(3)}\,100\% / \chi_{\mathrm{fs}}^{(3)}$
(^{i}Pr)$_3$Si—, —NO$_2$, —Si(^{i}Pr)$_3$, O$_2$N— **T1**	403 (34000)	0	100	2.4	39
(^{i}Pr)$_3$Si—, —NMe$_2$, —Si(^{i}Pr)$_3$, Me$_2$N— **T2**	459 (41000)	0	170	2.3	65
(^{i}Pr)$_3$Si—, —NMe$_2$, —Si(^{i}Pr)$_3$, O$_2$N— **T3**	468 (31200)	0	410	5.7	160
Me$_2$N—, —NMe$_2$, (^{i}Pr)$_3$Si—, —Si(^{i}Pr)$_3$ **T4**	428 (51100)	0	130	2.8	50
O$_2$N—, —NMe$_2$, (^{i}Pr)$_3$Si—, —Si(^{i}Pr)$_3$ **T5**	447 (19900)	8	28	0.39	11

Table 2.4. (Continued)

Compound	λ_{eg} (ε)	ε	γ (esu)	γ (SI)	$\dfrac{\chi^{(3)} 100\%}{\chi^{(3)}_{fs}}$
MeO—⬡—≡≡—⬡—OMe / Me₃Si—≡≡—Si(^{i}Pr)₃ **T6**	375 (36200)	0	55	0.77	26
MeO—⬡—≡≡—⬡—OMe / (^{i}Pr)₃Si—≡≡—⬡—NO₂ **T7**	428 (28900)	0	107	2.5	46
(^{i}Pr)₃Si—≡≡—⬡—NMe₂ / (^{i}Pr)₃Si—≡≡—⬡—NO₂ **T8**	471 (17000)	0	500	7.0	195
Me₂N—⬡—≡≡—⬡—NMe₂ / (^{i}Pr)₃Si—≡≡—⬡—NO₂ **T9**	461 (38299)	53	660	9.2	270
Me₂N—⬡—≡≡—⬡—NO₂ / O₂N—⬡—≡≡—⬡—NMe₂ **T10**	533 (22000)	1900	–	–	–
		115	365*	5.1*	160*
Me₂N—⬡—≡≡—⬡—NMe₂ / O₂N—⬡—≡≡—⬡—NO₂ **T11**	486 (40700)	430	1310	18.3	800
		0	700*	9.7*	300*
(^{i}Pr)₃Si—≡≡—⬡—NO₂ / Me₃Si—≡≡—Si(^{i}Pr)₃ **T12**	382 (28200)	0	41	0.57	17

Table 2.4. (Continued)

Compound	λ_{eg} (ε)	ε	γ (esu)	γ (SI)	$\chi^{(3)} 100\%$ / $\chi^{(3)}_{fs}$
T13 (^{i}Pr)$_3$Si, Me$_3$Si, Si(^{i}Pr)$_3$, NMe$_2$	433 (29300)	0	122	2.7	51
T14 (^{i}Pr)$_3$Si, O$_2$N, Si(^{i}Pr)$_3$, NO$_2$	456 (41600)	0	210	2.9	50
T15 (^{i}Pr)$_3$Si, Me$_2$N, Si(^{i}Pr)$_3$, NMe$_2$	486 (45100)	220	770	11	190
T16 (^{i}Pr)$_3$Si, O$_2$N, Si(^{i}Pr)$_3$, NMe$_2$	481 (31300)	0	1520	21	360
T17 OTBDMS, O$_2$N, TBDMSO, NO$_2$	372 (45300)	0	87	2.22	37
T18 OTBDMS, Me$_2$N, TBDMSO, NMe$_2$	380 (58400)	0	43	0.60	19
T19 OTBDMS, O$_2$N, TBDMSO, NMe$_2$	424 (25000)	0	251	3.5	108

Table 2.4. (Continued)

Compound	λ_{eg} (ε)	ε	γ (esu)	γ (SI)	$\dfrac{\chi^{(3)}\,100\%}{\chi^{(3)}_{fs}}$
T20	646 (171000)	27300	17500*	245*	1600*
T21	574 (144000)	12400	7020*	98*	470*

Our experimental results shown in Table 2.4 lead to a number of very interesting and fundamental conclusions [2.121]. First, the N,N-dimethylamino (donor) group is clearly superior in enhancing γ as compared to the nitro (acceptor) group. This holds for mono- and disubstituted TEE molecules (**T12–T13**, and **T1–T2**, respectively) as well as for the dimeric derivatives **T14–T15**.

Second, an increase in the donor strength leads to larger values of γ. This is illustrated in the comparison of molecules **T6** and **T4** as well as **T7** and **T9**, where replacement of methoxy by N,N-dimethylamino groups results in 2 and 6-fold increases, respectively, for γ. Similar observations have been made for monosubstituted stilbenes [2.83].

Third, two series (**T1–T3** and **T14–T16**) clearly show an increase of γ when changing from centrosymmetric (NO_2/NO_2 and NMe_2/NMe_2) to acentric (NO_2/NMe_2) substitution. These experimental results are in agreement with the work by Garito et al. [2.124] who theoretically predicted an enhancement of the nonresonant γ due to a lowering of symmetry and an increase in the number of contributing states, as well as with the simple three-level model for γ derived from static perturbation theory [2.56] (see Sect. 2.2): From (2.44) it follows that in noncentrosymmetric molecules an additional term proportional to $\Delta\mu^2$ contributes to γ. Theoretical and experimental investigations of pull–pull, push–push and push–pull polyenes also revealed an analogous behavior [2.125–127] and a similar trend was also found experimentally for benzene derivatives [2.21, 2.128]. Generally, one expects that an increase in the first-order hyperpolarizability, β, in donor/acceptor derivatives would lead to an increase in γ as well [2.83]. However, this is only valid within a certain range, since γ is predicted to be roughly zero when β is maximized (see Sect. 2.2.2.B) [2.58].

If we modify molecules **T1–T3** (to get molecules **T17–T19**) we obtain a different result. The acentric (NO_2/NMe_2) substitution still leads to the largest value of γ. However, we now have γ (NO_2/NO_2 substitution)$> \gamma$ (NMe_2/NMe_2 substitution). We attribute this change in comparison to **T1–T3** to the fact that molecules **T1–T3** (and **T14–T16**) are not really one-dimensional and that the triple bonds connecting, for example, the $Si(^iPr)_3$ groups also contribute to the third-order nonlinearity, whereas in molecules **T17–T19** this contribution is minor. The trend for molecules **T17–T19** is in agreement with results obtained in a series of biaryls and a series of elongated disubstituted diphenyl polyenes [2.127, 2.129]. It was also suggested that the importance of donor/donor vs. acceptor/acceptor substitution might depend on the conjugation length, where the *bis*-donor derivatives become superior after a critical conjugation length [2.130].

As was mentioned above, the increase of γ with increasing β_z is in agreement with the three-level model. As in the case of β_z γ also depends strongly on the longest wavelength electronic absorption maximum λ_{eg}. Since the TEE molecules described here are mostly two-dimensionally conjugated, no simple theoretical dependence can be deduced without knowledge of the energy states involved in the nonlinear optical processes. Figure 2.22 shows γ vs. λ_{eg} for the TEE molecules.

Fourth, we obtain large nonlinearities for donors/acceptors in *trans*- and *cis*-configurations due to the effect of favorable linear donor–acceptor conjugation, in contrast to substitution at the geminal position where only the weaker cross-conjugation is effective. This is nicely illustrated by comparing the γ value of molecule **T5** with the greatly enhanced values of molecules **T3** and **T8**. This effect is also demonstrated by comparison of molecules **T4** and **T5** where, in the absence of effective donor–acceptor conjugation, molecule **T5** shows a substantially lower γ than the electron richer **T4**.

Often, a strong increase in the third-order nonlinearity γ is associated with a sizeable increase in the longest wavelength electronic absorption maximum (λ_{eg}) [2.83]. This is not true for molecule **T11** as compared, for example, with molecule **T3**. Molecule **T11** is fully two-dimensionally conjugated; i.e.

it shows all six conjugation paths possible in a TEE molecule (Fig. 2.20). It exhibits a very large value of $18.3 \times 10^{-48}\,\mathrm{m^5/V^2}$ for γ and only a slight increase in λ_{eg}. To a first approximation we attribute this to a summation of the different possible conjugation paths in this molecule. We can clearly state that complete two-dimensional donor–acceptor conjugation is very important for increasing the third-order nonlinearity in our systems. Similar observations have been made by others as well [2.131].

In order to compare the influence of various donor–acceptor substitution patterns in TEE systems and to test the validity of the additivity approach an analogue of molecule **T11** was prepared in which the positions of the NMe$_2$ and NO$_2$ groups were changed, namely molecule **T10**. Since the influence of the *trans* conjugation is now reduced (no donor–acceptor conjugation) we expected a reduction of γ with respect to molecule **T12**. Due to the strong absorption of molecule **T10** at the third harmonic wavelength ($\varepsilon = 1900\,\mathrm{mol/(cm\ dm^3)}$), however, molecules **T10** and **T11** had to be also investigated at $\lambda = 2100\,\mathrm{nm}$ for a meaningful comparison using a flashlamp-pumped, Q-switched CrTmHo:YAG laser (Table 2.4). These measurements confirmed our expectation that the noncentrosymmetric molecule **T11** ($\gamma = 9.7 \times 10^{-48}\,\mathrm{m^5/V^2}$) is superior with respect to the third-order nonlinearity γ (γ (**T10**) $= 5.1 \times 10^{-48}\,\mathrm{m^5/V^2}$)) and that additional states contributing to γ are allowed due to the reduction of symmetry.

The importance of full two-dimensional donor–acceptor conjugation is also illustrated by comparison of the γ values of molecules **T11** and **T16**. The existence of four linear and two cross-conjugated paths (γ (**T11**) $= 18.3 \times 10^{-48}\,\mathrm{m^5/V^2}$) and the extension of one linear donor–acceptor conjugation path (γ (**T16**) $= 21 \times 10^{-48}\,\mathrm{m^5/V^2}$) both lead to a substantial increase in γ. However, due to the lower molecular volume molecule **T11** is superior as can be seen from the values of $\chi^{(3)}$, which were extrapolated from measured values of γ as an indication of the nonlinearity per unit volume. The comparison of $\chi^{(3)}$(**T11**) and $\chi^{(3)}$(**T16**) clearly demonstrates the more pronounced advantage of full two-dimensional conjugation.

Figure 2.22 shows plots of γ vs. wavelength of maximum absorption λ_{eg}. No straightforward dependence of γ on λ_{eg} can be found. For the molecules in Fig. 2.22a this can be expected: (i) the number of π electrons contributing to the third-order nonlinearity varies considerably from compound to compound, and (ii) since we have to deal with two-dimensional conjugation, multiple electron excitation is very likely to be considered which would lead to a much more complicated behavior of the dispersion of γ. For the molecules in Fig. 2.22b we have the following: The comparison between molecules **T1**–**T3** and **T17**–**T19** indicates that the triple bonds connecting the Si(^{i}Pr)$_3$ groups also contribute considerably to the third-order nonlinearity. In such a case no straightforward dependence of γ on λ_{eg} is expected. This is also illustrated by the compounds **T17**–**T19**, where the side groups are terminated by OTBDMS: we get smaller values of γ. However, not enough data points

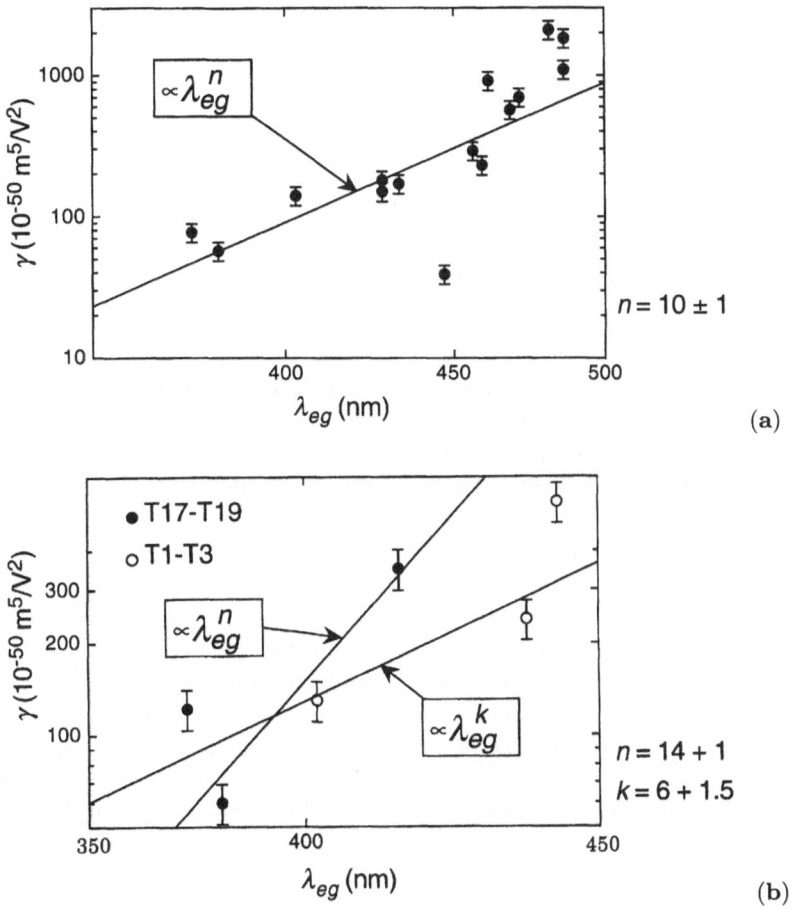

Fig. 2.22a,b. Measured values of γ vs. wavelength of maximum absorption λ_{eg}. (a) all tetraethynylethenes; (b) a comparison between molecules **T1–T3** and **T17–T19**. No straightforward dependence of γ on λ_{eg} can be found in either case

are available to confirm our expectations of a simple correlation between γ and λ_{eg}.

The last two molecules, so-called radialenes, are examples of more extended molecules based on the TEE backbone [2.132]. Huge third-order nonlinearities could be measured that are, however, very difficult to interpret since the compounds are strongly absorbing at the third-harmonic wavelength ($\lambda = 700\,\text{nm}$).

A detailed discussion of the UV/Vis absorption spectra of all compounds investigated here can be found in [2.122–123].

Hydrazone Derivatives. Unlike classical elongated π-conjugated chromophores [2.133], the hydrazone derivatives developed in our laboratory can basically

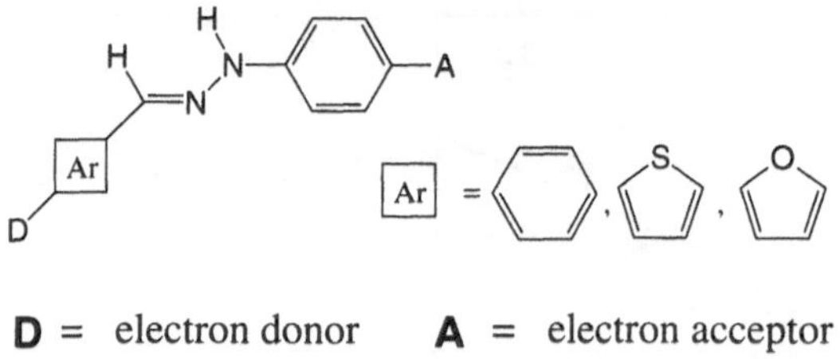

D = electron donor   **A** = electron acceptor

Fig. 2.23. Molecular structure of type I hydrazone derivatives. If donor (*D*) and acceptor (*A*) are exchanged we obtain different molecules: type II hydrazone derivatives

be divided into two groups since the hydrazone skeleton is not symmetrical (Fig. 2.23) [2.134]. Type I hydrazones are shown in Fig. 2.23. If the donor (D) and acceptor (A) are exchanged we obtain different molecules with different properties; type II hydrazone derivatives. These chromophores, which possess a bent hydrazone skeleton

$$(-CH=N\backslash_{NH-})$$

due to the non-rigid nitrogen–nitrogen single bond can be considered being formed by two dipolar chromophores e.g. (Donor–Ar–CH=N–) and (–NH–Ar–Acceptor) connected in a head-to-tail fashion. However, both experimentally and theoretically we have shown that the charge transfer process extends from the donor to the acceptor going through the entire hydrazone skeleton [2.134].

Type I hydrazones can be described as a *bis*-chromophoric system D-A'-D'-A where A' is equal to the azomethine double bond (C=N) and D' is equal to the central amino group, which suggests lower values of β in comparison with type II hydrazones (see below). Also, we expect increasing values of γ with an increase of β. This behavior is fully confirmed by our experimental results.

For the class of hydrazone derivatives discussed here, the probability of forming an acentric and efficient crystal is extremely high [2.60]. Seventeen out of the 24 4-nitrophenylhydrazone derivatives (71%) that have been synthesized in our laboratory are noncentrosymmetric and 25% of them exhibit a very strong second-harmonic signal in the powder test, comparable to 4-*N*, *N*-dimethylamino-4'-*N*'-methylstilbazolium tosylate (DAST) [2.87]. Such a high propensity is probably due to the bent and nonsymmetrical backbone of the hydrazone. Generally speaking, hydrazone derivatives can be crystallized very nicely in many cases. This is of great importance, since in a crystal a much higher density of active molecules can be obtained than in polymers, which results in much larger macroscopic nonlinearities. In addition, the macroscopic third-order nonlinearity can be further enhanced if the molecules are well oriented in the crystal.

Since noncentrosymmetry is not a requirement for third-order nonlinear optics, hydrazone derivatives are an interesting class of materials for third-order nonlinear optics due to their high crystallinity (Fig. 2.24) and the fact that the third-order nonlinearities of the two arrangements in Fig. 2.24 are expected to be similar (if intermolecular interactions are neglected).

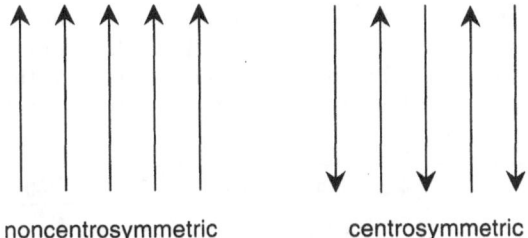

noncentrosymmetric   centrosymmetric

Fig. 2.24. Optimized noncentrosymmetric and centrosymmetric arrangements of polar molecules in a crystal

Table 2.5 shows a list of hydrazone derivatives that were investigated for their third-order nonlinear optical response. All third-harmonic generation measurements were performed in 1,4-dioxane and referenced against fused silica as described above. In addition to γ, $\chi^{(3)}$ values extrapolated from measured values of γ (assuming a linear dependence on concentration of our molecules) are also provided in Table 2.5 as an indication of the maximum macroscopic nonlinearity per unit volume. For molecules **H14** and **H15** we also calculated the macroscopic third-order susceptibility $\chi^{(3)}_{1111}$ based on (2.90), actual crystal data and refractive indices, and assuming that one single tensor element γ_{1111} dominates the molecular response. Values of more than $8000 \times \chi^{(3)}_{\mathrm{fs}}$ are estimated for molecule **H15**.

Compounds **H2–H3** are only moderately polar.[8] As expected, **H2** and **H3** have low values of γ since they are substituted by moderately weak donors on both sides. It is not understood yet why the molecule with the stronger donor (**H3**) has a lower value of γ.

By comparing **H2**, **H4** and **H5**, we see that the third-order nonlinearity continuously increases by first replacing the donor by an acceptor (type I, **H5**) and then exchanging donor and acceptor (type II, **H4**) with the largest value for the type II hydrazone. In a similar way, we have $\gamma(\mathbf{H6}) > \gamma(\mathbf{H7}) > \gamma(\mathbf{H3})$. Also, the stronger donor leads to larger values of γ (compare **H4–H6**, **H5–H7** and **H10**, **H12–H13**). The increase in γ is always associated with an increase in λ_{eg}.

---

[8] Despite being substituted with two acceptors **H1** is strongly polar due to a strong charge transfer from the NO_2 group through the thiopene ring to the nitrogen which acts as a moderate donor. Unfortunately γ could not be measured for this compound because it is very insoluble.

Table 2.5. Results of the third-harmonic generation experiments of hydrazone derivatives (γ(esu: 10^{-36}) and (SI: $10^{-48}\,\mathrm{m}^5/\mathrm{V}^2$)). $\chi^{(3)}$ was measured relative to $\chi^{(3)}$ of fused silica (fs), for which a value of $\chi^{(3)} = 2.62 \times 10^{-22}\,\mathrm{m}^2/\mathrm{V}^2$ (2.16×10^{-14} esu) was used. To compare our results with experimental results by other groups that used a different standard ($\chi_{\mathrm{fs}}^{(3)}(-3\omega,\omega,\omega,\omega) = 3.89 \times 10^{-22}\,\mathrm{m}^2/\mathrm{V}^2$ ($= 2.79 \times 10^{-14}$ esu)), our values of γ (and $\chi^{(3)}$) must be multiplied by a factor of 2.40. All measurements were carried out at $\lambda = 1907\,\mathrm{nm}$ in 1,4-dioxane except for **H13**, for which additional measurements were done in CHCl$_3$ and acetonitrile. The relative errors are estimated to 15%. λ_{eg} (nm) is the longest wavelength electronic absorption maximum and μ (10^{-29} Cm) is the static dipole moment. At the third-harmonic wavelength of $\lambda = 636\,\mathrm{nm}$ all molecules are completely transparent (except for **H14** ($\varepsilon \approx 3$) and **H15** ($\varepsilon \approx 13$))

Compound	λ_{eg}	μ	γ (esu)	γ (SI)	$\dfrac{\chi^{(3)}100\%}{\chi_{\mathrm{fs}}^{(3)}}$
H1	443	–	–	–	–
H2	364	–	18	0.25	13
H3	370	–	13	0.18	8.5
H4	470	2.3	136	2.9	89
H5	403	2.8	62	0.87	41

Table 2.5. (Continued)

Compound	λ_{eg}	μ	γ (esu)	γ (SI)	$\dfrac{\chi^{(3)}100\%}{\chi^{(3)}_{fs}}$
H6 (O_2N–thiophene–CH=N–NH–C$_6$H$_4$–OCH$_3$)	482	2.3	170	2.4	105
H7 (H_3CO–thiophene–CH=N–NH–C$_6$H$_4$–NO$_2$)	416	3.0	75	2.05	46
H8 (H_3CS–thiophene–CH=N–NH–C$_6$H$_4$–NO$_2$)	413	2.6	78	2.09	45
H9 (O_2N–C$_6$H$_4$–CH=N–NH–C$_6$H$_4$–CH$_3$)	423	2.0	82	2.14	55
H10 (H_3C–C$_6$H$_4$–CH=N–NH–C$_6$H$_4$–NO$_2$)	395	2.6	48	0.67	32
H11 (O_2N–C$_6$H$_4$–CH=N–NH–C$_6$H$_4$–OCH$_3$)	433	2.1	107	2.5	68
H12 (H_3CO–C$_6$H$_4$–CH=N–NH–C$_6$H$_4$–NO$_2$)	399	2.7	75	2.05	48

Table 2.5. (Continued)

Compound	λ_{eg}	μ	γ (esu)	γ (SI)	$\dfrac{\chi^{(3)}100\%}{\chi^{(3)}_{fs}}$
$(H_3C)_2N$—⟨benzene⟩—CH=N–N(H)—⟨benzene⟩—NO_2 **H13**	420 (dioxane)	2.9	131	2.8	78 1380[a] $(\chi^{(3)}_{2222})$
	429 (CHCl$_3$)		161	2.3	148 1720[a] $(\chi^{(3)}_{2222})$
	428 (CHCN)		238	3.3	88 2530[a] $(\chi^{(3)}_{2222})$
O_2N—⟨furan⟩—CH=CH–CH=N–N(H)—⟨benzene⟩—OCH_3 **H14**	474	2.1	249	3.5	150 6690[a] $(\chi^{(3)}_{1111})$
O_2N—⟨thiophene⟩—CH=CH–CH=N–N(H)—⟨benzene⟩—CH_3 **H15**	479	2.4	326	4.6	195 8270[a] $(\chi^{(3)}_{1111})$

[a] Values of $\chi^{(3)}_{iiii}$ based on (2.90), actual crystal data and refractive indices (see [2.134] for **H13** and $n^\omega \approx 2$, $n^{3\omega} \approx 2.4$ for **H14** and **H15**), and assuming that one single tensor element γ_{1111} dominates the molecular response

Molecule **H8** has a different donor with almost the same electron donating capacity as CH_3O, namely CH_3S. Our experimental results show that also the γ values are equal in this case (**H6** and **H8**).

Five-membered heteroaromatics such as thiophene and furan derivatives are known to have lower delocalization energies that phenyl derivatives. This often leads to larger values of the second-order polarizabilities β_z (see e.g. [2.84, 2.87, 2.135] and Sect. 2.4.2). As can be seen from a comparison of compounds **H4**–**H7** with **H9**–**H12** (except **H7**↔**H12**), replacing the phenyl ring by a thiophene ring also leads to larger values of the third-order polarizability γ.

Figure 2.25 shows the dependence of γ on λ_{eg}. Good agreement between experimental results of γ and λ_{eg} is obtained if the two lowest values are neglected (power law dependence which yields $\gamma \propto \lambda_{eg}^{5.5}$). If all data points are used we find a linear dependence of γ on λ_{eg}. In contrast to the TEE molecules the theoretical curves describe our experimental results rather well. This is probably due to the fact that, except for two cases, the same number

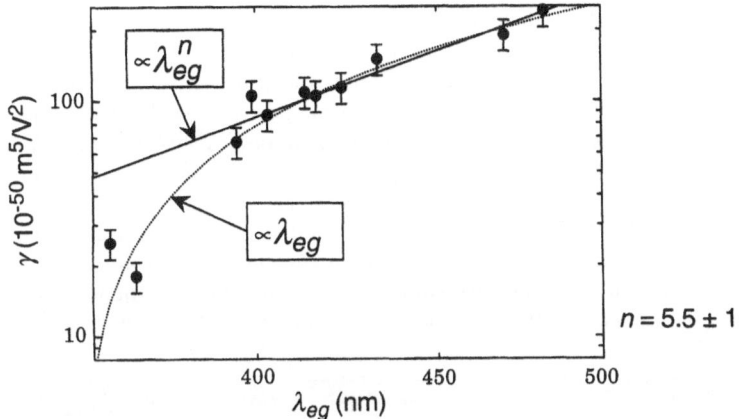

Fig. 2.25. Dependence of the rotational average γ of the hydrazone derivatives (**H2–H12**) on the wavelength of maximum absorption in 1,4-dioxane solutions. Solid line: power law dependence on λ_{eg} neglecting the two lowest values of γ; dotted line: linear dependence of γ on λ_{eg}

of π electrons are always involved in the third-order nonlinearity, since only different donors, acceptors and rings are used.

Stilbene Derivatives and Other Donor–Acceptor Molecules. Further experiments were carried out in order to (i) confirm our findings of larger third-order nonlinearities in noncentrosymmetric molecules compared with their centrosymmetric analogues and (ii) investigate molecules that form highly optimized noncentrosymmetric crystals or are easy to crystallize with favourable antiparallel orientation of the molecules for their third-order nonlinear optical response.

The stilbene derivatives shown in Fig. 2.26 were investigated (see Table 2.6). In contrast to the hydrazone derivatives an exchange of donor and acceptor leads to the same molecule.

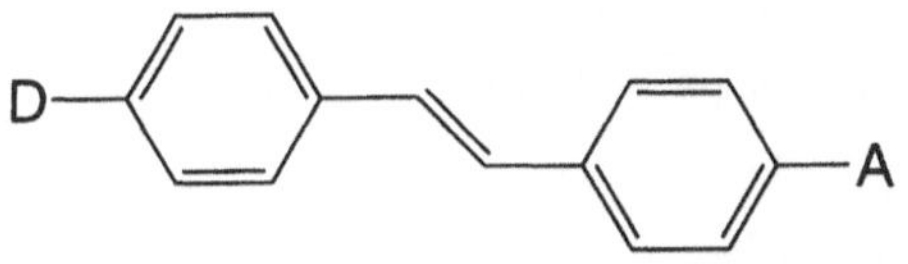

Fig. 2.26. Molecular structure of the prototype stilbene derivative

Unfortunately, stilbene derivatives are difficult to dissolve in many solvents since the backbone of the molecule is completely stiff. Therefore not all substances could be measured and a comparison was quite difficult. Never-

Table 2.6. Results of the third-harmonic generation experiments of stilbene derivatives and related molecules (γ (esu: 10^{-36}) and (SI: $10^{-48}\,\mathrm{m}^5/\mathrm{V}^2$)). $\chi^{(3)}$ was measured relative to $\chi^{(3)}$ of fused silica (fs), for which a value of $\chi^{(3)} = 2.62 \times 10^{-22}\,\mathrm{m}^2/\mathrm{V}^2$ (2.16×10^{-14} esu) was used. To compare our results with experimental results by other groups that used a different standard ($\chi^{(3)}_{fs}(-3\omega,\omega,\omega,\omega) = 3.89 \times 10^{-22}\,\mathrm{m}^2/\mathrm{V}^2$ ($= 2.79 \times 10^{-14}$ esu)), our values of γ (and $\chi^{(3)}$) must be multiplied by a factor of 2.40. The measurements were carried out at $\lambda = 1907\,\mathrm{nm}$ in solvents as indicated. The relative errors are estimated to 15%. λ_{eg} (nm) is the longest wavelength electronic absorption maximum. At the third-harmonic wavelength of $\lambda = 636\,\mathrm{nm}$ all molecules are completely transparent (except for **S5** (ε (methanol) ≈ 5, ε (ethanol) ≈ 10, ε (propanol) ≈ 33) and **S7** (ε (ethanol) ≈ 15, ε (propanol) ≈ 46))

Compound	λ_{eg} (solvent)	γ (esu)	γ (SI)	$\dfrac{\chi^{(3)}100\%}{\chi^{(3)}_{fs}}$
S1	334 (dioxane)	–	–	–
	347 (DMF)	57	0.79	33
S2	366 (dioxane)	–	–	–
	365 (DMF)	241	3.4	144
S3 (DANS)	426 (dioxane)	–	–	–
	443 (DMF)	208	2.9	124
	436 (CHCl3)	306	4.3	300
S4 (DAST)	478 (methanol)	382	5.3	97 / 8510[a]
S5	485 (methanol)	175	2.4	209
	512 (ethanol)	233	3.3	286
	524 (propanol)	223	3.1	274
S6	487 (methanol)	331	4.6	356
S7	484 (methanol)	294	4.1	258
	518 (ethanol)	182	2.6	160
	526 (propanol)	179	2.5	157

Table 2.6. (Continued)

Compound	λ_{eg} (solvent)	γ (esu)	γ (SI)	$\dfrac{\chi^{(3)}100\%}{\chi^{(3)}_{fs}}$
OH ... HO ... OCH₃ **S8**	298 (methanol)	<4	<0.06	<4
(stilbazolium structure) N⁺–C₇H₁₅ + OH ... HO ... OCH₃ **S9**	489 (methanol)	116	2.6	101 3750[a] $\chi^{(3)}_{1111}$
H₃CH₂C, HOH₂CH₂C N— —N=N— —NO₂ **DR1**	473 (dioxane) 502 (DMF) 481 (CHCl₃)	251 337 274	3.5 4.7 3.8	138 172 230
O₂N ... N ... OH **PNP**	376 (dioxane)	12.3	0.17	95 10.5[a] $\chi^{(3)}_{2222}$
1,4-dioxane	–	0.85	2.2×10^{-2}	2.73
CHCl₃	–	0.95	2.3×10^{-2}	2.07
DMF	–	2.08	2.9×10^{-2}	2.80
Acetonitrile	–	0.31	4.4×10^{-3}	0.82
Methanol	–	0.26	3.6×10^{-3}	0.83
Ethanol	–	0.48	6.7×10^{-3}	2.12
Propanol	–	0.61	8.5×10^{-3}	2.19

[a] Values of $\chi^{(3)}_{iiii}$ based on (2.90), actual crystal data and refractive indices (DAST [2.137, 2.138] , Mero-MDB [2.139] , PNP [2.140]), and assuming that one single tensor element γ_{1111} dominates the molecular response

theless, it was confirmed that the value of γ of the acentric molecule **S3** is clearly superior to that of molecule **S1** in the same solvent (DMF).[9]

Molecule **S5** (DAST) is an organic salt [2.136]. It was investigated since crystals of DAST are very well suited for second-order nonlinear optical and electro-optic applications because stilbazolium, as the nonlinear optically ac-

[9] A measurement with the molecule **S2** was not possible in 1,4-dioxane due to its limited solubility in that solvent.

tive part, is a very efficient organic chromophore. A crystal structure analysis showed that the angular deviation of the charge transfer axes of the stilbazolium chromophores from a completely aligned system is about 20°. Hence, if the chromophore also exhibits large third-order nonlinearities the good alignment of the chromophores in the crystal would also make DAST an interesting material for third-order nonlinear optics. DAST has a value of γ that is similar to **S3**. However, a direct comparison is difficult since not the same solvent was used (DAST does not dissolve well enough in 1,4-dioxane, DMF, CHCl$_3$ and others). **S3**, on the other hand, crystallizes very poorly. Therefore the large nonlinearity of the molecule **S3** cannot be used in a crystal, in contrast to DAST. Further results on DAST crystals will be presented in Sect. 2.5.

S5 is another molecule with excellent second-order nonlinear optical properties. However, it is difficult to crystallize. If **S5** is modified to **S6** and co-crystallized with **S8** noncentrosymmetric crystals of good optical quality with an optimized alignment of the nonlinear optically active chromophores (deviation < 1°) are obtained (4-{2-[1-(2-hydroxyethyl)-4-pyridylidene]-ethylidene}-cyclo-hexa-2,5-dien-1-one·methyl 2,4-dihydroxybenzoate (Mero-MDB)). **S5–S9** were investigated with third-harmonic generation in solution in order to estimate the third-order nonlinearity in the bulk. Surprisingly, **S6** and **S7** (analogue of **S5**, synthesized to increase the solubility) are superior to **S5** by a factor of two in the solvent methanol. A possible explanation is the following: in the solution we have an equilibrium of:

By varying R or R' we can change this equilibrium, and this can lead to different values of the second-order hyperpolarizabilities. The value of γ for **S8** is very small. The fact that **S9** (very similar to Mero-MDB), a solution of (almost) the actual co-crystal, shows considerably smaller nonlinearities than **S6** (by almost a factor of three) although **S8** has a very small value of γ, can be understood by looking at the total absorption spectrum: the solution contains not only **S9**, but also **S6** and **S8**. Therefore we have to consider a mixture of γ values, which obviously leads to reduction of the total γ.

In addition to γ, $\chi^{(3)}$ values extrapolated from measured values of γ (assuming a linear dependence on concentration of our molecules) are also provided in Table 2.6 as an indication of the maximum macroscopic nonlinearity per unit volume. Note, that in this case comparison between different molecules is only meaningful if the same solvent was used. For the compounds DAST (**S3**) and Mero-MDB (similar to **S9**) we also calculated the macroscopic third-order susceptibility $\chi_{1111}^{(3)}$ in the same way as for the two hydrazone derivatives **H14** and **H15**. Actual refractive indices as measured in these materials were used in this calculation. Values of more than $8500 \times \chi_{\text{fs}}^{(3)}$ are estimated for molecule **S4** (DAST).

Disperse Red 1 (DR1) is a molecule widely used in electro-optic polymers. It also crystallizes poorly. It is listed for comparison with the other molecules and because it is well suited as a standard material. Again its value of γ is quite similar to the other ones.

PNP is a pyridine derivative with large second-order nonlinearities in the bulk [2.71] and is listed for comparison and in order to demonstrate that generally small molecules only have moderate values of γ. If extrapolated to the expected macroscopic nonlinearity we obtain $\chi_{2222}^{(3)} \approx 11 \times \chi_{\text{fs}}^{(3)}$. In addition, the values of $\chi^{(3)}$ and γ of the solvents used are also given for completeness.

The possible contribution of cascading through the reaction field to γ will be discussed in Sect. 2.6.

2.4.2 Electric Field-Induced Second-Harmonic Generation

Electric field-induced second harmonic generation (EFISH) is the standard method for the determination of molecular second-order susceptibilities. The basic idea of the method is to measure the frequency doubling efficiency of the nonlinear optical molecules in dilute solutions in order to reduce inter-molecular effects. Because of the isotropic symmetry of a liquid, however, the macroscopic second-order susceptibility vanishes and no coherent frequency doubled light can be observed. This problem is solved by applying a static electric field which partially orients the dipolar molecules and thus breaks the symmetry of the liquid. With the induced macroscopic d-coefficient the observation of second-harmonic generation becomes possible (see also Sect. 2.2.3).

A Theory

A static electric field E^0 partially orients the molecules due to the coupling to the dipole moment. The solution becomes noncentrosymmetric and a laser beam E^ω incident to the solution produces a second-harmonic signal. The macroscopic polarization at the second-harmonic frequency is defined by

$$P_i^{2\omega} = \frac{1}{2}\varepsilon_0 \chi_{ijk}^{(-2\omega,\omega,\omega)}(E^0)E_j^\omega E_k^\omega = \varepsilon_0 d_{ijk}^{(-2\omega,\omega,\omega)}(E^0)E_j^\omega E_k^\omega . \qquad (2.120)$$

We choose the polarization of the incoming wave and the applied electric field in the x_3-direction. The frequency-doubled light is analyzed by a polarizer oriented along the x_3-direction. The only susceptibility component producing a second-harmonic signal is then the diagonal component along this direction, d_{333}.

The connection between the microscopic hyperpolarizability as defined in (2.9) and the macroscopically observable susceptibility was derived by Kielich [2.141–144] using classical statistical mechanics. He derived appropriate expressions for both weak and strong external fields applied to molecules with small intermolecular interactions. In practice, the condition for a weak electric field

$$\frac{\mu_g E^0}{kT} \ll 1 \tag{2.121}$$

is always satisfied. In this case, the orientational distribution function can be approximated by a linear function of the applied field and $d_{333}(E^0)$ becomes linearly dependent on the external field.

$$P_3^{2\omega} = \varepsilon_0 d_{333}(E^0)\left(E_3^\omega\right)^2 = \varepsilon_0 \Gamma_L \left(E_3^\omega\right)^2 E^0 , \tag{2.122}$$

where the third-order susceptibility Γ_L is connected to the molecular hyperpolarizabilities by (see also Table 2.2)

$$\Gamma_L = Nf\left\{\frac{3}{2}\bar{\gamma}(-2\omega,\omega,\omega,0) + \frac{1}{2}\frac{\mu_g \beta_z(-2\omega,\omega,\omega)}{5kT}\right\}, \tag{2.123}$$

where N is the number density of the molecules and f^0, f^ω and $f^{2\omega}$ are local field factors evaluated at the indicated frequencies. The vector part β_z of the hyperpolarizability tensor β_{ijk} along the direction of the permanent dipole moment and the scalar part of the second-order hyperpolarizability γ_{ijkl} have already been introduced in Sect. 2.2 ((2.23) and (2.24) respectively).

If we choose the dipole moment along the x_3-axis of the molecular coordinate system and assume Kleinman symmetry the microscopic quantities are given by (see also (2.24))

$$\bar{\gamma} = \frac{1}{5}\sum_{i,j}\gamma_{iijj} \tag{2.124}$$

$$\beta_z = \beta_{zzz} + \beta_{xxz} + \beta_{yyz} . \tag{2.125}$$

The quantity β_{zzz} is usually referred to as β_{CT} which corresponds to the nonlinearity along the charge transfer axis. EFISH does not allow the determination of single tensor elements but only of a combination of several components. For molecules with strong donor and acceptor groups, however, β_{CT} accounts for almost all of β_z.

The microscopic quantity that can actually be derived from the measurements is

$$\gamma' = \frac{3}{2}\bar{\gamma} + \frac{1}{2}\frac{\mu_g \beta_z}{5kT} . \tag{2.126}$$

This means that besides the nonlinearity Γ_L the permanent dipole moment μ_g also has to be determined. Using third-harmonic generation experiments the third-order hyperpolarizability $\gamma(-2\omega, \omega, \omega, 0)$ has been shown to contribute less than 10% to γ' for strongly conjugated molecules [2.135].[10] It is therefore usually neglected, which may lead to a slight overestimation of β_z.

In a two-component solution the macroscopically observed nonlinearity Γ_L arises from contributions of both the dissolved nonlinear optical chromophores (index p) and the pure solvent (s):

$$\Gamma_L = N_s f_s^0 \left(f_s^\omega\right)^2 f_s^{2\omega}\gamma_s' + N_p f_p^0 \left(f_p^\omega\right)^2 f_p^{2\omega}\gamma_p' \qquad (2.127)$$

In order to minimize solvent–solvent and solute–solute interactions, Singer and Garito [2.145] have developed an extrapolation procedure to infinite dilution where the quantity γ_p' is expressed in terms of the concentration dependencies of the nonlinearity, the dielectric constant, the specific volume and the refractive index of the solution.

Neglecting the terms containing the concentration dependence of the specific volume and of the refractive index because of their minor influence on γ_p' compared to the other terms, the microscopic nonlinearity can be expressed as

$$\gamma_p' = \frac{81\, M_p}{N_A (\varepsilon^s + 2)(n_s^2 + 2)^3} \left\{ v_s \left.\frac{\partial \Gamma_L}{\partial w}\right|_0 + v_s \Gamma_s \left[1 - \frac{1}{(\varepsilon^s + 2)} \left.\frac{\partial \varepsilon}{\partial w}\right|_0 \right] \right\},$$

$$(2.128)$$

where M_p is the molar weight of the solute and N_A is Avogadro's number. w represents the weight fraction of the solute ($w = m/(m + m_s)$; m: mass of solute, m_s: mass of solvent) and v_s, n_s and ε^s are the specific volume, the refractive index and the dielectric constant of the pure solvent. The concentration-dependent third-order susceptibility Γ_L can also be written as [2.30]

$$\Gamma_L = \frac{N_A d(C) f(c)}{1 + C} \left\{ \frac{\gamma_s'}{M_s} + C \frac{\gamma_p'}{M_p} \right\}, \qquad (2.129)$$

where $\gamma_{s,p}$ is the second-order hyperpolarizability of solvent and chromophore molecules, respectively, C is the concentration expressed as the total mass of chromophores divided by the total mass of the solvent, $f(C)$ contains all local field factors (Lorentz–Lorentz approximation), $d(C)$ is the density of the solution, N_A is Avogadro's number and $M_{p,s}$ are the molecular masses of the chromophore and solvent molecule, respectively. We have neglected the concentration dependence of the density and refractive indices, which proved to give small errors in comparison with other uncertainties for these very dilute solutions. The first-order hyperpolarizability β_z for the chromophore (p) and the solvent molecule (s) can then be evaluated.

The description of the theoretical curves used to evaluate the experimental results presented below is given in Sect. A.4.

---

[10] If the size of $\overline{\gamma}(-2\omega, \omega, \omega, 0)$ is estimated by measuring $\gamma(-3\omega, \omega, \omega, \omega)$ via third-harmonic generation, one must be careful concerning the frequency dispersion as well as the consistent definition of γ.

B Experimental Description

The EFISH measurements were performed with $\lambda = 1907\,$nm as the fundamental wavelength (see Sect. 2.4.2.B) [2.31]. The experimental setup is shown in Fig. 2.27. The generated second harmonic signal was detected with an IR sensitive photomultiplier, which was cooled to about $-6°$C to reduce the thermal noise.

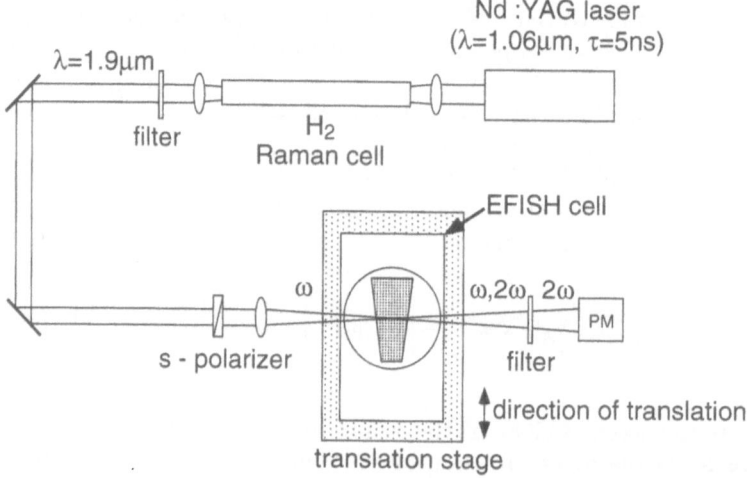

Fig. 2.27. Experimental setup for the EFISH measurements using a hydrogen-filled Raman cell to generate the fundamental radiation of $\lambda = 1907\,$nm

The measurements are performed by translating the wedge-shaped liquid cell containing the solutions to be investigated (Fig. 2.28) perpendicularly to the fundamental laser beam. Owing to the resulting variation of the liquid path length the typical Maker oscillations are observed when measuring the second-harmonic intensity. The liquid cell consisted of two glass windows (BK7) positioned between two stainless steel electrodes. The wedge angle α was $5.7°$ and the interelectrode distance was $5.0\,$mm. The high voltage applied to the glass cell was pulsed for a duration of about $1\,$ms in order to minimize the electrochemical degradation of the samples. The voltage was synchronized with the laser pulses with the Q-switch triggering the high voltage modulator. The voltage was typically $7\,$kV leading to a field strength of $E^0 = 14\,$kV/cm.

As was pointed out by Levine and Bethea [2.146], the liquid cell must fulfill the following requirement: the static field must be uniform over the whole region of the liquid in order to ensure that the full liquid nonlinearity $\Gamma_{\rm L}$ is achieved at the glass–liquid boundary. If the electrodes were immersed in a liquid without glass walls almost no second-harmonic signal would be produced.

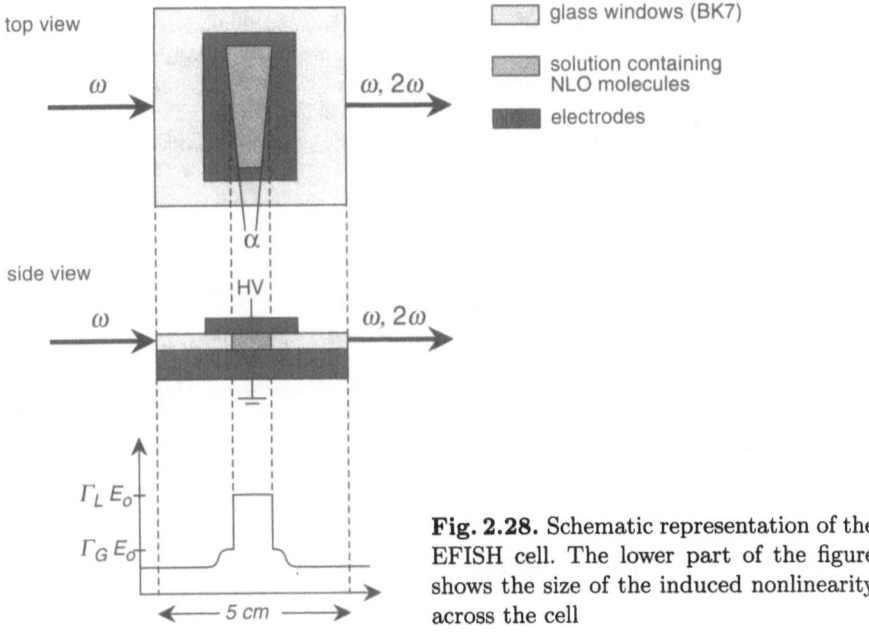

top view

glass windows (BK7)

solution containing NLO molecules

electrodes

ω

$\omega, 2\omega$

α

side view

HV

ω

$\omega, 2\omega$

$\Gamma_L E_o$

$\Gamma_G E_\sigma$

5 cm

Fig. 2.28. Schematic representation of the EFISH cell. The lower part of the figure shows the size of the induced nonlinearity across the cell

The β values were calibrated against a reference solution of 2-methyl-4-nitroaniline (MNA). The hyperpolarizability β of MNA was taken as $\beta_{MNA} = 44 \times 10^{-40}\,\mathrm{m^4/V}$ [2.31], as determined by an EFISH experiment with crystalline quartz ($d_{11} = 0.3\,\mathrm{pm/V}$ at $\lambda = 1064\,\mathrm{nm}$ plus Miller $- \delta \rightarrow d_{11} = 0.277\,\mathrm{pm/V}$ at $\lambda = 1907\,\mathrm{nm}$) [2.147] as a reference.

Tetraethynylethenes. The basic structure of the tetraethynylethenes is shown in Fig. 2.10. For each compound we measured at least two different concentrations and the pure solvent ($CHCl_3$). As was mentioned above, for conjugated donor–acceptor substituted systems, the contribution in (2.123) from the second-order hyperpolarizability $\overline{\gamma}(-2\omega, \omega, \omega, 0)$ is much smaller than that from $\beta_z(-2\omega, \omega, \omega)$, in our case less than 20% estimated from our third-harmonic generation data. As a consequence, $\overline{\gamma}(-2\omega, \omega, \omega, 0)$ was set to zero.

For all compounds, we used the calculated dipole moment (Table 2.7), which coincides well with the measured values for compounds **T3** and **T12**. It should be noted that the dipole moment is mainly determined by the $O_2NC_6H_4$-acceptor group (Table 2.7). To more accurately compare the hyperpolarizability β with measurements at other wavelengths and with the calculations, which always yield static values, we interpolated the measured value $\beta_z(-2\omega, \omega, \omega)$ to the zero frequency value β_0 by using the two level model. However, this is only a first approximation, since the two-level model may very likely be insufficient for accurately describing our two-dimensional molecules.

Table 2.7. Results of the EFISH experiments. All measurements were carried out in CHCl$_3$. Molecular structures, measured wavelengths of maximum absorption λ_{eg} (nm), calculated β_o, measured β_z and β_o (all $10^{-40}\,\mathrm{m}^4/\mathrm{V}$), calculated dipole moments μ_c and measured dipole moments μ_m (all $10^{-30}\,\mathrm{Cm}$). Precision of β_z is $\pm 10\%$. For comparison the values of γ ($10^{-48}\,\mathrm{m}^5/\mathrm{V}^2$) from Table 2.4 are given again

Compound	direction of dipole moment	λ_{eg}	Calc. β_o	Meas. β_z β_o	μ_c μ_m	γ
T3		468	–	430	21	5.7
			168	300	22	
T5		447	–	81	19	0.39
			82	61	–	
T8		471	–	150	17	7
			40	110	–	
T9		461	–	450	21	9.2
			180	320	–	

Table 2.7. (Continued)

Compound	direction of dipole moment	λ_{cg}	Calc. β_0	Meas. β_z β_0	μ_c μ_m	γ
Me$_2$N ... NMe$_2$ / O$_2$N ... NO$_2$ **T11**		486	– 200	510 360	24 29	18.3
(Pr^i)$_3$Si ... SiMe$_3$ / O$_2$N ... Si(^{i}Pr)$_3$ **T12**		382	– 36	51 42	16 –	0.57
(^{i}Pr)$_3$Si ... NMe$_2$ / Me$_3$Si ... Si(^{i}Pr)$_3$ **T13**		433	– 86	190 150	6.3 –	2.7
(Pr^i)$_3$Si ... (^{i}Pr)$_3$Si ... NMe$_2$ / O$_2$N ... Si(^{i}Pr)$_3$ ... Si(^{i}Pr)$_3$ **T16**		481	– 240	400 280	26 –	21

The EFISH measurements of donor–acceptor substituted TEEs showed significant values for the hyperpolarizability, as high as $\beta_z = (510 \pm 100) \times 10^{-40}\,\mathrm{m^4/V}$ (compound **T11**). The value for **T11** compares well with other highly nonlinear optical organic molecules such as DANPH ($\beta_z = 280 \times 10^{-40}\,\mathrm{m^4/V}$ [2.134]) or DANS ($\beta_z = 430 \times 10^{-40}\,\mathrm{m^4/V}$ [83]). Note that all hyperpolarizabilities were scaled for the new reference value for quartz of $d_{11} = 0.277\,\mathrm{pm/V}$ at $\lambda = 1907\,\mathrm{nm}$.

For comparison, and in order to get information about the different tensor elements of β, we calculated the molecular nonlinear optical properties of the TEEs (**T3**, **T5**, **T8–T9**, **T11–T13**, **T16**, Table 2.7), applying the semi-empirical finite field method [2.148] with the MOPAC6 [2.149] software and

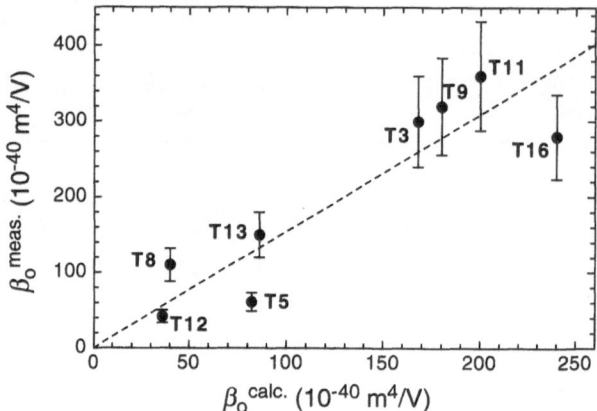

Fig. 2.29. Comparison of measured and calculated first-order hyperpolarizabilities of compounds **T3**, **T5**, **T8–T9**, **T11–T13**, **T16**, with an estimated overall error of 20%. The measured values were extrapolated to infinite wavelengths by the two-level model [2.21, 2.38]

the PM3 parametrization using the molecular geometry optimized within that calculation. With MOPAC, the hyperpolarizability tensor β in the coordinate system of the main axes of the moment of inertia tensor is obtained. MOPAC uses definitions of the molecular hyperpolarizabilities that differ from those defined by (2.9), and for comparison, the β_{ijk} tensor elements from MOPAC were corrected by

$$\beta_0^{\text{calc.}} = \frac{5}{6}\beta_{\text{vec.}}^{\text{MOPAC}} \tag{2.130}$$

$$\beta_{ijk}^{\text{calc.}} = \frac{1}{2}\beta_{ijk}^{\text{MOPAC}}. \tag{2.131}$$

To reduce the calculation time, the nonlinear inactive moieties $\text{Si}(^i\text{Pr})_3$ and SiMe_3 in all compounds were replaced by a single hydrogen atom. The electronic inactivity of the trialkylsilyl groups was determined experimentally by measuring the UV spectra for molecule **T5** and **T12** and its desilylated analogue. The spectra showed only minimal changes in the wavelength of maximum absorption, λ_{eg}, which is a good indication of the validity of this approximation.

In Fig. 2.29, the measured values for β_z, extrapolated to infinite wavelengths, are compared with the calculated ones. The straight line is a linear fit through the origin. With an estimated uncertainty of 20%, the measured and calculated values correlate well.

With the MOPAC calculations, in contrast to the EFISH measurements, values for all tensor elements β_{ijk} are obtained. Based on the observed correlation between $\beta_0^{\text{exp.}}$ and $\beta_0^{\text{calc.}}$ (Fig. 2.29), the calculated tensor elements $\beta_{ijk}^{\text{calc.}}$ should give realistic values. Therefore these calculations allow us to

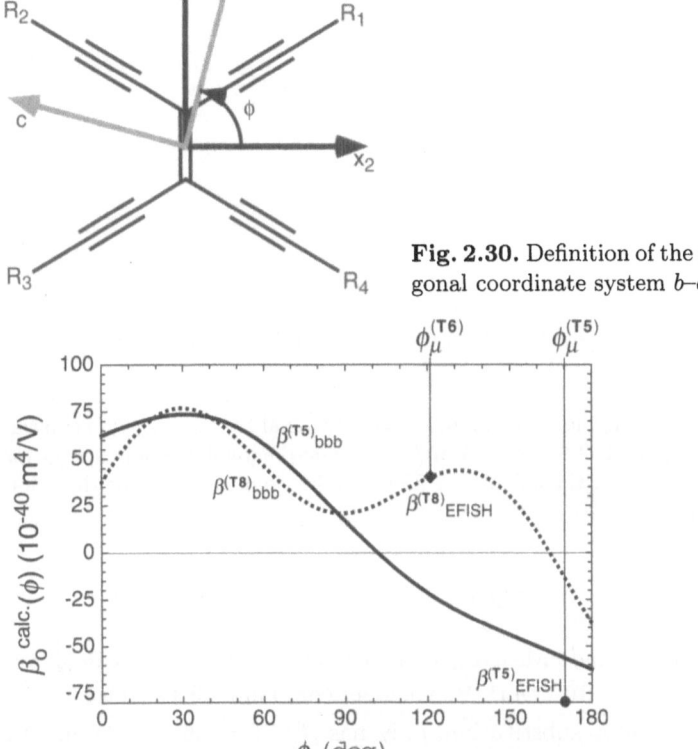

Fig. 2.30. Definition of the molecular orthogonal coordinate system b–c by the angle ϕ

Fig. 2.31. Calculated angular dependence of the diagonal first-order hyperpolarizability β_{bbb} for compounds **T5** and **T8**. β_{EFISH} is the calculated EFISH value, indicated by the direction of the dipole moment ϕ

accurately calculate the contribution of cascading through the reaction field (see Sect. 2.3) which will be discussed in Sect. 2.6.

To obtain the contributions which originate from the different conjugation paths in the molecule, we calculated the angular dependence of the diagonal element β_{bbb} (see Fig. 2.30 for the definition of the coordinate system) for compounds **T5**, **T8**–**T9** and **T11** (Figs. 2.31 and 2.32) and of the nondiagonal element β_{cbb} for compound **T11** (Fig. 2.32). If $e^b(\phi)$ and $e^c(\phi)$ are unit vectors in the b- and c-directions, respectively, then the angular dependence of the diagonal and nondiagonal element of compound **n** is given by (2.132) and (2.133), where the tensor β_{ijk} is given in the coordinate system (x_1, x_2, x_3) and summation over common indices is assumed

$$\beta_{bbb}^{(n)}(\phi) = \beta_{ijk}^{(n)} e_i^b(\phi) e_j^b(\phi) e_k^b(\phi) \tag{2.132}$$

$$\beta_{cbb}^{(n)}(\phi) = \beta_{ijk}^{(n)} e_i^c(\phi) e_j^b(\phi) e_k^b(\phi) \; . \tag{2.133}$$

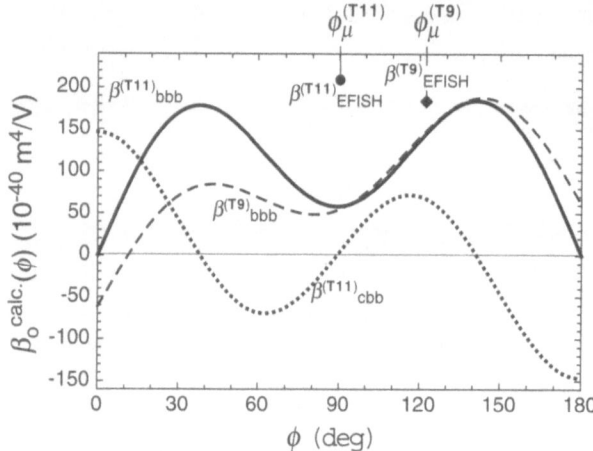

Fig. 2.32. Calculated angular dependence of the diagonal first-order hyperpolarizability β_{bbb} for compounds **T9** and **T11** and of the non-diagonal component β_{cbb} for compound **T12**. β_{EFISH} is the calculated EFISH value, indicated by the direction of the dipole moment ϕ

Let us now compare the experimental and theoretical results in more detail.

TEE **T13** with a single $Me_2NC_6H_4$-donor substitution has a calculated value of $\beta_0^{(T13)} = 86 \times 10^{-40}\,m^4/V$, whereas compound **T12**, with a single nitrophenyl acceptor substitution, only has $\beta_0^{(T12)} = 36 \times 10^{-40}\,m^4/V$. This is in contrast to donor–acceptor monosubstituted benzenes, where β_0 for nitrobenzene was higher than for N,N-dimethylaniline [2.83]. For the monosubstituted stilbenes with a longer conjugation length, almost identical values were measured for β_0, of both the $Me_2NC_6H_4$-donor and $O_2NC_6H_4$-acceptor substituted derivatives [2.83]. Compounds **T13** and **T12** are even more extensively conjugated. This indicates that by increasing the conjugation length, the effect of the donor group on β_0 becomes more important with respect to the acceptor group.

TEE **T3** is a combination of **T12** and **T13** and has a value of $\beta_0^{(T3)}$, which is higher than the sum $\beta_0^{(T12)} + \beta_0^{(T13)}$. This is the well-known charge-transfer effect between donor and acceptor [2.150].

Molecule **T5** shows a calculated $\beta_{bbb}^{(T5)}(30°)$ (direction of the donor group, Fig. 2.31), which corresponds well to that of compounds **T13**, **T8** and **T9** in the same direction. This suggests that this value is mainly determined by the single $Me_2NC_6H_4$-donor group. Furthermore, β_0 is relatively small, due to the inefficient cross-conjugation between donor and acceptor groups (see Sect. 2.4.2.B). For molecule **T8**, the $O_2NC_6H_4$-acceptor group and the $Me_2NC_6H_4$-donor group are located almost perpendicularly to each other. As a result, we get a value of $\beta_{bbb}^{(T8)}(30°) = 80 \times 10^{-40}\,m^4/V$ in the direction

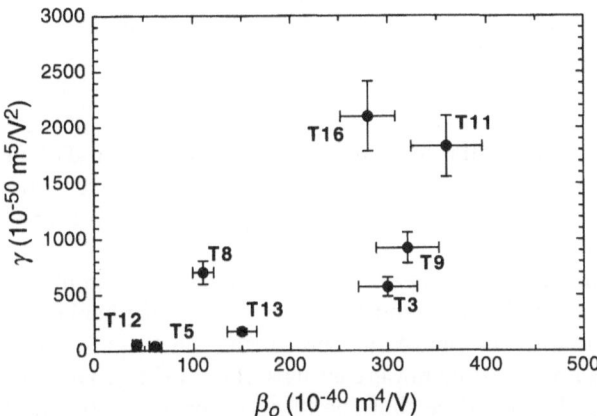

Fig. 2.33. Comparison of values of γ and β_0 for the TEE compounds. The values for γ were not extrapolated to infinite wavelength since no theory describing the dispersion of γ for these new compounds is currently available. There is a tendency for an increase of γ with increase of β_0. However, there is no obvious relationship visible, mainly due to the different averaging in EFISH and THG experiments

of the donor group and $\beta_{bbb}^{(T8)}(150°) = 32 \times 10^{-40}\,\mathrm{m^4/V}$ in the direction of the acceptor group, which coincides very well with that of compounds **T12** and **T13** (Fig. 2.31).

In addition, the measured value for **T8** (*cis*-configuration) is much lower than that measured for **T3** (*trans*-configuration). This is in contrast to the second-order hyperpolarizability γ, where comparable values for the *trans*- and *cis*-configured compounds were observed (see Sect. 2.4.2.B).

However, in the third-harmonic generation measurement of γ, one obtains the rotational average of γ_{ijkl} (2.31), whereas in the EFISH measurement one gets the average in (2.125) for β_z (and β_0), and the component along the dipole moment dominates. The discrepancy mentioned (see Fig. 2.33) might be a result of the different averaging in the EFISH and THG measurements. Still, a general trend is seen that can probably be associated with the $\Delta\mu^2$ term in the three-level model (2.44).

The value of compound **T9** in the direction of the push–pull charge transfer axis of $\beta_{bbb}^{(T9)}(150°)$ corresponds excellently with that of compound **T3** and **T11** in the same direction (Fig. 2.32 and Table 2.7).

The highest value ($\beta_z = (510 \pm 100) \times 10^{-40}\,\mathrm{m^4/V}$) of this series of compounds was measured for molecule **T12**. It can be seen that this value mainly originates from the nondiagonal tensor element β_{cbb}, as illustrated by (2.125) and Fig. 2.32 ($\beta_z \approx \beta_{bbb}^{(T11)}(90°) + \beta_{cbb}^{(T11)}(0°)$).

Hydrazone Derivatives. All the physical measurements were performed in dioxane. The ground state dipole moment, μ_g, was obtained from concentration-dependent dielectric measurements and the use of the Guggenheim equation [2.151]. The fundamental wavelength, $\lambda = 1.907\,\mu$m, was used in the EFISH experiment. $\overline{\gamma}(-2\omega, \omega, \omega, 0)$ was again set to zero. The β values were calibrated with a reference solution of MNA. The molecular hyperpolarizability β of MNA was again taken as $\beta_{\mathrm{MNA}} = 44 \times 10^{-40}\,\mathrm{m}^4/\mathrm{V}$.

The measured first-order molecular hyperpolarizability β_z, projected along the direction of the permanent dipole, was extrapolated to infinite wavelength (β_0) according to the two-level model, although it may lead to substantial errors because of multiple electron excitation [2.38].

We have carried out a comprehensive series of measurements to compare (i) type I and II hydrazone derivatives and (ii) the influence of phenyl and five-membered heteroaromatic (thiophene and furan) rings on β_z [2.87]. The acceptor was always NO_2 and the donors were either CH_3 or OCH_3. We have experimentally shown for the first time that the furan derived π-conjugated core can become the most effective π-conjugated bridge in enhancing the molecular hyperpolarizability when compared with the corresponding phenyl and thienyl counterparts if appropriate electron donors and acceptors are used.

Here we consider only the results for the molecules that were also investigated for their third-order nonlinearities (only one furan-based hydrazone derivative) which are summarized in Table 2.8.

In view of the electronic absorption spectra of the type I hydrazone derivatives, there is a significant bathochromic shift of the charge transfer absorption transitions, λ_{eg}, when the phenyl ring on the donor side is substituted by a five-membered heteroaromatic ring (compare e.g. H12 and H7). For the series bearing a methoxy group as a donating group (see e.g. H12 and H7), the molecular hyperpolarizabilities of the five-membered heteroaromatic derived hydrazones increase with increasing λ_{eg}. But for the series bearing a weak donor, the methyl group (see e.g. H10 and H5), the β_z of the thiophene-derived hydrazone drops below that of the phenyl counterpart.

In the case of the type II hydrazone derivatives, there is a dramatic red shift of the λ_{eg} ($\Delta\lambda = 22\,\mathrm{nm}$) in spite of the moderate or weak electron donor when the 4-nitrophenyl moiety of H9 or H11 is substituted by the 2-nitrofuran moiety (not shown in Table 2.8). Furthermore, the λ_{eg} shifts even further to longer wavelengths ($\Delta\lambda = 47 - 49\,\mathrm{nm}$) upon changing to the 2-nitrothiophene moiety (H4 or H6). However, the molecular hyperpolarizabilities do not increase in a similar manner to λ_{eg}. In turn, the furan-derived type II hydrazone skeleton capped with a moderate or weak electron donor and a strong electron acceptor (not shown in Table 2.8) is found to be the most effective π-conjugated system in enhancing the molecular hyperpolarizability among the series.

The approximation of a single excitation dominating the contribution to the molecular hyperpolarizability should not be valid for the hydrazone derivatives because there is more than one electron involved in the charge transfer process [2.134]. Thus, other excitations could contribute to a con-

Table 2.8. Results of EFISH experiments of the hydrazone derivatives. All measurements were carried out in dioxane. λ_{eg} (nm) is the longest wavelength electronic absorption maximum. μ (10^{-29} Cm) is the dipole moment derived from the Guggenheim equation. β_z (10^{-40} m^4/V) is the first-order molecular hyperpolarizability, projected along the direction of permanent dipole, determined by the EFISH experiment using fundamental wavelength at 1.907 μm with a precision of ±10%. β_o is the static hyperpolarizability extrapolated to infinite wavelength derived from the two-level model. For comparison the values of γ (10^{-48} m^5/V^2) from Table 2.5 are given again

Compound	λ_{eg}	μ	β_z β_o	γ
H4 (O$_2$N–thiophene–CH=N–N(H)–C$_6$H$_4$–CH$_3$)	470	2.3	205 145	2.9
H5 (H$_3$C–thiophene–CH=N–N(H)–C$_6$H$_4$–NO$_2$)	403	2.8	120 94	0.87
H6 (O$_2$N–thiophene–CH=N–N(H)–C$_6$H$_4$–OCH$_3$)	482	2.3	227 158	2.4
H7 (H$_3$CO–thiophene–CH=N–N(H)–C$_6$H$_4$–NO$_2$)	416	3.0	187 144	2.05
H8 (H$_3$CS–thiophene–CH=N–N(H)–C$_6$H$_4$–NO$_2$)	413	2.6	130 101	2.09

Table 2.8. (Continued)

Compound	λ_{eg}	μ	β_z β_0	γ
H9	423	2.0	150 115	2.14
H10	395	2.6	129 102	0.67
H11	433	2.1	210 170	2.5
H12	399	2.7	152 120	2.05
H13	420	2.9	277 212	2.8
H14	474	2.1	403 285	3.5
H15	479	2.4	433 305	4.6

siderable extent to the molecular hyperpolarizability of this class. This is consistent with our findings that the molecular hyperpolarizability is not directly correlated to λ_{eg}. The strongest red shift of λ_{eg} of the thiophene derived π-conjugated core among the series could be due to the participation of the highly polarizable sulphur atom of the thienyl ring.

With the addition of a double bond on the electron acceptor side of the type II hydrazone derivatives (see e.g. **H14** and **H15**), the λ_{eg} also shift to longer wavelengths with respect to those of the parent compounds (compare e.g. **H4** and **H15**). This is presumably due to the enhancement of π electron delocalization as observed in the case of donor–acceptor disubstituted stilbene series [2.135]. The β_0 values of the corresponding extended thiophene-derived hydrazones (e.g. **H15**) increase by a factor of two compared with those of their parent compounds (e.g. **H4**) and this extended type II hydrazone skeleton has become the most effective π-conjugated bridge in enhancing β_z in this class. The abrupt change in the effectiveness of the thiophene-derived extended π-conjugated core relative to its lower homologue in influencing β_z is probably due to the greater extent of participation of the highly polarizable sulfur atom in conjugation under the situation of high electron demand.

We can again try to compare the values of γ and β_0 of the hydrazone derivatives discussed here (Fig. 2.34). There is a tendency for an increase of γ with increase of β_0 (also noted by Cheng et al. for stilbene derivatives [2.83]). Although there is no simple relationship the general trend can probably be associated with the $\Delta\mu^2$ term in the three-level model.

Stilbene Derivatives and Other Donor–Acceptor Molecules. Organic salts cannot be measured with EFISH since the molecules have a charge and a current would flow on application of a static electric field. The merocyanine derivatives were also not investigated for their second-order nonlinear optical response with EFISH, since the applied electric field immediately broke down (due to an electric current flowing in the EFISH cell (mainly due to impurities in the solution that could not be removed)). Therefore we could only determine (or find) values for Disperse Red 1 (DR1), DANS and PNP (Table 2.9). DR1 and PNP were measured in our laboratory in dioxane, whereas the value for DANS was measured in $CHCl_3$ by Cheng et al. [2.83]. We can again notice an increase of γ with increasing values of β_0. Also very illustrative is the large increase in the molecular nonlinearities when extending the conjugation length (comparison of DANS and DR1 with PNP).

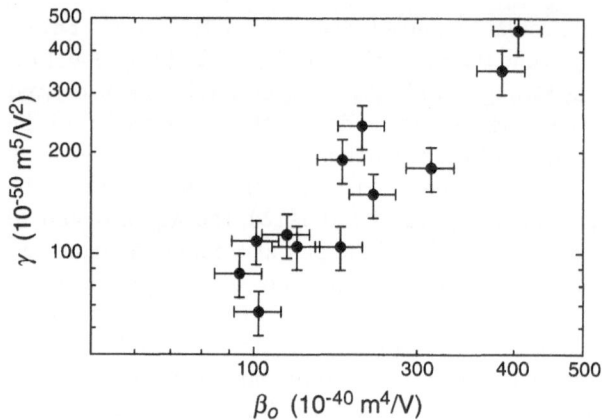

Fig. 2.34. Plot of γ vs. β_0 for the hydrazone derivatives shown in Table 2.8. The values for γ were not extrapolated to infinite wavelength since no theory describing the dispersion of γ for these new compounds is currently available. There is a tendency for an increase of γ with increase of β_0. However, there is no obvious relationship visible

Table 2.9. Results of EFISH experiments of DR1, DANS, and PNP. The measurements were carried out in the solvents indicated. λ_{eg} (nm) is the longest wavelength electronic absorption maximum. μ (10^{-29} Cm) is the dipole moment derived from the Guggenheim equation. β_z (10^{-40} m^4/V) is the first-order molecular hyperpolarizability, projected along the direction of permanent dipole, determined by the EFISH experiment using fundamental wavelength at 1.907 µm with a precision of ±10%. β_0 is the static hyperpolarizability extrapolated to infinite wavelengths derived from the two-level model. For comparison the values of γ (10^{-48} m^5/V^2) from Table 2.6 are given again

Compound	λ_{eg} (solvent)	μ	β_z β_0	γ
(H$_3$C)$_2$N— … —NO$_2$ **S3 (DANS)**	427 (CHCl3)	2.2	386[a] 294[a]	2.3[a] 4.3
H$_3$CH$_2$C, HOH$_2$CH$_2$C N— … N=N — NO$_2$ **DR1**	473 (dioxane) 481 (CHCl3)	3.0	540 382	2.4 3.9
O$_2$N … —OH **PNP**	376 (dioxane)		65 53	0.17

[a] Measurements from [2.83]

2.5   Nonlinear Optical Single Crystals

In this section experiments performed on single crystals and glasses are described. The two techniques used were third-harmonic generation and the z-scan technique. A major problem in nonlinear optics is that the exact intensities of the interacting waves are often not well known. Therefore most measurements are calibrated against a reference material with known nonlinearities. Part of this section is devoted to the investigation of reliable reference materials using the cascading of reference materials. As extensively discussed on the molecular scale, organic materials represent an important and promising class of materials for third-order nonlinear optics. In this section the first experimental results on single crystals of one of the most interesting organic materials, DAST, are reported.

2.5.1   Third-Harmonic Generation

The advantage of third-harmonic generation is that it is a direct measurement of the purely electronic third-order nonlinear optical response. In most other techniques a combination of electronic and other distortional effects that are often difficult or even impossible to resolve are determined. Most measurements are calibrated against a reference material with known nonlinearities. Third-harmonic generation in combination with cascaded second-harmonic and sum frequency generation is a very useful tool to obtain self-consistent reliable reference values.

The experimental setup is very similar to that described in Sect. 2.4.1.B (Fig. 2.18). The wavelengths used were $2.1\,\mu m$, $1.907\,\mu m$, $1.318\,\mu m$ and $1.064\,\mu m$. All measurements were carried out in either vacuum or air (in this case calibrated with measurements of fused silica in vacuum and air). Different geometrical conditions were selected in order to distinguish between cascaded second-harmonic and sum frequency generation and pure third-harmonic generation. The beams at frequency ω and 3ω were always s-polarized.

The description of the theoretical curves needed to evaluate the experimental results presented below is given in Sect. A.3.

A   Materials Description

Here we describe the important parameters required for the experiments below. We investigated fused silica, BK7 and SF59 (isotropic glasses); α-quartz (point group 32); KNbO$_3$ (point group $mm2$), KTaO$_3$ (point group $m3m$); and DAST (point group m).

Fused silica and BK7 are often used as reference materials in third-order nonlinear optical experiments. SF59 is a glass with a large refractive index and enhanced third-order nonlinearities with respect to fused silica [2.152–153]. It is a material that was investigated since it could very well serve as a reference material with enhanced third-order nonlinear optical response. BK7 is another glass that often serves as a reference material.

α-quartz is a well characterized trigonal crystal that is mainly used as a standard reference material for both $\chi^{(2)}$ and $\chi^{(3)}$.

KNbO$_3$ and KTaO$_3$ are both ferroelectric perovskite (ABO$_3$) compounds. Whereas KTaO$_3$ has cubic symmetry at room temperature, KNbO$_3$ is only cubic at high temperatures and transforms with decreasing temperature through the tetragonal phase to the orthorhombic phase at room temperature [2.154]. In Sect. 2.5.2 we will assume that the contribution to the nonlinear refractive n_2 arising from $\chi^{(3)}$ is the same for KTaO$_3$ and KNbO$_3$. With third-harmonic generation this assumption can be tested (see below).

In the DAST crystals the angular deviation of the charge transfer axes of the stilbazolium chromophores from a completely aligned system is about 20° [2.155]. Therefore large third-order susceptibilities are expected for light polarized along the polar axis a. Since DAST also has huge second-order nonlinear optical coefficients, significant cascaded second-order effects are expected as well.

The non-zero susceptibilities for third-harmonic generation for the point groups isotropic, 32, $mm2$, m, and $m3m$ (assuming Kleinman symmetry) are given in Table 2.10. For third-harmonic generation the following reduced notation simplifies the representation ($\chi_{ijkl} = \chi_{ip}$ ($i = 1\ldots3$, $p = 1\ldots$ $\ldots9$, 0)):

p	1	2	3	4	5	6	7	8	9	0	
jkl	111	222	333	233	223	133	113	122	112	123	213
				323	232	313	131	212	121	312	132
				332	322	331	311	221	211	231	321

For sum frequency generation the nonzero second-order susceptibilities for the point groups isotropic, 32, $mm2$, m and $m3m$ are given in Table 2.11.

Note that there is a prefactor that differs for sum frequency generation and frequency doubling (factor of two difference). This prefactor takes care of the fact that the low-frequency limit of all susceptibilities is the same for all frequency combinations. The polarization at the second harmonic for point group 32 (α-quartz) is, for example, given by

$$P_1^{2\omega} = \frac{1}{2}\varepsilon_0 \left\{ \begin{array}{l} \chi_{111}^{(2)}(-2\omega,\omega,\omega)E_1^\omega E_1^\omega + \chi_{122}^{(2)}(-2\omega,\omega,\omega)E_2^\omega E_2^\omega \\ +2\chi_{123}^{(2)}(-2\omega,\omega,\omega)E_2^\omega E_3^\omega \end{array} \right\} \quad (2.134)$$

$$P_2^{2\omega} = \varepsilon_0 \left\{ \chi_{213}^{(2)}(-2\omega,\omega,\omega)E_1^\omega E_3^\omega + \chi_{212}^{(2)}(-2\omega,\omega,\omega)E_1^\omega E_2^\omega \right\} \quad (2.135)$$

where $\chi_{111}^{(2)} = -\chi_{122}^{(2)} = -\chi_{212}^{(2)}$ and $\chi_{123}^{(2)} = -\chi_{113}^{(2)}$. Therefore no second-harmonic light is generated in an x-plate for perpendicular incidence. For

Table 2.10. Third-order nonlinear optical susceptibilities $\chi^{(3)}$ for third-harmonic generation. Notation: $\chi_{ijkl} = \chi_{ip}$ ($i = 1\ldots3$, $p = 1\ldots9, 0$)

Symmetry	χ_{ip}
Isotropic **(fused silica, BK7, SF59)**	
$m3m$ (cubic) **(KTaO$_3$)**	
32 (trigonal) **(α-quartz)**	
$mm2$ (orthorhombic) **(KNbO$_3$)**	
m (monoclinic, $m \perp x_2$) **(DAST)**	

- • $\quad \chi_{ip} = 0$
- ● $\quad \chi_{ip} \neq 0$
- ●—● $\quad \chi_{ip} = \chi'_{ip}$
- ●—○ $\quad \chi_{ip} = -\chi'_{ip}$
- ▲—● $\quad \chi_{ip} = 3\chi'_{ip}$

sum frequency generation we have

$$P_1^{\omega_3} = \varepsilon_0 \left\{ \begin{array}{l} \chi_{111}^{(2)}(-\omega_3, \omega_2, \omega_1) E_1^{\omega_1} E_1^{\omega_2} + \chi_{122}^{(2)}(-\omega_3, \omega_2, \omega_1) E_2^{\omega_1} E_2^{\omega_2} \\ +2\chi_{123}^{(2)}(-\omega_3, \omega_2, \omega_1) E_2^{\omega_1} E_3^{\omega_2} \end{array} \right\}$$

(2.136)

$$P_2^{\omega_3} = 2\varepsilon_0 \left\{ \chi_{213}^{(2)}(-\omega_3, \omega_1, \omega_2) E_1^{\omega_1} E_3^{\omega_2} + \chi_{212}^{(2)}(-\omega_3, \omega_1, \omega_2) E_1^{\omega_1} E_2^{\omega_2} \right\}.$$

(2.137)

Table 2.11. Second-order nonlinear optical susceptibilities $\chi^{(2)}$. Notation: $\chi_{ijk} = \chi_{ip}$ ($i = 1 \ldots 3$, $p = 1 \ldots 6$), where $p = j$ for $j = k$, and $p = 9 - (j + k)$ otherwise

Symmetry	χ_{ip}
Isotropic **(fused silica, BK7, SF59)**	Centrosymmetric All components are zero
$m3m$ (cubic) **(KTaO$_3$)**	Centrosymmetric All components are zero
32 (trigonal) **(α-quartz)**	$\begin{pmatrix} \bullet\!-\!\circ & \bullet & \cdot & \bullet\!\!\diagdown & \cdot & \cdot \\ \cdot & \cdot & \cdot & \cdot & \circ\!-\!\circ & \cdot \\ \cdot & \cdot & \cdot & \cdot & \cdot & \cdot \end{pmatrix}$
$mm2$ (orthorhombic) **(KNbO$_3$)**	$\begin{pmatrix} \cdot & \cdot & \cdot & \cdot & \bullet & \cdot \\ \cdot & \cdot & \cdot & \bullet & \cdot & \cdot \\ \bullet & \bullet & \bullet & \cdot & \cdot & \cdot \end{pmatrix}$
m (monoclinic, $m \perp x_2$) **(DAST)**	$\begin{pmatrix} \bullet & \bullet & \bullet & \cdot & \bullet & \cdot \\ \cdot & \cdot & \cdot & \bullet & \cdot & \bullet \\ \bullet & \bullet & \bullet & \cdot & \bullet & \cdot \end{pmatrix}$

- • zero modulus
- ● non zero modulus
- ●—● equal moduli
- ●—○ moduli numerically equal, but opposite in sign

B   Absolute Value of $\chi^{(3)}$

Third-harmonic generation experiments including cascaded second-harmonic and sum frequency generation summarized below were carried out for the following reasons.

(i) The reference value of $\chi_{111}^{(2)}$ of quartz has changed considerably over the last 15 years (from 1.0 pm/V down to 0.6 pm/V at $\lambda = 1064$ nm) [2.147, 2.156]. Kitamoto et al. [2.157] determined the nonlinear optical coefficient $d_{31} = 4.3 \pm 0.5$ pm/V of congruent LiNbO$_3$ with an absolute parametric fluorescence experiment at $\lambda = 532$ nm. A comparison with second-harmonic generation experiments relative to α-quartz (d_{11}) yields a value for α-quartz of $d_{11} = 0.3$ pm/V. In a similar way, Mito et al. [2.158] performed parametric fluorescence experiments with ADP (nonlinear optical coefficient d_{36}) at $\lambda = 632.8$ nm, which gave $d_{36} = (0.55 \pm 0.02)$ pm/V. Phase-matched frequency-doubling experiments at $\lambda = 1064$ nm with ADP yielded $d_{36} =$

(0.46 ± 0.03) pm/V, in perfect agreement with the value derived at this wavelength based on the Miller-δ [2.159],

$$\chi_{ijk}^{(-2\omega,\omega,\omega)} = \frac{\chi_i^{2\omega}\chi_j^{\omega}\chi_k^{\omega}}{\chi_i^{2\omega}\chi_j^{\omega}\chi_k^{\omega}}\chi_{ijk}^{(-2\omega,\omega,\omega)} = \varepsilon_0\delta_{ijk}\chi_i^{2\omega}\chi_j^{\omega}\chi_k^{\omega} \qquad (2.138)$$

which accurately describes the dispersion of second-order nonlinear optical coefficients in many inorganic materials. A subsequent Maker–Fringe experiment with α-quartz as a reference at $\lambda = 1064$ nm yielded $d_{11} = (0.30 \pm 0.02)$ pm/V. In addition, Maker–Fringe experiments at $\lambda = 632.8$ nm confirmed the validity of Miller's rule for ADP and α-quartz. Exact phase-matched frequency doubling in bulk KNbO$_3$ crystals and a comparison with Maker–Fringe experiments based on α-quartz also confirmed that $d_{11} = 0.30$ pm/V at $\lambda = 1064$ nm is a very realistic value [2.160].

(ii) The ratio of $\chi^{(3)}/\left[\chi^{(2)}\right]^2$ is not very well determined for quartz since $\chi^{(2)}$ of quartz is rather small. In the meantime high-quality crystals of KNbO$_3$ are available. KNbO$_3$ has much larger nonlinear optical susceptibilities $\chi^{(2)}$ with respect to its $\chi^{(3)}$ values and therefore allows a more precise determination of the ratio $\chi^{(3)}/\left[\chi^{(2)}\right]^2$.

(iii) It is of fundamental interest to know the dispersion of $\chi^{(3)}$, on the one hand to get a reliable standard for the dispersion of $\chi^{(3)}$, and on the other hand to get a better theoretical understanding of the dispersion behaviour in inorganic (and organic) materials. A generalization of Miller's rule (2.138) to third-order nonlinearities [2.161]

$$\chi_{ijkl}^{(-3\omega,\omega,\omega,\omega)} = \varepsilon_0\delta_{ijkl}\chi_i^{3\omega}\chi_j^{\omega}\chi_k^{\omega}\chi_l^{\omega} \qquad (2.139)$$

has not been as successful in predicting the wavelength dispersion as in second-order nonlinear optics. Wang [2.162] proposed a different relationship that seems to be more generally valid than Miller's rule. In the limit $\omega \to 0$ he obtains

$$\chi_i^{(3)} = Q\left[\chi^{(1)}\right]^2 . \qquad (2.140)$$

A third possibility is the formula of Boling, Glass and Owyoung [2.163]:

$$\chi_{iiii}^{(3)} = Q\left[n_i^2 + 2\right]^2\left[n_i^2 - 1\right]^2 . \qquad (2.141)$$

Q and Q' are quantities that are constant for many materials. These three dispersion relations will be compared for the cases of fused silica, α-quartz, BK7 and KNbO$_3$.

(iv) The influence of longitudinal second-harmonic generation and cascading through the local field on the third-harmonic signal is another important issue. The basic concept was presented in Sect. 2.3.3 and the influence on α-quartz and KNbO$_3$ will be described below.

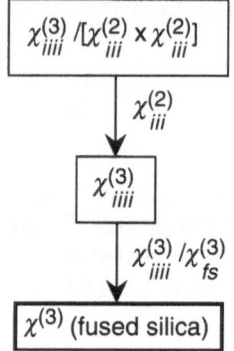

Fig. 2.35. Schematic diagram of the procedure to determine the third-order nonlinearity of fused silica based on the cascading of second-order nonlinearities in third-harmonic generation

As was mentioned in Sect. 2.3, a special feature of cascading is the fact that it often is a self-calibrating process. In the case of cascading in third-harmonic generation this means that a theoretical fit of an experimental curve that shows cascaded processes directly yields the ratio

$$\frac{\chi^{(3)}\left(-3\omega, \omega, \omega, \omega\right)}{\chi^{(2)}\left(-3\omega, 2\omega, \omega\right) \times \chi^{(2)}\left(-2\omega, \omega, \omega\right)} \tag{2.92}$$

without any required knowledge of the laser beam parameters. Therefore, if the values of $\chi^{(2)}$ are known, the value of $\chi^{(3)}$ can be determined. This principle has been applied by Meredith [2.7] to find the values of $\chi^{(3)}$ of crystalline quartz (α-quartz) and fused silica based on known values of $\chi^{(2)}$ of α-quartz (Fig. 2.35).

Typically third-harmonic Maker–Fringe curves based on a combination of $\chi^{(3)}$ and $\chi^{(2)} \times \chi^{(2)}$ show additional oscillations as compared to 'pure' third-harmonic generation curves (see Figs. 2.36 and 2.37). It is, however, also possible to have cascaded contributions to third-harmonic generation that do not show additional oscillations, as illustrated in Fig. 2.37b. In addition, so-called longitudinal second-harmonic and subsequent sum frequency generation can also contribute to the third-order nonlinearity: if we illuminate, for example a c-plate of $KNbO_3$ with an a-polarized fundamental beam, a longitudinal polarization can be generated through the nonlinear optical susceptibility $\chi^{(2)}_{311}$. The associated field cannot propagate but can couple to the field at frequency ω through $\chi^{(2)}_{113}$ and can also contribute to the signal at frequency 3ω. Since the intermediate field at frequency 2ω does not propagate, no additional interference effects appear in third-harmonic Maker–Fringe curves. The influence of longitudinal second-harmonic generation on the results in $KNbO_3$ will be discussed below.

At 1.907 μm and 1.318 μm measurements were performed with α-quartz and $KNbO_3$. At $\lambda = 1.064$ μm only α-quartz was investigated, since the third harmonic was absorbed in $KNbO_3$ too strongly. The data points between $\pm 5°$ were always neglected due to clearly visible multiple reflections that can influence the results.

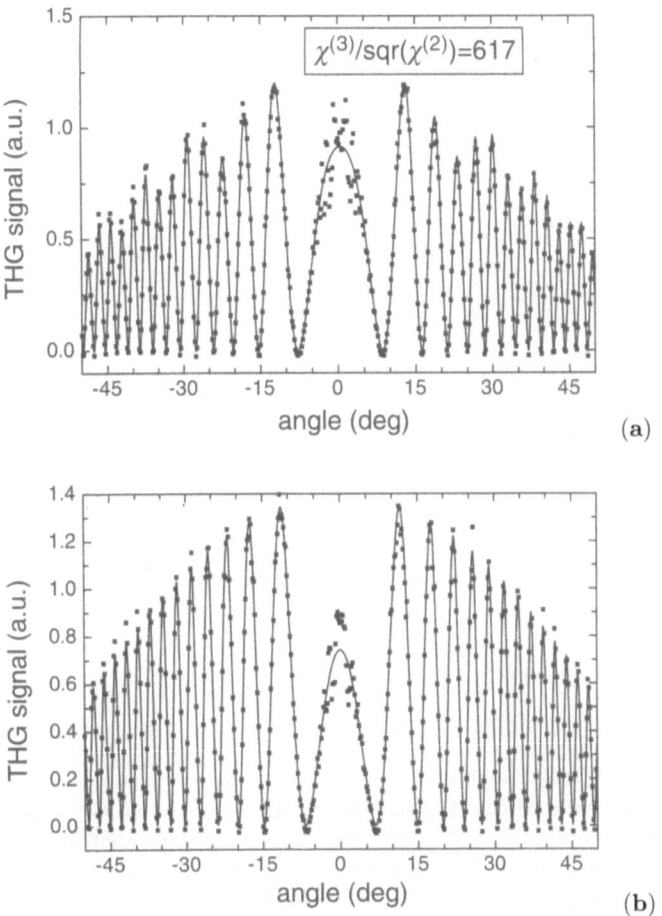

Fig. 2.36a,b. Third-harmonic Maker–Fringe curves of a y-plate of α-quartz at $\lambda = 1064\,\mathrm{nm}$. (**a**) Light polarization along x where cascaded second-order contributions appear. The interference effects are not very pronounced since the value of $\chi^{(2)}$ is quite small as compared to $\chi^{(3)}$. (**b**) Light polarization along z where no cascaded second-order contributions are present

The second-order nonlinear optical susceptibilities $\chi^{(2)}$ of α-quartz and KNbO$_3$ at different wavelengths were obtained from the Miller-δ (2.138) based on $\chi^{(2)}_{111} = 0.6\,\mathrm{pm/V}$ (α-quartz) at $\lambda = 1064\,\mathrm{nm}$ and measured values of $\chi^{(2)}_{333}$ of KNbO$_3$ [2.164]) at several wavelengths.

For the determination of absolute values for $\chi^{(3)}$ the ratios $\chi^{(3)}_{iiii}/\left[\chi^{(2)}_{iii}\right]^2$ and $\chi^{(3)}_{iiii}/\chi^{(3)}_{\mathrm{fs}}$ were first obtained from the experimental Maker–Fringe curves. The first ratio yields $\chi^{(3)}_{iiii}$ based on the known value of $\chi^{(2)}_{iii}$. From the second ratio $\chi^{(3)}_{\mathrm{fs}}$ can subsequently be determined (Fig. 2.35). This procedure was

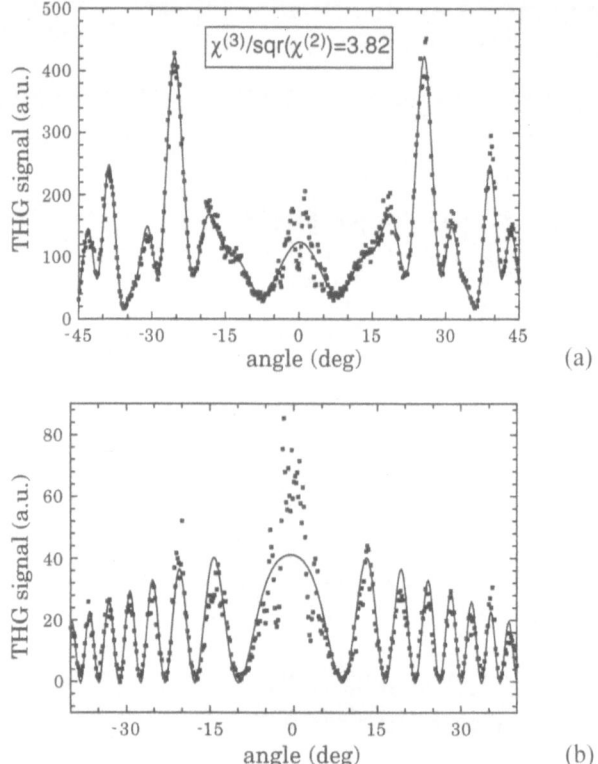

Fig. 2.37a,b. Third-harmonic Maker–Fringe curves of an a-plate of KNbO$_3$ at $\lambda =$ 1907 nm. (**a**) Light polarizations along c where cascaded second-order contributions appear. The interference effects are very pronounced since the value of $\chi^{(2)}$ is quite large compared to $\chi^{(3)}$. (**b**) Light polarizations along b where cascaded second-order contributions are also possible but cannot be detected due to an unfavourable combination of coherence lengths

carried out for α-quartz and KNbO$_3$. In the end, the results were weighted with the experimental errors. They are summarized in Table 2.12. The final values are the ones including the cascading through the local field. Inclusion of this effect barely influences the values for α-quartz and fused silica. Only in the case of KNbO$_3$ a reduction in the values of $\chi^{(3)}$ can be noted.

For α-quartz, x- and y-plates were investigated. For this crystal we have $\chi^{(3)}_{1111}(-3\omega,\omega,\omega,\omega) = \chi^{(3)}_{2222}(-3\omega,\omega,\omega,\omega)$ from symmetry considerations. As mentioned above, no second-harmonic signal and therefore no cascading is expected for s-polarized fundamental and third-harmonic beams. The x- and y-plates were rotated with their axes along the crystallographic y- and z-axes (x-plate) and x- and z-axes (y-plate). Only for the y-plate with rotation axis and light polarizations along the x-axis is cascading observed ($\chi^{(2)}_{111}(-3\omega,2\omega,\omega) \cdot \chi^{(2)}_{111}(-2\omega,\omega,\omega)$). For light polarized along the z-axis

Table 2.12. New absolute values of $\chi^{(3)}$ (in units of $10^{-22} \, \mathrm{m}^2/\mathrm{V}^2$) based on third-harmonic generation experiments of α-quartz and KNbO$_3$ with and without contributions from cascading through the local field. The ratios $\chi^{(3)}/(\chi^{(2)})^2$ are 618 for α-quartz and 3.57^a (3.88) for KNbO$_3$ when combining all experimental results

λ (nm)	$\chi^{(3)}_{1111}$	$\chi^{(3)}_{3333}$	$\chi^{(3)}_{\mathrm{fs}}$	$\chi^{(3)}_{1111}$ ^{a}	$\chi^{(3)}_{3333}$ ^{a}	$\chi^{(3)}_{\mathrm{fs}}$ ^{a}
	α-quartz	KNbO$_3$		α- quartz	KNbO$_3$	
1064	2.48 ± 0.15	–	1.99 ± 0.15	2.48 ± 0.15	–	1.99 ± 0.15
1318	2.22 ± 0.13	67.5 ± 2.5	–	2.21 ± 0.12	61.7 ± 2.5	–
1907	1.99 ± 0.08	51.5 ± 1.1	1.62 ± 0.06	1.99 ± 0.08	47.4 ± 1.1	1.62 ± 0.06

^{a} Including the contributions from cascading through the local field

(both x- and y-plate) only the interaction with $\chi^{(3)}_{3333} (-3\omega, \omega, \omega, \omega)$ produces a third-harmonic signal.

From a measurement of the x-plate with rotation axis and light polarization along the y-axis we expect the same value of $\chi^{(3)}$ as for the first configuration (y-plate with rotation axis and light polarization along x) since $\chi^{(3)}_{1111}(-3\omega, \omega, \omega, \omega) = \chi^{(3)}_{2222}(-3\omega, \omega, \omega, \omega)$ due to symmetry considerations. Figure 2.36 shows Maker–Fringe curves of a y-plate of α-quartz. Experimental results are summarized in Tables 2.12 and 2.13.

In KNbO$_3$ a-, b-, and c-plates were investigated. To determine the cascaded contributions the a- and b-plates were rotated around the crystallographic c-direction and the polarizer and analyzer were chosen along the same direction. This yields a signal at the third-harmonic wavelength that depends on $\chi^{(3)}_{3333} (-3\omega, \omega, \omega, \omega)$ and $\chi^{(2)}_{333} (-3\omega, 2\omega, \omega) \cdot \chi^{(2)}_{333} (-2\omega, \omega, \omega)$. For other geometries, e.g. a b-plate with light polarization along a, other cascaded processes are possible: $\chi^{(2)}_{323} (-3\omega, 2\omega, \omega) \cdot \chi^{(2)}_{233} (-2\omega, \omega, \omega)$. In our experiments no such contributions could be observed since the relevant coherence lengths $l_c (-3\omega, \omega, \omega, \omega)$ and $l_c (-3\omega, 2\omega, \omega)$ do not differ sufficiently. Figure 2.37 shows Maker–Fringe curves of an a-plate of KNbO$_3$ at $\lambda = 1907$ nm. Experimental results are summarized in Table 2.13.

Our new reference value for fused silica at $\lambda = 1907$ nm of $\chi^{(3)}_{\mathrm{fs}} = (1.62 \pm 0.06) \times 10^{-22} \, \mathrm{m}^2/\mathrm{V}^2$ derived from our experiments is a factor of 2.40 lower than the one currently used ($\chi^{(3)}_{\mathrm{fs}} = (3.89 \pm 0.15) \times 10^{-22} \, \mathrm{m}^2/\mathrm{V}^2$ [2.7]). By just adjusting the reference value for χ_{111} of α-quartz in [2.7] we calculate a factor of 2.46 which is in excellent agreement with the factor of 2.40 mentioned above. However, this agreement is quite fortuitous, for the following reasons: For α-quartz we obtained a ratio of $\chi^{(3)}/\left[\chi^{(2)}\right]^2 = 602$, Meredith $\chi^{(3)}/\left[\chi^{(2)}\right]^2 = 668$.[11] On the other hand, we determined $\chi^{(3)}_{\mathrm{fs}}/\chi^{(3)}_{1111} = 0.813$, whereas Meredith obtained $\chi^{(3)}_{\mathrm{fs}}/\chi^{(3)}_{1111} = 0.732$ [2.165], which is very close

---

[11] Note that $\chi^{(3)}/\left[\chi^{(2)}\right]^2$ (SI) $= (1/4\pi)\chi^{(3)}/\left[\chi^{(2)}\right]^2$(esu).

Table 2.13. Summary of all third-harmonic generation experiments ($\chi^{(3)}$ in units of $10^{-22}\,\mathrm{m^2/V^2}$) with [a] and without contributions from cascading through the local field

Crystal	λ (nm)	$\chi^{(3)}_{1111}$	$\chi^{(3)}_{2222}$	$\chi^{(3)}_{3333}$	$\chi^{(3)}_{1133}$
α-quartz	1064	2.48 ± 0.15	2.48 ± 0.15	2.57 ± 0.24	–
	1318	2.21 ± 0.15	2.21 ± 0.15	–	–
	1907	1.99 ± 0.8	1.99 ± 0.08	2.12 ± 0.11	–
KNbO$_3$	1318	–	–	62 ± 3	–
				56 ± 2^a	
	1907	44 ± 3	106 ± 8	51.4 ± 1.5	–
		44 ± 3^a	106 ± 8^a	47.3 ± 1.5^a	
KTaO$_3$	1907	67 ± 4	67 ± 4	67 ± 4	12.8 ± 2.3
					$(\chi^{(3)}_{\mathrm{eff}} = 53 \pm 3)$
SF59	1907	39.4 ± 0.3	–	–	–
BK7	1064	2.98 ± 0.26	–	–	–
	1907	2.38 ± 0.11	–	–	–
Fused silica	1064	1.99 ± 0.15	–	–	–
	1907	1.62 ± 0.06	–	–	–

to what we find for $\chi^{(3)}_{\mathrm{fs}}/\chi^{(3)}_{3333}$ (=0.746).[12] The product of $\chi^{(3)}/\left[\chi^{(2)}\right]^2$ and $\chi^{(3)}_{\mathrm{fs}}/\chi^{(3)}_{1111}$ gives the same value for both cases.

Mito et al. have recently carried out similar experiments with α-quartz. They obtained $\chi^{(3)}_{\mathrm{fs}} = (1.33 \pm 0.15) \times 10^{-22}\,\mathrm{m^2/V^2}$. The discrepancy from our data is mainly due to a different ratio of $\chi^{(3)}_{\mathrm{fs}}/\chi^{(3)}_{1111}$.

It is expected that the value of α-quartz at $\lambda = 1064\,\mathrm{nm}$ ($\chi^{(2)}_{111} = 0.6\,\mathrm{pm/V}$) that our results rely on is accurate. Nevertheless we give $\chi^{(3)}_{\mathrm{fs}}$ based on our experiments as a function of the square of χ_{111} of α-quartz at $\lambda = 1064\,\mathrm{nm}$ (using the Miller-δ) in Fig. 2.38. This equation allows us to calculate the value of $\chi^{(3)}_{\mathrm{fs}}$ at $\lambda = 1907\,\mathrm{nm}$ for any desired value of χ_{111} of α-quartz at $\lambda = 1064\,\mathrm{nm}$.

Figure 2.39 shows the dispersion of the third-order nonlinear optical susceptibilities of fused silica, α-quartz, BK7 and KNbO$_3$. The solid lines are a fit with the generalized Miller rule (2.139). This rule gave a much better, although not perfect, agreement with the experimental data than the relations given in (2.140) and (2.141).

---

[12] We speculate that in the α-quartz sample in [2.165] the axes y and z might have been exchanged.

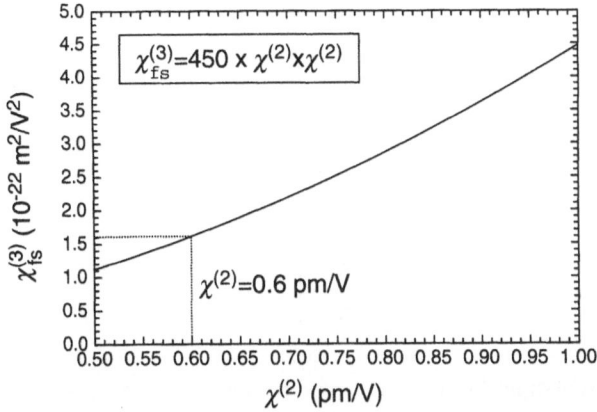

Fig. 2.38. Third-order nonlinear optical susceptibility $\chi^{(3)}(-3\omega,\omega,\omega,\omega)$ of fused silica at $\lambda = 1907\,\mathrm{nm}$ obtained from cascaded second-order processes in third-harmonic generation in α-quartz and KNbO$_3$ as a function of χ_{111} of α-quartz at $\lambda = 1064\,\mathrm{nm}$ ($\chi^{(3)}$ (SI) $= [4\pi/9 \times 10^8] \times \chi^{(3)}$ (esu))

C   Further Results

If we illuminate, for example a c-plate of KNbO$_3$ with an a-polarized fundamental beam, a longitudinal polarization can be generated through the nonlinear optical susceptibility $\chi^{(2)}_{311}$. The associated field cannot propagate but can couple to the field at frequency ω through $\chi^{(2)}_{113}$ and can also contribute to the signal at frequency 3ω. Since the intermediate field at frequency 2ω does not propagate, no additional interference effects appear in third-harmonic Maker–Fringe curves. Fortunately, this contribution is small. The experiments for the c-plate therefore directly yield $\chi^{(3)}_{1111}$ and $\chi^{(3)}_{2222}$ (Table 2.13).

KTaO$_3$ is cubic (point group $m3m$). The only two independent components are $\chi^{(3)}_{1111}(-3\omega,\omega,\omega,\omega) = \chi^{(3)}_{2222}(-3\omega,\omega,\omega,\omega) = \chi^{(3)}_{3333}(-3\omega,\omega,\omega,\omega)$ and $\chi^{(3)}_{2233}(-3\omega,\omega,\omega,\omega) = \chi^{(3)}_{1133}(-3\omega,\omega,\omega,\omega) = \ldots$. These two components can be determined through two independent measurements, e.g. by measuring the third-harmonic signal for light propagation along the crystallographic b direction for light polarized both (i) along the crystallographic a axis and (ii) at 45° to this direction (polarization and rotation always parallel to each other). In the first case one can determine $\chi^{(3)}_{1111}$; in the second case

$$\chi^{(3)}_{\mathrm{eff}} = 0.5\left\{\chi^{(3)}_{1111} + 3\chi^{(3)}_{1133}\right\} \text{ (Table 2.13)}.$$

If we assume the cubic (pure $\chi^{(3)}$) contributions to the third-order nonlinearities of KNbO$_3$ and KTaO$_3$ to be equal and if we neglect the temperature dependence of $\chi^{(3)}$ one can show that $\chi^{(3)}_{1111}$(KNbO$_3$) $= \chi^{(3)}_{3333}$(KNbO$_3$) $= \chi^{(3)}_{3333}$(KTaO$_3$) and $\chi^{(3)}_{2222}$(KNbO$_3$) $= \chi^{(3)}_{\mathrm{eff}}$(KTaO$_3$) (see paragraph above). Table 2.13 shows that the correspondence between $\chi^{(3)}_{3333}$(KNbO$_3$) and

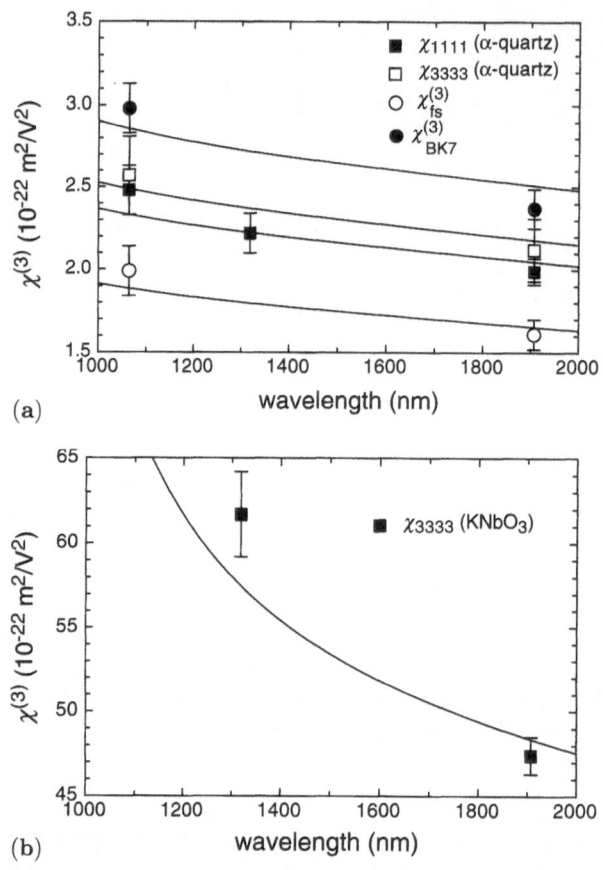

(a)

(b)

Fig. 2.39a,b. Dispersion of the third-order nonlinear optical susceptibilities in fused silica, α-quartz, BK7 and KNbO$_3$. The solid lines are a fit with the generalized Miller rule (2.139)

$\chi_{\text{eff}}^{(3)}$(KTaO$_3$) is excellent and also satisfactory for $\chi_{1111}^{(3)}$(KNbO$_3$). For $\chi_{2222}^{(3)}$ (KNbO$_3$), however, the simple model breaks down. Note that the b-axis is also the direction along which we have the largest linear refractive index.

SF59 has a third-order nonlinearity that is about 25 times as large as fused silica (Table 2.13). Although less nonlinear than, for example, KTaO$_3$, it is well suited as a reference material since (i) the nonlinearity is much larger than that of fused silica and (ii) it is a glass than can easily be cut and polished.

For DAST, b- and c-plates were used. At $\lambda = 1907$ nm only $\chi_{3333}^{(3)}$ could be determined due to the strong absorption of the third-harmonic in the material for light polarized along a. Figure 2.40 shows the Maker–Fringe curves of a b-plate of DAST at $\lambda = 1907$ nm. Unfortunately, $\chi_{1111}^{(3)}$ could also not be de-

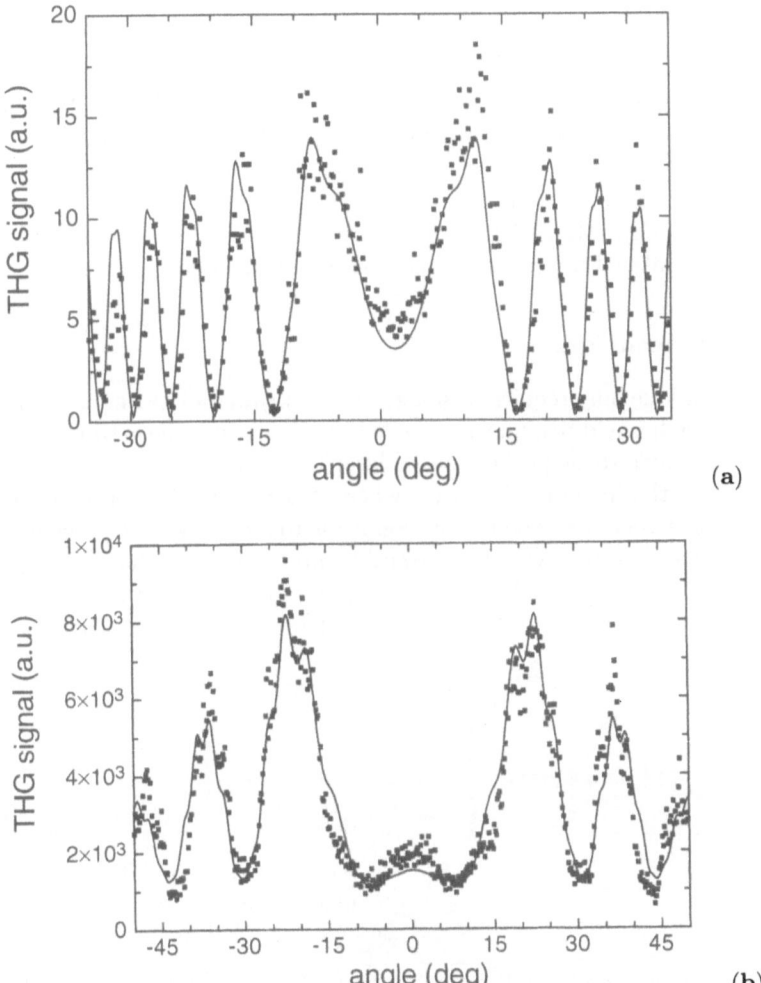

Fig. 2.40. (a) Third-harmonic Maker–Fringe curve of a b-plate of DAST at $\lambda =$ 1907 nm for light polarizations along c where cascaded second-order contributions appear. Interestingly, the best agreement is obtained for $\chi_{3333}^{(3)}(-3\omega,\omega,\omega,\omega) = 0$ and sqrt$(\chi_{333}^{(2)}(-3\omega,2\omega,\omega) \cdot \chi_{333}^{(2)}(-2\omega,\omega,\omega)) = 15.4 \pm 4\,\mathrm{pm/V}$. (b) Third-harmonic Maker–Fringe curve of a c-plate of DAST at $\lambda = 2100$ nm for light polarizations along a where cascaded second-order contributions appear. The theoretical analysis gave sqrt$(\chi_{111}^{(2)}(-3\omega,2\omega,\omega) \cdot \chi_{111}^{(2)}(-2\omega,\omega,\omega)) = 520\,\mathrm{pm/V}$, in good agreement with expected values based on measured nonlinear optical coefficients of DAST

termined at $\lambda = 2100$ nm since the product $\chi_{111}^{(2)}(-3\omega, 2\omega, \omega) \cdot \chi_{111}^{(2)}(-2\omega, \omega, \omega)$ is too large (Fig. 2.40): $\chi_{1111}^{(3)}$ could be varied from 0 to $5000 \times \chi_{fs}^{(3)}$ without significantly changing the experimental Maker–Fringe curves! This is exactly the opposite of the measurements with α-quartz where $\chi_{1111}^{(3)}$ is much larger than the cascaded contributions and which leads to a considerable uncertainty in the determination of the ratio $\chi^{(3)} / \left[\chi^{(2)}\right]^2$.

In the case of Mero-MDB the crystal quality was not sufficient to obtain third-harmonic Maker–Fringe curves that could be evaluated.

2.5.2   z-Scan Technique

The z-scan technique only requires a single focused beam and a circular aperture placed in the far field behind the sample [2.166]. The sample is translated along the optical axis through the focused laser beam (Fig. 2.41). The detector D1 measures the incoming intensity whereas detector D2 measures the intensity after the aperture. The translation leads to a change of the incident intensity and therefore to a variable refractive index change induced by n_2. By measuring the transmission through the circular aperture after the sample the sign as well as the magnitude of n_2 can be determined (Fig. 2.42a).

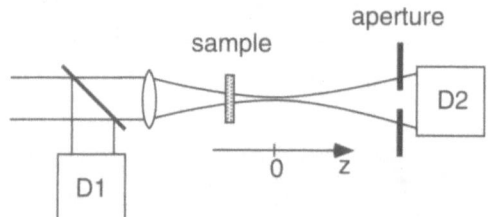

Fig. 2.41. Principle of the z-scan experiment

This can be understood as follows: we assume that $n_2 > 0$. For the sample placed far to the left of the focus the intensity and therefore also the effect of self-focusing is small. The transmission $T \propto D2/D1$ is therefore the same as that without a sample (ignoring absorption and reflection losses) and is usually normalized to one. If we move the sample towards the focus the self-focusing increases and the focus is moved towards $-z$. Therefore less energy is transmitted by the aperture. In the same way, the transmission increases if the sample is located to the right of the focal point. Finally, for large values of z the transmission becomes one again.

An analogous consideration leads to the dotted line for $n_2 < 0$ (Fig. 2.42a). Figure 2.42b shows the influence of two-photon absorption on the transmission. An identical z-scan experiment without the aperture allows the determination of the two-photon absorption coefficient α_2 (defined in (2.33)) (Fig. 2.43).

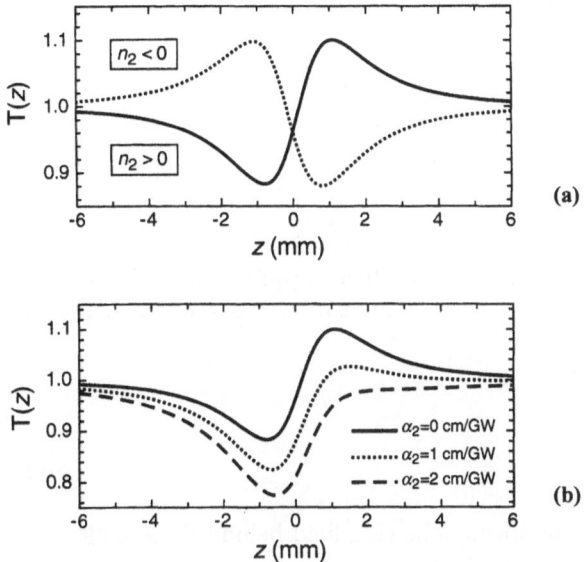

Fig. 2.42. (**a**) z-scan curves for positive (*solid line*) and negative (*dotted line*) n_2. (**b**) Influence of two-photon absorption on z-scan curves for $n_2 = 3.2 \times 10^{-5}\,\mathrm{cm}^2/\mathrm{GW}$

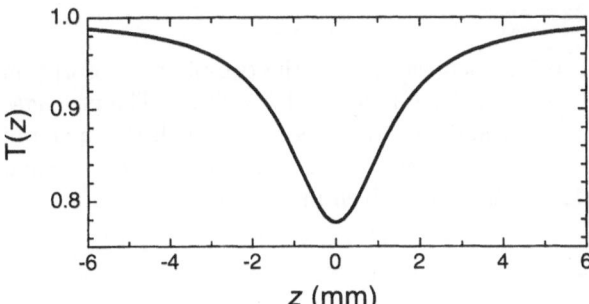

Fig. 2.43. Open aperture z-scan curve for the determination of the two-photon absorption coefficient α_2

A   Theory

Under the assumption that the sample is 'thin' the optical field behind the sample can be calculated. We assume a Gaussian $\mathrm{TEM}_{o,o}$ beam. The electric field in front of the sample is then given by [2.63]:

$$E(z,r,t) = E_0(t)\frac{w_0}{w(z)}\exp\left\{-\frac{r^2}{w^2(z)} - \frac{\mathrm{i}kr^2}{2R(z)}\right\}\exp\left(-\mathrm{i}\phi(z,t)\right)$$

$$r^2 = x^2 + y^2 \qquad\qquad z_0^2 = \frac{\pi w_0^2}{\lambda}$$

$$w^2(z) = w_0^2 \left(1 + \frac{z^2}{z_0^2} \right) \quad R(z) = z \left(1 + \frac{z_0^2}{z^2} \right) \qquad\qquad (2.142)$$

$$\phi(z,t) = kz - \eta(z) \qquad \eta(z) = \arctan \left(\frac{z}{z_0} \right) .$$

The assumption that the plate is thin means that the beam profile remains unchanged along the sample thickness L, which is equivalent to the condition $L < z_0$. The differential equations for the intensity and phase change are then given by

$$\frac{\mathrm{d}I}{\mathrm{d}z'} = -(\alpha + \alpha_2 I) I$$

$$\frac{\mathrm{d}\Delta\phi}{\mathrm{d}z'} = n_2 I k . \qquad\qquad (2.143)$$

Solving these equations leads to the following field behind the sample:

$$E_e(z,r,t) = E(z,r,t) \frac{\mathrm{e}^{-\alpha L/2}}{(1+q)^{1/2}} \exp \left\{ \mathrm{i} \frac{k n_2}{\alpha_2} \ln(1+q) \right\}$$

$$= E(z,r,t) \mathrm{e}^{-\alpha L/2} (1+q)^{\mathrm{i} k n_2/\alpha_2 - \frac{1}{2}} \qquad\qquad (2.144)$$

$$q(z,r,t) = I(z,r,t) \alpha_2 \frac{1 - \mathrm{e}^{-\alpha L}}{\alpha} ,$$

where α is the linear absorption coefficient, α_2 is the two-photon absorption coefficient, n_2 is the nonlinear refractive index and $k = 2\pi/\lambda$. The propagation of this field to the aperture and the transmission through the aperture has to be calculated. The solutions describing the final transmission through the aperture (see Figs. 2.42 and 2.43) are given in Sect. A.5.

B   Experimental Description

Either a mode-locked Nd:YAG laser (Quantronix 416, $\lambda = 1064\,\mathrm{nm}$) that seeds a regenerative amplifier (Continuum RGA 60-20, with a pulse duration of 100 ps, a repetition rate of 20 Hz, and pulse energies between 3 and 150 μJ, shown in Fig. 2.44) or a diode-pumped modelocked Nd:YLF laser (Lightwave, $\lambda = 1047\,\mathrm{nm}$) seeding a regenerative amplifier (Positive Light, with a pulse duration of 11 ps, a repetition rate of 10 Hz and pulse energies between 1 and 15 μJ) was used. Short pulses are needed to achieve the high intensities required to measure the nonlinearities. In addition, low repetition rates are preferred in order to exclude possible contributions due to thermal effects. A lens (focal length 120 mm) focused the beam to a minimum waist w_0 of around 20 μm. All measurements were referenced against a 1 mm thick cell filled with CS_2 ($n_2 = 3.2 \times 10^{-5}\,\mathrm{cm}^2/\mathrm{GW}$) [2.166]. The experimental setup is shown in Fig. 2.44.

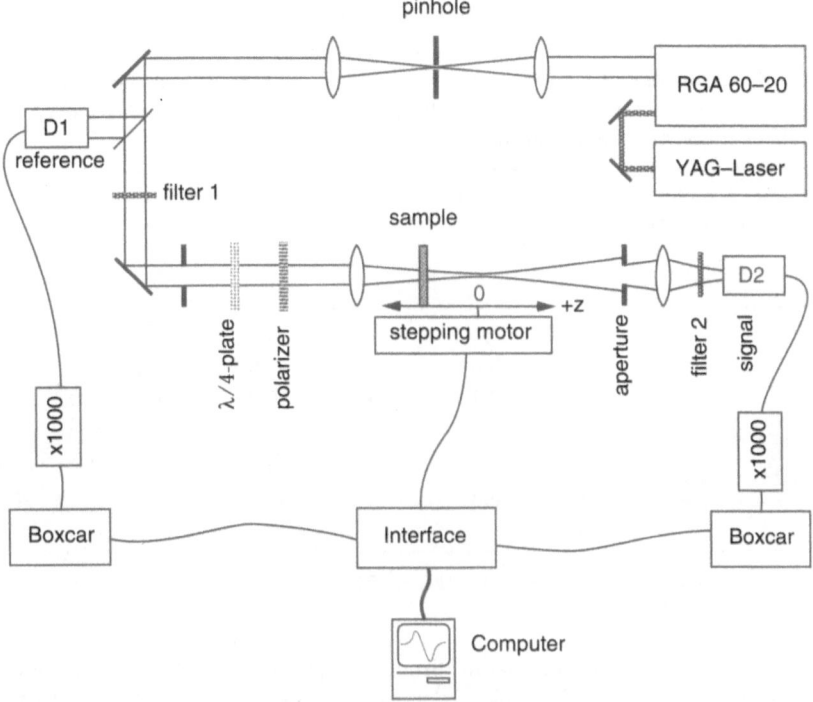

Fig. 2.44. Experimental z-scan setup for the measurement of nonlinear refractive indices and of two-photon absorption. The sample is translated along the optical axis through a focused laser beam. By measuring the transmission through the circular aperture after the sample, the sign as well as the magnitude of n_2 can be determined. Removing the aperture while performing the same experiment allows the determination of the two-photon absorption coefficient α_2. The filters allow easy adjustment of the incident intensity

C   Cascaded Optical Rectification and the Linear Electro-Optic Effect

The direct $\chi^{(3)}$ contribution to n_2 of KNbO$_3$ was determined from measurements on KTaO$_3$ [2.12]. Here we assume that the contribution to n_2 arising from $\chi^{(3)}$ is the same for KTaO$_3$ and KNbO$_3$ [2.68]. We have seen in the third-harmonic generation experiments that this assumption is justified for $\chi^{(3)}_{1111}$ and $\chi^{(3)}_{3333}$. In the cubic point group $m3m$ (assuming Kleinman symmetry [2.61]) there are only two independent tensor elements of $\chi^{(3)}$. We measured n_2 of KTaO$_3$ for light propagation along the crystallographic b-direction for both light polarized along the crystallographic a-axis and at 45° to this direction [2.12]. From the measured values of $+(7.7\pm1.5)\times10^{-6}$ cm^2/GW (prop. to $\chi^{(3)}_{1111}$) and $+(5.8\pm1.2)\times10^{-6}$ cm^2/GW (prop. to $1/2(\chi^{(3)}_{1111}+3\chi^{(3)}_{1133})$) we get $\chi^{(3)}_{1111}=+(1.3\pm0.3)\times10^{-20}$ m^2/V^2 and $\chi^{(3)}_{1133}=+(2.2\pm0.4)\times10^{-21}$ m^2/V^2 taking n(KTaO$_3$) = 2.2 (2.89). The nonlinear refractive indices of KNbO$_3$ in

Table 2.14. Comparison between the theory (2.115)) and experiment of KNbO$_3$ for measurements along the main axis of the indicatrix and for propagation along $\theta = 45°$ off the c-axis. The uncertainty of the theoretical as well as the measured values is 20%. n_2 is given in units of 10^{-6} cm^2/GW. n_2^{SHG} is based on $d_{111} = 0.3$ pm/V of α-quartz at $\lambda = 1064$ nm

Interacting second-order suscepti- bilities	Propagation direction	Pol.	Direct $\chi^{(3)}$ n_2^{direct}	Cascading n_2^{SHG}	Cascading n_2^{OR}	Theory n_2	Theory[a] n_2	Exp. n_2
$\chi_{113}^{(2)} \cdot \chi_{311}^{(2)}$	b	a	5.8	3.0	1.9	10.7	8.8	7.8
$\chi_{113}^{(2)} \cdot \chi_{311}^{(2)}$	c	a	5.8	–	1.9	7.7	7.7	4.5
$\chi_{223}^{(2)} \cdot \chi_{322}^{(2)}$	a	b	7.7	1.2	0.3	9.2	8.9	11
$\chi_{223}^{(2)} \cdot \chi_{322}^{(2)}$	c	b	7.7	–	0.3	8	8	9.0
$\chi_{333}^{(2)} \cdot \chi_{333}^{(2)}$	a/b	c	5.8	-1.9	4.4	8.3	3.9	5.6
$\chi_{eff}^{(2)} \cdot \chi_{eff}^{(2)}$	$45°$	b–c Plane	5.3	-1.5	19	22.8	–	26

[a] Including depolarization effects (see Sect. 2.3.5)

the orthorhombic phase arising from direct $\chi^{(3)}$ processes can then be calculated from a rotation of 45° of the $\chi^{(3)}$-tensor around the b-axis (Table 2.14) [2.68]. Note that it is not possible to measure n_2^{direct} in KNbO$_3$ at room temperature since a contribution from n_2^{OR} at least always exists. Experiments with KNbO$_3$ were performed with platelets cut normal to the crystallographic a-, b- and c-axes. In order to measure the largest contribution to n_2^{OR} arising from r_{232}, several propagation directions in the b–c-plane were selected and platelets cut at 37.5°, 45° and 71.4° with respect to the crystallographic c-axis were prepared.

No two-photon absorption was observed at $\lambda = 1064$ nm for both KNbO$_3$ and KTaO$_3$.[13]

Figure 2.45 shows an example of a z-scan curve for KNbO$_3$. For light propagating along the c-direction second-harmonic generation does not contribute to n_2 since the resulting polarization $P^{2\omega}$ is parallel to the propagation direction of the beam. For all other geometries this contribution has to be taken into account. Table 2.14 summarizes z-scan measurements for light propagation and polarization along dielectric main axes of the crystals together with theoretical predictions (see (2.115)). For KNbO$_3$, with its large dielectric constants we have $\varepsilon - 1 \approx \varepsilon + 2 \approx \varepsilon$ and therefore a negligible difference between (2.115) and (2.116). The values for the refractive indices and the nonlinear optical coefficients used for the calculation in Table 2.14 were taken from [2.164, 2.167]. For the 100 ps pulses used in these experiments the

---

[13] Note that strong two-photon absorption prohibits a determination of the nonlinear refractive indices of KNbO$_3$ and KTaO$_3$ at $\lambda = 532$ nm.

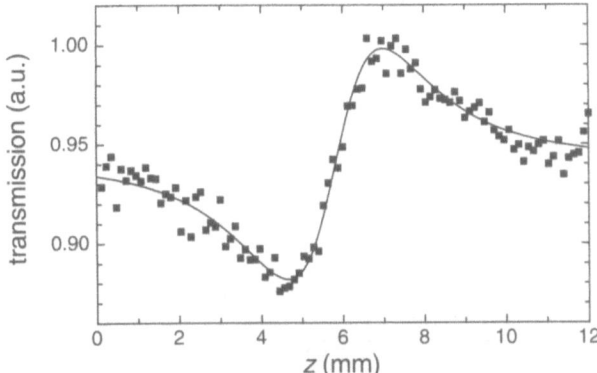

Fig. 2.45. Experimental z-scan curve using a KNbO$_3$ crystal (a-plate). $z = 0$ (sample in front of the focus) denotes the starting point of the scan. The decrease in transmission in front of the focus and the subsequent rise after the focus reveals that the nonlinear refractive index is positive

electro-optic response of the bound electrons and optical phonons shows up instantaneously, whereas the elasto-optic contribution is clamped. Therefore the strain-free electro-optic coefficients and dielectric constants were used in the analysis of n_2^{OR} [2.168].

We see from Table 2.14 that the correspondence between theory and experiment is remarkably good for b- and c-polarized light. The discrepancy between theory and experiment could possibly be attributed to a failure of our simple assumption that $\chi^{(3)}$ is exactly the same for KTaO$_3$ and KNbO$_3$. In addition, the precise effects of depolarization have to be further investigated. The largest clamped electro-optic coefficient in KNbO$_3$ is $r_{232}^s = 360 \pm 30$ pm/V which can contribute to n_2 for light propagation as well as polarization within the b-c-plane. In this plane we expect the largest value of n_2^{OR}; therefore we measured the nonlinear refractive index for several propagation directions, namely 37.5°, 45° and 71.4° off the crystallographic c-axis. Figure 2.46 shows the theoretically expected angular dependence of the different contributions to n_2 in KNbO$_3$, neglecting the effect of depolarization. n_2^{direct} is almost constant over the whole angular range. n_2^{SHG} is always negative for the configuration in Fig. 2.46 and is approximately constant. In contrast, the third term, n_2^{OR}, shows a pronounced angle dependence with a maximum value at around $\theta = 47°$.

Figure 2.47 shows the experimental results as well as the theoretical curve for all contributions. As expected a strong increase in the nonlinear refractive index is observed for the internal angles of incidence of 37.5° and 45°, where the contribution from r_{232}^s is more pronounced. The observed dependence of n_2 on the propagation direction and the increase in n_2 of more than a factor of three for an angle of $\theta = 47°$ can be explained only if the cascaded second-order effects due to optical rectification and the linear electro-optic effect are considered.

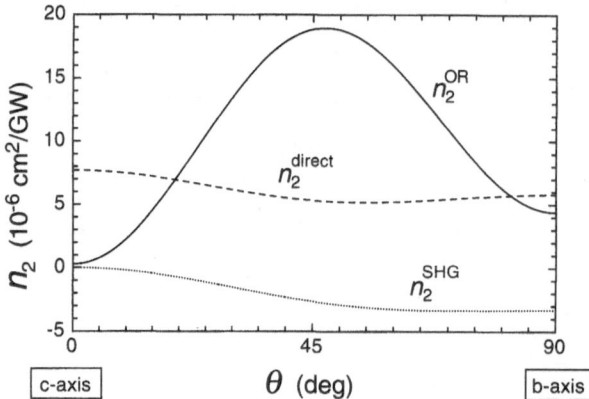

Fig. 2.46. Angular dependence of the different contributions to the nonlinear refractive index of KNbO$_3$ in the b–c-plane. θ is the internal angle of propagation measured from the crystallographic c-axis. The beam is polarized in the b–c-plane as well. n_2^{direct} is almost constant over the whole angular range. n_2^{SHG} is always negative for the configuration shown here and is approximately constant. In contrast, the third term, n_2^{OR}, shows a pronounced angle dependence with a maximum value of $1.9 \times 10^{-5}\,\text{cm}^2/\text{GW}$ at around $\theta = 47°$

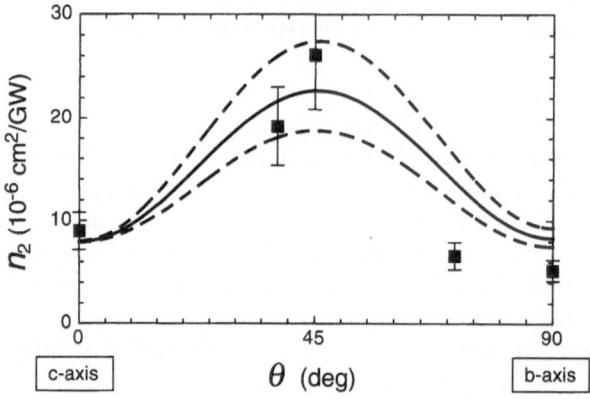

Fig. 2.47. Comparison of experimental and theoretical results of the nonlinear refractive index n_2 for KNbO$_3$ as a function of propagation direction in the b–c-plane with in-plane polarized beams. The theoretical curves contain all three contributions to the nonlinear refractive index. The dashed curves are lower and upper limits mainly determined by the uncertainties of the electro-optic coefficients and dielectric constants

D   Further Results

Table 2.15 summarizes the results from z-scan measurements for $KNbO_3$ [2.12], $KTaO_3$, SF59 and benzene. SF59 was investigated mainly to see whether it would be suitable as a better reference material than fused silica (low n_2) and CS_2 (liquid in cuvette with rather low damage threshold).

The third-harmonic generation results of SF59 gave $\chi^{(3)}/\chi_{\mathrm{fs}}^{(3)} = 24.8$, whereas the z-scan experiments yielded $\chi^{(3)}/\chi_{\mathrm{fs}}^{(3)} = 23$, an excellent agreement. Our results are also in good agreement with results from other groups that were obtained by other experiments [2.133, 2.152–153] although the exact definitions of the nonlinear optical susceptibilities and/or the reference values used are not always clear. We believe that SF59 is a very suitable reference material for third-order nonlinear optical experiments. $KTaO_3$ would be even more suitable due to the larger nonlinearities. However, $KTaO_3$ often shows stress-induced surface defects that make experiments involving the measurement of small phase shifts (as for example with the z-scan technique) very difficult.

It can generally be observed that the z-scan measurements with $KNbO_3$ and $KTaO_3$ yield larger values of $\chi^{(3)}$ in comparison with the THG experiments. This is not surprising since in third-harmonic generation one only determines electronic nonlinearities, whereas in z-scan measurements with 100 ps pulses optic phonons can also contribute to the effect.

Table 2.15. Summary of z-scan experiments for $KNbO_3$, $KTaO_3$, SF59, CS_2 and benzene at $\lambda = 1047$ nm or $\lambda = 1064$ nm (n_2 in units of 10^{-6} cm^2/GW, $\chi^{(3)}$ in units of 10^{-22} m^2/V^2). All values of n_2 were referenced against CS_2. The values of $\chi^{(3)}$ were calculated from (2.89)

Compound	Propagation direction	Polarization	$n_2/n_2^{CS_2}$	n_2	$\chi^{(3)}$
$KNbO_3$	b	a	0.24	7.8	$\chi_{1111}^{(3)} = 136$
	c	a	0.14	4.5	$\chi_{1111}^{(3)} = 79$
	a	b	0.34	11	$\chi_{2222}^{(3)} = 199$
	c	b	0.28	9	$\chi_{2222}^{(3)} = 162$
	a/b	c	0.18	5.6	$\chi_{3333}^{(3)} = 89$
	$45°$	b–c-plane	0.81	26	$\chi_{\mathrm{eff}}^{(3)} = 442$
$KTaO_3$	c	a	0.24	7.7	$\chi_{1111}^{(3)} = 132$
	c	$45°$	0.18	5.8	$\chi_{\mathrm{eff}}^{(3)} = 99$
					$\chi_{1113}^{(3)} = 22$
SF59	–	–	0.17	5.3	$\chi^{(3)} = 68$
CS_2	–	–	1	32	$\chi^{(3)} = 290$
Benzene	–	–	0.20	6.4	$\chi^{(3)} = 51$

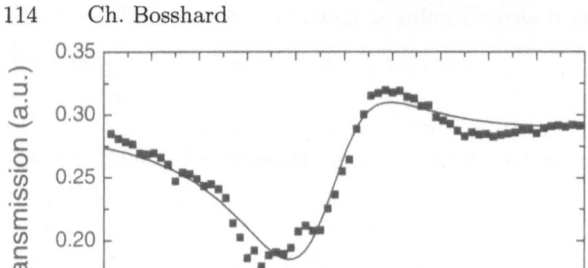

Fig. 2.48. Experimental z-scan curve using a DAST crystal (c-plate). The decrease in transmission in front of the focus and the subsequent rise after the focus reveals that the nonlinear refractive index is positive. Rather strong two-photon absorption is visible (compare with Fig. 2.42)

Note that our results with benzene are in excellent agreement with degenerate four-wave mixing experiments that were carried out at much lower intensities [2.169].

So far, only preliminary measurements have been carried out in single crystals of DAST (c-plates). There are several problems associated with such measurements. It seems that the nonlinear refractive index as well as two-photon absorption is strongly intensity dependent. In addition there is strong luminescence present that arises due to two-photon absorption. Furthermore, we have contributions due to cascaded second-order nonlinearities. The contribution due to cascaded optical rectification and the linear electro-optic effects can be calculated (neglecting depolarization effects) based on (2.115) or (2.116). This yields $n_2^{OR} = 1.0 \times 10^{-4}\,\mathrm{cm^2/GW}$ ($\chi_{1111}^{(3)} = 1730 \times 10^{-22}\,\mathrm{m^2/V^2}$ using (2.115)). Taking the effect of depolarization into account we estimate $n_2^{OR} = 4.8 \times 10^{-5}\,\mathrm{cm^2/GW}$ for the transverse case. The contribution arising from cascaded second-harmonic generation and difference frequency mixing, however, is very difficult to estimate since the second harmonic is strongly absorbed by the material.

Figure 2.48 shows an example of a z-scan for a 0.365 mm thick sample at $\lambda = 1047\,\mathrm{nm}$. The experiments revealed that the third-order nonlinearities (n_2 and α_2) are strongly dependent on the input intensity, suggesting fifth-order nonlinearities or excited state absorption [2.4]. The preliminary values of DAST are shown in Table 2.16.

2.6   Discussion and Conclusion

2.6.1   Second- and Third-Harmonic Generation

As mentioned earlier, an advantage of third-harmonic generation is that it is a direct measurement of the purely electronic third-order nonlinear optical

Table 2.16. Preliminary results of z-scan experiments in DAST at $\lambda = 1047$ nm. The experimental error is estimated to be 20%. The value of n_2 was referenced against CS_2. The value of $\chi^{(3)}$ was calculated from (2.89)

Propagation direction	Polari- zation	$n_2/n_2^{CS_2}$	$n_2(10^{-4}\,cm^2/GW)$	$\chi^{(3)}(10^{-22}\,m^2/V^2)$	$\alpha_2(cm/GW)$
c	a	7.5	2.4	$\chi^{(3)}_{1111} = 4200$	6. 7

response. Most other techniques measure a combination of electronic and other distortional effects that are often difficult or even impossible to resolve.

A   Single Crystals

Absolute Value of $\chi^{(3)}$ and Reference Materials. Based on third-harmonic generation measurements using cascaded second-harmonic and sum frequency generation we have determined a new reliable standard value for fused silica ($\chi^{(3)}_{fs} = 1.62 \times 10^{-22}\,m^2/V^2$ at $\lambda = 1907$ nm, $\chi^{(3)}_{fs} = 1.99 \times 10^{-22}\,m^2/V^2$ at $\lambda = 1064$ nm). At $\lambda = 1907$ nm this accurate determination was possible for the available high-quality crystals of $KNbO_3$. $KNbO_3$ has much larger nonlinear optical susceptibilities $\chi^{(2)}$ with respect to its $\chi^{(3)}$ values and therefore allows a more precise determination of the ratio $\chi^{(3)}/\left[\chi^{(2)}\right]^2$ than α-quartz.

In the search for a reliable reference material we propose the silicate glass SF59 as the best candidate at the moment. It combines sufficiently high third-order nonlinearities with excellent optical quality.

In comparison with previously published values of BK7 we find slightly lower values for BK7 ($\chi^{(3)} = 1.47 \times \chi^{(3)}_{fs}$ as compared to $\chi^{(3)} = 1.67 \times \chi^{(3)}_{fs}$ [2.165] at $\lambda = 1907$ nm). Also for $\chi^{(3)}_{1111}$ (α-quartz) our results differ: as explained in Sect. 2.5, we speculate that in the α-quartz sample in [2.165] the axes y and z might have been exchanged. An extension of the Miller-δ was proposed for third-order nonlinear optical susceptibilities [2.161] and also used in this work to describe the wavelength dispersion of $\chi^{(3)}$. However, this extension may not be sufficient: according to the Clausius–Mossotti equation the density ρ is proportional to $(n^2 - 1)/(n^2 + 2)$.[14] Accordingly, the Miller-δ defined in (2.139) is proportional to ρ^{-4}. A more appropriate 'Miller-δ' might therefore be [2.165]

$$\delta' = \delta \times \rho^3 \,. \tag{2.145}$$

Indeed, Table 2.17 shows that the quantity defined in (2.145) leads to much better consistency with our data and that the scaling of nonlinear optical properties is possible, at least for the examples presented here.

---

[14] For fused silica and α-quartz we have $\rho_{fs}/\rho_{quartz} = ((n^2 - 1)/(n^2 + 2))_{fs}/((n^2 - 1)/(n^2 + 2))_{quartz}$ [2.165].

Table 2.17. Relative values of the third-order nonlinearities. n is given by $n = (n^\omega + n^{3\omega})/2$

λ (nm)	Coefficient	ρ (g/cm^3)	n	$\chi^{(3)}/\chi^{(3)}_{\text{fs}}$	$\delta/\delta_{\text{fs}}$	$\delta'/\delta'_{\text{fs}}$
1907	$\chi^{(3)}_{\text{fs}}$	2.203	1.4482	1	1	1
	$\chi^{(3)}_{1111}$ (α-quartz)	2.648	1.5325	1.23	0.538	0.93
	$\chi^{(3)}_{3333}$ (α-quartz)	2.648	1.5412	1.31	0.531	0.92
1064	$\chi^{(3)}_{\text{fs}}$	2.203	1.4629	1	1	1
	$\chi^{(3)}_{1111}$ (α-quartz)	2.648	1.5494	1.24	0.543	0.94
	$\chi^{(3)}_{3333}$ (α-quartz)	2.648	1.5586	1.29	0.523	0.91

Dispersion of $\chi^{(3)}$. From the measurements of third-order nonlinear optical susceptibilities at different wavelengths we attempted to determine an appropriate dispersion relationship that can describe our results. Generally, the Miller-δ (2.139) gives the closest, although still not satisfying, agreement between theory and experiment. Further work is needed to more accurately describe the wavelength dispersion of $\chi^{(3)}$.

Organic Crystals. Based on measured second-order hyperpolarizabilities the macroscopic third-order nonlinearities were estimated using the oriented gas model. Extremely large values ($\chi^{(3)} > 8000 \times \chi^{(3)}_{\text{fs}}$; see Tables 2.5 and 2.6) are calculated due to the high degree of orientation and the large packing density of the molecules in the crystal lattice in comparison to disordered polymers.

So far, only measurements with DAST have been performed, in a configuration where cascaded contributions appear. The preliminary results showed that these cascaded contributions completely mask the third-order nonlinearity $\chi^{(3)} : \chi^{(3)}_{1111}$ could be varied from 0 to $5000 \times \chi^{(3)}_{\text{fs}}$ without significantly changing the experimental Maker–Fringe curves. Additional measurements with DAST and the most promising hydrazone derivatives are expected to show the potential of molecular crystals for third-order nonlinear optics.

Cascading Through the Local Field. The influence of cascading through the local field proved to be minor for the crystals investigated ($\leq 8\%$ change of $\chi^{(3)}$ with and without cascading through the local field). Larger second-order nonlinear optical coefficients are required to identify this effect clearly. An intriguing possibility is offered by organic crystals: if one single tensor element γ_{iiii} dominates the third-order response and intermolecular interactions can be neglected, the tensor elements of $\chi^{(3)}$ can be calculated for a known point group. If, in addition, the second-order nonlinear optical coefficients vary strongly for different input polarizations, a comparison of third-harmonic generation measurements with different experimental geometries may yield reliable quantitative results of the effect of local field cascading.

B   Structure–Property Relations on the Molecular Scale

Influence of Molecular Asymmetry. We have shown that important structure–property relationships exist in the novel nonlinear optical material classes presented here. Most importantly, we have clearly illustrated the importance of acentricity for large values of the third-order nonlinearity (the tetraethynylethenes; see Sects. 2.4.1 and 2.4.4). Moreover, a number of these compounds (e.g. **T1–T3**) can easily be polymerized after removal of the silyl-protecting groups, facilitating the fabrication of thin films.

Large values for the first-order hyperpolarizability β of donor–acceptor substituted TEEs, comparable to other highly nonlinear optical chromophores, have been measured. These compounds provide a large product $\mu \cdot \beta$, an important parameter for using the chromophores in a poled polymer guest–host or side-chain system. Values up to $15 \times 10^{-67}\,\mathrm{Cm^5/V}$ for compound **8**, compared with $9 \times 10^{-67}\,\mathrm{Cm^5/V}$ for the standard chromophore DANS, were obtained.

We showed that semi-empirical calculations are important in predicting trends in the nonlinearity of a series of structurally related molecules. From our calculations we estimated the entire tensor structure of the first-order hyperpolarizabilities, which is especially important for two- or more-dimensional conjugation. This knowledge leads to an understanding of the contributions from different donors and acceptors, as well as of different conjugation paths. This insight is crucial for the innovative design of improved chromophores.

As also observed by other groups [2.83] we find a clear trend of increasing values of γ with increasing β. Although no straightforward dependence could be deduced from the experimental data the general trend is probably associated with the $\Delta\mu^2$ term in the three-level model (2.44). In addition, there exists no simple correlation between γ and λ_{eg}. Theoretical calculations based on the sum-over-states approach to increase the understanding of the observed effects are currently under way.

The hydrazone derivatives show the same trend: acentricity increases the third-order nonlinearity. This could be nicely demonstrated in this materials class since an exchange of the donor and acceptor groups leads to a different molecule with different acentricity, offering additional flexibility. The interest in hydrazone derivatives stems from the fact that high quality crystals with optimized orientation of the chromophores can be grown. This leads to large values of $\chi^{(3)}$ due to the high degree of orientation and the large packing density of the molecules in the crystal lattice. As in the case of the tetraethynylethenes, the same arguments apply for the correlations between γ and β as well as γ and λ_{eg}.

It should be noted that eventually centrosymmetric molecules could have the largest second-order hyperpolarizabilities γ [2.19]. However, we are still quite far from this point although squaraine dyes have been identified as possible optimized centrosymmetric molecules for third-order nonlinear optics [2.60].

Two-Dimensional Conjugation. We have demonstrated the importance of full two-dimensional conjugation versus conjugation length. For the first time, the influence of donors and acceptors as well as of full two-dimensional conjugation on the third-order nonlinearity in one single class of materials is clearly shown (the tetraethynylethenes; see Sects. 2.4.1 and 2.4.4).

A further advantage of two-dimensionally conjugated chromophores such as **T11** is phase-matching: in a crystal composed of such chromophores, it is much more likely to find a phase-matching geometry for frequency doubling than in crystals of charge transfer molecules such as **T3** with one donor–acceptor conjugation path, because it is possible to use the large nondiagonal coefficients and the refractive indices often differ only slightly for different polarizations.

The formation of suitable noncentrosymmetric crystals as well as the incorporation of highly nonlinear optically active tetraethynylethene chromophores into polymers for solid state studies is being pursued.

Cascading Through the Reaction Field. One can argue that the increase of γ on the introduction of acentricity might be due to cascading through the reaction field. An estimation of this contribution is presented in the following for the example of the tetrasubstituted molecule **T11**.

We used the calculated values of all tensor elements of β_{ijk} (calculated as described in Sect. 2.4.2), calibrated them with the actual measured β_z (see Fig. 2.29), and took the dispersion of β_{ijk} into account using the two-level model. Molecule **T11** has point group symmetry $mm2$. Based on (2.99), which describes the rotational average $\langle \beta' \cdot \beta \rangle_{3333}$ with the polarization of the fundamental and the third- harmonic waves along the macroscopic 3-direction for this point group, we obtain $\langle \beta' \cdot \beta \rangle_{3333} = 7.41 \times 10^{-76}\,\mathrm{m^8/V^2}$. The reaction field factor $r^{2\omega}$ is calculated either from (2.71) or (2.73) using the refractive indices of the solvent ($CHCl_3$) or the refractive index of the solute n_i, which is estimated to be $n_i \approx 2$. The required cavity radius was obtained from either (2.75) or (2.77). Table 2.18 summarizes the results. The important quantity to compare is $2r^{2\omega}\,\langle \beta' \cdot \beta \rangle_{3333}\,/\,\langle \gamma \rangle_{\mathrm{eff}}$ (see (2.97)).

We see that for molecule **T11** the influence of cascading through the reaction field can be neglected. In the case of $n = 1.44$ and $n_i = 2$ the contribution of the cascading through the reaction field accounts for about 1% of the total effective hyperpolarizability. Even if we assume an (unrealistic[15]) value for r_{T11} of 2.8 Å, the contribution is still only around 6–7%. We can therefore conclude that cascading through the reaction field is not important for the extended molecules considered here and does not influence the interpretation of the effect of acentricity on γ.

---

[15] It seems unrealistic to assume a value of the spherical radius r for the large molecule **T11** that is smaller than the one for small molecules such as $CHCl_3$ ($r = 3.2$ Å).

Table 2.18. Estimated values for cascading through the reaction field for third-harmonic generation based on $\langle \beta' \cdot \beta \rangle_{3333} = 7.41 \times 10^{-76} \, \mathrm{m^8/V^2}$ (fundamental wavelength of $\lambda = 1907 \, \mathrm{nm}$, $\mathrm{CHCl_3}$: $M = 119.38 \, \mathrm{g}$, $\rho = 1.476 \, \mathrm{g/cm^3}$, $r = 3.2 \, \mathrm{Å}$, **T11**: $M = 604.7 \, \mathrm{g}$, $\rho = 1.37 \, \mathrm{g/cm^3}$, $r = 5.6 \, \mathrm{Å}$)

Refractive index of solvent	Refractive index of solute	Cavity radius (Å) (Eq.)	$2r^{2\omega} \langle \beta' \cdot \beta \rangle_{3333} / \langle \gamma \rangle_{\mathrm{eff}}$ (Eq.)	
1.44	1.44	7.2 (2.75)	0.006	(2.71)
			0.006	(2.73)
		6.1 (2.77)	0.011	(2.71)
			0.009	(2.73)
1.44	2	7.2 (2.75)	0.006	(2.71)
			0.006	(2.73)
		6.1 (2.77)	0.011	(2.71)
			0.011	(2.73)
1.44	1.44	4.4 (2.75)[a]	0.028	(2.71)
			0.025	(2.73)
		3.3 (2.77)[a]	0.067	(2.71)
			0.060	(2.73)

[a] Obtained by arbitrarily changing r_{solute} from 5.6 Å to 2.8 Å

2.6.2 Cascaded $\chi^{(2)} : \chi^{(2)}$ for the Optical Kerr Effect

A Experimental Determination of Cascaded Nonlinearities

Measurements of induced nonlinear phase shifts and n_2 in bulk materials are most often performed with the z-scan technique described in detail in this work. z-scan measurements in crystals of KTP and 4-(N, N-dimethylamino)-3-acetamidonitrobenzene (DAN) revealed a large n_2 that change sign as one goes through phase-matching (Fig. 2.49), therefore demonstrating the cascaded nature of the process [2.18, 2.107]. Values as large as $n_2 = \pm(2.5 \pm 0.5) \times 10^{-13} \, \mathrm{cm^2/W}$ near phase-matching were measured for DAN (Table 2.19).

Another possibility to measure third-order nonlinearities and cascaded nonlinearities is degenerate four-wave mixing. With this technique the anisotropy of $\chi^{(3)}$ $(-\omega, \omega, -\omega, \omega)$ can be measured and most importantly, a selection of propagation direction and polarization of the interacting waves allows a clear distinction between cascaded and direct $\chi^{(3)}$ effects, in contrast to the z-scan technique [2.12, 2.14]. The first experimental demonstration of cascaded optical rectification and the linear electro-optic effect in degenerate four-wave mixing was performed in polar $\mathrm{KNbO_3}$ crystals [2.14, 2.112]. Several geometries were investigated in order to evaluate the direct contribution of the third-order polarizabilities and the contribution from the second-order effects (cascading). Figure 2.50 illustrates the principle of the measurement [2.14, 2.112]. More than one order of magnitude increase in the diffracted signal due to cascading was obtained in these experiments. Contributions from

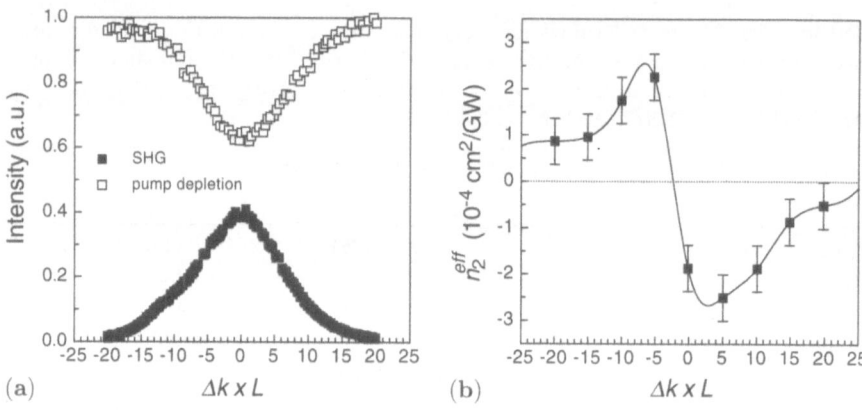

Fig. 2.49. (a) Measured pump depletion and second-harmonic generation efficiency, and (b) measured effective nonlinear refractive indices versus phase-mismatch for a DAN single crystal at $\lambda = 1064$ nm and at a peak intensity of $2\,\mathrm{GW/cm^2}$. The solid line is a guide to the eye (adapted from [2.18])

cascaded second-harmonic generation and difference frequency mixing can be neglected in this case.

It is important to note that values of $\chi^{(3)}\,(-\omega,\omega,-\omega,\omega)$ measured with the z-scan technique and degenerate four-wave mixing cannot be compared directly. The possibility of selecting the propagation directions and the polarizations of the interacting waves for the latter experiments can lead to different values of $\chi^{(3)}\,(-\omega,\omega,-\omega,\omega)$ compared with those determined in the single beam z-scan experiment [2.14].

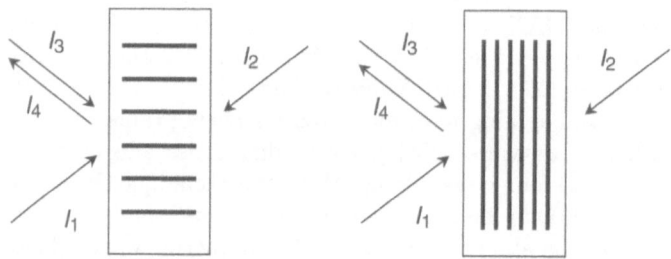

Fig. 2.50a,b. Principle of degenerate four-wave mixing in noncentrosymmetric crystals. Three laser pulses I_1, I_2, and I_3 simultaneously impinge on the crystal and generate a fourth beam I_4. The probe beam I_2 is delayed or advanced in time relative to I_1 and I_3. Two types of grating are produced: (a) a transmission grating generated by the beams I_1 and I_3, and (b) a reflection grating generated by I_2 and I_3. The signal of the diffracted beam I_4 is measured. A comparison with a reference material with known nonlinearity allows the determination of tensor elements $\chi^{(3)}_{ijkl}$

Table 2.19. Comparison of nonlinear refractive indices n_2. The materials are tabulated according to the different origins of n_2. Estimates are based on (2.110) for cascaded second-harmonic generation and difference-frequency mixing and on (2.115) for cascaded optical rectification and the linear electro-optic effect. For most materials the values were determined at a wavelength above at least twice the cut-off wavelength. The cases where this is not true are indicated below. For experimental geometries where quasi-phase matching was employed, the reduction of the nonlinear optical coefficients (typically $2/\pi$) was taken into account. All estimations for cascaded second-harmonic generation and difference frequency mixing are based on $\chi^{(2)}_{111}$ (α-quartz) $= 0.6$ pm/V at $\lambda = 1064$ nm.

Material	λ (nm)	Phase shift	n_2 $(10^{-16}\,\mathrm{cm^2/W})$	$|n_2/n_2^{\mathrm{fs}}|$	FM^a $(\mathrm{pm/V})^2$	Geometry	Experiment	Ref.
			Direct $\chi^{(3)}$					
Fused silica	1064	–	2.5	1	–	bulk	NDFWM^b	[2.234]
CS$_2$	1064	–	320	130	–	bulk	z-scan	[2.166]
AlGaAs	1550	–	1000–2000	400–800	–	waveguide	SPM^d	[2.235]
InGaAsP	1500	–	–32000	12800	–	waveguide	pump-probe	[2.221]
DANS (4-dimethylamino-4'-nitrostilbene)	1320	–	800	320	–	waveguide	interferometry	[2.172]
PPV (poly(p-phenylene-vinylene))	800	–	100 000	40 000	–	thin film	DFWM^b	[2.223]

Table 2.19. (Continued)

Material	λ (nm)	Phase shift	n_2 $(10^{-16}\,\mathrm{cm}^2/\mathrm{W})$	$\lvert n_2/n_2^{\mathrm{fs}}\rvert$	FM[a] $(\mathrm{pm/V})^2$	Geometry	Experiment	Ref.
PTS (poly[2,4-hexadiyne-1,6-diol-bis-p-toluene-sulfonate])	1600	–	22000	8800	–	bulk	z-scan	[2.4, 2.222]
Cascaded second-harmonic generation and difference frequency mixing								
KTP	1064	–	±200	80	1.9	bulk	z-scan	[2.107, 2.109, 2.156, 2.170]
	850	$> \pi$	480	190	3.9	waveguide[f]	SPM[d]	
	1587	0.45π	680	270	3.7	waveguide[f]	SPM[d]	
LiNbO$_3$	1320	0.53π	800	320	1.7	waveguide	interferometry	[2.170, 2.175]
KNbO$_3$	1064	–	±730	290	1.5	bulk	z-scan	[2.233]
DAN (4-(N,N-dimethylamino)-3-acetamidonitrobenzene)	625[e]	–	4.3×10^4	1.7×10^4	–	waveguide	SPM[d]	[2.18, 2.171]
	1064	–	±2500	1000	70	bulk	z-scan	
	1320	$-\pi/8$ to $-\pi/4$	–9000	3600	–	crystal	interferometry	
	1550	$\pi/2$	|6000|	2400	–	cored fiber	SPM[d]	

Compound								
DANPH (4-dimethylamino-benzaldehyde-4-nitrophenylhydrazone)	1500	—	$\pm 2.5 \times 10^5$	1.0×10^5	2530	bulk	estimation	[2.227]
DAST (4'-dimethylamino-N-methyl-4-stilbazolium tosylate)	1500	—	$\pm 3 \times 10^5$	1.2×10^5	3000	waveguide^f	estimation	[2.137, 2.225]
Optimized stilbene	1200	—	$\pm 63 \times 10^6$	25×10^6	510000	waveguide^f	estimation	this work
	1200	—	$\pm 9 \times 10^6$	3.7×10^6	76000	bulk	estimation	

Table 2.19. (Continued)

Material	λ (nm)	Phase shift	n_2 (10^{-16} cm^2/W)	$|n_2/n_2^{fs}|$	FM^a (pm/V)^2	Geometry	Experiment	Ref.
Alternating styrene-maleic-anhydrid copolymer (polyimide) +DR1	1500	—	$\pm 9.3 \times 10^3$	3700	100	waveguide^f	estimation	[2.236]
Cascaded optical rectification and the linear electro-optic effect								
KNbO$_3$	1064	—	260	100	2×10^4	bulk	z-scan	[2.12]
DAST (4'-dimethylamino-N-methyl-4-stilbazolium tosylate)	800^e	—	2540 820	1020 330	2.7×10^5 8.7×10^4	bulk bulk^g	estimation estimation	[2.137, 2.138]

Optimized stilbene	600^e	—	8.2×10^4 3.2×10^4	3.3×10^4 1.3×10^5	8.7×10^6 3.4×10^5	bulk $bulk^g$	estimation estimation	this work
Alternating styrene-maleic-anhydrid copolymer (polyimide) +DR1	1500	—	17	7	1810	waveguide	estimation	[2.236]

$$^a\; FM = \left\{ \begin{array}{l} d_{\text{eff}}^2/n^3 \quad \text{for SHG/DFM} \\ (n^3 r)_{\text{eff}}^2/(\varepsilon - 1) \quad \text{for OR/EO} \end{array} \right.$$

^{b} (N)DFWM: (non)degenerate four-wave mixing

^{c} Orientational contribution to n_2

^{d} SPM: self-phase modulation

^{e} Wavelength $\lambda < 2 \times$cut-off wavelength

^{f} Quasi-phase matching

^{g} Including depolarization effects (see (2.118))

For the characterization of waveguides other techniques are used. In the following we will describe self-phase modulation and interferometry in more detail.

Self-phase modulation: the easiest experiment to demonstrate phase shifts due to cascading is to show spectral broadening and modulation. From such experiments and with the help of numerical simulations the actual nonlinear phase shift can be deduced. The first demonstration of nonlinear phase shifts in a waveguide due to cascading was therefore performed with this method. The spectral modulation obtained on transmission through quasi-phase-matched KTP waveguides near 850 nm using a Kerr lens mode-locked Ti:Sapphire laser was measured and significant spectral broadening and modulation was observed slightly off phase-matching in the 854 to 857 nm range, with a typical power-dependent result shown in Fig. 2.51 [2.109]. This figure clearly shows the greater than threefold increase in spectral width and lobed modulation when going from low to high intensities in the waveguide. The low-intensity KTP waveguide transmission spectrum is indistinguishable from the input spectrum. Nonlinear peak phase shifts larger than π for 600 W peak power in a 2.8 mm long guide were observed (Table 2.19). Note that large second-harmonic conversion efficiencies were required to see this effect (60–70% average fundamental depletion for the largest phase modulation). The direct n_2 of KTP at 1.06 μm is $2.4 \times 10^{-15}\,\mathrm{cm^2/W}$ (n_2^{OR} was neglected!), which gives a phase shift almost two orders of magnitude less than that inferred here and confirms the cascaded nature of the observed effects. Similar results could be obtained near the interesting telecommunication wavelengths of 1.55 μm (peak nonlinear phase shifts of $\approx \pi/2$) [2.170]. Spectral broadening of 250 fs pulses due to self-phase modulation was also observed in DAN

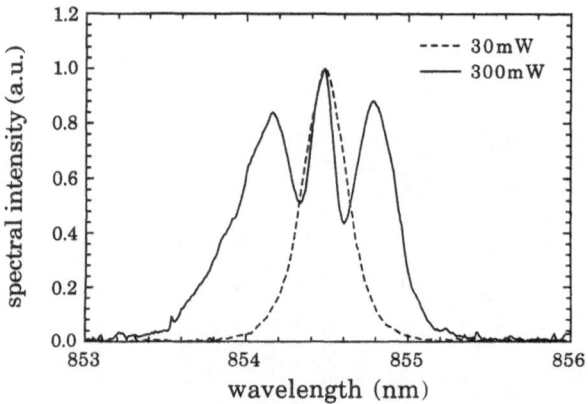

Fig. 2.51. Output fundamental spectra for two different incident powers (the 300 mW average power corresponds approximately to a peak intensity of 3 GW/cm^2) from spectral modulation measurements in quasi-phase-matched KTP waveguides (adapted from [2.109])

crystal cored fibers [2.171]. A comparison with a conventional fused silica fiber yielded a nonlinear refractive index n_2 that was 17 000 times higher. Later on similar experiments were performed at 1.55 μm with 600 fs pulses in a multimode 2.6 mm long crystal cored DAN fiber [2.18]. A phase shift of $\pi/2$ was measured and a corresponding value of $n_2 = |6 \pm 3| \times 10^{-13}$ cm^2/W was estimated for an input intensity of 0.3 GW/cm^2 (Table 2.19). Note, that spectral broadening was only observed if a strong second-harmonic signal was present. In the absence of a second-harmonic signal no broadening could be observed. Additional third-harmonic generation experiments on unpoled guest–host DAN/PMMA polymers (no macroscopic cascading possible) to estimate the direct $\chi^{(3)}$ of DAN were performed. These experiments further revealed that cascading was indeed the mechanism behind the large effective third-order nonlinearity in the DAN bulk crystals and crystal cored fibers.

Unfortunately, self-phase modulation only provides approximate values of the phase shift and no sign determination is possible.

Interferometric measurements: a nonlinear Mach–Zehnder interferometer can be used to measure directly the value as well as the sign of n_2 in waveguides with a remarkable resolution of $\pi/100$ [2.172]. With this method the nonlinear refractive index of DAN fibers [2.18, 2.173–174] and LiNbO$_3$ channel waveguides [2.175] were measured at $\lambda = 1319$ nm. In both cases the large effects that were observed could be clearly identified as coming from cascaded second-harmonic generation and difference frequency mixing. Whereas in the KTP self-phase modulation experiments the depletion of the fundamental was very large (60–70%), impressive phase shifts with low depletion could be demonstrated in LiNbO$_3$ [2.175].

Both the self-modulation and the interferometric experiments show that nonlinear phase shifts of sufficient magnitude for optical switching devices ($\geq \pi$) can be achieved. As a comparison, a push–pull Mach–Zehnder interferometer switch where phase shifts of opposite signs are induced in the two arms of the interferometer requires only a phase change of $\pm \pi/2$ in each arm.

B Effects Based on Cascaded Second-Order Processes

Most effects due to cascading have relied so far on the combined processes of second-harmonic generation and difference frequency mixing. The potential for cascaded optical rectification and the linear electro-optic effect has not yet been established. We will therefore primarily concentrate on the first process.

All-Optical Signal Processing. A substantial potential of cascading lies in all-optical signal processing applications that allow the manipulation of an output beam in a nonlinear material through the input signal or a separate control beam (Fig. 2.1) [2.2]. The potential of well-characterized $\chi^{(2)}$ materials for all-optical switching using cascading is already theoretically established [2.110, 2.176–179]. The processes are, however, very complex and there are several requirements for switching: undistorted fundamental pulse, low depletion and uniform phase profile across the pulse. It is not clear whether all these conditions can be met at the same time and, therefore, whether cascading is of interest for replacing the real n_2 for self-switching.

As discussed above large third-order effects could be observed in single-crystal cored fibers of DAN [2.171] and it could be proved from bulk and waveguide experiments that these effects originate from cascaded second-order nonlinearities [2.18, 2.174]. The interesting fact is that the effects observed in these waveguides (at $\lambda = 1.32\,\mu m$) are due to Cerenkov-type second-harmonic generation, that is, the second-harmonic wave is not guided in the waveguide core. Nevertheless nonlinear phase shifts as large as $-\pi/2$ could be measured at a fundamental wavelength of $1.32\,\mu m$. It was also predicted that high-contrast all-optical switching by using cascading in the same Cerenkov regime is possible [2.180–181].

A Mach–Zehnder push–pull interferometer based on cascaded phase shifts might also be a possible application [2.182]. The performance of such an interferometer relies on an optimized phase mismatch in one arm of the interferometer and an opposite phase mismatch in the other arm. This could lead to a phase shift of $\pi/2$ in one arm and one of $-\pi/2$ in the other arm therefore switching the light beam. The first experimental demonstration of all-optical switching based on cascaded second-order nonlinearities was performed in a nonlinear hybrid Mach–Zehnder interferometer using a LiNbO$_3$ channel waveguide. A contrast ratio of 7:1 was achieved in a regime with small fundamental depletion [2.183]. Shortly after that the first fully integrated asymmetric Mach–Zehnder interferometer based on the same material was demonstrated with a contrast ratio of 8:1 [2.184].

Novel all-optical approaches to signal processing can be exploited by coherent interactions between three waves in noncentrosymmetric materials [2.110, 2.178, 2.185–186]. As an example phase-controlled transistor action by cascading of second-order nonlinearities was demonstrated in KTP bulk crystals [2.187]. A switching ratio of 4.6 to 1 was achieved for a $1.064\,\mu m$ fundamental beam by modulating the relative input phase of the $0.532\,\mu m$ control beam. Much better results are expected for materials with higher nonlinear optical coefficients as well as in the case of waveguide geometries. There remains much theoretical work to be done to find proper operating conditions, since there exist almost infinite varieties of interactions.

Another interesting possibility for all-optical signal processing is the use of type II phase matching, since there is the additional freedom to adjust the strength of the two fundamental beams separately [2.178, 2.188–189]. In this way nonzero phase shifts can even be achieved with the phase-matching condition fulfilled as long as the two input intensities are different [2.190]. This kind of interaction is very similar to the seeded case for signal processing discussed above.

Spatial Solitons. One of the most interesting effects due to cascaded second-harmonic generation and difference frequency mixing is the mutual trapping

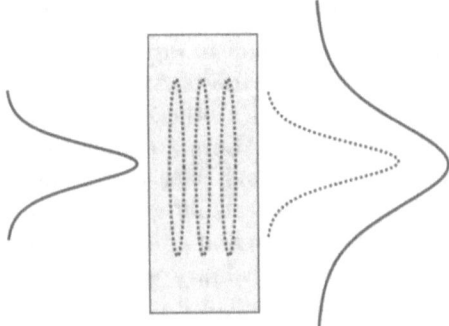

Fig. 2.52. Schematic drawing of a one-dimensional spatial soliton. Usually a beam diffracts as it propagates through a medium (*solid line*). If diffraction is exactly compensated by self-focusing the beam shape of the incident wave stays the same and we have a bright (one-dimensional) spatial soliton (*dotted line*). The lenses represent the self-focusing

of the fundamental and second-harmonic waves propagating in the material. The formation of such soliton-like states in a $\chi^{(2)}$ medium was first discussed by Karamzin [2.191–192]. Spatial optical solitons are light beams which propagate through a nonlinear medium without diffraction [2.193]. Figure 2.52 shows a schematic drawing of the principle for bright ($n_2 > 0$) spatial solitons: usually a beam diffracts as it propagates through a medium. If the diffraction is exactly compensated by self-focusing (for a material with positive n_2) the output beam has the same size as the incident beam and we have a bright (one-dimensional) spatial soliton. Stable self-trapping can occur only in a two-dimensional medium in which the beam can diffract along only one transverse direction. In a three-dimensional medium where light diffracts along two transverse directions self-trapping is not stable and can lead to catastrophic self-focusing [2.98]. Spatial solitons have attracted a great deal of attention since waveguides can be optically written and steered and beams can be guided in such channels [2.194–196]. This leads to a new type of reconfigurable optical device. Much theoretical work has also been done on solitons generated through cascaded nonlinearities [2.197–201]. In contrast to spatial solitons based on pure $\chi^{(3)}$ effects it could also be shown that in the case of cascading two-dimensional solitary waves are stable due to the strong nonlinear coupling which counteracts both diffraction and beam walk-off.

One-dimensional bright spatial solitons due to cascading were generated in Ti-diffused planar waveguides of $LiNbO_3$ [2.202]. The experiment was performed in a 47-mm long sample at a fundamental wavelength of 1.32 μm. An elliptical beam was coupled into the waveguide with a FWHM of 60–70 μm. Far from phase-matching the beam was diffracted to about 3 times the size of the input beam at the exit face of the waveguide. Near phase matching the output beam size at the exit face decreased to about the size of the input beam, forming a one-dimensional soliton due to cascaded second-order nonlinearities.

Very recently, two-dimensional spatial solitary waves in a quadratic medium, first predicted theoretically, were experimentally observed [2.203]. These solitary waves were generated in a 1 cm long KTP crystal cut for type II phase matching. A 1.064 μm beam was focused on the crystal to produce a waist of 20 μm (half-width at $1/e^2$ in air) at the entrance surface. At low intensities the three interacting beams moved away from each other in space and experienced diffraction at the output of the crystal. Above a certain intensity ($> 10\,\mathrm{GW/cm^2}$) the fundamental beam at the exit face had a stable waist of 12.5 μm demonstrating that diffraction and walk-off was overcome. In this intensity regime all three beams (two fundamental and one second-harmonic) are trapped in a cylindrically symmetric solitary wave. Experiments as well as numerical simulations showed that cascaded processes were responsible for the observed effect. Mutual trapping and dragging of such two-dimensional spatial solitary waves as well as all-optical switching based on the intensity-induced change of the propagation direction were experimentally realized [2.204–205]. These spatial solitons can also be used to reshape elliptical beams. Elliptical beams with an aspect ratio of 8:1 could be transformed into cylindrically symmetric beams through the formation of spatial solitons in KTP [2.206].

Optical Rectification and the Linear Electro-Optic Effect. Few experiments have been performed using the combined processes of optical rectification and the linear electro-optic effect. Early on, self-modulation of picosecond pulses in GaAs has been explained with this effect [2.13]. We have shown by a detailed z-scan analysis in $KTaO_3$ and $KNbO_3$ crystals and a continuous wave approach that this combined process can be expressed by a Kerr nonlinearity (n_2) [2.12]. Note that for a general propagation direction both types of cascading (n_2^{SHG} and n_2^{OR}) can contribute to the nonlinear phase shift. Thus the importance of the two contributions for a chosen experimental configuration must be carefully evaluated.

Another very interesting application of cascaded optical rectification and the linear electrooptic effect is degenerate four-wave mixing [2.14, 2.112, 2.115] (see above). More than one order of magnitude increase in the diffracted signal due to cascading was measured in polar $KNbO_3$ crystals. Contributions from cascaded second-harmonic generation and difference frequency mixing could be neglected in this case and direct $\chi^{(3)}$ and cascaded processes could be separated. In addition, an absolute determination of $\chi^{(3)}$ is possible by analogy with third-harmonic generation.

Other Effects. Besides the effects discussed so far there are many more interesting phenomena based on cascaded nonlinearities. Strong effective third-order nonlinearities were obtained through resonantly enhanced second-harmonic generation in a surface-emitting geometry [2.207–208]. Cascaded processes were also used in self-diffraction experiments in BBO [2.209–210]. Very recently, squeezed light was generated through cascaded nonlinearities [2.211–212]. This was done both by means of traveling-wave type -II phase-matched second-harmonic generation as well as using self-diffraction. Mode-locking of a cw Nd:YAG laser based on cascaded second-order nonlinearities was demonstrated with lithium borate. In contrast with Kerr lens mode locking this process was self-starting and it led to almost transform-limited laser pulses [2.213]. Optical bistability based on cascaded nonlinearities is another effect of interest that was mainly exploited in intracavity second-harmonic generation applications [2.214–216]. By placing a $KNbO_3$ crystal in an optical cavity in a type I phase-matched geometry, asymmetries in the temporal line shapes of the fundamental and second-harmonic light waves were measured [2.214]. Bistability in a monolithic $LiNbO_3$ resonator optimized for phase-matched second-harmonic generation at $\lambda = 1064$ nm could be interpreted in terms of cascaded nonlinearities [2.215]. In addition, bistability based on cascaded second-order nonlinearities was also observed by investigating the coupling into waveguide resonators [2.217]. By analogy with second-harmonic generation, nonlinear phase shifts of the fundamental in phase-matched third-harmonic generation based on cascaded third-order nonlinearities were recently predicted [2.218]. The analysis was performed for centrosymmetric materials and it was shown that the sign of the phase shift depends on the sign of the phase mismatch in third-harmonic generation (just as in the case of cascaded second-harmonic generation and difference frequency mixing).

C   Materials Consideration and Relevant Figures of Merit

We here compare the available materials and the parameters relevant for the optical Kerr effect. We first discuss the pure third-order materials glasses, semiconductors and polymers, before we concentrate on cascaded nonlinearities (Table 2.19). The figures of merit for the different materials classes relevant for applications will be discussed at the end of this section.

Glasses are one type of nonlinear optical material. Especially silica fibers have been extensively used for all-optical signal processing [2.102]. Despite the low nonlinearities large effects are possible due to the long interactions lengths that can be achieved in fibers.

Many experiments for all-optical signal processing are based on non-resonant nonlinearities that are used at photon energies one half below the bandgap [2.219]. In the telecommunications window around 1.5 μm Al_xGa_{1-x} As, also called *silicon of nonlinear optical materials*, is a suitable material [2.219]. Many devices, including X-junctions and Mach–Zehnder interferometers, were demonstrated [2.219].

Among the materials with the largest nonlinear refractive indices are active semiconductor amplifiers in a configuration where gain and absorption are balanced for the light-carrier interaction. Huge Kerr nonlinearities of $n_2 = -3.2 \times 10^{-12}$ cm^2/W were obtained for InGaAsP optical amplifiers [2.220]. The main problems in these materials are the large linear and nonlinear losses (see Table 2.20).

The most promising materials are currently active semiconductor amplifiers based on carrier nonlinearities [2.221]. Gain is first achieved by pumping

Table 2.20. Figures of merit for all-optical switching for a selection of materials. We assume an intensity of $1\,GW/cm^2$ which leads to a peak power of about $100\,W$ in the device. For a description of the organic compounds see Table 2.19

Material	λ (nm)	n_2 $(10^{-16}\,cm^2/W)$	α (cm^{-1})	W	T	Ref.
		Glasses				
SiO_2	1064	2.5	10^{-6}	$> 10^3$	$\ll 1$	[2.234]
RN (Corning)	1064	130	0.01	13	< 0.05	[2.34]
		Semiconductors				
AlGaAs[a]	15560	2000	0.1	13	< 0.15	[2.34]
AlGaAs[b]	810	-40000	18	2.7	0.45	[2.34]
InGaAsP	1500	-32000	30–50	0.5	2	[2.220]
		Organic materials				
PTS (crystal)	1600	22000	< 0.8	> 10	< 0.05	[2.222]
DANS (polymer)	1320	800	< 0.2	> 5	≈ 0.1	[2.172]
PPV (polymer)	800	> 100000	–	≥ 2	≤ 0.5	[2.223]

[a] For a bandgap at $\lambda = 750\,nm$ [b] For a bandgap at $\lambda = 790\,nm$

carriers into the conduction band. An incident beam can experience a nonlinear phase shift when electron transitions occur from the conduction to the valence bands. As an example, pulses of $3\,pJ$ in energy were enough to obtain a nonlinear phase shift of π in a $0.5\,mm$ long amplifier.

As extensively discussed in this work organic materials represent an important and promising class of materials for third-order nonlinear optics. The single crystal poly[2,4-hexadiyne-1,6-diol-*bis-p*-toluene-sulfonate] (PTS) is currently the best organic material in terms of nonlinearity and loss [2.222]. Values of $n_2 = 2.2 \times 10^{-12}\,cm^2/W$ were measured at $\lambda = 1600\,nm$. The drawback so far has been the scattering due to defects which limit the sample length that can be used.

Another material is the polymer poly(p-phenylenevinylene). It has huge nonlinearities at $\lambda = 800\,nm$ (see Table 2.19) [2.223]. A nice feature of PPV is the fact that the nonlinearity and two-photon absorption can be tuned by a change in the chemical composition. As in the case of PTS the major drawback so far has been the considerable optical losses.

Side-chain polymers have also been investigated for their third-order nonlinearity. One example is based on the (noncentrosymmetric) charge transfer molecule 4-dimethylamino-4'-nitrostilbene (DANS). At $\lambda = 1320\,nm$ nonlinearities of the same order as for AlGaAs were determined ($n_2 = 800 \times 10^{-14}\,cm^2/W$) [2.172].

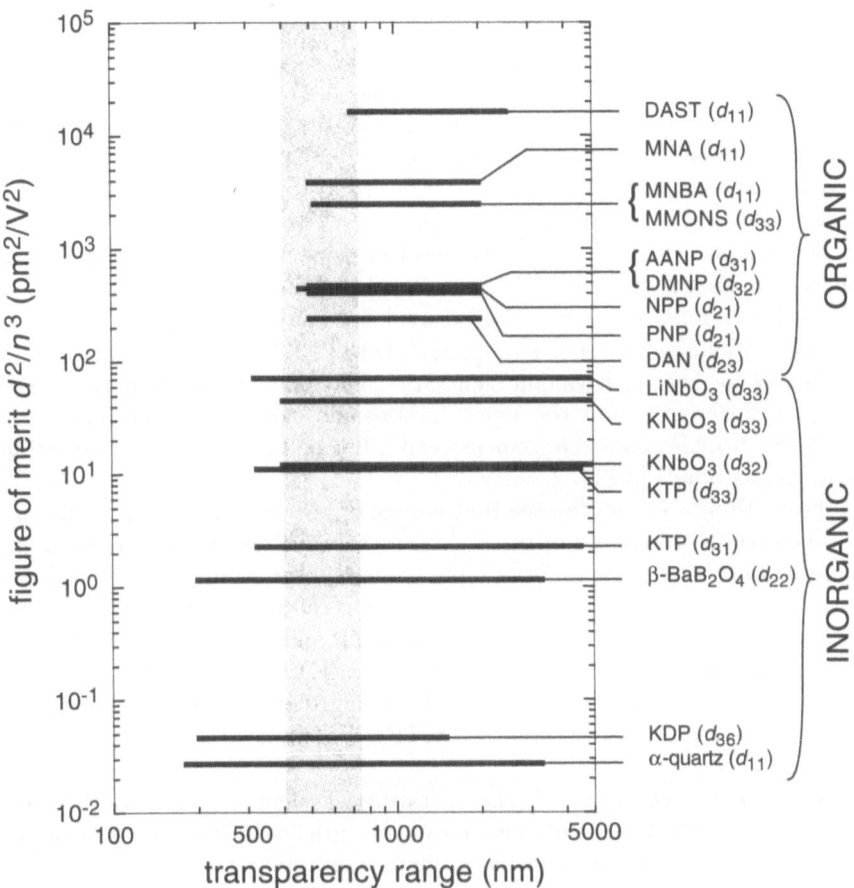

Fig. 2.53. Figures of merit d^2/n^3 for some inorganic and organic nonlinear optical materials. The curves are drawn over the approximate transparency range of the individual materials. The wavelength dispersion has been neglected. The shaded area indicates the visible part of the spectrum (400–750 nm). All values were calibrated with respect to $d_{111} = 0.3$ pm/V of α-quartz at $\lambda = 1064$ nm

We now concentrate on the cascaded nonlinearities. The requirements imposed on a material are different for the two different types of cascading. We will start with n_2^{SHG}. The most important condition is phase matching. In addition, one wants nonlinear optical coefficients to be as large as possible, since the induced nonlinear phase shift is proportional to the figure of merit d_{eff}^2/n^3. Figure 2.53 shows figures of merit for second-harmonic generation vs. the transparency range of a number of inorganic and organic single crystals where the wavelength dispersion of d_{eff}^2/n^3 was neglected. Also presented are most materials where cascaded processes have been experimentally observed (e.g. KTP [2.93, 2.107, 2.109, 2.170, 2.187, 2.203], LiNbO$_3$ [2.175, 2.202], BBO [2.210], KNbO$_3$ [2.12, 2.14, 2.112, 2.214],

DAN [2.18, 2.174], NPP [2.224]). We can clearly see that the potential of organics in terms of large figures of merit is superior as compared to inorganic materials. A problem to be solved is the phase-matching condition for molecular crystals: since large values of d_{eff}^2/n^3 are often connected with a diagonal tensor element (e.g. $d_{11} = 286\,\text{pm/V}$ for $\lambda = 1.542\,\mu\text{m}$ and $d_{11} = 1013\,\text{pm/V}$ for $\lambda = 1.318\,\mu\text{m}$ for DAST [2.225], no bulk phase matching using such coefficients is possible. There are, however, already single crystals with large nonlinear optical coefficients where birefringence phase matching is possible (e.g. N-4-nitrophenyl-L-prolinol (NPP)[2.226], 2-(N-prolinol)-5-nitropyridine (PNP) [2.140], 4-dimethylaminobenzaldehyde-4-nitrophenylhydrazone (DANPH) [2.227], DAST [2.225]. These materials are optimized for second-harmonic generation and have an angle between the charge-transfer axis and the polar crystal axis close to the optimum 54° [2.79]. Such molecular arrangements can often be favored by the formation of hydrogen bonds [2.134, 2.228–232].

From Table 2.19 one can see that a large d_{eff}^2/n^3 is not the only criterion. An increase of this figure of merit does not always lead to a corresponding increase in nonlinearity, but the input intensity, the sample length and the detuning from phase-matching play an important role as well. This can be seen by comparing the measured values of KTP and KNbO₃ at $\lambda = 1064\,\text{nm}$: the increase of n_2 from $\pm 200 \times 10^{-16}\,\text{cm}^2/\text{W}$ for KTP to $\pm 730 \times 10^{-16}\,\text{cm}^2/\text{W}$ for KNbO₃ is not due to an increase of the figure of merit but mainly arises from a larger crystal thickness (L (KTP) $= 1\,\text{mm}$, $L(\text{KNbO}_3) = 5\,\text{mm}$) [2.233].

Another important issue is the operating wavelength. For all-optical signal processing the telecommunication wavelengths at $1.3\,\mu\text{m}$ and $1.5\,\mu\text{m}$ are most interesting. This fact goes along our current research direction: we try to optimize the nonlinearities (nonlinear optical coefficient d_{ijk} and electro-optic coefficient r_{ijk}) of our crystals by pushing the wavelength of maximum absorption into the red spectral range (see Fig. 2.13) [2.71]. It can clearly be seen that an increase in λ_{eg} can lead to significant increases in d_{ijk} and therefore also larger values of the second-order nonlinearity in the near infrared. As an example we recently developed a new organic crystal with phase-matchable nonlinear optical coefficients in excess of $110\,\text{pm/V}$ at $1.5\,\mu\text{m}$ with a cut-off wavelength near $700\,\text{nm}$ [2.227]. This material is therefore almost optimized for cascaded second-harmonic generation and difference frequency mixing.

Table 2.19 shows a comparison of different inorganic and organic materials with results based on direct $\chi^{(3)}$ and processes using cascaded nonlinearities. It summarizes measured as well as estimated phase shifts and effective nonlinear refractive indices. Estimated values are based on (2.110) which gives n_2^{SHG} in the limit of low depletion of the fundamental beam. Whenever possible, d_{eff}^2/n^3 was corrected to the same reference value ($d_{111} = 0.3\,\text{pm/V}$ for α-quartz at $\lambda = 1064\,\text{nm}$). As discussed above, we assumed a phase mismatch of $\Delta kL \approx 3$ for optimum nonlinear refractive indices and a sample length of

1 mm for these estimates. Note again that these approximations are only valid for low conversion efficiencies since the validity of (2.110) quickly breaks down for larger efficiencies [2.16]. The estimates in Table 2.19 nevertheless provide a good guide to what could be expected.

Our estimates of the upper limits of these coefficients in molecular crystals indicate that orders of magnitude increases in obtainable phase shifts might be possible. For an optimized stilbene derivative with a nonlinear optical coefficient of 450 pm/V at $\lambda = 1200$ nm we estimate an n_2 of the order of 2×10^{-10} cm^2/W in the bulk. A large potential also exists for poled polymers. Large nonlinear optical coefficients of d_{22} of 115 pm/V ($\lambda = 1.064\,\mu$m) and $d_{22} = 23$ pm/V ($\lambda = 1.55\,\mu$m) were achieved in, for example, 4-dimethylamino-4'-nitrostilbene (DANS) side-chain polymers with poling fields of 300 V/μm and 200 V/μm, respectively, or $d_{33} = 32$ pm/V ($\lambda = 1.54\,\mu$m) for an alternating styrene-maleic-anhydrid copolymer (disperse red 1 polyimide side-chain polymer) [2.236–238].

We now look at n_2^{OR}. First of all we have $\Delta\phi^{\mathrm{OR}} \propto n_2^{\mathrm{OR}} I$, where $n_2^{\mathrm{OR}} \propto \left(n^3 r_{\mathrm{eff}}\right)^2 / (\varepsilon - 1)$ or $n_2^{\mathrm{OR}} \propto \left(n^3 r_{\mathrm{eff}}\right)^2 / (\varepsilon + 2)$. Here n is the linear refractive index, r_{eff} is the effective linear electrooptic coefficient and ε is the dielectric constant [2.12]. The phase shift is proportional to the square of the electro-optic figure of merit $\left(n^3 r_{\mathrm{eff}}\right)^2$ (see (2.115) and (2.116)). Again large nonlinearities, in this case large effective electro-optic coefficients are advantageous. Organic materials can be very promising in this regard: as in the case of second-harmonic generation an increase in λ_{eg} can lead to large increases in the electro-optic coefficients [2.71]. In contrast to n_2^{SHG}, however, large diagonal tensor elements can easily be used for this process. Therefore, a parallel alignment of all chromophores in the crystal lattice is optimum for cascaded optical rectification and the linear electro-optic effect using the transverse electro-optic effect. If the longitudinal electro-optic effect is to be used the same chromophore alignment as in the case of n_2^{SHG} is ideal. In addition, the combination of large second-order nonlinearities and low dielectric constants is another advantage of organic materials for cascaded optical rectification and the linear electro-optic effect. For the organic salt DAST we expect a contribution to n_2 of the order of 3×10^{-13} cm^2/W at $\lambda = 800$ nm. Table 2.19 summarizes measured values of n_2 using this combined effect and also gives estimates of possible values for optimized materials based on (2.115). Elements of the $\chi^{(3)}$ tensor can be calculated from n_2 using (2.89).

For all-optical signal processing (based on pure $\chi^{(3)}$ nonlinearities) the important quantity is the nonlinear refractive index n_2 and the relevant figures of merit are W (limit of linear absorption; see (2.35)), T (limit due to two-photon absorption; see (2.36)), and V (limit due to three-photon absorption; see (2.37)) with the requirements $W > 1$, $T^{-1} > 1$ and $V^{-1} > 1$. V has only been important for AlGaAs so far [2.239]. Table 2.20 lists W and T for various materials. Part of the data originate from [2.240].

We can see from Table 2.20 that several materials fulfill the requirements $W > 1$ and $T^{-1} > 1$. As an example, AlGaAs is such a material in which many device applications could be demonstrated [2.219]. However, due to the rather moderate nonlinearities, the switching energies for pulse lengths around 1 ps are just too high. On the other hand, the values of n_2 of, for example, PTS are very large, but the optical losses are too dominant. One can generally summarize by noting that up to now either the nonlinearities are not sufficiently high or that the sample quality is not good enough for a working device. Since cascading of second-order nonlinearities for all-optical signal processing is a very new approach the complications that might arise there are not yet known. Actual required nonlinear phase shifts and figures of merit for a selection of devices are tabulated in [2.34].

2.6.3   Final Remarks and Outlook

We have presented the first systematic investigation of the influence of acentricity and, at the same time, two-dimensional conjugation on the third-order nonlinear optical properties of molecules. We have shown that acentricity strongly enhances the third-order nonlinear optical properties of molecules. The underlying physical mechanisms responsible for the strong influence of acentricity is still under further investigation. On the one hand, additional molecular systems have to be investigated. On the other hand, nonlinear optical spectroscopic measurements are needed to identify the excited states that influence the molecular nonlinearities [2.59].

If crystals of the most promising molecules can be obtained with an optimized orientation in the lattice, huge pure third-order nonlinearities can be expected (see Tables 2.5 and 2.6). Much work is still needed to verify this potential of molecular crystals.

We have shown that there exists a large contribution to the nonlinear refractive index n_2 due to combined processes of optical rectification and the linear electro-optic effect. This contribution, which has been overlooked up to now, is always phase-matched since all interacting waves have either the same optical frequency or zero frequency. Another advantage of this new contribution to n_2 is that there is no net attenuation of the incident beam, in contrast to the contribution n_2^{SHG} where the fundamental is often [2.109], but not necessarily [2.175], depleted. It is also interesting to note that n_2^{OR} is always positive, also in contrast to n_2^{SHG} which can have either sign depending on the sign of the phase-mismatch. Moreover, n_2^{OR} can also be present for light propagation along the polar axis. In such a case the longitudinal electro-optic effect becomes active. For a general propagation direction in a noncentrosymmetric material both n_2^{SHG} and n_2^{OR} are important for induced nonlinear phase-shifts. It is, however, possible to find special orientations where one contribution is dominant, as was shown for $KNbO_3$. In the context of this type of cascading the influence of depolarization has to be further investigated.

Cascaded second-order nonlinearities offer an interesting and important alternative to direct $\chi^{(3)}$ nonlinearities. Larger third-order nonlinear optical effects are possible through cascading, therefore the development of materials with large second-order nonlinear optical susceptibilities $\chi^{(2)}$ may now pay off for cubic nonlinear optics. In addition to larger effective nonlinearities there also exist new effects, such as the control of the sign of n_2 when changing the sign of the phase mismatch. This is only possible through cascaded second-harmonic generation and difference frequency mixing. The effective nonlinear refractive indices obtained through cascaded processes (e.g. $n_2 = 6.8 \times 10^{-14}$ cm^2/W for KTP near 1.55 µm [2.170]) are already comparable to that of the pure $\chi^{(3)}$ effects in AlGaAs, which is a promising material for all-optical switching at telecommunication wavelengths [2.235]. Currently measured values of cascaded n_2 are, however, lower than those of the best materials based on direct $\chi^{(3)}$ (e.g. $n_2 = 2.2 \times 10^{-12}$ cm^2/W at $\lambda = 1600$ nm for PTS [2.222]). Our estimates show that this could easily be changed with either better organic crystals or poled polymers. Special attention will have to be paid to nonlinear effects near the interesting telecommunication wavelengths at around 1.3 µm and 1.5 µm.

Wavelength-dependent measurements should be carried out to determine spectroscopically the regions where the conditions imposed by the relevant figures of merit for all-optical signal processing are satisfied.

Besides basic characterization such as z-scan experiments, degenerate four-wave mixing, self-phase modulation and interferometry, many interesting effects based on cascading were observed: phase-controlled transistor action, squeezing of light, pulse compression, spatial optical soliton generation and others. The latter is of special interest for all-optical signal processing since two-dimensional bright spatial solitons as a result of cascading are stable, in contrast to the solitons resulting from direct $\chi^{(3)}$. Several additional possibilities for all-optical signal processing have been proposed and need to be demonstrated experimentally [2.110].

In the next few years the interplay of theory, organic synthesis, physics, materials science and device engineering may lead to exciting scientific results based on pure third-order and cascaded second-order effects.

Appendix

A.1   Definition of Nonlinear Optical Susceptibilities

The different conventions used in the definition of nonlinear optical susceptibilities were discussed in Sect. 2.1. Here we tabulate the nonlinear optical susceptibilities using the following consistent definition (closely following [2.92]): we express the macroscopic polarization P as a power series in the strength of the applied electric field E as

Table 2.21. List of nonlinear optical processes and the corresponding factors $K^{(n)}$ and frequency arguments (S: Stokes, AS: anti-Stokes, p: pump)

Process	Order n	Involved frequencies	$K^{(n)}$
Linear absorption/emission and refractive index	1	$-\omega, \omega$	1
Optical rectification	2	$0, \omega, -\omega$	1/2
Linear electro-optic effect	2	$-\omega, 0, \omega$	2
Second-harmonic generation	2	$-2\omega, \omega, \omega$	1/2
Sum and difference frequency mixing, parametric oscillation and amplification	2	$-\omega_3, \omega_1, \pm\omega_2$	1
Quadratic electro-optic effect (dc Kerr effect)	3	$-\omega, 0, 0, \omega$	3
Electric field-induced second-harmonic generation	3	$-2\omega, 0, \omega, \omega$	3/2
Third-harmonic generation	3	$-3\omega, \omega, \omega, \omega$	1/4
General four-wave mixing	3	$-\omega_4, \omega_1, \omega_2, \omega_3$	3/2
Sum and difference frequency mixing	3	$-\omega_4, \pm\omega_1, \omega_2, \omega_2$	3/4
Coherent anti-Stokes Raman scattering	3	$-\omega_{AS}, \omega_p, \omega_p, -\omega_S$	3/4
Optical Kerr effect (optically induced birefringence), cross-phase modulation, stimulated Raman scattering, stimulated Brillouin scattering	3	$-\omega_S, \omega_p, -\omega_p, \omega_S$	3/2
Cross-phase modulation, degenerate four-wave mixing[a]		$-\omega, \omega, -\omega, \omega$	3/2
Intensity-dependent refractive index, optical Kerr effect (self-induced and cross-induced birefringence), self-focusing, self-phase modulation	3	$-\omega, \omega, -\omega, \omega$	3/4
Two-photon absorption/ionisation/emission	3	$-\omega_1, -\omega_2, \omega_2, \omega_1$ or $-\omega, -\omega, \omega, \omega$	3/4
nth harmonic generation	n	$-n\omega, \omega, \ldots, \omega$	2^{1-n}

[a] Note that cross-phase modulation and degenerate four-wave mixing are listed in the wrong row in [2.92]

$$P_i = P_{i,0} + \varepsilon_0 \left\{ K^{(1)} \chi_{ij}^{(1)} E_j + K^{(2)} \chi_{ijk}^{(2)} E_j E_k + K^{(3)} \chi_{ijkl}^{(3)} E_j E_k E_l + \ldots \right\},$$

(2.146)

where the Einstein summation convention has to be applied. $\boldsymbol{P}_0$ is the spontaneous polarization and $\chi^{(n)}$ is the nth order susceptibility tensor. ε_0 is the vacuum permittivity. $K^{(n)}$ is a numerical factor which makes sure that the low frequency limit of all susceptibilities is the same for all frequency combinations. Table 2.21 summarizes various nonlinear optical processes and the corresponding factors $K^{(n)}$.

A.2   Conversion Between SI, cgs and Atomic Units

A   SI and cgs Units

In the field of nonlinear optics, especially in the characterization of molecular hyperpolarizabilities, the cgs system of units is frequently used. Apart from the different units for the fundamental quantities (length and mass), the cgs system differs from the SI system in the definition of electric quantities, thus introducing factors of c (velocity of light, $c = 3 \times 10^8$ throughout this work). We have

$$\left.\begin{array}{ll} P^{(n)} = \varepsilon_0 \chi^{(n)} E^n & \mathrm{C\,m^{-2}} \\ \chi^{(n)} & (\mathrm{m\,V^{-1}})^{n-1} \\ E & \mathrm{V\,m^{-1}} \end{array}\right\} \mathrm{SI} \quad \left.\begin{array}{ll} P^{(n)} = \chi^{(n)} E^n & \mathrm{statvolt\,cm^{-1}} \\ \chi^{(n)} & (\mathrm{cm\,statvolt^{-1}})^{n-1} \\ E & \mathrm{statvolt\,cm^{-1}} \end{array}\right\} \mathrm{esu}$$

The units of the nonlinear susceptibilities are not usually stated explicitly in the Gaussian system of units: rather one simply states that the value is given in electrostatic units (esu).

Electric field strength	$E(\mathrm{esu}) = \dfrac{10^4}{c} E(\mathrm{SI})$	$= \dfrac{1}{3 \times 10^4} E(\mathrm{SI})$
Macroscopic polarization	$P(\mathrm{esu}) = \dfrac{c}{10^3} P(\mathrm{SI})$	$= 3 \times 10^5\, P(\mathrm{SI})$
Linear susceptibility	$\chi^{(1)}(\mathrm{esu}) = \dfrac{1}{4\pi}\chi^{(1)}(\mathrm{SI})$	
Second-order susceptibility $\chi^{(2)}$, d	$\chi^{(2)}(\mathrm{esu}) = \dfrac{c}{4\pi \times 10^4}\chi^{(2)}(\mathrm{SI})$	$= \dfrac{3 \times 10^4}{4\pi}\chi^{(2)}(\mathrm{SI})$
Third-order susceptibility $\chi^{(3)}$, Γ	$\chi^{(3)}(\mathrm{esu}) = \dfrac{c^2}{4\pi \times 10^8}\chi^{(3)}(\mathrm{SI})$	$= \dfrac{9 \times 10^8}{4\pi}\chi^{(3)}(\mathrm{SI})$[a]
nth-order susceptibility $\chi^{(n)}$	$\chi^{(n)}(\mathrm{esu}) = \dfrac{(10^{-4}c)^{n-1}}{4\pi}\chi^{(n)}(\mathrm{SI})$	$= \dfrac{(3 \times 10^4)^{n-1}}{4\pi}\chi^{(n)}(\mathrm{SI})$
Electro-optic coefficient	$r(\mathrm{esu}) = \dfrac{c}{10^4}r(\mathrm{SI})$	$= 3 \times 10^4\, r(\mathrm{SI})$
Molecular polarization p, dipole moment μ	$\mu(\mathrm{esu}) = 10^3\, c\, \mu(\mathrm{SI})$	$= 3 \times 10^{11}\, \mu(\mathrm{SI})$
Linear polarizability $\alpha(\mathrm{esu})$	$\alpha(\mathrm{esu}) = \dfrac{10^6}{4\pi}\alpha(\mathrm{SI})$	$= 79577\, \alpha(\mathrm{SI})$
First-order hyperpolarizability	$\beta(\mathrm{esu}) = \dfrac{10^2 c}{4\pi}\beta(\mathrm{SI})$	$= \dfrac{3 \times 10^{10}}{4\pi}\beta(\mathrm{SI})$
Second-order hyperpolarizability	$\gamma(\mathrm{esu}) = \dfrac{c^2}{4\pi \times 10^2}\gamma(\mathrm{SI})$	$= \dfrac{9 \times 10^{14}}{4\pi}\gamma(\mathrm{SI})$
nth-order hyperpolarizability[b]	$\alpha^{(n)}(\mathrm{esu}) = \dfrac{(10^{-4}c)^{n-1}}{4\pi \times 10^{-6}}\alpha^{(n)}(\mathrm{SI})$	$= \dfrac{(3 \times 10^4)^{n-1}}{4\pi \times 10^{-6}}\alpha^{(n)}(\mathrm{SI})$

[a] Note that $\chi^{(3)}/\left[\chi^{(2)}\right]^2$ (SI) $= (1/4\pi)\chi^{(3)}/\left[\chi^{(2)}\right]^2$ (esu)
[b] We have e.g. $\alpha^{(2)} = \beta$ and $\alpha^{(3)} = \gamma$.

In practice, the following relations are very useful:

$$\chi^{(2)}(\text{pm/V}) \qquad = 0.4192\,\chi^{(2)}(10^{-9}\,\text{esu}) \qquad 1\,\text{pm/V} \;= 2.386(10^{-9}\,\text{esu})$$

$$\chi^{(3)}(10^{-22}\,\text{m}^2/\text{V}^2) = 1.396\chi^{(3)}(10^{-14}\,\text{esu}) \qquad 10^{-14}\,\text{esu} = 0.7163(10^{-22}\,\text{m}^2/\text{V}^2)$$

$$\mu(10^{-29}\,\text{Cm}) \qquad = \tfrac{1}{3}\mu(\text{D})^{16} \qquad\qquad\;\; 10^{-29}\,\text{Cm} = 3\,\text{D}$$

$$\beta(10^{-40}\,\text{m}^4/\text{V}) \quad = 4.192\beta(10^{-30}\,\text{esu}) \qquad 10^{-30}\,\text{esu} = 2.386(10^{-40}\,\text{m}^4/\text{V})$$

$$\gamma(10^{-48}\,\text{m}^5/\text{V}^2) = 1.396 \times 10^{-2}\gamma(10^{-36}\,\text{esu}) \quad 10^{-36}\,\text{esu} \;= 71.63(10^{-48}\,\text{m}^5/\text{V}^2)$$

B Atomic Units

For theoretical calculations of molecular dipole moments, polarizabilities and hyperpolarizabilities atomic units are frequently used. In this system of units all quantities are expressed as a multiple of an appropriate combination the three quantities $\hbar$, e and m_e which symbolize Planck's constant, the electron charge and mass, respectively. Since the equations defining electric quantities are the same as for the cgs system, the conversion between atomic units and the cgs system can readily be made by inserting the cgs values of the fundamental constants:

$$\hbar = 1.05457 \times 10^{-27}\ \text{erg sec}$$

$$e = 4.80324 \times 10^{-10}\ \text{g}^{1/2}\,\text{cm}^{3/2}\,\text{sec}^{-1}$$

$$m_e = 9.10939 \times 10^{-28}\ \text{g}\,.$$

The unit of length is Bohr's radius:

$$[\ell] = \frac{\hbar^2}{me^2} \equiv a_0 = 0.52918 \times 10^{-8}\,\text{cm} \cong 0.529\,\text{Å}\,.$$

With this definition, most quantities can be expressed with the electron mass and a_0:

---

[16] 1 D = 1 Debye = 10^{-18} esu.

Quantity	Units	1 a.u. = ...	
Energy	$[W] = \dfrac{e^2}{a_0}$	$= 4.360 \times 10^{-11}\,\text{erg}$	$= 4.360 \times 10^{-18}\,\text{J}$ $= 27.2\,\text{eV}$
Electric field strength	$[E] = \dfrac{e}{a_0^2}$	$= 1.715$ $\times 10^7\,\text{g}^{1/2}\,\text{cm}^{-1/2}\,\text{sec}^{-1}$	$= 5.140 \times 10^{11}\,\text{V/m}$
Voltage	$[V] = \dfrac{e}{a_0}$	$= 9.077$ $\times 10^{-2}\,\text{g}^{1/2}\,\text{cm}^{1/2}\,\text{sec}^{-1}$	$= 27.2\,\text{V}$
Dipole moment	$[\mu] = ea_0$	$= 2.542$ $\times 10^{-18}\,\text{g}^{1/2}\,\text{cm}^{5/2}\,\text{sec}^{-1}$ $= 2.542\,\text{D}$	$= 0.847 \times 10^{-29}\,\text{Cm}$
Linear polarizability	$[\alpha] = a_0^3$	$= 1.482 \times 10^{-25}\,\text{cm}^3$	$= 1.862 \times 10^{-30}\,\text{m}^3$
First-order hyperpol.	$[\beta] = \dfrac{a_0^5}{e}$	$= 8.639$ $\times 10^{-33}\,\text{g}^{-1/2}\,\text{cm}^{7/2}\,\text{sec}^1$	$= 3.622$ $\times 10^{-42}\,\text{m}^4/\text{V}$
Second-order hyperpol.	$[\gamma] = \dfrac{a_0^7}{e^2}$	$= 5.037$ $\times 10^{-40}\,\text{g}^{-1}\,\text{cm}^4\,\text{sec}^2$	$= 7.033$ $\times 10^{-54}\,\text{m}^5/\text{V}^2$

There are again some practically useful relations:

$\mu\ (\text{a.u.}) = 0.393\,\mu\ (\text{D}) \qquad = 1.18\,\mu\ (10^{-29}\,\text{C m})$

$\alpha\ (\text{a.u.}) = 67.48\,\alpha\ (10^{-23}\ \text{esu}) = 53.7\,\alpha\ (10^{-28}\,\text{m}^3)$

$\beta\ (\text{a.u.}) = 115.8\,\beta\ (10^{-30}\ \text{esu}) = 27.6\,\beta\ (10^{-40}\,\text{m}^4/\text{V})$

$\gamma\ (\text{a.u.}) = 1985\,\gamma\ (10^{-36}\ \text{esu}) = 142200\,\gamma\ (10^{-48}\,\text{m}^5/\text{V}^2)$

C n_2, $\overline{n_2}$, α_2 and $\chi^{(3)}$

The nonlinear refractive index n_2 is defined as (see also Sect. 2.2)

$$n = n_0 + \overline{n_2}\left\langle E(t)^2 \right\rangle = n_0 + \frac{\overline{n_2}}{2}\left| E(\omega) \right|^2 = n_0 + n_2 I, \qquad (2.88)$$

where n_0 is the linear refractive index and I is the intensity of the incoming beam. In the following we list the appropriate conversion factors between n_2, $\overline{n_2}$, and $\chi^{(3)}$ (always $c = 3 \times 10^8$, except where specially indicated).

$$n_2\left(m^2W^{-1}\right) = \frac{1}{\varepsilon_0 n_0 c}\overline{n}_2\left(m^2V^{-2}\right) \qquad = \frac{10^{-8}}{3\varepsilon_0 n_0}\overline{n}_2\left(m^2V^{-2}\right)$$

$$n_2\left(m^2W^{-1}\right) = \frac{40\pi}{n_0 c}\overline{n}_2\;(esu) \qquad\qquad = \frac{4\pi 10^{-7}}{3n_0}\overline{n}_2\;(esu)$$

$$n_2\left(m^2W^{-1}\right) = \frac{3}{4\varepsilon_0 n_0^2 c}\,Re\!\left(\chi^{(3)}\right)\left(m^2V^{-2}\right) = \frac{10^{-8}}{4\varepsilon_0 n_0^2}\,Re\;\left(\chi^{(3)}\right)\left(m^2V^{-2}\right)$$

$$\overline{n}_2\;(esu) \qquad = \frac{3\pi}{n_0}\,Re\!\left(\chi^{(3)}\right)(esu) \qquad = \frac{3\pi}{n_0}\,Re\!\left(\chi^{(3)}\right)(esu)$$

$$n_2\left(m^2W^{-1}\right) = \frac{120\pi^2}{n_0^2 c}\,Re\!\left(\chi^{(3)}\right)(esu) \qquad = \frac{40\pi^2 10^{-8}}{n_0^2}\,Re\!\left(\chi^{(3)}\right)(esu)$$

$$\alpha_2\;(mW^{-1}) \quad = \frac{3\pi}{\varepsilon_0 n_0^2 c\lambda}\,Im\!\left(\chi^{(3)}\right)\left(m^2V^{-2}\right) = \frac{10^{-8}\pi}{\varepsilon_0 n_0^2 \lambda}\,Im\!\left(\chi^{(3)}\right)\left(m^2V^{-2}\right)$$

$$\alpha_2\;(mW^{-1}) \quad = \frac{480\pi^3}{n_0^2 c\lambda}\,Im\!\left(\chi^{(3)}\right)(esu) \qquad = \frac{16\times 10^{-9}\pi^3}{n_0^2 \lambda}\,Im\!\left(\chi^{(3)}\right)(esu)$$

Again there are some practically useful relations:

$$n_2\left(10^{-16}\,cm^2\,W^{-1}\right) = \frac{3.765}{n_0}\overline{n}_2\left(10^{-22}\,m^2V^{-2}\right)$$

$$n_2\left(10^{-16}\,cm^2\,W^{-1}\right) = \frac{4.189}{n_0}\overline{n}_2\left(10^{-13}\,esu\right)$$

$$n_2\left(10^{-16}\,cm^2\,W^{-1}\right) = \frac{2.824}{n_0^2}\,Re\left(\chi^{(3)}\right)\left(10^{-22}\,m^2V^{-2}\right)$$

$$\overline{n}_2\left(10^{-13}\,esu\right) \qquad = \frac{0.9425}{n_0}\,Re\left(\chi^{(3)}\right)\left(10^{-14}\,esu\right)$$

$$n_2\left(10^{-16}\,cm^2\,W^{-1}\right) = \frac{3.948}{n_0^2}\,Re\left(\chi^{(3)}\right)\left(10^{-14}\,esu\right)$$

$$\alpha_2\;\left(cm\,GW^{-1}\right) \quad = \frac{35.48}{n_0^2\lambda}\,Im\left(\chi^{(3)}\right)\left(10^{-22}\,m^2V^{-2}\right) \quad (\lambda\;in\;nm)$$

$$\alpha_2\;(cm\,GW^{-1}) \quad = \frac{49.61}{n_0^2\lambda}\,Im\left(\chi^{(3)}\right)\left(10^{-14}\,esu\right) \qquad (\lambda\;in\;nm)$$

$$n_2\left(m^2\,W^{-1}\right) \qquad = 10^3\,n_2\;(esu)$$

$$SI:\;\;I = \frac{1}{2}\varepsilon_0 n_0 c\,|E|^2 \quad (Wm^{-2}) \qquad E\;in\;V/m,\;c = 3\times 10^8\,m/s$$

$$esu:\;I = \frac{n_0 c}{8\pi}\,|E|^2 \quad (esu) \qquad\qquad E\;in\;statvolt/cm,\;c = 3\times 10^{10}\,cm/s$$

A.3   Theoretical Description of Third-Harmonic Generation

A   Third-Harmonic Generation in Solution

By solving the wave equation and applying the electromagnetic boundary conditions, the field at the third-harmonic can be calculated. We assume that the fundamental and the third-harmonic waves are both s-polarized (e.g. along the z-axis) and that the sample is rotated around the same axis (Fig. 2.54).

In this case the total field at frequency 3ω after the sample is given by

$$|E_{\text{tot}}|^2 = (\text{Re } E)^2 + (\text{Im } E)^2 \tag{2.147}$$

with

$$\text{Re } E = \begin{bmatrix} (E_0 + E_1) \times \cos\left(k_G^{3\omega} L_{G1} + k_L^{3\omega} L_L + k_G^{3\omega} L_{G2}\right) \\ + (E_2 + E_3) \times \cos\left(k_G^{\omega} L_{G1} + k_L^{3\omega} L_L + k_G^{3\omega} L_{G2}\right) \\ + (E_4 + E_5) \times \cos\left(k_G^{\omega} L_{G1} + k_L^{\omega} L_L + k_G^{3\omega} L_{G2}\right) \\ + (E_6 + E_7) \times \cos\left(k_G^{\omega} L_{G1} + k_L^{\omega} L_L + k_G^{\omega} L_{G2}\right) \end{bmatrix} \times (E^{\omega})^3 \tag{2.148}$$

$$\text{Im } E = \begin{bmatrix} (E_0 + E_1) \times \sin\left(k_G^{3\omega} L_{G1} + k_L^{3\omega} L_L + k_G^{3\omega} L_{G2}\right) \\ + (E_2 + E_3) \times \sin\left(k_G^{\omega} L_{G1} + k_L^{3\omega} L_L + k_G^{3\omega} L_{G2}\right) \\ + (E_4 + E_5) \times \sin\left(k_G^{\omega} L_{G1} + k_L^{\omega} L_L + k_G^{3\omega} L_{G2}\right) \\ + (E_6 + E_7) \times \sin\left(k_G^{\omega} L_{G1} + k_L^{\omega} L_L + k_G^{\omega} L_{G2}\right) \end{bmatrix} \times (E^{\omega})^3 \ , \tag{2.149}$$

where E^{ω} is the external field at frequency ω. The wave vectors are written as

$$k_G^{\omega} = \frac{6\pi}{\lambda} n_G^{\omega} \cos\theta_G^{\omega} \qquad k_G^{3\omega} = \frac{6\pi}{\lambda} n_G^{3\omega} \cos\theta_G^{3\omega} \tag{2.150}$$

$$k_L^{\omega} = \frac{6\pi}{\lambda} n_L^{\omega} \cos\theta_L^{\omega} \qquad k_L^{3\omega} = \frac{6\pi}{\lambda} n_L^{3\omega} \cos\theta_L^{3\omega} \ , \tag{2.151}$$

where n_G (n_L) is the refractive index of the glass (liquid), L_{G1} and L_{G2} are the thicknesses of the two glass walls, L_L is the thickness of the liquid compartment, and θ_G (θ_L) is the angle inside the medium (glass or liquid). The factors E_0–E_7 containing the transmission factors, the factors resulting from the electromagnetic boundary conditions, and the nonlinear optical

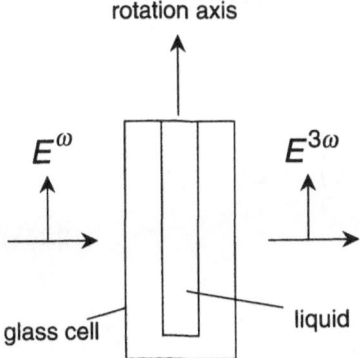

Fig. 2.54. Schematic of the experimental sample geometry. The sample is in air

susceptibilities are given by

$$E_0 = t_{G/L}^{3\omega} \times t_{L/G}^{3\omega} \times t_{G/A}^{3\omega} \times t_{A/G}^{3\omega} \times \chi_{\text{air}}^{(3)} \times \exp\left(-\frac{3\alpha^{3\omega} L_L}{2\cos\theta_L^{\omega}}\right) \tag{2.152}$$

$$E_1 = -t_{G/L}^{3\omega} \times t_{L/G}^{3\omega} \times t_{G/A}^{3\omega} \times b_{A/G} \times \left(t_{A/G}^{\omega}\right)^3 \times \frac{\chi_G^{(3)}/4}{(n_G^{\omega})^2 - (n_G^{3\omega})^2}$$
$$\times \exp\left(-\frac{3\alpha^{3\omega} L_L}{2\cos\theta_L^{\omega}}\right) \tag{2.153}$$

$$E_2 = t_{L/G}^{3\omega} \times t_{G/A}^{3\omega} \times f_{G/L} \times \left(t_{A/G}^{\omega}\right)^3 \times \frac{\chi_G^{(3)}/4}{(n_G^{\omega})^2 - (n_G^{3\omega})^2}$$
$$\times \exp\left(-\frac{3\alpha^{3\omega} L_L}{2\cos\theta_L^{\omega}}\right) \tag{2.154}$$

$$E_3 = -t_{L/G}^{3\omega} \times t_{G/A}^{3\omega} \times b_{G/L} \times \left(t_{A/G}^{\omega} \times t_{G/L}^{\omega}\right)^3 \times \frac{\chi_L^{(3)}/4}{(n_L^{\omega})^2 - (n_L^{3\omega})^2}$$
$$\times \exp\left(-\frac{3\alpha^{3\omega} L_L}{2\cos\theta_L^{\omega}}\right) \tag{2.155}$$

$$E_4 = t_{G/A}^{3\omega} \times f_{L/G} \times \left(t_{A/G}^{\omega} \times t_{G/L}^{\omega}\right)^3 \times \frac{\chi_L^{(3)}/4}{(n_L^{\omega})^2 - (n_L^{3\omega})^2}$$
$$\times \exp\left(-\frac{3\alpha^{\omega} L_L}{2\cos\theta_L^{\omega}}\right) \tag{2.156}$$

$$E_5 = -t_{G/A}^{3\omega} \times b_{L/G} \times \left(t_{A/G}^{\omega} \times t_{G/L}^{\omega} \times t_{L/G}^{\omega}\right)^3 \times \frac{\chi_G^{(3)}/4}{(n_G^{\omega})^2 - (n_G^{3\omega})^2}$$
$$\times \exp\left(-\frac{3\alpha^{\omega} L_L}{2\cos\theta_L^{\omega}}\right) \tag{2.157}$$

$$E_6 = f_{G/A} \times \left(t_{A/G}^{\omega} \times t_{G/L}^{\omega} \times t_{L/G}^{\omega}\right)^3 \times \frac{\chi_G^{(3)}/4}{(n_G^{\omega})^2 - (n_G^{3\omega})^2}$$
$$\times \exp\left(-\frac{3\alpha^{\omega} L_L}{2\cos\theta_L^{\omega}}\right) \tag{2.158}$$

$$E_7 = -b_{G/A} \times \left(t_{A/G}^{\omega} \times t_{G/L}^{\omega} \times t_{L/G}^{\omega} \times t_{G/A}^{\omega}\right)^3 \times \chi_{\text{air}}^{(3)}$$
$$\times \exp\left(-\frac{3\alpha^{\omega} L_L}{2\cos\theta_L^{\omega}}\right). \tag{2.158}$$

The transmission factors for the fundamental and third-harmonic waves are (θ: external angle of incidence)

$$t^\omega_{A/G} = \frac{2\cos\theta}{n^\omega_G \cos\theta^\omega_G + \cos\theta} \qquad t^\omega_{G/L} = \frac{2n^\omega_G \cos\theta^\omega_G}{n^\omega_G \cos\theta^\omega_G + n^\omega_L \cos\theta^\omega_L}$$

$$(2.159)$$

$$t^\omega_{L/G} = \frac{2n^\omega_L \cos\theta^\omega_L}{n^\omega_G \cos\theta^\omega_G + n^\omega_L \cos\theta^\omega_L} \qquad t^\omega_{G/A} = \frac{2n^\omega_G \cos\theta^\omega_G}{n^\omega_G \cos\theta^\omega_G + \cos\theta}$$

$$(2.160)$$

$$t^{3\omega}_{A/G} = \frac{2\cos\theta}{n^{3\omega}_G \cos\theta^{3\omega}_G + \cos\theta} \qquad t^{3\omega}_{G/L} = \frac{2n^{3\omega}_G \cos\theta^{3\omega}_G}{n^{3\omega}_G \cos\theta^{3\omega}_G + n^{3\omega}_L \cos\theta^{3\omega}_L}$$

$$(2.161)$$

$$t^{3\omega}_{L/G} = \frac{2n^{3\omega}_L \cos\theta^{3\omega}_L}{n^{3\omega}_G \cos\theta^{3\omega}_G + n^{3\omega}_L \cos\theta^{3\omega}_L} \qquad t^{3\omega}_{G/A} = \frac{2n^{3\omega}_G \cos\theta^{3\omega}_G}{n^{3\omega}_G \cos\theta^{3\omega}_G + \cos\theta}. \qquad (2.162)$$

The factors f and b result from the electromagnetic boundary conditions at the different interfaces:

$$f_{G/L} = \frac{n^\omega_G \cos\theta^\omega_G + n^{3\omega}_G \cos\theta^{3\omega}_G}{n^{3\omega}_G \cos\theta^{3\omega}_G + n^{3\omega}_L \cos\theta^{3\omega}_L} \qquad f_{L/G} = \frac{n^\omega_L \cos\theta^\omega_L + n^{3\omega}_L \cos\theta^{3\omega}_L}{n^{3\omega}_G \cos\theta^{3\omega}_G + n^{3\omega}_L \cos\theta^{3\omega}_L}$$

$$(2.163)$$

$$f_{G/A} = \frac{n^\omega_G \cos\theta^\omega_G + n^{3\omega}_G \cos\theta^{3\omega}_G}{n^{3\omega}_G \cos\theta^{3\omega}_G + \cos\theta} \qquad (2.164)$$

$$b_{A/G} = \frac{n^\omega_G \cos\theta^\omega_G + \cos\theta}{n^{3\omega}_G \cos\theta^{3\omega}_G + \cos\theta} \qquad b_{G/L} = \frac{n^\omega_L \cos\theta^\omega_L + n^{3\omega}_G \cos\theta^{3\omega}_G}{n^{3\omega}_G \cos\theta^{3\omega}_G + n^{3\omega}_L \cos\theta^{3\omega}_L}$$

$$(2.165)$$

$$b_{L/G} = \frac{n^\omega_G \cos\theta^\omega_G + n^{3\omega}_L \cos\theta^{3\omega}_L}{n^{3\omega}_G \cos\theta^{3\omega}_G + n^{3\omega}_L \cos\theta^{3\omega}_L}. \qquad (2.166)$$

B   Third-Harmonic Generation in Single Crystals

We again assume that the fundamental, the second-harmonic (in the case of cascading), and the third-harmonic waves are all s-polarized (e.g. along the z-axis) and that the sample is rotated around the same axis (Fig. 2.55).[17] The nonlinear optical susceptibilities are defined as in Table 2.21 where $\chi^{(2)}$ ($\chi^{(2)\prime}$) represents the process of frequency doubling (sum frequency generation).

---

[17] The theory of [2.7] is here extended to Maker–Fringe experiments.

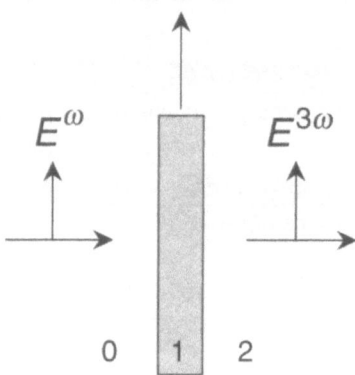

rotation axis

E^ω $E^{3\omega}$

0   1   2

Fig. 2.55. Schematic of the experimental sample geometry (0, 2: air; 1: crystal)

In this case the total field at frequency 3ω after the sample is given by

$$
\begin{aligned}
|E_{\text{tot}}|^2 ={}& C_{\text{air}} + (E_{b1} - E_{b2})^2 + (E'_{b1} - E'_{b2})^2 + 2\,(E'_{b1} - E'_{b2})\,(E_{b1} - E_{b2}) \\
&+ 4 E_{b1}\,[E_{b2} + E'_{b2}]\sin^2\left(\frac{\Delta k^{3\omega} L}{2}\right) \\
&+ 4 E'_{b1}\,[E_{b2} + E'_{b2}]\sin^2\left(\frac{\Delta k^{321\omega} L}{2}\right) \\
&- 4 E_{b1} E'_{b1}\sin^2\left(\frac{\Delta k^{3\omega\prime} L}{2}\right)\,.
\end{aligned}
\tag{2.167}
$$

The different quantities are described in the following. The contribution from air is contained in

$$
\begin{aligned}
C_{\text{air}} ={}& (E_{\text{air\_1}} - E_{\text{air\_2}})^2 - 2\,(E_{\text{air\_1}} - E_{\text{air\_2}})\,(E_{b2} + E'_{b2}) \\
&+ 2\,(E_{b1} + E'_{b1})\,(E_{\text{air\_1}} - E_{\text{air\_2}}) \\
&- 4\sin^2\left(\frac{\Delta k^{3\omega} L}{2}\right)[E_{b1} E_{\text{air\_1}} + E_{b2} E_{\text{air\_2}} \\
&\hspace{3.5cm} + E'_{b2} E_{\text{air\_2}} - E_{\text{air\_1}} E_{\text{air\_2}}] \\
&- 4\sin^2\left(\frac{\Delta k^{321\omega} L}{2}\right) E'_{b1} E_{\text{air\_1}} + 4\sin^2\left(\frac{\Delta k^{3\omega\prime} L}{2}\right) E'_{b1} E_{\text{air\_2}}\,.
\end{aligned}
\tag{2.168}
$$

The wave vector mismatches are written as

$$
\begin{aligned}
\Delta k^{3\omega} &= k_b^{3\omega}\cos\theta^\omega - k^{3\omega}\cos\theta^{3\omega} = \frac{6\pi}{\lambda}\left(n^\omega\cos\theta^\omega - n^{3\omega}\cos\theta^{3\omega}\right) \\
\Delta k^{3\omega\prime} &= k_b^{3\omega}\cos\theta^\omega - k^{2\omega}\cos\theta^{2\omega} - k^\omega\cos\theta^\omega \\
&= \frac{4\pi}{\lambda}\left(n^\omega\cos\theta^\omega - n^{2\omega}\cos\theta^{2\omega}\right) \\
\Delta k^{321\omega} &= k^{3\omega}\cos\theta^{3\omega} - k^{2\omega}\cos\theta^{2\omega} - k^\omega\cos\theta^\omega \\
&= \frac{2\pi}{\lambda}\left(3n^{3\omega}\cos\theta^{3\omega} - 2n^{2\omega}\cos\theta^{2\omega} - n^\omega\cos\theta^\omega\right)\,,
\end{aligned}
\tag{2.169}
$$

where n is the refractive index of the crystal, L is the sample thickness, and θ^ω $(\theta^{2\omega}, \theta^{3\omega})$ is the internal angle of incidence of the wave with the appropriate frequency. The further quantities are

$$E_{\text{air\_1}} = t_{0/1}^{3\omega} t_{1/2}^{3\omega} \chi_{\text{air}}^{(3)} \qquad E_{\text{air\_2}} = \left(t_{0/1}^{\omega} t_{1/2}^{\omega}\right)^3 \chi_{\text{air}}^{(3)} \tag{2.170}$$

$$E_{b1} = T_1^{1/2} E_b^{3\omega} \qquad E_{b2} = t_{1/2}^{3\omega} T_2^{0/1} E_b^{3\omega} \tag{2.171}$$

$$E_{b1}' = \left(T_1^{1/2}\right)' E_b^{3\omega'} \qquad E_{b2}' = t_{1/2}^{3\omega} \left(T_2^{0/1}\right)' E_b^{3\omega'} . \tag{2.172}$$

In the case of cascading through the local field induced by transverse waves at the second-harmonic frequency we have (where E^ω is the external field at frequency ω)

$$E_b^{3\omega} = \left(t_{0/1}^\omega\right)^3 \frac{1}{(n^\omega)^2 - (n^{3\omega})^2} \times \left(\frac{1}{4}\chi^{(3)} + \frac{1}{2}\chi^{(2)} \left(\chi^{(2)}\right)' \frac{1}{(n^\omega)^2 - (n^{2\omega})^2}\right.$$
$$\left. + \frac{1}{2}\chi^{(2)} \left(\chi^{(2)}\right)' \varepsilon_0 \frac{L^{2\omega}}{f^{2\omega}}\right) (E^\omega)^3 . \tag{2.173}$$

In the case of cascading through the local field induced by a longitudinal field at frequency 2ω and no macroscopic cascading (as in the case of a c-plate of KNbO$_3$, see Sect. 2.5) we have

$$E_b^{3\omega} = \left(t_{0/1}^\omega\right)^3 \frac{1}{(n^\omega)^2 - (n^{3\omega})^2} \times \left(\frac{1}{4}\chi^{(3)} - \frac{1}{2}\chi_{\text{long}}^{(2)} \left(\chi_{\text{long}}^{(2)}\right)' \frac{1}{\left(n_{\text{long}}^{2\omega}\right)^2}\right.$$
$$\left. + \frac{1}{2}\chi_{\text{long}}^{(2)} \left(\chi_{\text{long}}^{(2)}\right)' \varepsilon_0 \frac{L^{2\omega}}{f_{\text{long}}^{2\omega}}\right) (E^\omega)^3 . \tag{2.174}$$

Here $\chi_{\text{long}}^{(2)}$ represents the second-harmonic that is generated along the propagation direction of the fundamental (e.g. $\chi_{311}^{(2)}$ for perpendicular incidence on a c-plate of KNbO$_3$ with light polarized along a). Note that in this case the field generated at 2ω does not propagate itself. Likewise n_{long} is the refractive index along the propagation direction

$$L = \frac{1}{3\varepsilon_0} \qquad f^{2\omega} = \frac{(n^{2\omega})^2 + 2}{3} \qquad f_{\text{long}}^{2\omega} = \frac{\left(n_{\text{long}}^{2\omega}\right)^2 + 2}{3} . \tag{2.175}$$

$E_b^{3\omega'}$ is given by

$$E_b^{3\omega'} = -\left(t_{0/1}^\omega\right)^3 \frac{9}{(n^\omega)^2 + 4(n^{2\omega})^2 + 4n^\omega n^{2\omega}\cos(\theta^{2\omega} - \theta^\omega) - 9(n^{3\omega})^2}$$
$$\times T_{vv}\frac{1}{2}\chi^{(2)'}\chi^{(2)} \frac{1}{(n^\omega)^2 - (n^{2\omega})^2} (E^\omega)^3 . \tag{2.176}$$

The transmission factors for the fundamental and third-harmonic waves are given by (θ: external angle of incidence)

$$t^{\omega}_{0/1} = \frac{2\cos\theta}{n^{\omega}\cos\theta^{\omega} + \cos\theta} \qquad t^{\omega}_{1/0} = \frac{2n^{\omega}\cos\theta^{\omega}}{n^{\omega}\cos\theta^{\omega} + \cos\theta} \qquad (2.177)$$

$$t^{3\omega}_{0/1} = \frac{2\cos\theta}{n^{3\omega}\cos\theta^{3\omega} + \cos\theta} \qquad t^{3\omega}_{1/0} = \frac{2n^{3\omega}\cos\theta^{3\omega}}{n^{3\omega}\cos\theta^{3\omega} + \cos\theta} . \qquad (2.178)$$

The factors resulting from the electromagnetic boundary conditions at the different interfaces are expressed as

$$T_{vv} = \frac{\cos\theta + n^{\omega}\cos\theta^{\omega}}{\cos\theta + n^{2\omega}\cos\theta^{2\omega}} \qquad (2.179)$$

$$T^{0/1}_2 = \frac{n^{\omega}\cos\theta^{\omega} + \cos\theta}{n^{3\omega}\cos\theta^{3\omega} + \cos\theta}$$

$$\left(T^{0/1}_2\right)' = \frac{\frac{1}{3}\left(2n^{2\omega}\cos\theta^{2\omega} + n^{\omega}\cos\theta^{\omega}\right) + \cos\theta}{n^{3\omega}\cos\theta^{3\omega} + \cos\theta} \qquad (2.180)$$

$$T^{1/2}_1 = \frac{n^{3\omega}\cos\theta^{3\omega} + n^{\omega}\cos\theta^{\omega}}{n^{3\omega}\cos\theta^{3\omega} + \cos\theta}$$

$$\left(T^{1/2}_1\right)' = \frac{\frac{1}{3}\left(2n^{2\omega}\cos\theta^{2\omega} + n^{\omega}\cos\theta^{\omega}\right) + n^{3\omega}\cos\theta^{3\omega}}{n^{3\omega}\cos\theta^{3\omega} + \cos\theta} . \qquad (2.181)$$

Incoherent multiple reflections were taken into account in the form of

$$|E_{\text{tot}}|^2_{\text{final}} = R_{\text{mult}} \times |E_{\text{tot}}|^2$$

$$= \left(\frac{1}{1 - (r^{3\omega})^4}\right) \times \left(\frac{1}{1 - (r^{\omega})^{12}}\right) \times \left(1 + (r^{3\omega})^2 (r^{\omega})^6\right) \times |E_{\text{tot}}|^2 \qquad (2.182)$$

with

$$r^{\omega} = t^{\omega}_{0/1} - 1 \qquad r^{3\omega} = t^{3\omega}_{0/1} - 1. \qquad (2.183)$$

This correction reduced the resulting value of $\chi^{(3)}$ by about 1%. However, it did not influence the ratio $\chi^{(3)}/\left(\chi^{(2)}\right)^2$.

A.4   Theoretical Description
of Electric Field-Induced Second-Harmonic Generation

With the type of configuration introduced in Sect. 2.4.2.B the generated second-harmonic intensity also contains contributions from the induced glass nonlinearity Γ_G and can be expressed as [2.38]

$$I^{2\omega}_{\text{L}} = \frac{128}{\varepsilon_0 c\lambda^2} \eta_{\text{L}} \left[T_G\Gamma_G l_{c;G} - T_L\Gamma_L l_{c;L}\right]^2 \left(E^0\right)^2 \left(I^{\omega}\right)^2 f(L), \qquad (2.184)$$

where I^ω is the intensity of the fundamental beam and $l_{c;G}$ and $l_{c;L}$ are the coherence lengths of glass and of the liquid:

$$l_{c;L} = \frac{\lambda}{4(n_L^{2\omega} - n_L^\omega)}, \quad l_{c;G} = \frac{\lambda}{4(n_G^{2\omega} - n_G^\omega)} . \tag{2.185}$$

The factor $\eta_L = (t^\omega)^4 (t^{2\omega})^2$ contains Fresnel transmission factors of the fundamental and the second-harmonic wave at the air/glass and the glass/air interface, respectively, and the factors T_G and T_L result from the electromagnetic boundary conditions [2.38] at the glass/liquid and liquid/glass interfaces:

$$t^\omega = \frac{2}{1 + n_G^\omega} \qquad\qquad t^{2\omega} = \frac{2n_G^{2\omega}}{1 + n_G^{2\omega}} \tag{2.186}$$

$$T_G = \frac{n_L^{2\omega} + n_G^\omega}{(n_L^{2\omega} + n_G^{2\omega})(n_G^{2\omega} + n_G^\omega)} \qquad T_L = \frac{1}{(n_L^{2\omega} + n_G^{2\omega})} \left(\frac{2n_G^\omega}{n_L^\omega + n_G^\omega} \right)^2 . \tag{2.187}$$

E_0 is the static field strength, n_G^ω ($n_G^{2\omega}$) is the refractive index of the glass at the fundamental (second-harmonic), and n_L^ω ($n_L^{2\omega}$) is the refractive index of the liquid at the fundamental (second-harmonic). Γ_G (Γ_L) is the glass (liquid) nonlinearity.

The dependence of the second-harmonic intensity on the length L of the liquid path is given by the factor

$$f(L) = \frac{1}{2} \exp\left[-\left(\alpha^\omega + \frac{\alpha^{2\omega}}{2} \right) \right]$$
$$\times \left\{ \cosh\left[-\left(\alpha^\omega + \frac{\alpha^{2\omega}}{2} \right) \right] - \cos\left(\frac{\pi L}{l_{c;L}} \right) \right\}, \tag{2.188}$$

where α^ω and $\alpha^{2\omega}$ are the absorption coefficients at the corresponding wavelengths. In the case of negligible absorption ($\alpha^\omega = \alpha^{2\omega} = 0$) the factor $f(L)$ reduces to

$$f(L) = \sin^2\left(\frac{\pi L}{2 l_{c;L}} \right) . \tag{2.189}$$

From the experimental curves the amplitude of the fringes and the fringe spacing are determined. The coherence length $l_{c;L}$ is related to the separation of the fringe minima Δx via

$$l_{c;L} = \frac{\Delta x}{2} \sin\alpha \tag{2.190}$$

where α is the wedge angle of the liquid cell.

A.5   Theoretical Description of the z-Scan Technique

The field given by (2.144) has to propagate to the aperture. We here describe two possible ways to calculate this propagation. First, by means of Fresnel integration [2.241] and, second, by means of a decomposition of the field into Gaussian beams that individually propagate to the aperture and which are summed up there [2.166].

A   Fresnel Integration

In this case the field at the aperture is given by $(r = r_s)$

$$E_A(z, r_a, t) = \frac{k}{d_a} \int_0^\infty dr_s E_e(z, r_s, t) r_s J_0\left(\frac{k r_s r_a}{d_a}\right) \exp\left\{i\frac{k}{2 d_a} r_s^2\right\}, \quad (2.191)$$

where r_s is the radial dimension perpendicular to the propagation direction, d_a is the distance between the sample output and the aperture, and J_0 is the zeroth-order Bessel function. The energy transmitted by the aperture is then given by

$$I_a(z) = \int_{-\infty}^\infty dt \int_0^{r_a} dr_s 2\pi r_s |E_A(z, r_s, t)|^2, \quad (2.192)$$

where r_a is the radius of the aperture.

B   Gaussian Decomposition

In this case the field at the aperture is written as

$$E_a(r, t) = E(z, r = 0, t) e^{-\frac{\alpha L}{2}} \sum_{m=0}^\infty \frac{(i\Delta\phi_0(z, t))^m}{m!} \prod_{n=1}^m \left(1 + i(2n - 1)\frac{\alpha_2 \lambda}{4\pi n_2}\right)$$

$$\times \frac{w_{m0}}{w_m} \exp\left(-\frac{r^2}{w_m^2} - \frac{ikr^2}{2R_m} + i\Theta_m\right) \quad (2.193)$$

with

$$\Delta\phi_0(z, t) = \frac{\Delta\phi_0(t)}{1 + z^2/z_0^2} \qquad d_m = \frac{k w_{m0}^2}{2}$$

$$\Delta\phi_0(t) = k\Delta n_0(t) \qquad w_m^2 = w_{m0}^2\left(g^2 + \frac{d^2}{d_m^2}\right)$$

$$L_{\text{eff}} = \left(1 - e^{\alpha L}\right)/\alpha \qquad R_m = d\left(1 - \frac{g}{g^2 + d^2/d_m^2}\right)^{-1} \quad (2.194)$$

$$g = 1 + \frac{d}{R(z)} \qquad \Theta_m = \tan^{-1}\left(\frac{d/d_m}{g}\right)$$

$$w_{m0}^2 = \frac{w^2(z)}{2m + 1} \qquad w^2(z) = w_0^2\left(1 + \frac{z^2}{z_0^2}\right).$$

The sum in (2.193) converges for $\alpha_2 I L_{\text{eff}} < 1$. The intensity at the aperture is then

$$I_a(r, t) = \frac{1}{2}\varepsilon_0 c E_a(r, t) E_a^*(r, t). \quad (2.195)$$

The power is

$$P_a(t) = 2\pi \int_0^{r_a} r I_a(r,t) \mathrm{d}r. \tag{2.196}$$

This leads to the normalized transmission

$$T(z) = \frac{\int_{-\infty}^{\infty} P_a(t)\mathrm{d}t}{S \int_{-\infty}^{\infty} P_{in}(t)\mathrm{d}t} \quad \text{with} \quad S = 1 - \mathrm{e}^{-2r_a^2/w^2}. \tag{2.197}$$

$P_{in}(t)$ is the incident power and S is the part of the energy that passes the aperture in the case that there are no nonlinearities (sample far from focal point). Carrying out the integration (assuming a Gaussian pulse in time) leads to

$$T(z) = \frac{2\mathrm{e}^{-\alpha L}}{S \pi w_0^2 (1 + z^2/z_0^2)} \sum_{m=0}^{\infty} \sum_{p=0}^{\infty} \sqrt{\frac{1}{m+p+1}} \frac{(\mathrm{i}\Delta\phi_0(z))^m}{m!} \frac{(-\mathrm{i}\Delta\phi_0(z))^p}{p!}$$

$$\times \frac{w_{m0} w_{p0}}{w_m w_p} \prod_{n=1}^{m} \left(1 + \mathrm{i}(2n-1)\frac{\alpha_2 \lambda}{4\pi n_2}\right)$$

$$\times \prod_{v=1}^{p} \left(1 - \mathrm{i}(2v-1)\frac{\alpha_2 \lambda}{4\pi n_2}\right) \frac{\pi}{\left(\frac{1}{w_m^2} + \frac{1}{w_p^2} + \mathrm{i}\left(\frac{\pi}{\lambda R_m} - \frac{\pi}{\lambda R_p}\right)\right)}$$

$$\times \left(1 - \exp\left\{-a^2 \left(\frac{1}{w_m^2} + \frac{1}{w_p^2} + \mathrm{i}\left(\frac{\pi}{\lambda R_m} - \frac{\pi}{\lambda R_p}\right)\right)\right\}\right). \tag{2.198}$$

Remark: we assume that the laser pulses are Gaussian both in time and space (TEM$_{o,o}$):

$$I(r,t) = I_0(z)\exp\left(-\frac{t^2}{\tau^2}\right)\exp\left(-\frac{2r^2}{w(z)^2}\right), \tag{2.199}$$

where τ is the temporal half-width at $1/e$ of the maximum intensity, $w(z)$ is the beam radius at $1/e^2$ of the maximum intensity, and z is the distance from the beam waist. In the focus the energy of a single pulse is expressed as

$$E = 2\pi I_0 \int_{-\infty}^{\infty} \mathrm{d}t \exp\left(-\frac{t^2}{\tau^2}\right) \int_0^{\infty} \mathrm{d}r\, r \exp\left(-\frac{2r^2}{w_0^2}\right) = \frac{1}{2}\pi^{3/2} w_0^2 \tau I_0. \tag{2.200}$$

A.6   List of Symbols and Abbreviations

A   Constants

c	velocity of light	2.99792458×10^8 ms^{-1}
ε_0	vacuum permeability	8.85418×10^{-12} As V^{-1}m^{-1}
h	Planck's constant	6.6262×10^{-34} Js
$\hbar$	$h/2\pi$	1.0546×10^{-34} Js
N_A	Avogadro's number	6.022×10^{23} mol^{-1}

B Important Symbols

α	linear absorption coefficient
α_2	two-photon absorption coefficient
α_3	three-photon absorption coefficient
α_{ij}	molecular first-order polarizability
β_{ijk}	molecular first-order hyperpolarizability (or second-order polarizability)
β_z	vector part of β_{ijk} along the direction of the dipole moment
β_{zzz}	component of β_{ijk} along the charge transfer direction
β^{MOPAC}	molecular first-order hyperpolarizability as calculated with MOPAC
β_0	static value of the hyperpolarizability
c	velocity of light
C	concentration
$\Delta\phi^{\text{NL}}$	nonlinear phase shift
Δk	wave vector mismatch
d_{ijk}	second-order susceptibility (describing second-harmonic generation)
d_{eff}	effective nonlinear optical coefficient
$E^\omega, E^{2\omega}$	Fourier components of the electric field strength
E^0	static electric field
$\mathcal{E}$	molar extinction coefficient
ε	dielectric constant
f	local field factor
FM	figure of merit
γ_{ijkl}	molecular second-order hyperpolarizability
$\overline{\gamma}$	scalar part of the second-order hyperpolarizability
γ	rotational average of γ_{ijkl}
Γ	third-order susceptibility
I	light intensity
k	wave vector in vacuum ($= 2\pi/\lambda$)
M	molar weight
μ	dipole moment
μ_{eg}	transition dipole moment
$\Delta\mu$	difference between the excited and the ground state dipole moments
n	linear refractive index
n_2	nonlinear refractive index
N	number density of molecules
λ	wavelength
λ_{eg}	wavelength of maximum absorption
l_{c}	coherence length
ω	frequency of light

ω_{eg}	resonance frequency
P	macroscopic polarization
$p^{2\omega}, p^{3\omega}$	Fourier components of the molecular polarisation
r_{ijk}	linear electro-optic coefficient
R_{ijkl}	quadratic electro-optic coefficient
ρ	density
T	figure of merit for all-optical switching (determined by two-photon absorption)
$\chi^{(n)}$	nth order susceptibility
V	figure of merit for all-optical switching (determined by three-photon absorption)
V_π	half-wave voltage
w	weight fraction
W	figure of merit for all-optical switching (determined by linear absorption)
ω_0	resonance frequency

C Abbreviations and Names

MOPAC	Semiempirical quantum chemistry program package
PM3	Parametrization method 3
TEE	tetraethynylethenes

D Materials

DAN	4-(N,N-dimethylamino)-3-acetamidonitrobenzene
DANPH	4-dimethylaminobenzaldehyde-4-nitrophenylhydrazone
DANS	4-dimethylamino-4'-nitrostilbene
DAST	4'-dimethylamino-N-methyl-4-stilbazolium tosylate
DR1	Disperse Red 1
Mero-MDB	4-{2-[1-(2-hydroxyethyl)-4-pyridylidene]-ethylidene}-cyclo-hexa-2,5-dien-1-one·methyl 2,4-dihydroxybenzoate
MNA	2-methyl-4-nitroaniline
PNP	2-(N-prolinol)-5-nitropyridine
PPV	poly(p-phenylenevinylene)
PTS	poly[2,4-hexadiyne-1,6-diol-bis-p-toluene-sulfonate]

References

2.1 D. S. Chemla, J. Zyss, "Nonlinear Optical Properties of Organic Molecules and Crystals", in *Quantum Electronics – Principles and Applications*, edited by P. F. Liao and P. Kelley (Academic Press, Inc., Orlando, 1987), Vol. 2.

2.2 G. I. Stegeman, A. Miller, in *Photonics in Switching, Vol. I, Background and Components*, edited by J. E. Midwinter, Quantum Electronics. Principles and Applications Vol. I (Academic Press, Inc., Boston, 1993), p. 81.

2.3 C. Sauteret, J.-P. Hermann, R. Frey, F. Pradere, J. Ducuing, R. H. Baughman, R. R. Chance, Phys. Rev. Lett. **36**, 956 (1976).

2.4 B. Lawrence, W. E. Torruellas, M. Cha, M. L. Sundheimer, G. I. Stegeman, J. Meth, S. Etemad, G. Baker, Phys. Rev. Lett. **73**, 597 (1994).

2.5 C. Flytzanis, N. Bloembergen, Prog. Quant. Electr. **4**, 271 (1976).

2.6 E. Yablonovitch, C. Flytzanis, N. Bloembergen, Phys. Rev. Lett. **29**, 865 (1972).

2.7 G. R. Meredith, Phys. Rev. B **24**, 5522 (1981).

2.8 L. A. Ostrovskii, JETP Lett. **5**, 272 (1967).

2.9 D. N. Klyshko, B. F. Polkovnikov, Sov. J. Quant. Electron. **3**, 324 (1974).

2.10 J.-M. R. Thomas, J.-P. E. Taran, Opt. Commun. **4**, 329 (1972).

2.11 S. A. Akhmanov, A. I. Kovrygin, A. P. Sukhorukov, in *Quantum Electronics: A Treatise, Nonlinear Optics, Part B*, edited by H. Rabin and C. L. Tang, Vol. 1 (Academic Press, New York, 1975), p. 476.

2.12 Ch. Bosshard, R. Spreiter, M. Zgonik, P. Günter, Phys. Rev. Lett. **74**, 2816 (1995).

2.13 T. K. Gustafson, J.-P. E. Taran, P. L. Kelley, R. Y. Chiao, Opt. Commun. **2**, 17 (1970).

2.14 M. Zgonik, P. Günter, J. Opt. Soc. Am. B **13**, 570 (1996).

2.15 B. Wedding, D. Jäger, Appl. Phys. Lett. **41**, 1028 (1982).

2.16 G. I. Stegeman, M. Sheik-Bahae, E. Van Stryland, G. Assanto, Opt. Lett. **18**, 13 (1993).

2.17 G. Assanto, G. Stegeman, M. Sheik-Bahae, E. Van Stryland, Appl. Phys. Lett. **62**, 1323 (1993).

2.18 D. Y. Kim, Torruellas W. E., J. Kang, Ch. Bosshard, G. I. Stegeman, P. Vidakovic, J. Zyss, W. E. Moerner, R. Twieg, G. Bjorklund, Opt. Lett. **19**, 868 (1994).

2.19 M. Kuzyk, C. W. Dirk, Phys. Rev. **A 41**, 5098 (1990).

2.20 J. F. Nye, *Physical Properties of Crystals* (Clarendon, Oxford, 1967).

2.21 J. L. Oudar, D. S. Chemla, J. Chem. Phys. **66**, 2664 (1977).

2.22 C. C. Teng, A. F. Garito, Phys. Rev. Lett. **50**, 350 (1983).

2.23 C. C. Teng, A. F. Garito, Phys. Rev. **B 28**, 6766 (1983).

2.24 D. Li, T. J. Marks, M. A. Ratner, J. Phys. Chem. **96**, 4325 (1992).

2.25 G. D. Boyd, D. A. Kleinman, J. Appl. Phys. **39**, 3597 (1968).

2.26 D. M. Burland, C. A. Walsh, F. Kajzar, C. Sentein, J. Opt. Soc. Am. B **8**, 2269 (1991).

2.27 A. Willetts, J. E. Rice, D. M. Burland, D. P. Shelton, J. Chem. Phys. **97**, 7590 (1992).

2.28 A. Buckingham, B. Orr, Q. Rev. Chem. Soc **21**, 195 (1967).

2.29 S. Kielich, IEEE J. Quantum Electron. **QE-4**, 744 (1968).

2.30 F. Kajzar, I. Ledoux, J. Zyss, Phys. Rev. **A 36**, 2210 (1987).

2.31 Ch. Bosshard, G. Knöpfle, P. Prêtre, P. Günter, J. Appl. Phys. **71**, 1594 (1992).

2.32 M. S. Paley, J. M. Harris, H. Looser, J. C. Baumert, G. C. Bjorklund, D. Jundt, R. J. Twieg, J. Org. Chem. **54**, 3774 (1989).

2.33 N. Bloembergen, Y. R. Shen, Phys. Rev. **A 133**, 37 (1964).

2.34 G. I. Stegeman, in *Proceedings of Nonlinear optical properties of advanced materials*, **SPIE 1852**, edited by S. Etemad, SPIE – The International Society for Optical Engineering, Bellingham, Washington (1993), p. 75.

2.35 J. F. Ward, Rev. Mod. Phys. **37**, 1 (1965).
2.36 L. Pauling, *The Nature of the Chemical Bond* (Cornell University Press, Ithaca, New York, 1960).
2.37 D. S. Chemla, J. L. Oudar, J. Zyss, Echo Rech. **103**, 3 (1981).
2.38 J. L. Oudar, J. Chem. Phys **67**, 446 (1977).
2.39 J. L. Oudar, J. Zyss, Phys. Rev. **A 26**, 2016 (1982).
2.40 P. W. Atkins, *Molecular Quantum Mechanics* (Oxford University Press, Oxford, 1983).
2.41 K. Jain, J. I. Crowley, G. H. Hewig, Y. Y. Cheng, R. J. Twieg, Optics and Laser Technology, **297** (1981).
2.42 J. L. Oudar, D. S. Chemla, Opt. Commun. **13**, 164 (1975).
2.43 J. Zyss, D. S. Chemla, in *Quantum Electronics – Principles and Applications*, edited by P. F. Liao and P. Kelley, Vol. 1 (Academic Press, Inc., Orlando, 1987), p. 23.
2.44 M. Joffre, Y. Yaron, R. J. Silbey, J. Zyss, J. Chem. Phys. **97**, 5607 (1992).
2.45 J. Zyss, J. Chem. Phys. **98**, 6583 (1993).
2.46 T. Yoshimura, Phys. Rev. B **40**, 6292 (1989).
2.47 D. R. Kanis, M. A. Ratner, T. J. Marks, Chem. Rev. **94**, 195 (1994).
2.48 J. W. Wu, J. R. Heflin, R. A. Norwood, K. Y. Womg, O. Zamini-Khamiri, A. F. Garito, P. Kalyanaraman, P. Sounik, J. Opt. Soc. Am. B **6**, 707 (1989).
2.49 B. Pierce, in *Proceedings of Nonlinear optical properties of organic materials*, **SPIE 1560**, edited by K. Singer, SPIE – The International Society for Optical Engineering, Bellingham, Washington (1991), p. 148.
2.50 K. C. Rustagi, J. Ducuing, Opt. Commun. **10**, 258 (1974).
2.51 J. P. Hermann, J. Ducuing, J. Appl. Phys. **45**, 5100 (1974).
2.52 S. R. Marder, J. W. Perry, G. Bourhill, C. B. Gorman, B. G. Tiemann, K. Mansour, Science **261**, 186 (1993).
2.53 B. J. Orr, J. F. Ward, Mol. Phys. **20**, 513 (1971).
2.54 J. L. Brédas, C. Adant, P. Tackx, A. Persoons, B. M. Pierce, Chem. Rev. **94**, 243 (1994).
2.55 F. Kajzar, J. Messier, in *Conjugated Polymers*, edited by J. L. Brédas and R. Silbey, Kluwer Academic Publishers, 1991), p. 509.
2.56 C. W. Dirk, L.-T. Cheng, M. G. Kuzyk, Int. J. Quant. Chem. **43**, 27 (1992).
2.57 F. Meyers, S. R. Marder, B. M. Pierce, J. L. Brédas, J. Am. Chem. Soc. **116**, 10703 (1994).
2.58 S. R. Marder, C. B. Gorman, F. Meyers, J. W. Perry, G. Bourhill, J. L. Brédas, B. M. Pierce, Science **265**, 632 (1994).
2.59 J. H. Andrews, J. D. V. Khaydarov, K. D. Singer, D. L. Hull, K. C. Chuang, J. Opt. Soc. Am. B **12**, 2360 (1995).
2.60 M. G. Kuzyk, U. C. Paek, C. W. Dirk, Appl. Phys. Lett. **59**, 902 (1991).
2.61 D. A. Kleinman, Phys. Rev. **126**, 1977 (1962).
2.62 S. H. Wemple, M. DiDomenico, Jr., J. Appl. Phys. **40**, 720 (1969).
2.63 A. Yariv, *Quantum Electronics* (John Wiley & Sons, New York, 1975).
2.64 F. S. Chen, Proc. IEEE, **58**, 1440 (1970).
2.65 P. Günter, in *Proceedings of Electro-optics/Laser International '76*, edited by H. G. Jerrard, IPC Science and Technology (1976), p. 121.
2.66 F. Brehat, B. Wyncke, J. Phys. B **22**, 1891 (1989).
2.67 B. Wyncke, F. Brehat, J. Phys. **B 22**, 363 (1989).

2.68 S. H. Wemple, M. DiDomenico, Jr., in *Advances in Materials and Device Research*, edited by R. Wolfe, Applied Solid State Science Vol. 3 (Academic, New York, 1972), p. 263.

2.69 P. Günter, Ferroelectrics **24**, 35 (1980).

2.70 J. Fousek, Ferroelectrics **20**, 11 (1978).

2.71 Ch. Bosshard, K. Sutter, R. Schlesser, P. Günter, J. Opt. Soc. Am. B **10**, 867 (1993).

2.72 W. Baumann, in *Physical Methods of Chemistry*, edited by B. W. Rossiter and J. F. Hamilton, Vol. IIIb (John Wiley & Sons, New York, 1986).

2.73 C. J. Boettcher, *Theory of Electric Polarization* (Elsevier, Amsterdam, 1973).

2.74 A. T. Amos, B. L. Burrows, Advan. Quantum Chem. **7**, 289 (1973).

2.75 L. Onsager, J. Am. Chem. Soc. **58**, 1486 (1936).

2.76 G. R. Meredith, B. Buchalter, J. Chem. Phys. **78**, 1938 (1983).

2.77 M. W. Wong, M. J. Frisch, K. B. Wiberg, J. Am. Chem. Soc. **113**, 4776 (1991).

2.78 J. G. Bergman, G. R. Crane, J. Chem. Phys. **60**, 2470 (1974).

2.79 J. Zyss, J. L. Oudar, Phys. Rev. **A 26**, 2028 (1982).

2.80 Ch. Bosshard, K. Sutter, Ph. Prêtre, J. Hulliger, M. Flörsheimer, P. Kaatz, P. Günter, *Organic Nonlinear Optical Materials* (Gordon & Breach Science Publishers, Amsterdam, 1995).

2.81 J. Zyss, D. S. Chemla, in *Nonlinear Optical Properties of Organic Molecules and Crystals*, edited by D. S. Chemla and J. Zyss, Vol. 1 (Academic Press, Inc., Orlando, 1987), p. 3.

2.82 C. J. Boettcher, *Theory of Electric Polarization* (Elsevier, Amsterdam, 1952).

2.83 L. T. Cheng, W. Tam, S. H. Stevenson, G. R. Meredith, G. Rikken, S. Marder, J. Phys. Chem. **95**, 10631 (1991).

2.84 A. K.-Y. Jen, V. P. Rao, K. Y. Wong, K. J. Drost, J. Chem. Soc., Chem. Commun. **1993**, 90 (1993).

2.85 V. P. Rao, A. K.-Y. Jen, K. Y. Wong, K. J. Drost, Tetrahedron Lett. **34**, 1747 (1993).

2.86 V. P. Rao, A. K.-Y. Jen, K. Y. Wong, K. J. Drost, J. Chem. Soc., Chem. Commun. **1993**, 1118 (1993).

2.87 M. S. Wong, U. Meier, F. Pan, V. Gramlich, Ch. Bosshard, P. Günter, Adv. Mater. **7**, 416 (1996).

2.88 G. F. Lipscomb, A. F. Garito, R. S. Narang, Appl. Phys. Lett. **38**, 663 (1981).

2.89 J. D. Bierlein, L. K. Cheng, Y. Wang, W. Tam, Appl. Phys. Lett. **56**, 423 (1990).

2.90 G. Knöpfle, Ch. Bosshard, R. Schlesser, P. Günter, IEEE J. Quantum. Electron. **30**, 1303 (1994).

2.91 R. Morita, N. Ogasawara, S. Umegaki, R. Ito, Jap. J. Appl. Phys. **26**, L1711 (1987).

2.92 P. N. Butcher, D. Cotter, *The Elements of Nonlinear Optics* (Cambridge University Press, Cambridge, 1990).

2.93 R. DeSalvo, M. Sheik-Bahae, A. A. Said, D. J. Hagan, E. W. Van Stryland, Opt. Lett. **18**, 194 (1993).

2.94 R. W. Boyd, *Nonlinear Optics* (Academic Press, Inc., San Diego, 1992).

2.95 R. A. Fisher, *Optical Phase Conjugation* (Academic Press, Inc., Orlando, 1983).

2.96 Y. R. Shen, Prog. Quant. Electr. **4**, 1 (1975).

2.97 R. Y. Chiao, E. Garmire, C. H. Townes, Phys. Rev. Lett. **13**, 479 (1964).

2.98 P. L. Kelley, Phys. Rev. Lett. **15**, 1005 (1965).

2.99 J. H. Marburger, Prog. Quant. Electr. **4**, 35 (1975).

2.100 H. M. Gibbs, *Optical Bistability* (Academic Press, Inc., Orlando, 1985).

2.101 L. F. Mollenauer, R. H. Stolen, J. P. Gordon, Phys. Rev. Lett. **45**, 1095 (1980).

2.102 G. V. Agrawal, *Nonlinear Fiber Optics* (Academic Press, Inc., San Diego, 1989).

2.103 A. Barthelemy, S. Maneuf, C. Froehly, Opt. Commun. **55**, 201 (1985).

2.104 V. E. Zakharov, A. B. Shabat, Sov. Phys. JETP **34**, 62 (1972).

2.105 G. R. Meredith, Chem. Phys. Lett. **92**, 165 (1982).

2.106 D. Bedeaux, N. Bloembergen, Physics **69**, 57 (1973).

2.107 R. DeSalvo, D. J. Hagan, M. Sheik-Bahae, G. Stegeman, E. W. Van Stryland, H. Vanherzeele, Opt. Lett. **17**, 28 (1992).

2.108 R. C. Eckhardt, J. Reintjes, IEEE J. Quantum Electron. **20**, 1178 (1984).

2.109 M. L. Sundheimer, Ch. Bosshard, E. W. Van Stryland, G. I. Stegeman, J. D. Bierlein, Opt. Lett. **18**, 1397 (1993).

2.110 G. Assanto, G. I. Stegeman, M. Sheik-Bahae, E. Van Stryland, IEEE J. Quantum Electron. **31**, 673 (1995).

2.111 M. L. Sundheimer, "Cascaded second-order nonlinearities in waveguides", PhD thesis, University of Arizona, 1994.

2.112 M. Zgonik, P. Günter, Ferroelectrics **126**, 33 (1992).

2.113 M. Bass, P. A. Franken, J. F. Ward, G. Weinreich, Phys. Rev. Lett. **9**, 446 (1962).

2.114 C. Kittel, *Einführung in die Festkörperphysik* (R. Oldenbourg Verlag, München, 1991), p. 420.

2.115 P. Unsbo, J. Opt. Soc. Am. B **12**, 43 (1995).

2.116 H. Hopf, M. Kreutzer, P. G. Jones, Chem. Ber. **124**, 1471 (1991).

2.117 J. Anthony, A. M. Boldi, Y. Rubin, M. Hobi, V. Gramlich, C. B. Knobler, P. Seiler, F. Diederich, Helv. Chim. Acta **78**, 13 (1995).

2.118 M. Schreiber, J. Anthony, F. Diederich, M. E. Spahr, R. Nesper, M. Hubrich, F. Bommeli, L. Degiorgi, P. Wachter, P. Kaatz, Ch. Bosshard, P. Günter, M. Colussi, U. W. Suter, C. Boudon, J.-P. Gisselbrecht, M. Gross, Adv. Mater. **6**, 786 (1994).

2.119 "X-ray analysis of compounds **2** and **3** shows a < 14° deviation from planarity across the π framework including the aromatic substituents. A variety of related TEE derivatives show completely planar solid-state structures. All compounds shown here have been fully characterized by infrared (IR), electronic absorption (UV/VIS), and ^{1}H–^{13}C-nuclear magnetic resonance spectroscopy as well as by mass spectrometry and elemental analyses."

2.120 H. Meier, Angew. Chem. Int. Ed. Eng. **31**, 1339 (1992).

2.121 Ch. Bosshard, R. Spreiter, P. Günter, R. R. Tykwinski, M. Schreiber, F. Diederich, Adv. Mater. **8**, 231 (1996).

2.122 R. R. Tykwinski, M. Schreiber, R. Pérez Carlón, F. Diederich, V. Gramlich, Helv. Chim. Acta **79**, 2249 (1996).

2.123 R. R. Tykwinski, M. Schreiber, V. Gramlich, P. Seiler, F. Diederich, Adv. Materials **8**, 226 (1996).

2.124 A. F. Garito, J. R. Heflin, K. Y. Wong, O. Zamani-Khamiri, in *Organic Materials for Non-linear Optics: Royal Society of Chemistry Special Publication*

No. 69, edited by R. A. Hann and D. Bloor, The Royal Society of Chemistry, Burlington House, London, Oxford (1989), p. 16.

2.125 F. Meyers, J. L. Brédas, in *Organic Materials for Nonlinear Optics III: Royal Society of Chemistry Special Publication No. 137*, edited by G. J. Ashwell and D. Bloor, The Royal Society of Chemistry, Burlington House, London, Oxford (1993), p. 1.

2.126 G. Puccetti, M. Blanchard-Desce, I. Ledoux, J.-M. Lehn, J. Zyss, J. Phys. Chem. **97**, 9385 (1993).

2.127 C. W. Spangler, K. O. Havelka, M. W. Becker, T. A. Kelleher, L.-T. A. Cheng, in *Proceedings of Nonlinear Optical Properties of Organic Materials*, **SPIE 1560**, edited by K. Singer, SPIE – The International Society for Optical Engineering, Bellingham, Washington (1991), p. 139

2.128 J. L. Oudar, D. S. Chemla, E. Batifol, J. Chem. Phys. **67**, 1626 (1977).

2.129 C. Combellas, G. Mathey, A. Thiebault, F. Kajzar, Nonlinear Optics **12**, 251 (1995).

2.130 M. Blanchard-Desce, J.-M. Lehn, M. Barzoukas, C. Runser, A. Fort, G. Puccetti, I. Ledoux, J. Zyss, Nonlinear Optics **10**, 23 (1995).

2.131 H. S. Nalwa, Adv. Mater. **5**, 341 (1993).

2.132 M. Schreiber, R. R. Tykwinski, F. Diederich, R. Spreiter, U. Gubler, Ch. Bosshard, I. Poberaj, P. Günter, C. Boudon, J.-P. Gisselbrecht, M. Gross, U. Jonas, H. Ringsdorf, Adv. Mater. **9**, 339 (1997).

2.133 I. Thomazeau, J. Ethcepare, G. Grillon, A. Migus, Opt. Lett. **10**, 223 (1985).

2.134 Ch. Serbutoviez, Ch. Bosshard, G. Knöpfle, P. Wyss, P. Prêtre, P. Günter, K. Schenk, E. Solari, G. Chapuis, Chem. Mater. **7**, 1198 (1995).

2.135 L. T. Cheng, W. Tam, S. R. Marder, A. E. Stiegman, G. Rikken, C. W. Spangler, J. Phys. Chem. **95**, 10643 (1991).

2.136 S. R. Marder, J. W. Perry, W. P. Schaeffer, Science **245**, 626 (1989).

2.137 G. Knöpfle, R. Schlesser, R. Ducret, P. Günter, Nonlinear Optics **9**, 143 (1995).

2.138 F. Pan, G. Knöpfle, Ch. Bosshard, S. Follonier, R. Spreiter, M. S. Wong, P. Günter, Appl. Phys. Lett. **69**, 13 (1996).

2.139 M. S. Wong, F. Pan, M. Boesch, R. Spreiter, Ch. Bosshard, V. Gramlich, P. Günter, J. Opt. Soc. Am. **B 15**, 426 (1998).

2.140 K. Sutter, Ch. Bosshard, W. S. Wang, G. Surmely, P. Günter, Appl. Phys. Lett. **53**, 1779 (1988).

2.141 S. Kielich, Acta Phys. Polonica **XXXVI**, 621 (1969).

2.142 S. Kielich, IEEE J. Quantum Electron. **5**, 562 (1969).

2.143 S. Kielich, J. Opto-Electron. **2**, 5 (1970).

2.144 S. Kielich, Acta Phys. Polonica **A37**, 205 (1970).

2.145 K. D. Singer, A. F. Garito, J. Chem. Phys. **75**, 3572 (1981).

2.146 B. F. Levine, C. G. Bethea, J. Chem. Phys. **63**, 2666 (1975).

2.147 R. C. Eckardt, H. Masuda, Y. X. Fan, R. L. Byer, IEEE J. Quant. Electron. **26**, 922 (1990).

2.148 J. Zyss, J. Chem. Phys. **70**, 333 (1979).

2.149 H. A. Kurtz, J. J. P. Stewart, K. M. Dieter, J. Comp. Chem. **11**, 82 (1990).

2.150 B. F. Levine, Chem. Phys. Lett. **37**, 516 (1976).

2.151 E. A. Guggenheim, Trans. Faraday Soc. **45**, 714 (1949).

2.152 D. W. Hall, M. A. Newhouse, N. F. Borelli, W. H. Dumbaugh, D. L. Weidman, Appl. Phys. Lett. **54**, 1293 (1989).

2.153 S. R. Friberg, P. W. Smith, IEEE J. Quantum Electron. **23**, 2089 (1987).
2.154 G. Shirane, H. Danner, A. Pavlovic, R. Pepinsky, Phys. Rev. **93**, 672 (1954).
2.155 S. R. Marder, J. W. Perry, C. P. Yakymyshyn, Chem. Mater. **6**, 1137 (1994).
2.156 D. A. Roberts, IEEE J. Quantum Electron. **28**, 2057 (1992).
2.157 A. Kitamoto, T. Kondo, I. Shoji, R. Ito, Opt. Rev. **2**, 280 (1995).
2.158 A. Mito, K. Hagimoto, C. Takahashi, Nonlinear Optics **13**, 3 (1995).
2.159 R. C. Miller, Appl. Phys. Lett. **5**, 17 (1964).
2.160 T. Pliska, F. Mayer, D. Fluck, P. Günter, D. Rytz, J. Opt. Soc. Am. B **12**, 1878 (1995).
2.161 J. J. Wynne, Phys. Rev. **178**, 1295 (1969).
2.162 C. C. Wang, Phys. Rev. **B 2**, 2045 (1970).
2.163 N. Boling, A. J. Glass, A. Owyoung, IEEE J. Quantum Electron. **14**, 601 (1978).
2.164 J.-C. Baumert, J. Hoffnagle, P. Günter, in *Proceedings of 1984 European Conference on Optics, Optical Systems, and Applications*, **SPIE 492**, edited by B. Bolger and H. A. Ferwerda, Soc. Photo-Opt. Intrum. Eng. (1984), p. 374
2.165 B. Buchalter, G. R. Meredith, Appl. Opt. **21**, 3221 (1982).
2.166 M. Sheik-Bahae, A. A. Said, W. Tai-Huei, D. J. Hagan, E. W. Van Stryland, IEEE J. Quantum Electron. **26**, 760 (1990).
2.167 B. Zysset, I. Biaggio, P. Günter, J. Opt. Soc. Am. B **19**, 380 (1992).
2.168 M. Zgonik, R. Schlesser, I. Biaggio, E. Voit, J. Tscherry, P. Günter, J. Appl. Phys. **74**, 1287 (1993).
2.169 N. Tang, J. P. Partanen, R. W. Hellwarth, R. J. Knize, Phys. Rev. B **48**, 8404 (1993).
2.170 M. L. Sundheimer, A. Villeneuve, G. I. Stegeman, J. D. Bierlein, Electron. Lett. **30**, 1400 (1994).
2.171 M. Yamashita, K. Torizuka, T. Uemiya, Appl. Phys. Lett. **57**, 1301 (1990).
2.172 D. Y. Kim, M. Sundheimer, A. Otomo, G. Stegeman, W. H. G. Horsthuis, G. R. Möhlmann, Appl. Phys. Lett. **63**, 290 (1993).
2.173 P. Vidakovic, J. Zyss, D. Kim, W. Torruellas, G. Stegeman, Nonlinear Optics **10**, 239 (1995).
2.174 W. E. Torruellas, G. Krijnen, D. Y. Kim, R. Schiek, G. I. Stegeman, P. Vidakovic, J. Zyss, Opt. Commun. **112**, 122 (1994).
2.175 R. Schiek, M. L. Sundheimer, D. Y. Kim, Y. Baek, G. I. Stegeman, Opt. Lett. **19**, 1949 (1994).
2.176 R. Schiek, J. Opt. Soc. Am. B **10**, 1848 (1993).
2.177 R. Schiek, Opt. Quantum Electron. **26**, 415 (1994).
2.178 G. Assanto, I. Torelli, Opt. Commun. **119**, 143 (1995).
2.179 G. I. Stegeman, D. J. Hagan, L. Torner, J. Opt. Quantum Electron. **28**, 1691 (1996).
2.180 G. J. M. Krijnen, W. Torruellas, G. I. Stegeman, H. J. W. M. Hoekstra, P. V. Lambeck, IEEE J. Quantum Electron. **32**, 729 (1996).
2.181 G. J. M. Krjinen, J. W. M. Hoekstra, G..I. Stegeman, W. Torruellas, Opt. Lett. **21**, 851 (1996).
2.182 C. N. Ironside, J. S. Aitchison, J. M. Arnold, IEEE J. Quantum Electron. **29**, 2650 (1993).
2.183 Y. Baek, R. Schiek, G. I. Stegeman, Opt. Lett. **20**, 2168 (1995).
2.184 Y. Baek, R. Schiek, G. I. Stegeman, G. Krjinen, I. Baumann, W. Sohler, Appl. Phys. Lett. **68**, 2055 (1996).

2.185 L. Lefort, A. Barthelemy, Opt. Commun. **119**, 163 (1995).
2.186 P. St. J. Russell, Electron. Lett. **29**, 1228 (1993).
2.187 D. J. Hagan, Z. Wang, G. Stegeman, E. W. Van Stryland, M. Sheik-Bahae, G. Assanto, Opt. Lett. **19**, 1305 (1994).
2.188 G. Leo, G. Assanto, W. E. Torruellas, Opt. Lett. **22**, 7 (1997).
2.189 A. Kobyakov, U. Peschel, F. Lederer, Opt. Commun. **124**, 184 (1996).
2.190 A. L. Belostotsky, A. S. Leonov, A. V. Meleshko, Opt. Lett. **19**, 856 (1994).
2.191 Y. N. Karamzin, A. P. Sukhoruv, JETP Lett. **20**, 339 (1974).
2.192 Y. N. Karamzin, A. P. Sukhoruv, Sov. Phys. JETP. **41**, 414 (1976).
2.193 Y. Silverberg, in *Optical solitons*, edited by J. Satsuma, Springer Series on Wave Phenomena, Springer-Verlag, Berlin (1992).
2.194 B. Luther-Davies, Y. Xiaoping, Opt. Lett. **17**, 1755 (1992).
2.195 Ch. Bosshard, P. V. Mamyshev, G. I. Stegeman, Opt. Lett. **19**, 90 (1994).
2.196 P. V. Mamyshev, A. Villeneuve, G. I. Stegeman, J. S. Aitchison, Electron. Lett. **30**, 726 (1994).
2.197 K. Hayata, M. Koshiba, Phys. Rev. Lett. **71**, 3275 (1993).
2.198 A. G. Kaloscai, J. W. Haus, Phys. Rev. **A 49**, 574 (1994).
2.199 L. Torner, C. R. Menyuk, G. I. Stegeman, Opt. Lett. **19**, 1615 (1994).
2.200 L. Torner, C. R. Menyuk, G. I. Stegeman, J. Opt. Soc. Am. **B 12**, 889 (1995).
2.201 C. R. Menyuk, R. Schiek, L. Torner, J. Opt. Soc. Am. **B 11**, 2434 (1994).
2.202 R. Schiek, Y. Baek, G. I. Stegeman, Phys. Rev. **A 53**, 1138 (1996).
2.203 W. E. Torruellas, Z. Wang, D. J. Hagan, E. W. Van Stryland, G. I. Stegeman, L. Torner, C. R. Menyuk, Phys. Rev. Lett. **74**, 5036 (1995).
2.204 W. E. Torruellas, Z. Wang, L. Torner, G. I. Stegeman, Opt. Lett. **20**, 1949 (1995).
2.205 W. E. Torruellas, G. Assanto, B. L. Lawrence, R. A. Fuerst, G. I. Stegeman, Appl. Phys. Lett. **68**, 1449 (1996).
2.206 R. A. Fuerst, B. L. Lawrence, W. E. Torruellas, G. I. Stegeman, Opt. Lett. **22**, 19 (1997).
2.207 J. B. Khurgin, Y. J. Ding, Opt. Lett. **19**, 1016 (1994).
2.208 S. J. Lee, J. B. Khurgin, Y. J. Ding, J. Opt. Soc. Am. **B 12**, 275 (1995).
2.209 A. Berzanskis, R. Danielus, A. Dubietis, A. Piskarskas, A. Stabimis, Appl. Phys. **B 60**, 421 (1995).
2.210 R. Danielus, P. Di Trapani, A. Dubietis, A. Piskarskas, D. Podenas, Opt. Lett. **18**, 574 (1993).
2.211 A. Berzanskis, K.-H. Feller, A. Stabinis, Opt. Commun. **118**, 438 (1995).
2.212 Ruo-Ding Li, P. Kumar, Opt. Lett. **18**, 1961 (1993).
2.213 G. Cerullo, S. De Silvestri, A. Monguzzi, D. Segala, V. Magni, Opt. Lett. **20**, 746 (1995).
2.214 Z. Y. Ou, Opt. Commun. **124**, 430 (1996).
2.215 A. G. White, J. Mlynek, S. Schiller, Europhys. Lett. **35**, 425 (1996).
2.216 C. Richy, K. I. Petsas, E. Giacobino, C. Fabre, L. Lugiato, J. Opt. Soc. Am. **B 12**, 456 (1995).
2.217 R. Reinisch, M. Nevière, E. Popov, Opt. Lett. **20**, 2472 (1995).
2.218 S. Saltiel, S. Tanev, A. D. Boardman, Opt. Lett. **22**, 148 (1997).
2.219 G. I. Stegeman, A. Villeneuve, J. Kang, J. S. Aitchison, C. N. Ironside, K. Al-hemyari, C. C. Yang, C.- H. Lin, H.-H. Lin, G. T. Kennedy, R. S. Grant, W. Sibbett, Int. J. Nonlinear Opt. Phys. **3**, 347 (1994).
2.220 K. L. Hall, A. M. Darwish, E. P. Ippen, U. Koren, G. Raybon, Appl. Phys. Lett. **62**, 1320 (1993).

2.221 M. J. Adams, D. A. O. Davies, M. C. Tatham, M. A. Fisher, Opt. Quant. Electron. **27**, 1 (1995).

2.222 B. Lawrence, M. Cha, J. U. Kang, W. E. Torruellas, G. I. Stegeman, G. Baker, J. Meth, S. Etemad, Electron. Lett. **30**, 447 (1994).

2.223 A. Samoc, M. Samoc, M. Woodruff, B. Luther-Davies, Opt. Lett. **20**, 1241 (1995).

2.224 Z. Wang, D. J. Hagan, E. W. Van Stryland, J. Zyss, P. Vidakovik, W. E. Torruellas, J. Opt. Soc. Am. **B 14**, 76 (1997).

2.225 U. Meier, M. Boesch, Ch. Bosshard, F. Pan, P. Günter, J. Phys. Lett. **83**, 3486 (1998).

2.226 I. Ledoux, C. Lepers, A. Périgaud, J. Badan, J. Zyss, Opt. Commun. **80**, 149 (1990).

2.227 S. Follonier, Ch. Bosshard, G. Knopfle, U. Meier, C. Serbutoviez, F. Pan, P. Günter, J. Opt. Soc. Am. **B 14**, 593 (1997).

2.228 Ch. Bosshard, M. S. Wong, F. Pan, R. Spreiter, S. Follonier, U. Meier, P. Günter, in *Electrical and Related Properties of Organic Solids*, edited by R. W. Munn, A. Miniewicz, and B. Kuchta, NATO ASI Series 3. High Technology Vol. 24 (Kluwer Academic Publishers, Dordrecht, The Netherlands, 1997), p. 279.

2.229 M. C. Etter, G. M. Frankenbach, D. A. Adsmond, Mol. Cryst. Liq. Cryst. **187**, 22 (1990).

2.230 F. Pan, M. S. Wong, V. Gramlich, Ch. Bosshard, P. Günter, Chem. Commun. **1996**, 1557 (1996).

2.231 F. Pan, M. S. Wong, V. Gramlich, Ch. Bosshard, P. Günter, J. Am. Chem Soc. **118**, 6315 (1996).

2.232 J. Zyss, J. F. Nicoud, M. Coquillay, J. Chem. Phys. **81**, 4160 (1984).

2.233 J. R. DeSalvo, "On nonlinear refraction and two-photon absorption in optical media", PhD thesis, CREOL, University of Central Florida, 1993.

2.234 R. Adair, L. L. Chase, S. A. Payne, J. Opt. Soc. Am. **B 4**, 875 (1987).

2.235 C. C. Yang, A. Villeneuve, G. I. Stegeman, G.-H. Lin, H.-H. Lin, Electron. Lett. **29**, 37 (1992).

2.236 P. Prêtre, P. Kaatz, A. Bohren, P. Günter, B. Zysset, M. Ahlheim, M. Stähelin, F. Lehr, Macromolecules **27**, 5476 (1994).

2.237 Ch. Bosshard, P Günter, in *Nonlinear Optics of Organic Molecular and Polymeric Materials*, edited by S. Miyata and H. S. Nalwa, CRC Press, Inc., Boca Raton (1997), p. 391.

2.238 A. Otomo, G. I. Stegeman, W. H. G. Horsthuis, G. R. Möhlmann, Appl. Phys. Lett. **65**, 2389 (1994).

2.239 J. U. Kang, A. Villeneuve, M. Sheik-Bahae, G. I. Stegeman, Appl. Phys. Lett. **68**, 147 (1994).

2.240 G. I. Stegeman, W. E. Torruellas, Phil. Trans. R. Soc. Lond. **A 354**, 745 (1996).

2.241 J. W. Goodman, *Introduction to Fourier Optics*, (McGraw Hill, New York, 1968).

3 Second-Order Nonlinear Optical Organic Materials: Recent Developments

Ch. Bosshard, M. Bösch, I. Liakatas, M. Jäger, and P. Günter

This chapter presents an overview of the current status of second-order non-linear optical organic materials for their use in photonic applications such as optical frequency converters and electro-optic modulators. Special emphasis is placed on the material aspects and a thorough discussion of presently available substances. The potential of organic materials for third-order nonlinear optical applications has been presented in detail in Chap. 2. Earlier reviews of organic nonlinear optical materials covering different physical and material aspects can be found in [3.1–8].

Section 3.1 presents a brief introduction to important nonlinear optical and electro-optic effects, while Sect. 3.2 discusses basic material issues relevant for second-order effects. In Sect. 3.3 we discuss the nonlinearities on a molecular scale presenting comprehensive tables. Section 3.4 is devoted to an overview on the macroscopic properties of single crystals and polymers. The advantages and disadvantages of both material systems are discussed and critically confronted with inorganic dielectrics and semiconductors. Section 3.5 deals with the potential utilization of the presented materials and effects: we present a selection of the most promising applications. Some of the organic devices have already matured to a point of commercial availability. In Sect. 3.6 we discuss the issue of stability of the materials, focusing on optical damage in single crystals and orientational relaxation of poled polymers. Finally in Sect. 3.7 we summarize our findings and give an outlook on future work and materials development.

3.1 Nonlinear Optical and Electro-Optic Effects

Linear and nonlinear optical effects can be described in terms of the linear polarization P^L and the nonlinear polarization P^{NL} induced by the electric field

$$P_i = P_i^L + P_i^{NL} = \varepsilon_0 \chi_{ij}^{(1)} E_j + \varepsilon_0 \chi_{ijk}^{(2)} E_j E_k + \cdots \quad . \tag{3.1}$$

This series is truncated because the terms higher than second-order are beyond the scope of this chapter. Some of the higher-order effects are discussed in Chap 2. (3.1) also defines the susceptibility tensors $\chi^{(n)}$, which contain all the information about the optical properties of the respective material.

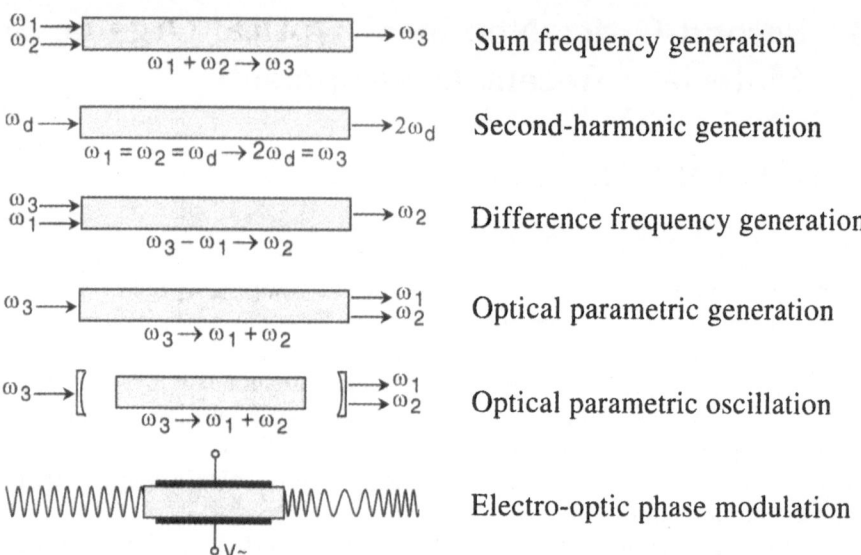

Fig. 3.1. Schematic representation of important nonlinear optical and electro-optic effects

It is important to note that the magnitude of the nonlinear susceptibilities depends on the definition of the electric field amplitude

$$\boldsymbol{E}(\boldsymbol{r},t) = \frac{1}{2}\sum_{k,\omega}\left[\boldsymbol{E}(\boldsymbol{k},\omega)\mathrm{e}^{\mathrm{i}(\boldsymbol{kr}-\omega t)} + \text{c.c.}\right].$$ (3.2)

Interestingly, all the effects discussed in this chapter are generally governed by (3.1). However, based on the characteristics of the involved electric fields several nonlinear and electro-optic effects can be distinguished (see Fig. 3.1). Some of the more important ones even have their own notation and will be briefly introduced below.

3.1.1 Sum Frequency Generation and Optical Frequency Doubling

Sum frequency generation is the mixing of two incident light waves of frequencies ω_1 and ω_2 creating a wave of $\omega_3 = \omega_1 + \omega_2$. This situation is represented by the nonlinear polarization

$$P_i^{\omega_3} = \varepsilon_0\chi_{ijk}^{(2)}\left(-\omega_3,\omega_1,\omega_2\right)E_j^{\omega_1}E_k^{\omega_2}.$$ (3.3)

Optical frequency doubling or second-harmonic generation (SHG) is just a special case of sum-frequency generation. Only one light wave of frequency ω is incident, which is "mixing with itself", thus generating a wave of twice the frequency. The nonlinear polarization for SHG can also be expressed using the nonlinear optical coefficient d_{ijk} which is often used for the nonlinear optical characterization of macroscopic samples.

$$P_i^{\omega_3} = \frac{1}{2}\varepsilon_0\chi_{ijk}^{(2)}\left(-2\omega,\omega,\omega\right)E_j^\omega E_k^\omega = \varepsilon_0 d_{ijk}\left(-2\omega,\omega,\omega\right)E_j^\omega E_k^\omega \ . \tag{3.4}$$

Sum frequency and second harmonic generation are standard techniques to create a new coherent output from existing laser systems and especially to access the short wavelength range towards the ultraviolet region.

3.1.2 Difference Frequency Generation and Optic Parametric Oscillation/Generation

Difference frequency generation (DFG) is characterized as the interaction of two input beams of frequencies ω_3 and ω_1 resulting in an optical field with the frequency $\omega_2 = \omega_3 - \omega_1$, i.e. the difference of the two. The nonlinear polarization for DFG can be written as

$$P_i^{\omega_2} = \varepsilon_0\chi_{ijk}^{(2)}\left(-\omega_2,\omega_3,-\omega_1\right)E_j^{\omega_3}\left(E_k^{\omega_1}\right)^* \ . \tag{3.5}$$

The beam with frequency ω_3 is typically the strongest and therefore referred to as the pump beam. Optic parametric generation is a special case of difference frequency generation, where only the pump beam is incident on the nonlinear material generating two beams at the frequencies ω_1 and ω_2. These frequencies are selected based on the phase-matching condition (3.7) to be discussed in the next section. In order to enhance the efficiency of either process, the nonlinear medium can be placed inside a cavity with highly reflecting mirrors for the frequencies ω_1 and/or ω_2.

In contrast to sum frequency generation, difference frequency generation is well suited to achieving coherent light sources at longer wavelengths, i.e. the near- and mid-infrared region. Another application is parametric amplification where the strong pump beam at frequency ω_3 tranfers energy to amplify an optical signal at frequency ω_1. In this case the output at frequency ω_2 remains unused and is thus often referred to as idler. Finally optical parametric generation and oscillation are of particular importance because they allow to turn a single frequency laser into a broadly tunable laser system by adjusting the phase-matching condition (3.7) using, for example, angle or temperature tuning.

3.1.3 Conservation of Energy and Momentum

All the nonlinear processes previously discussed have in common that they conserve energy, which was already implicitly assumed:

$$\omega_3 = \omega_1 + \omega_2 \quad \text{(energy conservation)}. \tag{3.6}$$

Another joint feature is that efficient nonlinear interaction can only occur if momentum is conserved. This requirement also referred to as phase-matching is manifested in the following condition (see e.g. [3.9]):

$$\mathbf{k}_{\omega_3} = \mathbf{k}_{\omega_1} + \mathbf{k}_{\omega_2} \quad \text{(momentum conservation or phase matching)}. \quad (3.7)$$

For collinear parametric interactions, where all wave vectors $\mathbf{k}_i$ are parallel to one another, the phase-matching condition simplifies to

$$n(\omega_3)\,\omega_3 \; = \; n(\omega_1)\,\omega_1 + \; n(\omega_2)\,\omega_2 \;, \quad (3.8)$$

where $n(\omega_i)$ are the refractive indices of the waves of frequency ω_i. The nonlinear polarization for general directions in the crystal can be written as

$$|P^{\omega_3}| = 2\varepsilon_0 d_{\text{eff}} |E^{\omega_1}||E^{\omega_2}| \quad \text{(sum frequency generation)} \quad (3.9)$$

$$d_{\text{eff}} = \sum_{ijk} d_{ijk}^{(\omega_3,\omega_1,\omega_2)} \cos\left(\alpha_i^{\omega_3}\right) \cos\left(\alpha_j^{\omega_1}\right) \cos\left(\alpha_k^{\omega_2}\right) \;, \quad (3.10)$$

where d_{eff} is the effective nonlinear optical coefficient and α_i^ω are the angles between the electric field vector $\mathbf{E}^\omega$ of the wave with frequency ω and the main axis i of the indicatrix [3.10, 11]. Note, that the walk-off angle ρ between the wave vector and the Poynting vector must be taken into account to calculate the electric field vectors and α_i^ω.

3.1.4   Linear Electro-Optic Effect

Electro-optic effects describe the deformation and rotation of the optical indicatrix if an electric field is applied to a noncentrosymmetric sample [3.12]. The linear electro-optic effect can be also expressed using the nonlinear $\chi^{(2)}$ tensor

$$P_i^\omega = 2\varepsilon_0 \chi_{ijk}^{(2)} \left(-\omega, \omega, 0\right) E_j^\omega E_k^\circ \;. \quad (3.11)$$

However, it is not considered a nonlinear optical effect because one of the involved fields E_k^0 is not an optical but a static electric field. Typically the linear electro-optic effect is described in terms of the change of the optical indicatrix

$$\Delta\left(\frac{1}{n^2}\right)_{ij} = \Delta\left(\frac{1}{\varepsilon}\right)_{ij} = r_{ijk}E_k \;. \quad (3.12)$$

The above equation is also the defining equation of the electro-optic tensor r_{ijk}. For small changes the linear refractive index change can be approximated by [3.9]

$$\Delta n_i \cong -\frac{n_i^3 r_{ijk}E_k}{2} \quad \text{(for } i = j\text{)} \;. \quad (3.13)$$

The linear electro-optic effect is widely used in electro-optic modulators. These devices employ the induced phase change of an optical wave and convert it to a change in intensity (see Sect. 3.5.3). Therefore the optical intensity can be controlled by an electrical signal, a frequent task in telecommunications.

3.2   Material Considerations

While (3.1) governs nonlinear effects on a macroscopic scale, one may also consider this problem on the molecular level. The dipole moment of the molecule p consists of its ground state dipole moment μ_{g} and the induced contribution. The corresponding expansion

$$p_i = \mu_{\mathrm{g},i} + \varepsilon_0 \alpha_{ij} E_j + \varepsilon_0 \beta_{ijk} E_j E_k + \dots \tag{3.14}$$

defines the molecular coefficients: the linear polarizability α_{ij} and the first-order hyperpolarizability β_{ijk}. The task of linking the macroscopic coefficients to the molecular ones is a nontrivial problem because of interactions between neighboring molecules. However, most often the macroscopic second-order nonlinearities of organic materials can be well explained by the nonlinearities of the constituent molecules (oriented gas-model [3.13]). For example the nonlinear optical and electro-optic coefficients can be expressed assuming only one dominant tensor element β_{zzz} (see Chap. 2):

$$d_{ijk}\left(-2\omega,\omega,\omega\right) = \frac{1}{2}\chi_{ijk}^{(2)}\left(-2\omega,\omega,\omega\right) \tag{3.15}$$

$$= \frac{1}{2} N \frac{1}{n(g)} f_i^{2\omega} f_j^{\omega} f_k^{\omega} \sum_{s}^{n(g)} \sum_{mnp}^{3} \cos\left(\theta_{im}^s\right) \cos\left(\theta_{jn}^s\right)$$

$$\times \cos\left(\theta_{kp}^s\right) \beta_{mnp}\left(-2\omega,\omega,\omega\right)$$

$$= g(n) \times \sum_{s}^{n(g)} \cos\left(\theta_{iz}^s\right) \cos\left(\theta_{jz}^s\right) \cos\left(\theta_{kz}^s\right) \beta_{zzz}\left(-2\omega,\omega,\omega\right)$$

and

$$r_{ijk}\left(-\omega,\omega,0\right) = -\frac{2}{n_i^2\left(\omega\right) n_j^2\left(\omega\right)} \chi_{ijk}^{(2)}\left(-\omega,\omega,0\right) \tag{3.16}$$

$$= -N \frac{1}{n(g)} \frac{2 f_i^{(\omega)} f_j^{(\omega)} f_k^{(0)}}{n_i^2\left(\omega\right) n_j^2\left(\omega\right)} \sum_{s}^{n(g)} \sum_{mnp}^{3}$$

$$\times \cos\left(\theta_{im}^s\right) \cos\left(\theta_{jn}^s\right) \cos\left(\theta_{kp}^s\right) \beta_{mnp}\left(-\omega,\omega,0\right)$$

$$= -g'(n) \times \sum_{s}^{n(g)} \cos\left(\theta_{iz}^s\right) \cos\left(\theta_{jz}^s\right) \cos\left(\theta_{kz}^s\right) \beta_{zzz}\left(-\omega,\omega,0\right) \,,$$

where θ_{im}^s are the angles between the dielectric and molecular axes i and m, N is the number of molecules per unit volume, $n(g)$ is the number of equivalent positions in the unit cell, s denotes a site in the unit cell, f_i^{ω} are local field corrections (mostly in the Lorentz approximations), and β_{mnp} is the molecular first-order hyperpolarizability.

3.2.1   Dispersion of the Nonlinear and Electro-Optic Coefficients

As an important consequence of the oriented gas model, the wavelength-dependence of the nonlinear optical coefficients can be approximated by the wavelength-dependence of the molecular hyperpolarizability

$$\beta_{zzz}(-\omega_3, \omega_1, \omega_2) = \frac{1}{2\varepsilon_0 \hbar^2} \frac{\omega_{eg}^2 \left(3\omega_{eg}^2 + \omega_1\omega_2 - \omega_3^2\right)}{\left(\omega_{eg}^2 - \omega_1^2\right)\left(\omega_{eg}^2 - \omega_2^2\right)\left(\omega_{eg}^2 - \omega_3^2\right)} \Delta\mu\, \mu_{eg}^2 \,,$$

(3.17)

which was discussed in detail in Chap. 2.

In (3.17) only electronic contributions to the optical material response have been included which is valid for all mentioned nonlinear optical effects. However, the electro-optic effect also involves an applied static or quasi-static electric field with a frequency much lower than the optical frequencies. This allows other effects of much slower response time, such as acoustic and optic phonons, to contribute to the electro-optic effect. In polar materials, e.g. in $KNbO_3$, these contributions can be very large [3.14, 15]. In contrast, in organic materials electro-optic effects are assumed to be of purely electronic origin in the visible spectral range. Thus the electro-optic coefficient is virtually independent of the modulation frequency of the applied electric field over a very large range and up to extremely high frequencies, an important advantage for the application in electro-optic modulators. Furthermore, organic materials often have low dielectric constants which yield a large frequency bandwidth and a low electrical power consumption. A more detailed discussion of these aspects can be found in Chap. 2 and in Sect. 3.5.3.

3.2.2   Symmetry Considerations
for Second-Order Nonlinear Optical Materials

All of the second-order effects described by the nonlinear $\chi^{(2)}$ tensor can only occur in noncentrosymmetric materials. This important fact can be shown by symmetry considerations [3.9] and holds from the molecular level up to the macroscopic level. As a consequence, the search for suitable second-order nonlinear materials starts with the search for molecules with large first-order hyperpolarizabilities β only found in noncentrosymmetric molecules. Section 3.3 is devoted to a detailed discussion of these molecular engineering issues.

Large second-order nonlinear coefficients require a noncentrosymmetric packing of the organic molecules. A variety of technological approaches have been demonstrated, including the polar arrangement of the molecules in single crystals or Langmuir–Blodgett films. Alternatively the organic molecules can be introduced into suitable hosts such as polymers or sol–gels. In this case the noncentrosymmetric arrangement is typically achieved in a separate poling process. Organic single crystals and poled polymers, as the two most promising technologies, will be further investigated in Sect. 3.4.

3.3   Organic Nonlinear Optical Molecules

The basic design of nonlinear optical molecules is based on π bond systems. π bonds are regions of delocalized electronic charge distribution resulting from the overlap of π orbitals. This delocalization leads to a high mobility of the electron density. The electron distribution can be distorted by substituents at both sides of the π bond system. The extent of the redistribution is measured by the dipole moment, and the ease of redistribution in response to an externally applied field by the hyperpolarizability. The optical nonlinearity of organic molecules can be increased by either increasing the conjugation length or by using appropriate electron donor and electron acceptor groups (Fig. 3.2) The addition of the appropriate functionality at the ends of the π system can enhance the asymmetric electronic distribution in either or both the ground state and excited state configurations (Fig. 3.3).

The organic molecules are used mainly in four kinds of materials showing a macroscopic nonlinear optical response [3.16]:

- organic crystals
- polymer systems
- Langmuir–Blodgett (LB) films
- films formed by molecular beam epitaxy (MBE)

Although the relation between macroscopic nonlinearity and molecular structure is not yet completely understood and no control over the arrangement of the individual molecules can be obtained, the mechanisms leading to large microscopic effects are well understood (see also Chap. 2). A straightforward connection between microscopic and macroscopic nonlinearities is

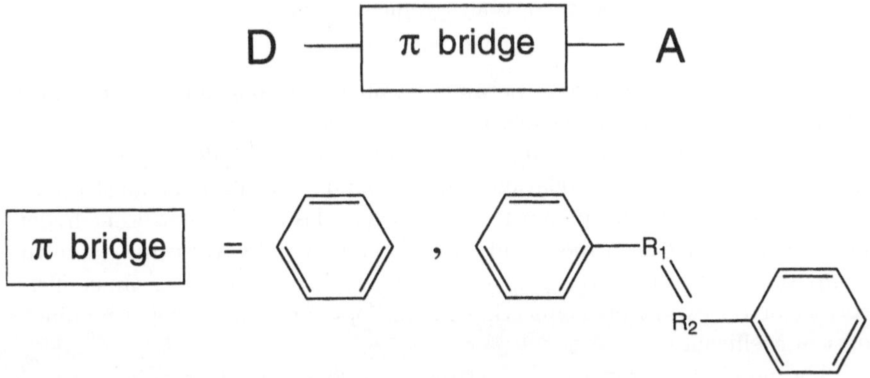

Fig. 3.2. Typical organic molecules for second-order nonlinear optical effects. The electron donor group (D) is connected to the electron acceptor group (A) through a π electron system. The most common systems are those containing one benzene ring (benzene analogs) and those containing two benzene rings (stilbene analogs). R_1 and R_2 are usually carbon or nitrogen

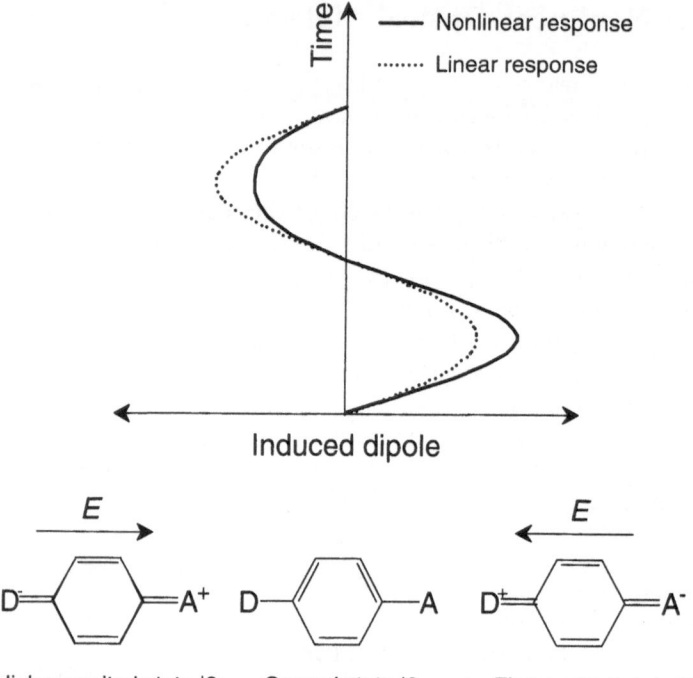

Fig. 3.3. Simple picture of the physical mechanisms of the nonlinearity of conjugated molecules. $|0\rangle$ neutral state, $|1\rangle$ situation where one electron is transferred from the donor to the acceptor (maximum charge transfer state), $|2\rangle$ situation where one electron is transferred from the acceptor to the donor (very unlikely). As a result, we get an asymmetric electronic response of the polarization on application of an optical field for molecules with donor and/or acceptor groups in contrast to the symmetric electronic response for centrosymmetric molecules

not always possible as several parameters play an important role. The measurement of molecular hyperpolarizabilities is, however, of great importance. On the one hand, a direct comparison of different molecules is possible and useful structure–property relations can be found. The effects of substituents, conjugation length, and planarity of the molecules on the first-order hyperpolarizability have been thoroughly investigated [3.3]. On the other hand, the comparison between microscopic and macroscopic nonlinearities can confirm, correct or reject models connecting microscopic and macroscopic nonlinear optical coefficients.

The nonlinear optical and electro-optic properties of organic materials will be interpreted mainly in terms of the oriented gas model. This model [3.13] is based on the assumption that the interactions among molecules are satisfactorily described by the dipole approximation, which is true in rarefied gases where the intermolecular distance is large compared with molecular dimensions. The model applied to solids and liquids, except for local field corrections, takes into account only intramolecular contributions to the optical nonlinearity. Based on this model measured first-order hyperpolarizability (β) values could be reasonably well compared with measured nonlinear optical coefficients (d) and electro-optic coefficients (r).

In Table 3.1, some of the most important categories of organic nonlinear optical molecules are presented.

3.3.1   Measurement Techniques

Various techniques have been employed in the past for the measurement of molecular hyperpolarizabilities. The techniques that are mostly used and which are presented in this chapter are electric field-induced second-harmonic generation (EFISH) and the hyper-Rayleigh scattering (HRS). They are both well established and widely used. Apart from them, the method of solvatochromism and the technique of measuring nonlinear optical properties in the vapor phase are only occasionally used [3.17, 18].

For the definition of the nonlinear optical susceptibilities alternative conventions are frequently used (see Chap. 2). This has led to some confusion in the literature concerning the comparison of experimentally determined values obtained with different techniques as well the comparison of calculated and experimental values. The fact that most often the precise definition in use is not clearly stated complicates the comparison of optical hyperpolarizabilities. In this chapter we follow the power series convention (see Table 2.1 and (2.2) from Chap. 2) where the molecular polarizabilities are defined by the expansion of the molecular dipole moment p as given in (3.14).

As absolute measurements of molecular polarizabilities are very difficult to perform, the signals measured are usually referenced to the one of a quartz crystal. Several absolute values for the nonlinear optical susceptibility d_{11} of quartz varing from 0.3 to 0.5 pm/V have been reported [3.19, 20]. Although values of 0.4 and 0.5 pm/V have been broadly used in the past, the one that is currently mostly accepted and used throughout this chapter is $d_{11} = 0.3$ pm/V measured at 1064 nm (see [3.20]).

A   Electric Field-Induced Second-Harmonic Generation

Electric field-induced second-harmonic generation (EFISH) is the most frequently used method to measure nonlinear optical properties of molecules (see also Chap. 2). It is based on measuring the frequency-doubled light generated in a solution under the influence of a static electric field used to break the isotropy of the liquid. EFISH is usually performed using a Maker–Fringe technique where the intensity of the second-harmonic wave $I^{2\omega}$ generated in the sample shows oscillations due to different phase velocities of the fundamental and frequency-doubled beams in the material [3.17]. A solution of the material to be investigated is placed in a wedged cell which consists of two

Table 3.1. Categories of organic nonlinear optical molecules. The references correspond to the molecule entry number in Table 3.2

	Structure	Reference
π bond system		
Benzenes		[3.1, 2]
Stilbenes		[3.4, 5, 13]
Azo-stilbenes		[3.7, 8, 15, 16]
Tolanes		[3.11]
Phenyl-thiophenes	$n,m=1,2,...$	[3.17–23, 30]
Thiophenes	$n=0,1,2,...$	[3.24, 26, 27]
Polyenes	$n=1,2,...$	[3.32, 33]
Carotenoids		[3.34, 35]
Donors		
Amino	H_2N-	[3.1, 2, 10]
Dialkylamino	$H_{2n+1}C_n$ $\diagdown$ $N-$ $H_{2m+1}C_m$ $\diagup$ $n,m=1,2,...$	[3.4, 11–14, 16, 17, 19] [3.20, 22, 23, 25, 26] [3.29–31, 36, 38–41, 45]
Diphenylamino		[3.5, 8, 15, 18, 21, 24]
Methoxy	H_3CO-	[3.6, 9]
Ketone dithioacetal		[3.27, 28]
Julolidinyl		[3.32–35]

Table 3.1. (Continued)

	Structure	Reference
Acceptors		
Nitro	$-NO_2$	[3.1–12, 45]
Cyano	$-CN$	[3.36]
Dicyanoethenyl		[3.13, 14, 40]
Tricyanoethenyl		[3.15–28, 37, 38]
3-(dicyanomethylidene)-2,3-dihydrobenzothiophen-2-ylidene-1,1-dioxide		[3.30, 31]
N, N'-diethyl-thiobarbituric acid		[3.32, 34, 41]
3-phenyl-5-isoxazolone		[3.33, 35]

glass windows positioned between two stainless steel electrodes where the static electric field is applied (Fig. 3.4). By translating the wedged liquid cell across the beam Maker–fringe amplitude oscillations of the generated second harmonic are obtained (Fig. 3.5).

This method is easy to apply and can be performed quite fast. Its drawback, though, is that only the product $\mu_g\beta_z$ can be determined, where β_z is the vector part of the hyperpolarizability tensor β_{ijk} along the direction of the permanent dipole moment μ_g. This means that the dipole moment must also be independently determined. Moreover, due to solvent/solute effects care must be taken when comparing EFISH results from measurements performed in different solvents.

In the following paragraphs the EFISH method will be described and the determination of the molecular first-order hyperpolarizability β_z will be

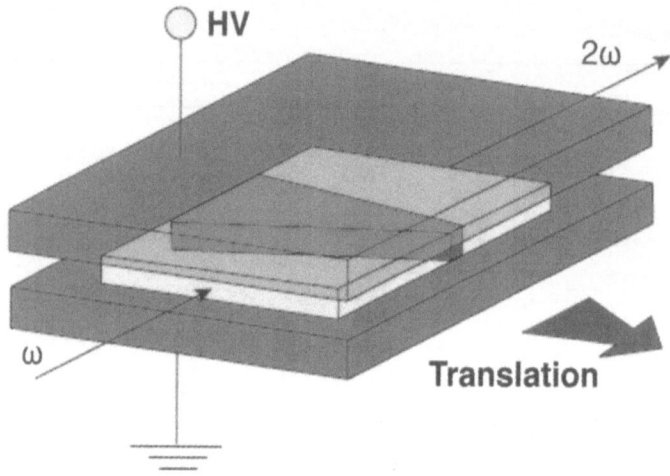

Fig. 3.4. EFISH cell

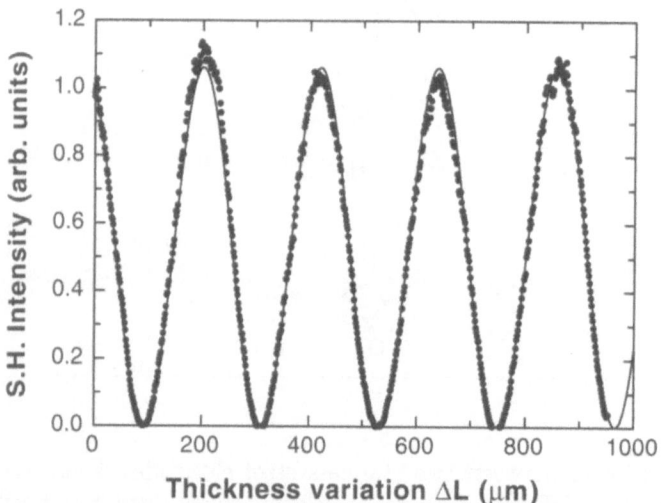

Fig. 3.5. Example of an EFISH measurement performed with pure chloroform at 1907 nm. The solid line is the theoretical curve described by (3.22)

presented. The macroscopic polarization $P^{2\omega}$ induced in a solution by an incident laser field E^ω is described by

$$P_I^{2\omega} = \varepsilon_0 d_{IJK}(E^0) E_J^\omega E_K^\omega \,, \tag{3.18}$$

where the components d_{IJK} of the nonlinear optical susceptibility tensor are dependent on the strength of the applied dc field E^0 and ε_0 is the vacuum permittivity. Assuming both the dc field and the polarization of the fundamental laser field to be parallel to the 3-axis, the only susceptibility component producing a second-harmonic signal is $d_{333}(E^0)$ which for weak fields becomes proportional to the external field. If the molecular z-axis is chosen to lie parallel to the ground state dipole moment μ_g, then $P_3^{2\omega}$ is written as

$$P_3^{2\omega} = \varepsilon_0 \Gamma_L (E_3^\omega)^2 E_3^0 ,\tag{3.19}$$

with

$$\Gamma_L = \frac{d_{333}(E_3^0)}{E_3^0} = N f^0 (f^\omega)^2 f^{2\omega} \left(\frac{3}{2}\gamma + \frac{1}{2}\frac{\mu_g \beta_z}{5kT} \right),\tag{3.20}$$

where N is the number density of the molecules and f^0, f^ω and $f^{2\omega}$ are local field factors evaluated at the indicated frequency. If Kleinman symmetry [3.21] is assumed, the microscopic quantities γ and β_z are given by

$$\gamma = \frac{1}{5}\gamma_{iijj} \quad \text{and} \quad \beta_z = \beta_{zzz} + \beta_{xxz} + \beta_{yyz} ,\tag{3.21}$$

where β_z is the vector part of the hyperpolarizability tensor β_{ijk} along the direction of the permanent dipole moment μ_g.

The intensity of the generated second harmonic signal after the liquid cell is given by

$$I_L^{2\omega}(L) = (I_L^{2\omega})_E \, f(L) = K\eta_L [T_{G,L}\Gamma_G l_c^G - T_L \Gamma_L l_c^L]^2 (E^0)(I^\omega)^2 f(L) ,\tag{3.22}$$

where I^ω and $I_L^{2\omega}(L)$ are the intensities of the fundamental and of the generated second-harmonic wave, respectively, $(I_L^{2\omega})_E$ is the envelope of the interference fringes (if no absorption is present), and E^0 is the static field strength. The factor $\eta_L = (t^\omega)^4 (t^{2\omega})^2$ contains Fresnel transmission factors of the fundamental and the second-harmonic wave at the air/glass and glass/air interfaces, respectively. The factors $T_{G/L}$ and T_L result from the electromagnetic boundary conditions at the glass/liquid and liquid/glass interfaces. l_c^G and l_c^L are the coherence lengths of the glass and the liquid, respectively, and are given by

$$l_c^G = \frac{\lambda}{4(n_G^{2\omega} - n_G^\omega)} \quad \text{and} \quad l_c^L = \frac{\lambda}{4(n_L^{2\omega} - n_L^\omega)} .\tag{3.23}$$

The constant K is

$$K = \frac{128}{\varepsilon_0 c \lambda^2} \, (\text{SI}) \quad K = \frac{8192\pi^3}{c\lambda^2} \, (\text{esu}) .$$

Generally $f(L)$ is given by

$$f(L) = \frac{1}{2}\exp\left[-\left(\alpha^\omega + \frac{\alpha^{2\omega}}{2} \right) L \right]$$
$$\times \left\{ \cosh\left[\left(\alpha^\omega - \frac{\alpha^{2\omega}}{2} \right) L \right] - \cos\left(\frac{\pi L}{l_c^L} \right) \right\}\tag{3.24}$$

describing the well-known oscillations as the distance traveled in the liquid, L, is varied. α^ω and $\alpha^{2\omega}$ are the absorption coefficients at the corresponding wavelengths. In the case of negligible absorption $f(L)$ is reduced to

$$f(L) = \sin^2\left(\frac{\pi L}{2l_c^L}\right) . \tag{3.25}$$

Figure 3.5 shows an example of an EFISH measurement performed with pure chloroform at $\lambda = 1907\,\mathrm{nm}$. The curve has been analyzed using a least squares fit of the form

$$I^{2\omega} = a_1 \sin^2\left(\frac{\pi L}{2a_3} + \frac{a_4}{2}\right) + a_2 , \tag{3.26}$$

where the parameters a_3 and a_4 are the coherence length and the phase offset, respectively. a_2 is the fringe minimum and a_1 is the fringe amplitude. By fitting these parameters and by referencing to the signal of a quartz crystal or a reference solution, the macroscopic third-order nonlinearity Γ_L can be deduced for a specific solution.

For a two component solution, Γ_L is the sum of solute (index 1) and solvent (index 0) contributions expressed as [3.22]

$$\Gamma_L = N_0 f_0^0 (f_0^\omega)^2 f_0^{2\omega} \gamma_0' + N_1 f_1^0 (f_1^\omega)^2 f_1^{2\omega} \gamma_1' , \tag{3.27}$$

where the microscopic third-order nonlinearity γ' is defined as

$$\gamma' = \left(\frac{3}{2}\gamma + \frac{1}{2}\frac{\mu_g \beta_z}{5kT}\right) . \tag{3.28}$$

Mostly pure Lorenz–Lorentz type local field factors with $n^\omega \cong n^{2\omega}$ are applied.

In order to minimize solvent–solvent and solute–solute interactions, an extrapolation procedure to infinite dilution is used. The nonlinearity Γ_L is measured for different concentrations (again using a reference for calibration) and using (3.27) and (3.28) the product $\mu_g \beta_z$ can be deduced, neglecting the term γ.

B   Hyper-Rayleigh Scattering

Elastic second-harmonic light scattering in macroscopically isotropic media was discovered as early as in 1965 by Terhune et al. [3.23] and has been described in a series of publications [3.24–27]. Recently it was used for determining the polarizability of nonlinear optical molecules and became known as hyper-Rayleigh scattering (HRS) [3.28].

The observation of coherent second-harmonic generation from solutions is prevented by the absence of any macroscopic second-order susceptibility $\chi^{(2)}$ due to the isotropic symmetry of the liquid. This macroscopic isotropy, however, is a temporally and spatially averaged property. Fluctuations in the liquid caused by the random movement of the molecules lead to local deviations of $\chi^{(2)}$ from zero and are responsible for second-harmonic light scattering.

The hyper-Rayleigh scattering method has been widely used lately as it allows the determination of nonlinearities of molecules that cannot be measured with EFISH. Octupolar molecules [3.29], for example, may have no ground state dipole moment, in which case they cannot orient themselves to the static electric field applied in EFISH. Ionic molecules, too, cannot be measured with EFISH, as a flowing current due to the ions creates an electrical breakdown between the electrodes. Another advantage of the HRS method is that not only can the vector part of the hyperpolarizability tensor along the direction of the permanent dipole moment, β_z, be determined but also individual components of the tensor. This method is also attractive because of the simplicity of the experimental setup (no voltage source and translation stage needed) and because no local field factors and knowledge of the dipole moment are necessary for the evaluation.

The drawbacks of this method are the low signals due to the incoherent nature of the scattered light that has to be detected and the good beam quality required. The longer the wavelength of the fundamental light, needed to avoid absorption resonances, the lower the scattering signal is ($I_{\mathrm{sc}} \propto \omega^4$). Moreover, care must be taken to avoid broadband fluorescence, which can lead to erroneous results.

In the following paragraphs the HRS method will be described and the derivation of the molecular first-order hyperpolarizability β from this method will be presented. Let us consider the case of Fig. 3.6, where a liquid cell is illuminated by a fundamental laser beam. Each molecule acts as an independent, incoherent source of light generated at the second-harmonic frequency. The intensity of the second-harmonic scattered light $I_{ij}^{2\omega}$ is given by

$$I_{ij}^{2\omega} = GN(f^\omega)^4(f^{2\omega})^2 \sum_{ijklmn} \langle \beta_{ijk}\beta_{lmn}^* \rangle \left(I_{ij}^\omega \right)^2 , \tag{3.29}$$

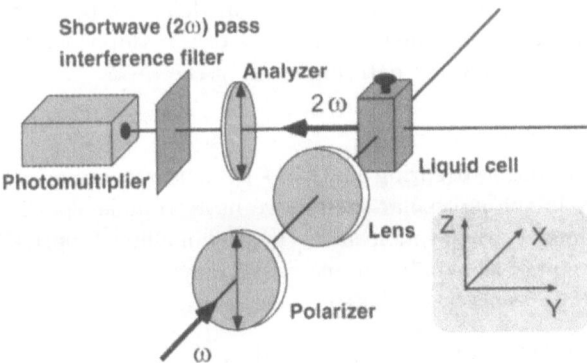

Fig. 3.6. Experimental setup for hyper-Rayleigh scattering

where G is an experimental geometry factor, N is the number density of molecules, f^ω and $f^{2\omega}$ are local field factors (for which the Lorentz local field factors $f = (n^2+2)/3$ are usually used), I^ω is the intensity of the fundamental light and

$$\sum_{ijklmn} \langle \beta_{ijk} \beta_{lmn}^* \rangle \qquad (3.30)$$

is the orientational average of the square of the first-order hyperpolarizability tensor of the molecules. In the general case, at most five invariants of the molecular hyperpolarizabilities may be determined, depending on the molecular symmetry as well as the polarization of the fundamental and the detected light. In the special and most common case where both polarizations lie along the Z-axis (see Fig. 3.6), the two intensities are related as follows

$$I_Z(2\omega) \propto N \, \langle \beta_{ZZZ}^2 \rangle \, I^2(\omega) \,, \qquad (3.31)$$

where the average $\langle \beta_{ZZZ}^2 \rangle$ is a combination of β_{ijk} tensor elements and its expression depends on the symmetry of the molecules. i, j and k are the molecule-fixed Cartesian coordinates, while X, Y, Z refer to the laboratory frame. From a measurement of the dependence of the second-harmonic intensity as a function of the fundamental intensity, the factor $A = G(f^\omega)^4(f^{2\omega})^2 N$ $\times \langle \beta_{ZZZ}^2 \rangle$ relating the two intensities can be determined. This can be done, for example, by rotating a polarization rotator between two crossed polarizers. The resulting data points can be analyzed using a least squares fit of the form [3.30]

$$I_Z(2\omega) = A \, \sin^4 \alpha \,, \qquad (3.32)$$

where α is the rotation angle. By fitting the parameter A for solutions of different concentrations the factor $\langle \beta_{ZZZ}^2 \rangle$ can be determined for both the molecule under investigation and the one used as reference. An example of such curves is presented in Fig. 3.7, where the second-harmonic intensity scattered from 4-dimethylaminobenzaldehyde-4-nitrophenylhydrazone (DANPH) molecules dissolved in dioxane was recorded while the angle of the polarization rotator was changed. Figure 3.8 shows a linear fit of the amplitudes A for different concentrations allowing the determination of the ratio

$$\langle \beta_{ZZZ}^2 \rangle_{\text{molecule}} / \langle \beta_{ZZZ}^2 \rangle_{\text{reference}} \,. \qquad (3.33)$$

In order to express the terms occuring in (3.31) by tensor components of the hyperpolarizability in the molecular system we have to make specific assumptions on the symmetry of the molecules. Typical nonlinear optical molecules can be approximated as axially symmetric with symmetry C_{2v}. In this case the averages $\langle \beta_{ZZZ}^2 \rangle$ and $\langle \beta_{XZZ}^2 \rangle$ can be expressed as

$$\langle \beta_{ZZZ}^2 \rangle = (1/7) \, \beta_{zzz}^2 + (9/35) \, \beta_{zxx}^2 + (6/35) \, \beta_{zzz} \beta_{zxx}$$
$$\langle \beta_{XZZ}^2 \rangle = (1/35) \, \beta_{zzz}^2 + (11/105) \, \beta_{zxx}^2 - (2/105) \, \beta_{zzz} \beta_{zxx} \,, \qquad (3.34)$$

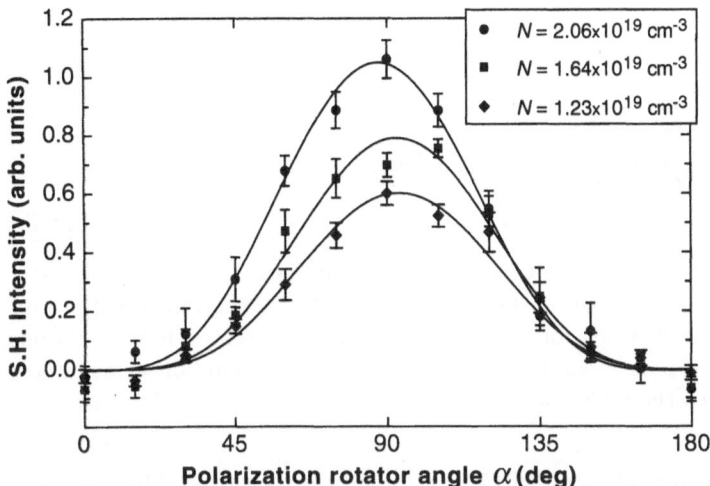

Fig. 3.7. Second-harmonic scattered light intensity for DANPH in dioxane for a fundamental wavelength of 1542 nm, at different number densities N. The solid lines are theoretical curves (3.32)

where the tensor elements β_{zzz} and β_{zxx} are given in the molecule's coordinate system. For the most interesting molecules used for nonlinear optics, showing strong nonlinear optical hyperpolarizabilities, one assumes that one single tensor element β_{zzz} is dominant and therefore the last two terms of (3.34) are neglected.

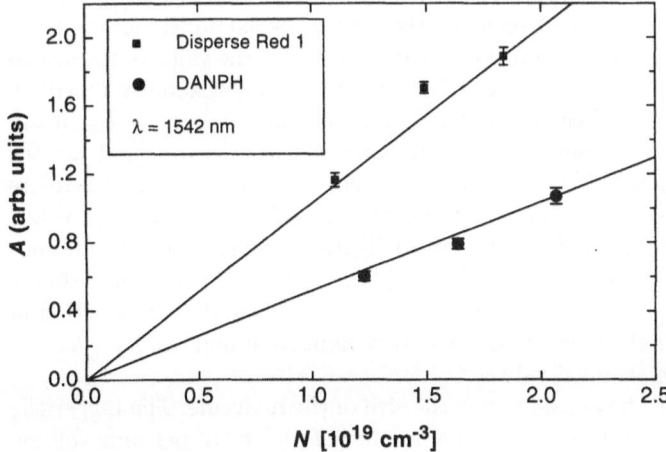

Fig. 3.8. Amplitude of hyper-Rayleigh scattered intensity A vs. number density N of molecules obtained for DANPH (from Fig. 3.7) and Disperse Red 1 (DR1). The solid lines are theoretical curves assuming a linear dependence of A on concentration with $A = 0$ for zero concentration

When the measurement is performed with the analyzer oriented along the X-axis, the quantity $\langle \beta_{XZZ}^2 \rangle$ can be determined. The ratio

$$D_p = \langle \beta_{XZZ}^2 \rangle / \langle \beta_{ZZZ}^2 \rangle \qquad (3.35)$$

known as the depolarization ratio, is a function of the ratio

$$H = \beta_{zxx} / \beta_{zzz} \ , \qquad (3.36)$$

which reflects the anisotropy of the in-plane nonlinearity. Once D_p is measured, H can be determined. Performing the same experiment with combinations of polarizations for the incident and the detected light, relations between various tensor components can be derived assuming a symmetry corresponding to the molecule's structure.

3.3.2   Discussion of Second-Order Nonlinear Optical Molecules

During the last two decades, a large number of nonlinear optical molecules have been synthesized and investigated allowing scientists to gain insight into the chemistry and physics of optical hyperpolarizabilities. Improvement of the size of the hyperpolarizabilities, by using new electron donor and acceptor groups, led to new nonlinear optical materials. Molecules assemble in crystals, Langmuir–Blodgett films and polymers. Measurements of microscopic and macroscopic nonlinearities of these systems reveal new relationships between structure and nonlinear optical properties. Table 3.2 contains a selection of nonlinear optical molecules presenting different approaches towards optimized properties as well as the state of the art in this area.

Special care has to be taken when comparing different molecules. First of all, the measurement method must be the same. For example, electric field-induced second-harmonic generation, EFISH, provides the value of the vector part $\beta_z = \beta_{zzz} + \beta_{xxz} + \beta_{yyz}$ of the first-order hyperpolarizability tensor β, whereas hyper-Rayleigh scattering, HRS, provides the component β_{zzz} itself. Depending on the wavelength of the fundamental laser beam used for the experiments, considerable enhancement of the values can occur. Therefore it is necessary to compare the dispersion-free β_o values (see (3.17)). When comparing molecules for poled polymer applications the product $\mu \beta_o$ and not just β_o is of importance, since the chromophores have to have a large dipole moment to achieve a good poling efficiency (see Sect. 3.4.3). The selection of the solvent used for the measurements is very important and can have a large influence on the determined values.

What has also to be considered is the size of the molecule. The bigger the charge transfer length, the smaller the number of molecules per unit volume that can be achieved in the bulk. It is not possible to measure the length of the molecule; therefore the molecular weight is used as a size parameter instead. Therefore the figure of merit for nonlinear optical molecules is $\mu \beta_o / M$, where M is the molecular weight.

Table 3.2. Selected chromophores for second-order nonlinear optics. λ_{eg} is the absorption peak of the charge transfer band, β is the first order hyperpolarizability at the wavelength measured, β_0 is the first order hyperpolarizability extrapolated to infinite wavelength, μ is the ground state dipole moment, MW is the molecular weight of the molecule and T_d is the temperature of onset of decomposition. EFISH stands for electric-field-induced second harmonic generation, HRS for hyper-Rayleigh scattering and EOAM for electrooptical absorption measurements. All values are given in SI units. For the determination of β the convention given in (3.20) is used and for the calibration $d_{11} = 0.3$ pm/V at 1064 nm of quartz is adopted, unless indicated otherwise. To achieve a common calibration, in first approximation values of β were multiplied by the appropriate factor (e.g. by $0.3/0.5 = 0.6$ if data using a reference value of $d_{11} = 0.5$ pm/V at 1064 nm was reported)

No	Structure	λ_{eg} (nm) (solvent)	μ (10^{-29}Cm)	β (10^{-40}m^4/V)	β_o (10^{-40}m^4/V)	$\mu\beta$ (10^{-69}m^{5}C/V)	$\mu\beta_o$ (10^{-69}m^{5}C/V)	$\mu\beta_o$/MW (10^{-69}m^{5}C/V)	Method for β at λ(nm) (solvent for β)	T_d (°C)	Ref.
1	H_2N–⬡–NO_2 p - NA	358 (Dioxane)	—	80	40	—	—	—	EFISH 1064 (Dioxane)	379	[3.31]
		376 (Acetone)	2.1	62	50	130	105	0.8	EFISH 1907 (Acetone)	—	[3.32]
2	H_3C,H_2N–⬡–NO_2 MNA	361 (Dioxane)	2.5	78	36	194	92	0.6	EFISH 1064 (Dioxane)	—	[3.31]
		361 (Dioxane)	2	46	38	92	75	0.5	EFISH 1907 (Dioxane)	—	[3.32]

Table 3.2. (Continued)

No	Structure	λ_{eg} (nm) (solvent)	μ (10^{-29} Cm)	β (10^{-40} m^4/V)	β_0 (10^{-40} m^4/V)	$\mu\beta$ (10^{-69} m^{5}C/V)	$\mu\beta_0$ (10^{-69} m^{5}C/V)	$\mu\beta_0$/MW (10^{-69} m^{5}C/V)	Method for β at λ (nm) (solvent for β)	T_d (°C)	Ref.
3	COANP	361 (Dioxane)	2.3	64	59	147	136	0.57	EFISH 1907 (Dioxane)	—	[3.17]
4	DANS	427 (Chloroform)	2.2	380	300	840	640	2.4	EFISH 1907 nm (Chloroform)	289	[3.32]
5		436 —	1.6	—	160	—	250	0.6	EFISH[a] —	358	[3.33]
6	MPNP5	435 (Dioxane)	2.3	1460	400	3410	940	3.7	EFISH[b] 1064 (Dioxane)		[3.34]

Compound									Method		[Ref]
7 DR1	509 (DMSO)	2.9	290	110	850	320	1	EFISH[b] 1356 (DMSO)	308	[3.35]
	455 (Dioxane)	2.3	260	190	600	440	1.4	EFISH 1907 (Dioxane)	—	[3.32]
8	473 (Dioxane)	3	570	403	1700	1207	3.8	EFISH 1907 (Dioxane)	—	[3.36]
	486	2	—	440	—	880	2.2	EFISH 1907 (Chloroform)	393	[3.37]
9 MNBA	380	2.9	88	74	254	205	0.68	EFISH 1907 (Dioxane)	—	[3.38]

(from dispersion of β)

Table 3.2. (Continued)

No	Structure	λ_{eg} (nm) (solvent)	μ (10^{-29} Cm)	β (10^{-40} m^4/V)	β_o (10^{-40} m^4/V)	$\mu\beta$ (10^{-69} m^{5}C/V)	$\mu\beta_o$ (10^{-69} m^{5}C/V)	$\mu\beta_o$/MW (10^{-69} m^{5}C/V)	Method for β at λ (nm) (solvent for β)	T_d (°C)	Ref.
10	ANDS (H$_2$N–⟨⟩–S–⟨⟩–NO$_2$)	341 (Acetone)	1.9	720	380	1390	730	3	EFISH^b 1064 (Acetone)	—	[3.39]
11	(H$_3$C)$_2$N–⟨⟩–≡–⟨⟩–NO$_2$	402 (Dioxane)	2.4	840	310	1970	730	2.7	EFISH^a 1064 (Dioxane)	—	[3.40]
		415 (Chloroform)	3	240	80	720	240	0.9	EFISH 1907 (Chloroform)	—	[3.41]
12	(H$_3$C)$_2$N–⟨⟩–≡–⟨⟩–≡–⟨⟩–NO$_2$	384 (Chloroform)	3.3	500	210	1670	710	1.9	EFISH^{b,c} 1064 (Chloroform)	287	[3.42]
13	DADS (H$_3$C)$_2$N–⟨⟩–CH=CH–⟨⟩–CH=C(CN)$_2$	492 (DMSO)	2.7	760	310	2080	860	2.9	EFISH 1356 (DMSO)	—	[3.35]

	λmax							Method		Ref.
14 (structure)	508 (Chloroform)	2.9	1030	690	2990	1990	6.3	EFISH 1907 (Chloroform)	307	[3.43]
15 (structure)	602	2.4	—	1090	—	2600	5.8	EFISH 1907 (Chloroform)	364	[3.37]
16 (structure, TCV)	582 (DMSO)	3.5	920	360	3210	1270	3.6	EFISH 1580 (DMSO)	—	[3.35]
17 (structure)	640 (Dioxane)	—	—	—	—	4220	11.8	EFISH[a,b] 1907 nm (Dioxane)	240	[3.44]
18 (structure)	601 (Dioxane)	—	—	—	4540	2470	5.4	EFISH[a,b] 1907 (Dioxane)	315	[3.45]

Table 3.2. (Continued)

No	Structure	λ_{eg} (nm) (solvent)	μ (10^{-29} Cm)	β (10^{-40} m^4/V)	β_0 (10^{-40} m^4/V)	$\mu\beta$ (10^{-69} m^{5}C/V)	$\mu\beta_0$ (10^{-69} m^{5}C/V)	$\mu\beta_0$/MW (10^{-69} m^{5}C/V)	Method for β at λ (nm) (solvent for β)	T_d (°C)	Ref.
19		662 (Dioxane)	—	—	—	12700	5800	14.5	EFISH[a,b] 1907 (Dioxane)	—	[3.44]
20		653 (Dioxane)	—	—	—	10340	4850	10.4	EFISH[a,b] 1907 (Dioxane)	—	[3.44]
21		611 (Dioxane)	—	—	—	5510	2910	5.4	EFISH 1907 (Dioxane)	—	[3.46]
22		665 (Dioxane)	—	—	—	14740	6650	13.4	EFISH 1907 (Dioxane)	—	[3.46]
23		655 (Dioxane)	—	—	—	19960	9300	19.9	EFISH 1907 (Dioxane)	—	[3.46]

Compound	λ (nm) (Dioxane)							Method		Ref.
24	665 (Dioxane)	—	—	—	14250	6430	14	EFISH[a,b] 1907 (Dioxane)	265	[3.45]
25	684 (Dioxane)	—	—	—	18170	7680	16.1	EFISH[a,b] 1907 (Dioxane)	—	[3.47]
	686 (Dioxane)	—	6310	2440	—	—	—	HRS[b,d] 1064 (Dioxane)	—	[3.48]
26	654 (Dioxane)	4	—	3150	—	12600	40.6	EOAM[b,e] (Dioxane)	286	[3.49]

Table 3.2. (Continued)

No	Structure	λ_{eg} (nm) (solvent)	μ (10^{-29}Cm)	β $(10^{-40}\text{m}^4/\text{V})$	β_o $(10^{-40}\text{m}^4/\text{V})$	$\mu\beta$ $(10^{-69}\text{m}^5\text{C}/\text{V})$	$\mu\beta_o$ $(10^{-69}\text{m}^5\text{C}/\text{V})$	$\mu\beta_o$/MW $(10^{-69}\text{m}^5\text{C}/\text{V})$	Method for β at λ(nm) (solvent for β)	T_d (°C)	Ref.
27		594 (Dioxane)	3	—	3000	—	9000	22.7	EOAM[b,e] (Dioxane)	—	[3.49]
28		570 (Dioxane)	—	—	—	3070	1800	4.8	EFISH[a,b] 1907 (Dioxane)	310	[3.50]
29		524 (Dioxane)	13	240	150	3070	1990	4.6	EFISH 1907 (Dioxane)	207	[3.51]
30		744 (Dichloromethane)	—	—	—	20960	6990	12.6	EFISH[a,b] 1907 (Dichloromethane)	—	[3.52]
31		826 (Chloroform)	2.7	37000	7500	102000	21000	40	EFISH 1907 (Chloroform)	197	[3.53]

No.	Structure								Method			Ref.
32	(structure)	686 (Chloroform)	2.9	11000	4700	32500	14000	30	EFISH 1907 (Chloroform)	190		[3.54]
33	(structure)	640 (Chloroform)	3.3	5100	2500	16600	8100	19	EFISH 1907 (Chloroform)	—		[3.54]
34	(structure)	680 (Chloroform)	—	—	—	60000	25500	45	EFISH 1907 (Chloroform)	—		[3.54]
35	(structure)	647 (Chloroform)	5.3	9100	4300	49000	23000	43.8	EFISH 1907 (Chloroform)	—		[3.54]
36	(structure)	330 (Chloroform)	—	250	140	—	—	—	EFISH^b 1064 (Chloroform)	—		[3.55]

Table 3.2. (Continued)

No	Structure	λ_{eg} (nm) (solvent)	μ (10^{-29} Cm)	β (10^{-40} m^4/V)	β_0 (10^{-40} m^4/V)	$\mu\beta$ (10^{-69} m^{5}C/V)	$\mu\beta_0$ (10^{-69} m^{5}C/V)	$\mu\beta_0$/MW (10^{-69} m^{5}C/V)	Method for β at λ (nm) (solvent for β)	T_d (°C)	Ref.
37	DADC	459 (NMP)	2.8	400	290	1130	820	1.4	EFISH[a,b,f] 1907 (Chloroform)	375	[3.56]
38	DADB	500 (NMP)	3.2	670	450	2170	1460	2.4	EFISH[a,b,f] 1907 (Chloroform)	371	[3.56]
39		513 (Dioxane)	—	—	—	1820	1200	2.2	EFISH[a,b] 1907 (Dioxane)	354	[3.57]
40	D1 - DTT-A2	562 (Dichloromethane)	—	—	—	5590	3280	6.5	EFISH[a,b] 1907 (Dichloromethane)	344	[3.58]
41	D1-DTT-A3	616 (Dichloromethane)	—	—	—	6990	3630	5.7	EFISH[a,b] 1907 (Dichloromethane)	252	[3.58]

#										Method	Ref.
42	430	—	2.5	480	140	1190	340	1.1	—	EFISH 1064 (Dioxane)	[3.37]
43	670 (Dichloromethane)	—	—	1340	470	—	—	—	—	HRS [b,g] 1064 (Dichloromethane)	[3.59]
44	462	—	4.3	1110	—	—	—	—	—	Solvatochromic [b]	[3.60]
45	450 and 680 (Chloroform)	—	4.1	20680	3350	—	13750	14.7	—	HRS [b,h] 1064 (Chloroform)	[3.61]

[a] Calibration reference not reported
[b] Convention for definition of β not reported
[c] For comparison: β (DR1 at 1064 nm) = $698 \times 10^{-40}\,\mathrm{m^4/V}$
[d] Reference: β (p-NA at 1064 nm) = $70.84 \times 10^{-40}\,\mathrm{m^4/V}$. Differs from value for molecule 1 in this table
[e] Reference: β_0 (p-NA) = $181 \times 10^{-40}\,\mathrm{m^4/V}$. Differs from value for molecule 1 in this table
[f] For Comparison: β_0 (DR1 at 1907 nm) = $733 \times 10^{-40}\,\mathrm{m^4/V}$. Differs from value for molecule 7 in this table
[g] Reference: β_0 (p-NA at 1064 nm) = $91 \times 10^{-40}\,\mathrm{m^4/V}$. Differs from value for molecule 1 in this table
[h] Reference: β_0 (p-NA at 1064 nm) = $96 \times 10^{-40}\,\mathrm{m^4/V}$. Differs from value for molecule 1 in this table

In the next paragraphs some of the most important types of nonlinear optical molecules are discussed. The numbers in bold type refer to the molecule entry numbers in Table 3.2. The various molecular classes were described in Table 3.1.

A Benzenes

Benzenes are the simplest nonlinear optical molecules and their thorough investigation gave rise for the first time to structure–property relations and especially on the effect of donor and acceptor groups on the nonlinearity. Molecule **1**, para-nitroaniline (PNA), was one of the first nonlinear optical molecules to be investigated and together with 2-methyl-4-nitroaniline (MNA), **2**, has been the base for research on donor–acceptor substitutes for many years [3.31, 3.32, 3.35]. They are still very often used as a reference for EFISH and HRS measurements [3.17, 3.48].

B (Azo-)stilbenes

(Azo-)stilbenes are a substantial improvement of the benzenes and their investigation has revealed a direct relation between second-order hyperpolarizability and charge transfer transition wavelength (absorption maximum). A similar relation also holds for benzenes. Disperse Red 1 (DR1), **7**, and DANS, **4**, are the main representatives of azostilbenes and stilbenes, respectively. Due to their high nonlinearity they have been widely used as a point of reference and in investigations of guest–host polymer systems.

C Polyenes and Carotenoids

Polyenes have been investigated for the influence of the number of double bonds (or molecule's length) on the molecular nonlinearity. It was found that the $\mu\beta_0$ product increases with increased length but at the expense of a red shift of the absorption maximum. The highest $\mu\beta_0$ values ever reported were achieved using extended polyene π bridge systems (polyenes **31–33** and carotenoids **34–35**) with very strong electron acceptors, like thiobarbituric acid, in order to reach the bond length alternation at which the hyperpolarizability is maximized [3.53, 3.54]. The drawback of this approach, however, is the extended length of the molecules making them thermally unstable (decomposition temperatures in the range of 175–235°C).

D Thiophenes

Until 1993 the most common π-conjugated bridges consisted of phenylene moieties (e.g. molecules **1**, **2**, **4**, **7** in Table 3.2). Thiophene rings seemed to be a good alternative as they have a lower delocalization energy upon charge separation. The use of benzene–thiophene systems with a tricyanovinyl acceptor **17, 19, 20** by Rao et al. [3.44] led to dramatically enhanced nonlinearities. It was shown that an extension of the conjugation length by an vinyl **19** or a thienylvinyl **20** moiety leads to higher nonlinearities. Vinyl units were also shown to be superior to thienylvinyl ones (compare **19** and **20**). Since then several variations have been investigated. It was shown that the thiophene–thiophene systems **24** are better than the benzene–thiophene analogs **18** by a factor of 2.5 for $\mu\beta_o$, but with a 50°C lower decomposition temperature (265°C) (which is nevertheless rather good). Molecule **24** was mixed (15 wt.%) with a high T_g (265°C) polyquinoline (PQ-100) and was measured to have an electro-optic coefficient $r_{33} = 13$ pm/V at 1.3 µm. The combination of one benzene and two thiophene rings **21–23** has also been investigated [3.46] resulting in a very high $\mu\beta_o$ product when using a double bond connection **23**. Shu et al. [3.47] used a thiophene ring and a triene as conjugating moiety **25** and achieved higher $\mu\beta_o$ and better stability than the analog without the triene. Molecule **25** was used to form thin films in the polymer PQ-100 (same as for molecule **24** and at the same chromophore loading) and was measured to have an electro-optic coefficient of $r_{33} = 33$ pm/V at 1.3 µm.

It was observed that the incorporation of benzene rings into the polyene-based donor–acceptor systems limits or saturates the molecular nonlinearity while enhancing the thermal stability. Moreover, bithiophenes were found to have limited thermal stability due to thermally driven *cis-trans* isomerization of the olefinic linkage. Therefore Rao et al. [3.50] suggested in 1994 the use of fused thiophenes instead of olefinic bonds. The resulting thienothiophene **28** has a nonlinearity twice as high as that of the simple thiophene analog and also higher thermal stability. Accordingly, very high nonlinearities have been obtained by Kim et al. [3.58] by using three fused thiophenes **40, 41**.

E Λ-(lambda)-shape Molecules

Λ-shape molecules are interesting for organic crystals as they tend to crystallize into noncentrosymmetric space groups. This happens because Λ-shape molecules stack easily along one direction, which means that all the molecular dipoles are parallel to the crystal axis. Of all the investigated compounds having a Λ-shape conformation, more than 75% crystallize lacking a center of symmetry [3.62]. This type of molecule has also been used, however, in polymer systems. In 1995 Ermer et al. [3.56] synthesized a series of lambda-shaped donor–acceptor-donor molecules. They have two charge transfer axes lying close to each other in energy which leads to low wavelengths of maximum absorption. Molecule **37** (DADC) has a nonlinearity of the order of DR1 but with a lower wavelength of maximum absorption and a higher decomposition temperature (375°C). Rao et al. [3.57] synthesized lambda-shaped molecules with thiophene rings. They replaced the most reactive CN in the tricyanovinyl acceptor group with a a benzene–thiophene ring with a diethylamino donor **39** in order to increase the solubility in host polyimides. Molecule **39** was mixed (20 wt. %) with polyamic acid and was measured to have an electro-optic coefficient $r_{33} = 12$ pm/V at 830 nm.

F   Electron–Acceptor Groups

Knowing the importance of the appropriate donor and acceptor selection for large nonlinearities, many combinations have been investigated. The most common acceptor is the nitro group (NO_2), which has been used almost explicitly until recently. A major breakthrough came in 1987 when Katz et al. [3.63] used for the first time tricyanovinyl acceptors in nonlinear optical molecules. The use of tricyano acceptor compounds by Rao et al. [3.44] attracted again the interest of the scientific community in 1993. Comparing, for example, molecules **8** and **15** [3.37] we see that the tricyanovinyl acceptor leads to a three times higher $\mu\beta_o$ product compared with the nitro one. The price to be paid is a red shift of 116 nm in the wavelength of maximum absorption and a 30°C decrease of the decomposition temperature. This is, however, not so important as **8** has one of the highest differential scanning calorimetry (DSC) decomposition temperatures ever observed (393°C). The use of heterocyclic acceptors with two strong electron withdrawing groups (dicyanomethylidene and sulfone) by Ahlheim et al. in 1996 [3.52] led to molecules like **30** with very high nonlinearities. An analog of **30** but with the thiophene ring replaced by a double bond and with a 20% lower nonlinearity was mixed (20 wt. %) with a low $T_g (\approx 80°C)$ polycarbonate and was measured to have an impressive electro-optic coefficient $r_{33} = 55$ pm/V at 1.3 μm. Thiobarbituric acid (**32**, **34**, **41**) is a very strong acceptor and has been sucessfully used in polyenes giving rise to very large nonlinearities.

G   Electron–Donor Groups

The main donors used until recently were the amino and the dialkylamino group. One advantage of the dialkylamino group is that by changing the length of the alkyl chains, the solubility of the molecule can be changed. An important improvement came with the use of the diphenylamino group (see e.g. **8**). Comparing **18** [3.45] to its analog with a diethylamino donor 17 [3.44] we notice an almost 50% decrease of the nonlinearity followed, however, by a 75°C increase of the decomposition temperature. Julolidinyl is a strong donor that has been used in molecules with large nonlinearities (**32–35**).

Although a large number of nonlinear optical molecules have been synthesized, a considerable insight into the physics and chemistry of molecular hyperpolarizabilities has been gained and basic structure–property relations have been established, no ideal molecule yet exists. The reason is that, depending on the application, different requirements – and often more than one – have to be fulfilled. For frequency doubling, a transparency at the wavelength of the second-harmonic light is needed. For organic crystals, a noncentrosymmetric packing and a favorable orientation for either frequency doubling or electro-optics is a strict condition. For electro-optic polymers, the demand for molecules with high nonlinearity, limited size and high thermal and photochemical stability is a real challenge for synthetic chemists. On the other hand, the unlimited possibilities of structures to be synthesized form a fertile ground for challenges like this to be met. Most efforts until now have concentrated on the "largest" nonlinearities "at all costs". The focus of future work, however, should be the word "compromise". Effort should be focused on developing molecules for a *specific application*, meeting *all the necessary requirements* and making an *optimized compromise* between them.

3.4   Nonlinear and Electro-Optic Single Crystals and Polymers

There are several possibilities to arrange molecules in a macroscopic way. The most often used forms are single crystals [3.1, 3.3–3.5, 3.8], polymers [3.4, 3.6], and Langmuir–Blodgett films [3.64]. Besides these there exist other interesting possibilities, such as molecular beam deposition for the preparation of thin films. Crystals and polymers will be discussed in detail in this chapter. In this section we will first discuss measurement techniques for nonlinear optical and electro-optic properties in general and then discuss organic single crystals and poled polymers in detail. At the end we present a few inorganic materials to compare their properties with those of organic materials.

3.4.1   Measurement Techniques

A   Nonlinear Optical Measurement Techniques

There exists a wide variety of techniques used for the determination of nonlinear optical properties. Since there is extensive literature discussing these methods we will simply list these, giving a very short description and the corresponding references. A discussion of the advantages and disadvantages of the different techniques can be found in e.g. Bosshard et al. [3.3].

(i) The second-harmonic *powder test* [3.19] allows a rough estimate of the nonlinear optical response. Using this technique a powder sample is irradiated by the fundamental laser beam and the generated second-harmonic light is detected with a photomultiplier. Often a parabolic mirror is used to collect more of the randomly scattered light. A reference sample with known nonlinearity is also measured to calibrate the second-harmonic signal.

(ii) The *wedge technique*: a wedged sample is moved perpendicular to the incident laser beam and the resulting second-harmonic intensity is measured [3.65]. Since the fundamental and the second-harmonic wave usually have different phase velocities in the materials the change in thickness due to the wedge leads to a change of the interference pattern resulting in an oscillation of the observed second-harmonic signal. A reference crystal (often α-quartz) with known nonlinearity is also measured to calibrate the second-harmonic signal.

(iii) The *Maker–Fringe method*: a plane-parallel sample is rotated perpendicular to the laser beam and the resulting second-harmonic intensity is recorded [3.66]. For the same reason as in the case of the wedge technique an oscillation of the observed signal is detected. The Maker–Fringe method is the most widely used method and is especially suited to organic crystals, which tend to grow with plane-parallel faces. As in the above case, a reference crystal (often α-quartz) with known nonlinearity is also measured to calibrate the second-harmonic signal.

(iv) *Phase-matched* second-harmonic and sum frequency generation. Here the fundamental wave is incident at an angle of incidence where phase matching is present: the fundamental and the newly generated waves travel at the same speed in the material. This leads to a strong increase in the detected signal and allows measurement of absolute conversion efficiencies without a reference sample. The drawback is that a precise knowledge of the laser beam parameters is required (see e.g. [3.20]).

(v) *Parametric fluorescence* [3.67–68]. This technique also allows absolute measurements but only requires the knowledge of the laser power, not the intensity! A laser beam with frequency ω_3 is incident on the sample in a geometry where phase matching is achieved. Through spontaneous parametric emission the two frequencies ω_1 and ω_2 are generated. The power of one of these waves (e.g. P_1) is measured with a calibrated detector. From the ratio of the powers P_1/P_3 the nonlinear optical coefficient can absolutely be determined.

B Electro-Optic Measurement Techniques

Alseo in the area of electro-optics there exist many different measurement techniques. An overview of the different methods to determine electro-optic coefficients can be found in e.g. Bosshard et al. [3.3–4]. In the following we will briefly describe the various techniques.

(i) Interferometric techniques such as a *Michelson interferometer*: the sample is placed in one arm of a Michelson interferometer. With a piezoelectric mirror in the reference arm the output of the interferometer is set to 50%. When applying an electric field to the sample the electro-optic effect changes the refractive index of the materials and therefore the optical path length in this arm of the interferometer is changed. This leads to a shift of the intensity pattern at the output of the interferometer that can be analyzed. If the refractive index and the thickness of the sample, the applied voltage, and the maximum and the minimum output of the interferometer are known, the electro-optic coefficient can be determined. Changing the polarization of the incident laser beam allows the measurements of different electro-optic coefficients. This method is the most widely used one for single crystals.

(ii) *Field-induced birefingence* using crossed polarizers. The sample is placed between crossed polarizers in such a way that the incident polarization is at 45° to one of the main dielectric axis of the crystal. With an additional birefringent plate the output is set to 50% transmisson. The application of an electric field now leads to a change of the polarization state at the output of the crystal and therefore to a change of the transmission after the analyzer. This change can be measured. If the same parameters as in the case of the Michelson interferometer are known an effective electro-optic coefficient can be determined. It is often only an effective coefficient since a linear combination of electro-optic coefficients is usually involved.

(iii) *Reflection technique.* This method is most often applied to polymeric thin films. The input laser beam is polarized 45° with respect to the plane of incidence, whereas a polarizer at –45° is used to analyze the reflected beam [3.69–70]. The reflected intensity is set to 50% with the use of a Soleil–Babeil compensator in front of the analyzer. A phase difference is acquired between the parallel (p) and perpendicular (s) components of the beam as they pass through the polymer film when the modulating field E is applied. The phase difference is due to the different effective electro-optic coefficients experienced by the s- and p-polarized waves and the resulting difference in the optical path length L between the two waves in the polymer film. If the intensity contrast, the modulation intensity, the layer thickness, the refractive index, and the angle of incidence are known, effective electro-optic coefficients can be calculated.

(iv) *Attenuated total reflection.* This is another widely used method for the determination of electro-optic coefficients in thin films. The method is based on the Kretschmann method for the observation of surface plasmon waves at metallic interfaces [3.71]. The sample geometry usually consists of a combination of prism/electrode (often gold)/nonlinear optical film/electrode (semitransparent, e.g. ITO)/substrate. The beam incident on one face of the prism is either s- or p-polarized. The reflected beam is now measured as a function of angle of incidence. Various waveguide modes can be excited as a function of θ which leads to sharp dips in the reflected signal. Electro-optic coefficients can now be determined by measuring the shift in coupling angles of the waveguide modes with applied voltage.

(v) *Resonant reflection mode Fabry–Perot microcavities:* with Fabry–Perot microcavities electrorefraction, electroabsorption, and possible converse piezoelectric contributions to the electro-optic effects can be measured. This concept is ideally suited for polymeric thin films [3.72]. A beam with defined polarization is weakly focused on a resonant mode Fabry–Perot microcavity and the reflected optical signal is recorded as a function of angle of incidence.

This experiment is performed with and without application of an electrical modulation field. Analysis of the two measurements then yields the electro-optic coefficients.

(vi) *Pulsed electro-optic* measurements. With this technique the electro-optic response is determined in the time domain. Usually a Michelson interferometer, as described above, is used. The difference lies in the shape of the electric field, which is applied in the shape of a step function (rise time of typically 1 ns or shorter). An analysis of the time-dependent electro-optic response allows the distinction between the acoustic phonon contribution and the clamped electro-optic contribution [3.15, 3.73] since the acoustic phonons only build up after some μs.

(vii) *Second-harmonic generation.* The nonlinear optical coefficients are first determined by second-harmonic generation using one of the techniques described above. These values are obviously of purely electronic origin. Using the two-level model [3.74] the electronic contribution to the electro-optic coefficient, r^e, can be calculated by using the approximate wavelength dispersion of $\chi^{(2)}$.

3.4.2   Single Crystals

There have been significant advances in understanding and optimizing classical π-conjugated donor–acceptor chromophores with large first-order molecular hyperpolarizabilities in the area of organic nonlinear optics in the last few years [3.1–4, 3.6, 3.32, 3.41, 3.54]. However, there are only few chromophores with very large molecular hyperpolarizabilities such as donor–acceptor stilbenes and tolanes that form potentially useful crystalline materials. The interest in molecular crystals stems from the fact that the potential upper limits of macroscopic nonlinearities and long-term orientational stability as well as the optical quality of molecular crystals are significantly superior to those of polymers. In addition to a large molecular hyperpolarizability, the second-order macroscopic nonlinearities are strongly dependent on the relative arrangement and orientation of the π-conjugation chromophores in the crystalline solid. To be an efficient as well as useful second-order nonlinear optical crystalline material, the orientation of the chromophores in the bulk also needs to be optimized.

We have calculated upper limits with regard to electro-optic and nonlinear optical coefficients (Chap. 2). This calculation showed that the macroscopic susceptibilities of crystalline materials based on highly extended π-conjugated donor–acceptor chromophores, e.g. donor–acceptor disubstituted stilbenes have by far not reached the upper limit yet.

Crystal growth is the prototype of self-assembly in nature [3.167]. However, crystallization of large organic molecules with desired optical properties is still a challenging topic. There are several routes to achieve optimized nonlinear optical organic crystals that are summarized here (see e.g. [3.168]).

(i) *Use of molecular asymmetry.* Molecules tend to undergo shape simplification during crystal growth, which gives rise to dimers and then high-order aggregates in order to adapt a close-packing in the solid state [3.169]. The high tendency of achiral molecules to crystallize centrosymmetrically could be due to such a close-packing driving force. Therefore if the symmetry of the chromophores is reduced dimerization and subsequent aggregation are no longer of advantage to the close packing and increase the probability of acentric crystallization. This symmetry reduction can be accomplished by either the introduction of molecular (structural) asymmetry or the incorporation of steric (bulky) substituents into the chromophore. These two approaches were widely and successfully applied to benzenoid chromophores (see Fig. 3.9). Tsunekawa and co-workers found that an introduction of a substituent at the 3-position of 4'-nitrobenzylidene 4-donor-substituted-aniline can induce a favorable non-centrosymmetric packing for large optical nonlinearities [3.170,171]. This led to the discovery of 4'-nitrobenzylidene-3-acetamino-4-methoxyaniline, MNBA, which shows a large SHG efficiency that is 230 times of that of urea. Another example using this approach, developed by Tam and co-workers is 3-methyl-4-methoxy-4'-nitrostilbene, MMONS, which shows an SHG powder efficiency of 1250 times that of urea [3.32, 3.147, 3.172].

(ii) The *use of strong Coulomb interactions* can help to override the weak dipole–dipole interactions to induce a noncentrosymmetric packing. Meredith proved the validity of this concept for the case of 4-dimethylamino-N-methylstilbazolium salts. This led to the discovery of 4-dimethylamino-N-methylstilbazolium methylsulfate, DMSM, which shows an SHG efficiency of 220 times of that of urea [3.173]

Subsequently, Nakanishi and co-workers found that the 4-toluenesulfonate anion was an effective counterion to induce the noncentrosymmetric packing of stilbazolium chromophores, which led to the development of 1-methyl-4-(2-(4-hydroxyphenyl)vinyl)pyridium (or 4-hydroxy-N-methylstilbazolium) 4-toluene-sulfonate, MC-PTS, which exhibits an SHG signal of 14 times of that of urea at 1.06 μm [3.151–153].

Marder and co-workers adopted the same strategy to perform an extensive investigation by means of varying the counterions of various stilbazolium chromophores including 2-N-methylstilbazolium and 4-N-methylstilbazolium cations [3.174–175]. They found that whereas rod-shaped 4-N-methylstilbazolium cations can often be forced to crystallize noncentrosymmetrically this is not true for non-rod-shaped 2-N-methylstilbazolium cations, a result also found by others (our work, unpublished). 4-dimethylamino-N-methylstilbazolium 4-toluene-sulfonate, DAST, was shown to exhibit the largest powder SHG efficiency (1000× urea) at 1.9 μm. As in several other stilbazolium-based acentric crystals, a polar ionic sheet-packing motif was evidenced in the crystal packing of DAST. However, we have found that by either incorporating a non-planar or bulky donating group or replacing the phenyl ring with a

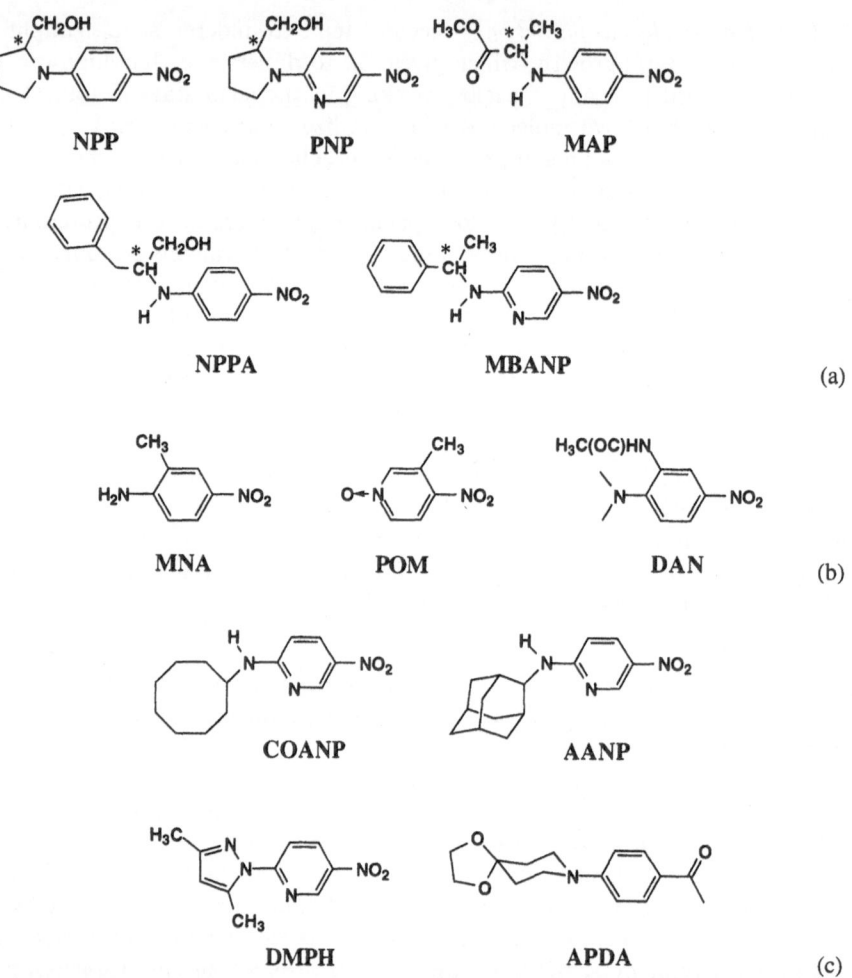

Fig. 3.9. Engineering strategies for inducing an acentric packing of benzenoid chromophores and examples: (**a**) introduction of chirality (* refers to a chiral center), (**b**) use of molecular asymmetry, (**c**) incorporation of steric (bulky) substituents

heteroaromatic ring into the skeleton of 4-*N*-methylstilbazolium cations, the probability of getting acentric crystals went down significantly. This suggests that the polar ionic sheet is very sensitive to the structural change of the stilbazolium cation. The ease of twinning and cracking of ionic crystals such as DAST are often detrimental for growing a large and good optical quality bulk crystal. Nevertheless, very recently we have accomplished the growth of such large, high-quality bulk crystals of sizes up $18 \times 18 \times 4\,\text{mm}^3$ by controlled temperature lowering technique [3.176].

(iii) *Use of non-rod-shaped m-conjugated cores.* In contrast to donor–acceptor disubstituted stilbene derivatives, hydrazone derivatives generally adopt a bent, non-rod-shaped conformation in the solid state because of the non-rigid nitrogen–nitrogen single bond ($-CH=N_{\backslash NH}-$). We have found that donor-substituted (hetero)aromatic aldehyde-4-nitrophenylhydrazones show an overwhelmingly high propensity for a noncentrosymmetric packing [3.36, 3.62, 3.177]. Of particular importance is that the majority of these acentric crystals exhibit very strong SHG signals that are at least two orders of magnitude greater than that of urea. Furthermore, most of the hydrazone crystals developed show very good crystallinity and high thermal stability. Note, however, that the flexibility of the hydrazone backbone poses a problem of polymorphism; however, with proper control of the growth conditions such as careful choices of solvent and method of crystal growth, the desirable acentric bulk crystal phase can be selectively grown [3.165].

The best example in this class is 4-dimethylaminobenzaldehyde-4-nitrophenylhydrazone, DANPH, which exhibits a very strong SHG signal that is comparable to that of DAST. Another potential candidate is 5-(methylthio)-thiophenecarboxaldehyde-4-nitrophenylhydrazone, MTTNPH, which also shows the same order of powder efficiency as DANPH. However, its molecular hyperpolarizability, β_0 (usually equal to β_μ, the first-order molecular hyperpolarizability, projected along the direction of the permanent dipole, usually determined by electric field-induced second-harmonic generation experiments (see Sect. 3.4.1.A)) is only half that of DANPH [3.36, 3.164]. However, MTTNPH molecules have a large off-diagonal tensor component which is one-third of the diagonal tensor component, according to the finite-field calculations using AM 1 parameters in MOPAC 6. These components lead to large off-diagonal elements of the nonlinear optical susceptibility. A third example is the newly developed 3,4-DHNPH, with an excellent alignment of the chromophores in the crystal lattice and a molecular hyperpolarizability comparable to DANPH [3.166].

(iv) *Supramolecular synthetic approach.* This is the design of molecular or ionic aggregates or assemblies to favor the desired crystal packing. It offers more design feasibility as one or both molecules can be tailor-made or modified to fit one another to acquire the desirable molecular properties in the solid state. Furthermore, the physical properties such as melting point and solubility, as well as the crystal properties, such as crystallinity and ease of crystal growth of the co-crystals, can usually be improved compared with those of its starting components. Etter and co-workers first demonstrated the induction of a net dipole moment with a complimentary host-guest pair of 4-aminobenzoic acid and 3,5-dinitrobenzoic acid; however, the SHG signal generated by this co-crystal is in the order of the urea standard [3.178].

We have found that the co-crystals formed from the merocyanine dyes (Mero-1 and Mero-2) and the class I phenolic derivatives in which the electron acceptor is *para*-related to the phenolic functionality together with a

Substituent

Mero-1 : R=CH₃

Mero-2 : R=CH₂CH₂OH

Class I phenolic derivatives

Fig. 3.10. Chemical structures of Mero-1, Mero-2 and Class I phenolic derivatives

substituent either in the *ortho-* or *meta-*position (Fig. 3.10) show the highest tendency of forming acentric co-crystals. In addition, a large fraction of acentric co-crystals (25%) based on Mero-2 and the class I phenolic derivatives exhibit strong second-harmonic signals that are at least two orders of magnitudes larger than that of urea. Their packing motifs can be distinctively divided into two groups.

The type I co-crystal is generally characterized by anionic and cationic assemblies or arrays. A fascinating example in this class is the co-crystal Mero-2·DBA (DBA = 2,4-dihydroxy-benzaldehyde) [3.162]. Mero-2·DBA contains a water molecule and packs noncentrosymmetrically with space group $P1$ and point group 1. The anionic assembly is constructed by the co-aggregation of two DBA molecules in which one of the molecules gives up a proton and bonds to another by a hydrogen bond. Additionally, Mero-2 acquires a proton and co-aggregates in antiparallel fashion with another Mero-2 by a short hydrogen bond constituting a cationic assembly. The interesting fact is now that, although the net dipole moment almost vanishes in this arrangement, the Mero-2·DBA co-crystal exhibits a large second-harmonic signal in the powder test. This can be explained by the asymmetric position of the hydrogen-bonded proton between the two Mero-2 dyes (Fig. 3.11) which results in a positive reinforcement of molecular hyperpolarizabilities within the cationic assembly since Mero-2 has a negative sign of the hyperpolarizability and the protonated form of Mero-2 ([Mero-2-H⁺]) has a positive sign of the hyperpolarizability. As a consequence, the co-crystal Mero-2·DBA is a potential candidate for linear electro-optic effects because of its perfectly parallel alignment of molecular hyperpolarizabilities in the solid state.

Type II co-crystals are formed by linear molecular aggregates. One of the representative examples in this class is the co-crystal Mero-2·DAP which exhibits a very strong SHG signal that is three orders of magnitudes larger than that of urea [3.160]. The molecular aggregate is assembled by the highly electronegative oxygen of Mero-2 and the acidic proton of the phenolic derivative through a short hydrogen bond. Then the rod-like aggregates connect laterally by hydrogen bonds, resulting in a staircase-like polar chain. These polar chains align in a parallel fashion constituting a two-dimensional acen-

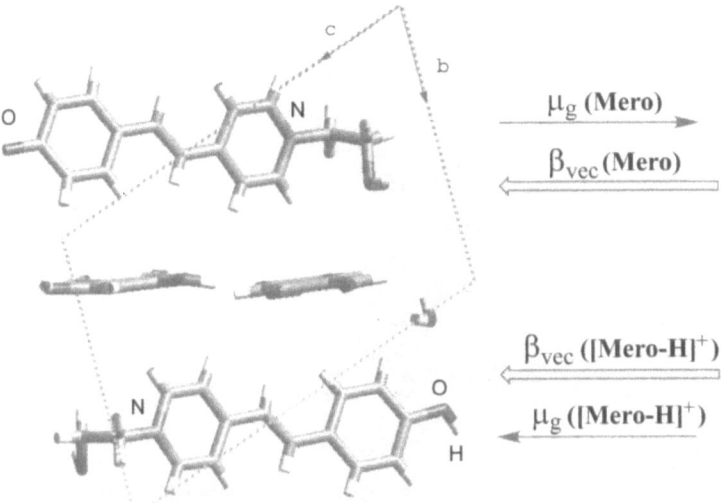

μ_g (Mero)

β_{vec} (Mero)

β_{vec} ([Mero-H]$^+$)

μ_g ([Mero-H]$^+$)

Fig. 3.11. Crystal structure of Mero-DBA co-crystal

tric layer which is found to be the common and key feature of all the highly noncentrosymmetric co-crystals in this class. Since the charge transfer axis of Mero-2 is inclined by an angle of about 70° to the polar direction of the crystal, this co-crystal is an attractive candidate for nonlinear optical effects such as frequency-doubling. In addition, we have found in this newly developed system that the orientation of the merocyanine dye can be changed and tuned within the crystal lattice by a "careful" selection of a guest molecule–phenolic derivative, provided that the linear molecular aggregate and the acentric layer packing motifs are maintained. Although Mero-2 only exists in an amorphous state by itself, both types of co-crystals formed show greatly improved crystal and physical properties compared to its constituted components. Another interesting type II co-crystal shown in Fig. 3.12 is Mero-2-MDB which is optimized for electro-optic applications due to the parallel alignment of the nonlinear optical chromophores [3.161].

Other crystal engineering approaches include the design of molecules with a deliberately low dipole moment. Due to negligible dipole–dipole interactions an almost parallel alignment of the molecules (ideal for electro-optics) can be achieved [3.146], with the drawback, however, that the conjugation is interrupted. No data on nonlinear optical experiments is available yet.

Furthermore, octupolar molecules having no dipole moment were synthesized in order to induce an asymmetric crystal packing due to the absence of dipole–dipole interactions [3.179]. Initial expectations of a higher occurrence of noncentrosymmetric crystalline materials could not yet be confirmed, likely due to the strong influence of steric forces (see above).

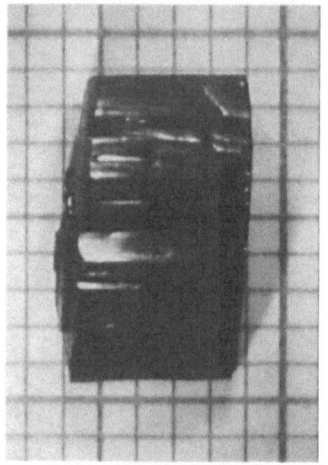

(a)

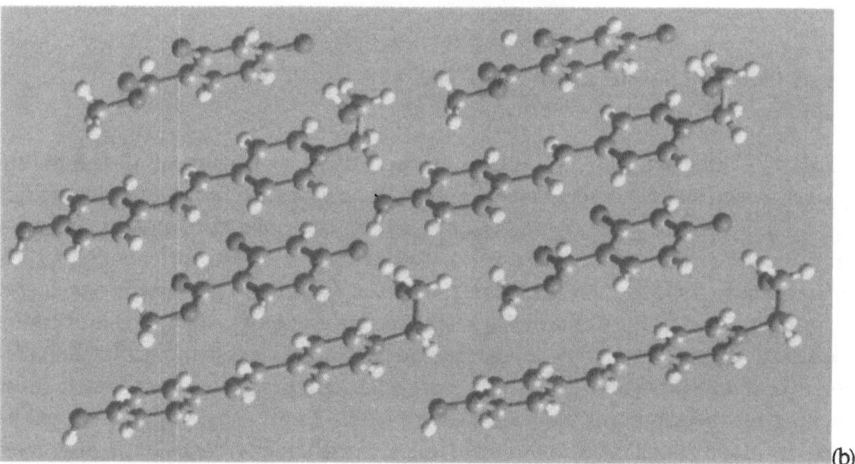

(b)

Fig. 3.12. (a) An as grown crystal of Mero-2-MDB. (b) X-ray structure of the hydrogen bond-directed acentric layer structure of the co-crystal Mero-2 · MDB

Another route is the incorporation of highly nonlinear optical molecules (with associated large dipole moments) into a lattice host. An example is perhydrotriphenylene, which serves as a channel-like framework for highly active molecules [3.180, 3.181]. A dilution of the chromophore density and a reduction of the nonlinear optical activity must be taken into account in this case. However, an optimum polar alignment of the molecules can be achieved in this way. Nonlinear optical measurements on crystals must be performed in order to know whether this approach leads to competitive materials. Other solutions for the design of nonlinear optical crystals have been proposed using assemblies of organic chromophores and inorganic salts [3.182].

Table 3.3 lists examples of electro-optic and/or nonlinear optical organic crystals. Only materials that are noncentrosymmetric and where the crystal structure is known are included. With the exception of a few classical examples, only crystals on which results in 1990 or later were reported are included in Table 3.3. The crystals are organized with respect to the cut-off wavelength λ_c, where λ_c was taken from literature and not adjusted for a consistent definition:

- $\lambda_c = 300\,\mathrm{nm}$
- $300\,\mathrm{nm} \leq \lambda_c \leq 450\,\mathrm{nm}$
- $450\,\mathrm{nm} \leq \lambda_c \leq 550\,\mathrm{nm}$
- $\lambda_c \geq 550\,\mathrm{nm}$

The values of the first-order hyperpolarizability β and the nonlinear optical coefficients d were adjusted to the same reference value (e.g. $d_{11}(\alpha\text{-quartz})=0.3\,\mathrm{pm/V}$ ($\lambda=1064\,\mathrm{nm}$) and $=0.28\,\mathrm{pm/V}$ ($\lambda=1907\,\mathrm{nm}$) see Chap. 2). θ_p is the angle between the molecular dipole axis and the polar crystal axis. The full names of the crystals related to the abbreviations can also be found in Table 3.3. As a general remark an increase of the nonlinearity is often accompanied by an increase of the cut-off wavelength.

A $\lambda_c \leq 300\,\mathrm{nm}$

These materials are all colorless. HMTA is the first material on this list because it is the first electro-optic organic material [3.75, 3.76]. Urea was the first organic crystal in which interesting applications such as optic parametric oscillation (conversion efficiency up to 20%) and UV light generation by frequency doubling (down to 213 nm) were demonstrated [3.79, 3.183–185]. In L-PCA UV, generation down to 266 nm was demonstrated by phase-matched frequency doubling [3.82]. The crystal has a high melting point and large crystals up to $2.5 \times 1.5 \times 1.5\,\mathrm{cm}^3$ in size and 17.5g in weight were grown [3.186]. HFB is the last material in the group with $\lambda_c \leq 300\,\mathrm{nm}$ [3.83]. It is a semiorganic crystal with the advantage of good mechanical and thermal properties. It is easy to grow and phase-matched parametric processes between 300 nm and 1300 nm have been predicted.

Table 3.3. Examples of molecular crystals that have been investigated for their nonlinear optical and/or electro-optic response. λ_{eg} is the wavelength of maximum absorption, β is the molecular first-order hyperpolarizability (mostly given as the vector part of β_{ijk}), β_o is β extrapolated to infinite wavelengths, λ_c is the cut-off wavelength in the bulk, d is the nonlinear optical coefficient, r is the electro-optic coefficient, n is the refractive index, ε is the low-frequency dielectric constant, θ_p is the angle between the molecular dipole axis and the polar crystal axis, and T_m is the melting point. PM means that phase-matched second-harmonic generation is possible. Apart from some special cases only materials on which new results were published in 1990 or later are considered

Material	λ_{eg} (nm) solvent	$\beta(\beta_o)$ $(10^{-40}\,m^4/V)$ λ (nm) solvent	Point group	λ_c of bulk (nm)	$d, r(pm/V),$ n, ε	θ_p (deg) $T_m(°C)$	Ref.
Cut-off wavelength $\lambda_c \leq 300$ nm							
HMTA (hexamethylenetetramine)	–	–	$\bar{4}3m$	–	$n(547.5\,nm) = 1.591$ $r_{41}(546\,nm) = 0.71$–0.8	–	[3.75–76]
Urea	< 200	0.76 1064 DMSO	$\bar{4}2m$	200	$d_{14}(480\text{–}640\,nm)=1.0$ $d_{14}(1064\,nm)=1.1$ $r_{41}(632.8\,nm)=1.9$ $r_{63}(632.8\,nm)=0.8$ $n_o(632.8\,nm)=1.474$ $n_e(632.8\,nm)=1.590$ PM	$T_m=133\text{-}135$	[3.77–81]
L-PCA (L-pyrrolidone-2-carboxylic acid)	–	–	222	260	$d_{14}(532\,nm)=0.32$ reference: $d_{eff}=1.62$ of BBO (type I PM) $n_1(632.8\,nm)=1.49$ $n_2(632.8\,nm)=1.52$ $n_3(632.8\,nm)=1.63$ PM	$T_m=162$	[3.82]

Compound					
HFB (L-histidine tetrafluoroborate)	—	—	2	250 d_{eff}(1064 nm)$\approx$2 PM	T_m=205 [3.83]

Cut-off wavelength 300 nm $\leq \lambda_c \leq$ 450 nm

Compound					
MHBA (3-methoxy-4-hydroxy-benzaldehyde)	—	—	2	370 d_{13}(1064 nm)=13 d_{11}(1064 nm)=9.8 n_1(546.1 nm)=1.558 n_2(546.1 nm)=1.700 n_3(546.1 nm)=1.893	T_m=82–83 [3.84-85]
IAPU (isopropyl-4-acetylphenylurea)	293 CHCl$_3$	42 (27) by MOPAC/AM1 calculations	2	380 d_{22}(1064 nm)=30.5 n_2(1064 nm)=1.779 n_2(532 nm)=1.700 n_3(1064 nm)=1.634 n_3(532 nm)=1.610 PM	θ_p=33.4 T_m=150 [3.86]
FMA (2-furyl methacrylic anhydride)	—	—	4 mm	380 d_{33}(1064 nm)=18 d_{31}(1064 nm)=12 n_e(632.8 nm)=1.887 n_o(632.8 nm)=1.641 PM	T_m=135 [3.87]

Table 3.3. (Continued)

Material	λ_{eg} (nm) solvent	$\beta(\beta_o)$ $(10^{-40} \text{m}^4/\text{V})$ λ (nm) solvent	Point group	λ_c of bulk (nm)	$d, r(\text{pm/V})$, n, ε	θ_p (deg) T_m (°C)	Ref.
				Cut-off wavelength 300 nm $\leq \lambda_c \leq$ 450 nm			
APDA (8-(4'-acetylphenyl)-1,4-dioxa-8-azaspiro[4.5]decane)	—	—	mm2	384	d_{31} (1064 nm)=7, d_{33} (1064 nm)=50, reference not specified, n_a (532 nm)=1.679, n_b (532 nm)=1.565, n_c (532 nm)=1.702, PM	T_m=123–124	[3.88–90]
ABP (4-aminobenzophenone)	—	—	2	400	d_{eff} (954 nm)=6.8 (type I), d_{eff} (954 nm)=1.4 (type II), reference: d_{eff}=1.62 of BBO (type I PM), n_1 (477 nm)=1.65, n_2 (477 nm)=1.67, n_3 (477 nm)=2.07, PM	T_m=122.3	[3.91–92]
5-NU (5-nitrouracil)	295 ethanol; 284 CCl$_4$	—	222	410	d_{14} (1064 nm)=5.2, PM	—	[3.93]

						Ref.
2A5NPDP (2-amino-5-nitropyridinium dihydrogen phosphate) H_2N—⟨NO₂ pyridinium⟩—N^+ H^+ $H_2PO_4^-$	347 2A5NP in acetone	85 (44) 1064 2A5NP in acetone	$mm2$	420	d_{15} (1064 nm)=5.4 d_{24} (1064 nm)=0.96 n_1 (546 nm)=1.642 n_2 (546 nm)=1.656 n_3 (546 nm)=1.772 PM $\theta_p=36.7$	[3.94–96]
DMACB (dimethylaminocyanobiphenyl) $(H_3C)_2N$—⟨biphenyl⟩—CN	345 CHCl₃	260 (135) 1064 CHCl₃ reference not given	m	420	d_{11} (1064nm)=276 r_{11} (1064nm)=55 estimated from oriented gas model $\theta_p=0$	[3.55, 3.97]
PCNB (4·(isopropylcarbomoyl)nitro-benzene) $(H_3C)_2HCOOCHN$—⟨benzene⟩—NO₂	—	—	$mm2$	420	n_1 (632.8 nm)=1.63 n_2 (632.8 nm)=1.58 n_3 (632.8 nm)=1.74 PM —	[3.98]
BMC (4-Br-4'-methoxychalcone) H_3CO—⟨benzene⟩—C(=O)—CH=CH—⟨benzene⟩—Br	—	—	m	420	d_{13} (1064 nm)=16.4 d_{33} (1064 nm)=3.8 n_x (532 nm)=1.58 n_y (532 nm)=1.50 n_z (532 nm)=1.92 PM —	[3.99]

Ch. Bosshard et al.

Table 3.3. (Continued)

Material	λ_{eg} (nm) solvent	$\beta(\beta_o)$ ($10^{-40}\,\mathrm{m^4/V}$) λ (nm) solvent	Point group	λ_c of bulk (nm)	d, r (pm/V), n, ε	θ_p (deg) T_m (°C)	Ref.
Cut-off wavelength $300\,\mathrm{nm} \leq \lambda_c \leq 450\,\mathrm{nm}$							
EMC (4-ethoxy-4'-methoxychalcone)	–	–	$mm2$	430	d_{eff} (1064 nm)=4.3, n_x (532 nm)=1.493, n_y (532 nm)=1.710, n_z (532 nm)=1.983, PM	–	[3.100]
MTP (3-(4-methylphenyl)-1-(2-thienyl)-2-propen-1-one)	–	–	2	430	d_{eff} (1064 nm)=5.6, n_x (532 nm)=1.602, n_y (532 nm)=1.665, n_z (532 nm)=1.713, PM	–	[3.101]
BBCP (2,5-bis(benzylidene)-cyclopentanone)	–	–	222	435 (pol. along a) 445 (pol. along b) 435 (pol. along c)	d_{14} (1064 nm)=4.2, n_a (532 nm)=1.596, n_b (532 nm)=1.850, n_c (532 nm)=1.827, PM	T_m=198	[3.102]

Compound	λ/solvent	Point group	λ (nm)	Optical properties	T_m	Ref.
MBBCH (2,6-bis(p-methylbenzylidene)-4-tert-butylcyclohexanone)	—	$mm2$	435 (pol. along a) 445 (pol. along b) 435 (pol. along c)	d_{31} (1064 nm)=9 d_{32} (1064 nm)=7.2 d_{33} (1064 nm)=2.4 n_a (532 nm)=1.471 n_b (532 nm)=1.762 n_c (532 nm)=1.711 PM	T_m=157	[3.102]
DIVA (ortho-dicyanovinyl-anisole)	365 solvent not given	2	440	d_{eff} (888 nm)=81 (type I) Ref. 13.7 pm/V of KNbO$_3$ PM d_{22} (1064 nm)=6 d_{eff} (1064 nm)=12 n_1 (450 nm)=1.58 n_2 (450 nm)=1.79 n_3 (450 nm)=2.37 PM	—	[3.103–104]
DMNP (3,5-dimethyl-1-(4-nitrophenyl)-pyrazole)	—	$mm2$	450	d_{21} (950 nm)=90 d_{22} (950 nm)=29 reference not specified n_x (884 nm)=1.513 n_y (884 nm)=1.793 n_z (884 nm)=1.696 PM	T_m= 102.5	[3.105]

Table 3.3. (Continued)

Material	λ_{eg} (nm) solvent	$\beta(\beta_0)$ (10^{-40} m^4/V) λ (nm) solvent	Point group	λ_c of bulk (nm)	d, r (pm/V), n, ε	θ_p (deg) T_m (°C)	Ref.
				Cut-off wavelength 450 nm $\leq \lambda_c \leq$ 550 nm			
MNA (2-methyl-4-nitroaniline)	361 dioxane	44 (37), 46 (38), 1907 dioxane	m	480	d_{11} (1064 nm) = 150, d_{12} (1064 nm) = 23, r_{11} (632.8 nm) = 67, n_1 (632.8 nm) = 2.00	θ_p = 21, T_m = 133	[3.17, 3.32, 3.106–111]
			1	480		T_m = 132	
Mixed MNA/DPA (2-methyl-4-nitroaniline/2,4-dinitrophenyl-L-alanine)	–	–	2	–	–	T_m = 138	[3.111]
Mixed p-NA/NENA (para-nitroaniline/N-ethyl-4-nitroaniline)	354 p-NA in dioxane, 376 p-NA in acetone	24 (19.5), 1907 p-NA in acetone	–	≈500	d (average, 1064 nm) = 71	–	[3.41, 3.112, 3.113]

Compound				Measurements		
MNMA (2-methyl-4-nitro-N-methyl-aniline)	—	$mm2$	500	d_{33} (1064 nm)=2.0 d_{31} (1064 nm)=9.8 r_{33} (632.8 nm)=7.6 r_{13} (632.8 nm)=8.0 n_1 (632.8 nm)=2.178 n_3 (632.8 nm)=1.520	T_{m}=139.5 —	[3.114]
DBNMNA (2,6-dibromo-N-methyl-4-nitroaniline)	360 CNCH$_3$	—	$mm2$	500	r_{42} (514.5 nm)=86 $n^3 r$ (514.5 nm)=367 $n^3 r$ (632.8 nm)=79 $n^3 r$ (810 nm)=40 n_1 (632.8 nm)=1.90 n_2 (632.8 nm)=1.62 n_3 (632.8 nm)=1.48	—	[3.115]
DAN (4-(N,N-dimethylamino)-3-acetamidonitrobenzene)	360 dioxane	55 (45) experimental details not specified dioxane	2	485	d_{23} (1064 nm)=38 r_{32} (632.8 nm)=13 n_1 (632.8 nm)=1.539 n_2 (632.8 nm)=1.682 n_3 (632.8 nm)=1.949 PM	θ_{p}=70.8 T_{m}=165.7	[3.116,117]

Table 3.3. (Continued)

Material	λ_{eg} (nm) solvent	$\beta(\beta_o)$ $(10^{-40}\,m^4/V)$, λ (nm) solvent	Point group	λ_c of bulk (nm)	d, r(pm/V), n, ε	θ_p(deg) T_m(°C)	Ref.
			Cut-off wavelength $450\,nm \leq \lambda_c \leq 550\,nm$				
mNA (meta-nitroaniline)	–	–	$mm2$	500	d_{31} (1064 nm)=11.7 d_{33} (1064 nm)=12.3 r_{33} (632.8 nm)=16.7 n_1 (632.8 nm)=1.751 n_2 (632.8 nm)=1715 n_3 (632.8 nm)=1.665 ε_1 (3 kHz)=3.9 ε_2 (3 kHz)=4.2 ε_3 (3 kHz)=4.6 PM	T_m=112	[3.118–120]
POM (3-methyl-4-nitropyridine-1-oxide)	320 from crystal data	21 (12.6) 1064 from crystal data	222	460	d_{14} (1064 nm)=6.0 d_{eff} (1064 nm)=5.7 (type I) d_{eff} (1064 nm)=2.6 (type II) r_{41} (632.8 nm)=3.6 r_{52} (632.8 nm)=5.2 r_{63} (632.8 nm)=2.6 n_1 (632.8 nm)=1.711 n_2 (632.8 nm)=1.920 n_3 (632.8 nm)=1.641 PM	T_m=136	[3.121–123]

Compound	λ_{max} (nm)	$\mu\beta$ (nm)	point group		d-coefficients, refractive indices	properties	Ref.
MAP (methyl-(2,4-dinitrophenyl)-aminopropanoate) CH_3 $H-C-NH-$ (ring with NO_2, NO_2) CH_3O_2C	360 from crystal data	55 (26.4) 1064 acetone	2	500	d_{21} (1064 nm) $=10.0$ d_{22} (1064 nm) $=11.1$ n_x (532 nm) $=1.557$ n_y (532 nm) $=1.710$ n_z (532 nm) $=2.035$ PM	$T_m=69$	[3.124]
NPP (N-(4-nitrophenyl)-(L)-prolinol) CH_2OH (ring with N, NO_2)	391 ethanol 397 acetone 390 $CHCl_3$	106 (40) 1064 acetone 113 (45) 1064 $CHCl_3$	2	500	d_{21} (1064 nm) $=51$ d_{22} (1064 nm) $=16.8$ assuming d_{11} (quartz) $=0.3\,\mathrm{pm/V}$ n_x (632.6 nm) $=2.066$ n_y (632.8 nm) $=1.876$ n_z (632.8) $=1.478$ PM	$\theta_p=58.6$ $T_m=116$	[3.125] [3.126]
NPAN (N-(4-nitrophenyl)-N-methyl-aminoacetonitrile) H_3C $N-$ (ring with NO_2) $N\equiv C-CH_2$	358 ethanol 360 acetone	60 (23) 1064 acetone	$mm2$	500	d_{32} (1064 nm) $=57$ d_{33} (1064 nm) $=27$ PM	$\theta_p=58.6$ $T_m=114$	[3.127,128]

Table 3.3. (Continued)

Material	λ_{eg} (nm) solvent	$\beta(\beta_o)$ ($10^{-40}\,\mathrm{m}^4/\mathrm{V}$) λ (nm) solvent	Point group	λ_c of bulk (nm)	$d, r\,(\mathrm{pm/V})$, n, ε	θ_p(deg) T_m(°C)	Ref.
				Cut-off wavelength 450 nm $\leq \lambda_c \leq$ 550 nm			
DAD ((−)-1-(4-dimethylaminophenyl)-2-(2-hydroxypropylamino)cyclobutene-3,4-dione)	388 dioxane	720 by solvatochromic method	1	500	d_{11} (1064 nm)=150 n_1 (1064 nm)=1.81 n_2 (1064 nm)=1.60 n_3 (1064 nm)=1.56	θ_p=10	[3.129]
NPNa (4-nitrophenol sodium salt hydrate)	–	–	mm2	≈500	d_{eff} (1064 nm)=5.0 n_1 (632.8 nm)=1.414 n_2 (632.8 nm)=1.979 n_3 (632.8 nm)=1.582 PM	–	[3.130–131]
DPNa (4-nitrophenol sodium salt hydrate (deuterated water))	–	–	mm2	≈500	d_{eff} (1064 nm)=5.5 n_1 (632.8 nm)=1.414 n_2 (632.8 nm)=1.979 n_3 (632.8 nm)=1.582 PM	–	[3.131–132]

Compound					Properties		References
PNP (2-(N-prolinol)-5-nitropyridine)	376 dioxane	66 (53) 1907 dioxane	2	490	d_{21} (1064 nm)=51 d_{22} (1064 nm)=16.2 r_{22} (514.5 nm)=28.3 r_{22} (632.8 nm)=12.8 n_1 (632.8 nm)=1.990 n_2 (632.8 nm)=1.788 n_3 (632.8 nm)=1.467 PM	θ_p=59.6 T_m=83	[3.17, 3.133–135]
	370 CHCl₃	83 (38) 1064 CHCl₃					[3.126]
COANP (2-cyclooctylamino-5-nitro-pyridine)	361 dioxane	64 (59) 1907 dioxane	mm2	490	d_{32} (1064 nm)=32 r_{33} (514.5 nm)=28 r_{33} (632.8 nm)=15 n_1 (632.8 nm)=1.672 n_2 (632.8 nm)=1.781 n_3 (632.8 nm)=1.647 ε_1 (3 kHz)=2.8 ε_2 (3 kHz)=2.7 ε_3 (3 kHz)=2.6 PM	θ_p=61.8 T_m=72.8	[3.17, 3.133, 3.136, 3.137]
(-)MBANP ((-)-2-(α-methylbenzylamino)-5-nitropyridine)	359 ethanol	63 (30) reference not specified 1064	2	450	d_{22} (1064 nm)=36 r_{eff} (488 nm)=31.4 r_{eff} (514.5 nm)=26.6 r_{eff} (632.8 nm)=18.2 PM	θ_p=33.2 T_m=83	[3.138–141]

Table 3.3. (Continued)

Material	λ_{eg} (nm) solvent	$\beta(\beta_o)$ (10^{-40} m^4/V) λ (nm) solvent	Point group	λ_c of bulk (nm)	d, r (pm/V), n, ε	θ_p (deg) T_m (°C)	Ref.
				Cut-off wavelength 450 nm $\leq \lambda_c \leq$ 550 nm			
(±)MBANP ((±)2-(α-methylbenzyl-amino)-5-nitropyridine)	—	—	$mm2$	440 (pol. along a) 430 (pol. along c)	d_{31} (1064 nm)=6.8 d_{32} (1064 nm)=4.7 d_{33} (1064 nm)=0.84 n_1 (632.8 nm)=1.78 n_3 (632.8 nm)=1.61	θ_p=77.8 T_m=124	[3.141]
AANP (2-adamantylamino-5-nitro-pyridine)	—	—	$mm2$	460	d_{31} (1064 nm)=48 d_{33} (1064 nm)=36 n_1 (532 nm)=1.77 n_2 (532 nm)=1.61 n_3 (532 nm)=1.86 PM	θ_p=58 T_m=167	[3.142–143]
NPPA (N-(4-nitro-2-pyridinyl)-phenyl-aninol)	345 from crystal data	85 (42) 1064 from crystal data	2	500	d_{23} (1064 nm)=23 d_{23} (1318 nm)=21 n_1 (632.8 nm)=1.524 n_2 (632.8 nm)=1.694 n_3 (632.8 nm)=1.907 PM	θ_p=74.8	[3.144]

Compound							Ref.
NPLO (L-N-(5-nitro-2-pyridyl)leucinol) $H_3C{-}CH$ H_3C HOH_2C N N NO_2	359 dichloro-methane	—	2	480	d_{eff} (1064 nm)=33 (type I PM) d_{eff} (1064 nm)=2.7 (type II PM) n_1 (632.8 nm)=1.457 n_2 (632.8 nm)=1.631 n_3 (632.8 nm)=1.933 PM	θ_p=76.4 T_m=118	[3.145]
Azine derivative Br CH_3 $N{-}N$ H_3C OCH_3	—	—	2	—	—	$\theta_p \approx 0$	[3.146]
MMONS (3-methyl-4-methoxy-4'-nitro-stilbene) H_3C H_3CO NO_2	366 dioxane	138 (114) 1907 dioxane	$mm2$	515	d_{33} (1064 nm)=112 d_{24} (1064 nm)=43 r_{33} (632.8 nm)=40 n_1 (632.8 nm)=1.569 n_2 (632.8 nm)=1.693 n_3 (632.8 nm)=2.129 PM	θ_p=34	[3.147]

Table 3.3. (Continued)

Material	λ_{eg} (nm) solvent	$\beta(\beta_0)$ (10^{-40} m^4/V) λ (nm) solvent	Point group	λ_c of bulk (nm)	d, r (pm/V), n, ε	θ_p (deg) T_m (°C)	Ref.
		Cut-off wavelength 450 nm $\leq \lambda_c \leq$ 550 nm					
MNBA (4'-nitrobenzylidene-3-acetamino-4-methoxyaniline)	380 dioxane	88 (74) 1907 dioxane	m	520	d_{11} (1064 nm)=131 r_{11} (532 nm)=50 r_{11} (632.8 nm)=29 n_1 (632.8 nm)=2.024 n_2 (632.8 nm)=1.648 n_3 (632.8 nm)=1.583 ε_1 (3 kHz)=4.0 ε_2 (3 kHz)=2.9 ε_3 (3 kHz)=3.2	θ_p=18.7	[3.38]
NMBA (4-nitro-4'-methylbenzylidene aniline)	351 dioxane	40 (33) 1907 dioxane	m	500	d_{11} (1064 nm)=83 r_{11} (488 nm)=37.2 r_{11} (514.5 nm)=36.4 r_{11} (632.8 nm)=25.2 n_1 (632.8 nm)=2.081 n_2 (632.8 nm)=1.695 n_3 (632.8 nm)=1.524 PM	$\theta_p \approx$18	[3.32, 3.148–149]
MBPO (4-(4'-methoxybenzylidene)-2-phenyloxazolin-5-one)	437 acetonitrile	100 1064 dioxane reference not given	222	$\approx$500	d_{14} (1064 nm)=4.6 reference not given n_3 (600 nm)=1.527	T_m=159–161	[3.150]

MC-PTS (4-hydroxy-N-methyl-stilbazolium-4-toluene sulfonate)	392 methanol	985 (389) 1064 methanol	1	510	d_{11} (1064 nm)$=314$	$\theta_p=0$ [3.151–153]

H_3C—N$^{\oplus}$ ... HO— ... SO_3^- ... H_3C—

Cut-off wavelength $\lambda_c \geq 550$ nm

DR1 (Disperse Red 1)	473 dioxane	570 (403)	–	–	–	$T_m=160–162$ [3.36]

H_3CH_2C, HOH_2CH_2C'N— —N=N— —NO$_2$

DCNP (3-(1,1-dicyanoethenyl)-1-phenyl-4,5-dihydro-1H-pyrazole)	466 CHCl$_3$	330 (190) calculated	m	≈630	r_{33} (632.8 nm)$=87$; n_3 (632.8 nm)$=2.7$	$T_m=194$ [3.154]

CN CN N—N—

Table 3.3. (Continued)

Material	λ_{eg} (nm) solvent	$\beta(\beta_o)$ ($10^{-40}\,\mathrm{m}^4/\mathrm{V}$), λ (nm) solvent	Point group	λ_c of bulk (nm)	$d, r\,(\mathrm{pm/V})$, n, ε	θ_p(deg) $T_m(^\circ\mathrm{C})$	Ref.
		Cut-off wavelength 450 nm $\leq \lambda_c \leq$ 550 nm					
DAST (4'-dimethylamino-N-methyl-4-stilbazolium tosylate)	475 methanol	—	m	700	d_{11} (1318 nm)=1010 d_{11} (1542 nm)=290 d_{26} (1542 nm)=39 r_{11} (720 nm)=92 r_{11} (1313 nm)=53 r_{11} (1535 nm)=47 n_1 (720 nm)=2.519 n_2 (720 nm)=1.720 n_3 (720 nm)=1.635 ε_1 (3 kHz)=5.2 ε_2 (3 kHz)=4.1 ε_3 (3 kHz)=3.0 PM	θ_p=20 T_m=256	[3.155–157]
DMSM (SPCD) (4'-dimethylamino-N-methyl-4-stilbazolium methylsulfate)	—	≈ 600	$mm2$	620	r_{33} (632.8 nm)=430 n_3 (632.8 nm)=1.55	θ_p=34	[3.158–159]

DAST structure: H_3C-N^+ ... $N(CH_3)_2$; H_3C- ... SO_3^-

DMSM structure: H_3C-N^+ ... $N(CH_3)_2$; $MeSO_4^-$

Name									
Mero-2-MDB (4-{2-[1-(2-hydroxyethyl)]-4-pyridylidene]-ethylidene}-cyclo-hexa-2,5-dien-1-one-methyl-2,4-dihydroxybenzoate)	487 methanol	420–4200	m	experimental details not known	phase II: 680 phase I: 615	phase II: d_{11} (1318 nm)=267 r_{11} (1313 nm)=34 n_1 (700 nm)=2.20 phase I: d_{11} (1318 nm)=108 r_{11} (318 nm)=24 n_1 (700 nm)=2.07	$\theta_\mathrm{p}=0$ $T_\mathrm{m}=185$	[3.160–161]
Mero-2-DBA (4-{[1-2(hydroxy-1-ethyl)-4((1H)-pyridinylidene]ethylidene]} -2,5-cyclohexadien-1-one · 2,4-dihydroxy-benzaldehyde)	487 methanol	420–4200	1	experimental details not known	—	—	$\theta_\mathrm{p}=0$ —	[3.162]

Table 3.3. (Continued)

Material	λ_{eg} (nm) solvent	$\beta(\beta_o)$ (10^{-40} m^4/V) λ (nm) solvent	Point group	λ_c of bulk (nm)	d, r(pm/V), n, ε	θ_p(deg) T_m(°C)	Ref.
		Cut-off wavelength 450 nm $\leq \lambda_c \leq$ 550 nm					
Mero-2-DAP (4-{2-[1-(2-hydroxyethyl)-4-pyridylidene]ethylidene}-cyclohexa-2,5-dien-1-one·2,4-dihydroxyacetophenone)	487 methanol	420–4200 experimental details not known	m	–	–	θ_p=70 T_m=193	[3.160]
DANPH (4-dimethylaminobenzaldehyde-4-nitrophenylhydrazone)	420 dioxane	277 (212) 1907 dioxane	m	670	d_{12} (1542 nm)=190 d_{12} (1542 nm)=140 n_1 (820 nm)=2.19 n_2 (820 nm)=2.16 n_3 (820 nm)=1.451 PM	θ_p=50 T_m=186	[3.163]
MTTNPH (5-(methylthio)-thiophenecarboxaldehyde-4-nitro-phenylhydrazone)	412 dioxane	130 (101) 1907 dioxane	mm2	620	d_{32} (1313 nm)=32 n_1 (720 nm)=2.519 n_2 (720 nm)=1.720 n_3 (720 nm)=1.635 PM	θ_p=55 T_m=172	[3.164]

NTMPH (5-nitro-2-thiophenecarbox-aldehyde-4-methylphenylhydrazone)	470 dioxane	205 (145) 1907 dioxane	2/m (I) mm2 (II) 2/m (III)	—	—	$\theta_p=51.4$ (II) $T_m=152$ (I) $T_m=157$ (II) $T_m=156$ (II)	[3.165]
3,4-DHNPH (3,4-dihydroxybenzaldehyde-4-nitrophenyl-hydrazone)	403 dioxane	340 (230) 1542 dioxane	m	730	—	$\theta_p=20$ $T_m=230$	[3.166]

B   300 nm $\leq \lambda_c \leq$ 450 nm

All materials in this class are also colorless. The first one, MHBA, is a useful material for frequency doubling Ti:Sapphire laser light [3.84]. An all solid state blue laser based on intracavity frequency doubling of MHBA was fabricated. Furthermore, big crystals with sizes up to $50 \times 25 \times 10$ mm^3 were grown from solution. IAPU is a promising candidate for second-harmonic generation into the blue spectral range [3.86]. It is transparent down to 380 nm and has a large diagonal nonlinear optical coefficient of $d_{22} = 30.5$ pm/V ($\lambda = 1064$ nm). Therefore it cannot be phase-matched in the bulk but is suitable for waveguide nonlinear optics [3.86, 3.187]. Phase-matched frequency doubling in a four-layer waveguide by mode conversion was predicted [3.86]. In actual experiments mode-to-mode conversion and the Čerenkov radiation scheme were both employed, with a lowest reported second-harmonic wavelength of 420 nm [3.187]. FMA is another 'white' crystal [3.87]. Both collinear and noncollinear (with several experimental configurations) phase-matched second-harmonic generation were demonstrated.

Single crystals of APDA can be grown from the melt with sizes up to a diameter of 12 mm and a length of 28 mm [3.90]. They are stable and transparent down to 384 nm. Due to the excellent optical quality, intracavity cw frequency doubling (810 nm → 405 nm, noncritical type II phase-matching) could be demonstrated [3.88]. In fact, APDA was the first organic crystal with UV emission. In the UV experiments, Fresnel losses were eliminated by coating the crystal with a copolymer of tetrafluoroethylene and 2,2-*bis*-trifluoromethyl-4,5-difluoro-1,3-dioxole (TEFRON AF-2400). ABP crystals can be grown in large size ($33 \times 50 \times 24$ mm^3) [3.91]. They are transparent down to 400 nm. Phase-matched frequency doubling down to 460 nm and 433 nm were experimentally measured with substantially larger effective nonlinear optical coefficients than e.g. BBO. 5-NU is a nonlinear optical crystal with reasonable nonlinearity and an excellent transparency in the blue spectral range. In addition, strong hydrogen bonds increase the laser damage threshold as well as the stability towards air [3.93, 3.188].

2A5NPDP was designed for large nonlinear optical effects by incorporating highly polarizable organic molecules arranged in polar order between phosphate polyanion sheets [3.94]. The ionic nature of the crystal strongly influences the nonlinearities and leads to a dramatic failure of the one-dimensional oriented gas model [3.95]. Type II phase-matching and thermo-optic tuning possibilities were demonstrated [3.189] and near-infrared optic parametric oscillation experiments were performed [3.190]. DMACB is a crystal based on a biphenyl chromophore. It has an optimized alignment for electro-optics combined with an excellent absorption cut-off of 420 nm [3.97]. This low absorption edge is thought to be due to excitonic coupling. Unfortunately, it is extremely difficult to grow crystals large enough for optical characterization. PCNB crystals can be grown in large sizes ($5.5 \times 10 \times 10$ mm^3) [3.98]. It has been demonstrated that laser light at $\lambda = 930$ nm can be frequency-doubled to

465 nm using noncritical phase-matching with large acceptance angles. BMC [3.99], EMC [3.100] and MTP [3.101] are chalcone-based crystals with cut-off wavelengths between λ_c=420 nm and 430 nm. Chalcone-based crystals have the advantage of rather easy crystal growth and a transparency in the blue spectral range. Using MTP intracavity second-harmonic generation pumped at $\lambda = 1064$ nm was successfully demonstrated [3.101]. BBCP and MBBCH are two organic crystals developed for frequency-doubling into the blue spectral range [3.102]. They both have high melting points and are chemically stable. Large size crystals with dimensions up to $50 \times 10 \times 3$ mm^3 (BBCP) and $20 \times 3 \times 0.5$ mm^3 (MBBCH) could be grown. DIVA is another crystal that is phase-matchable (type I) into the blue spectral range [3.104]. At a wavelength of $\lambda = 888$ nm an effective nonlinear coefficient $5.9 \times d_{32}$ of KNbO$_3$, one of the best inorganic crystals for frequency conversion into the blue, was measured making DIVA a very promising material. DMNP can be grown into cored fibers [3.105]. It combines a short cut-off wavelength with large nonlinear optical coefficients. In a 16 mm long fiber a conversion efficiency of 0.96% was reached for an input of 16.6 mW cw power at $\lambda = 884$ nm.

C   450 nm $\leq \lambda_c \leq$ 550 nm

In this class all materials have colors varying between yellow and orange. In this wavelength range the largest class of materials can be found. MNA exhibits both large electro-optic and nonlinear optical effects [3.106, 3.107]. The molecules are favorably aligned for electro-optics (angle between the molecular dipole axis and the polar crystal axis of $\theta = 21°$). Unfortunately, the growth of large size crystals is extremely difficult due to inherent cleavage planes. In a prototype transverse phase modulator, modulation frequencies up to 94 GHz were obtained [3.191]. In 1 μm thin samples ultrabroadband second-harmonic generation was demonstrated [3.110]. Mixed MNA/DPA crystals were investigated to improve the morphology of the pure MNA crystals [3.111]. Besides a DPA:MNA crystal complex, a triclinic MNA polymorph with point group $P1$ was found. Both new structures exhibited large second-harmonic signals in the powder test. Both p-NA and NENA crystals both are nonlinear optically inactive. The crystalline mixture p-NA/NENA, however, exhibits large second-harmonic signals in the powder test [3.113]. Using the second-harmonic evanescent wave technique [3.192], a large averaged nonlinear optical coefficient of 71±12 pm/V was determined. However, no crystal structure data is available yet. MNMA, of which single crystals can be grown easily, was developed by a modification of MNA. The birefringence is unusually high and can be directly related to the anisotropy of the molecules [3.114]. DBNMNA, similar to MNA, is another crystal with large electro-optic coefficients [3.115].

High-quality crystals of DAN can easily be grown from dimethylsulfoxide solutions [3.116, 3.117]. DAN combines large nonlinear optical coefficients with interesting phase-matching possibilities in bulk crystals and in crystal cored fibers [3.193]. Intracavity phase-matched second-harmonic generation has been demonstrated [3.194]. A large potential for the cascading of second-order nonlinearities was experimentally verified [3.195].

mNA is another yellow material with large nonlinear optical and electro-optic coefficients and interesting phase-matching possibilities [3.118–120]. It is one of the few organic materials in which the dielectric constants, important for fast electro-optic modulation, were determined. The low dielectric constants confirmed the almost pure electronic origin of the linear optical properties of organic materials. POM was deliberately designed with a vanishing dipole moment in order to generate an acentric crystal structure due to the absence of dipole–dipole interactions [3.121–123, 3.196]. It can be grown in large sizes (cut and polished samples larger than 1 cm^3). It is the only organic crystal in which the acoustic and acousto-optic properties have been thoroughly investigated [3.196]. MAP is a successful example where chirality was introduced to obtain an acentric crystal packing (see also Fig. 3.9) [3.124].

NPP is one of the materials that is farthest developed with respect to applications. The angle of $\theta_p = 58.6°$ makes it a material optimized for parametric interactions. Low-threshold parametric oscillation with large conversion efficiencies was demonstrated [3.96, 3.197–198]. Recently, a detailed study of the crystal growth behavior of NPP in different solvents led to high-quality large crystals up to $2 \times 0.8 \times 0.8$ cm^3 in size [3.199].

In NPAN the first temperature-tuned phase-matching experiments were performed in an organic crystal [3.128]. It is an efficient frequency doubler for Nd:YAG and cw semiconductor lasers.

DAD is a crystal with a large diagonal tensor element d_{11} due to the small angle of $\theta_p = 10°$ [3.129] that could be favorable in waveguide geometries.

NPNa and DPNa (deuterated form of NPNa) are very hard ionic crystals that are available in large size [3.130–132]. Because of the high-quality crystals efficient intracavity second-harmonic generation experiments could be performed in DPNa [3.132].

PNP is the pyridine analog of NPP [3.134–135] with almost the same angle θ_p. The replacement of the benzene ring by the pyridine ring leads to a well-known shift of the cut-off wavelength towards the blue. This shift is expected to be accompanied by a slight reduction of the nonlinearity. This is the case for the molecular hyperpolarizability, however, not for the macroscopic nonlinear optical coefficients [3.126]. This discrepancy may possibly be due to the uncertain reference used in the experiments concerning NPP (see Table 3.3) [3.125]. The unexpectedly large difference in the values of β could either be due to the different solvents used in the experiments (acetone for NPP, dioxane for PNP) or due to a different definition of β. COANP is another nitropyridine derivative. The bulky substituent favored a noncentrosymmetric crystal packing [3.136–137]. Also COANP is optimized for second-order nonlinear optics. The different values of the nonlinear optical coefficients in comparison to PNP can be explained by the slightly different value of θ_p (using the oriented gas model). In both PNP and COANP the influence of acoustic phonons on the linear electro-optic effect could be nicely investigated and explained in terms of the boundary conditions [3.133].

Both racemic ($\pm$) and enantiomorphic ($-$) crystals of the chiral nitropyridine derivative MBANP were investigated with respect to their linear and nonlinear optical properties [3.141]. Although the cut-off wavelength of ($\pm$) MBANP is blue-shifted with respect to ($-$) MBANP, surprisingly the same nonlinear optical coefficients were determined for both crystals showing the validity of the oriented gas model in second-order nonlinear optics when all intermolecular interactions are neglected. Crystals of ($\pm$)MBANP are more stable and harder than ($-$) crystals. Detailed damage threshold experiments were carried out with ($-$) MBANP crystals showing that fluence (energy/area) is the main criterion determining the damage threshold in organic single crystals ([3.200]; see also Sect. 3.6). A further nitropyridine derivative is AANP, developed based on the molecular asymmetry concept (bulky substituent) with $\theta_p = 58°$ [3.142–143]. It combines large nonlinear optical coefficients with a high thermal stability. Crystals of AANP can be grown by the novel technique of indirect laser-heated pedestal growth [3.143]. The last two nitropyridine derivatives in Table 3.3 are NPPA [3.144] and NPLO [3.145]. They have a very similar molecular structure and almost the same angle θ_p. As expected from the oriented gas model these similarities also lead to a close correspondence in the macroscopic nonlinear optical coefficients.

The azine derivative in Table 3.2 is interesting since the molecules (although probably not very nonlinear) have a perfect parallel alignment in the crystal lattice [3.146]. However, nonlinear optical data are not available at the moment.

MMONS is the first stilbene derivative in Table 3.3 [3.147]. Its angle of $\theta_p = 34°$ makes it a crystalline material optimized for neither electro-optic nor nonlinear optical applications. Nevertheless it has large nonlinear optical and electro-optic coefficients due to the efficient chromophore. In MMONS the concept of molecular asymmetry was successfully applied.

MNBA is another example of the successful implementation of molecular asymmetry [3.38]. The crystal has an angle of $\theta_p = 18.7°$ leading to both large diagonal elements d_{111} and r_{111}. NMBA is very similar to MNBA in molecular structure except that there are no side groups and that the donor is expected to be slightly less efficient [3.148]. Having almost the same chromophoric orientation in the crystal lattice it is not surprising that the macroscopic nonlinearities are very close to each other (compare e.g. r_{111}).

MBPO, an azlactone derivative, is an example of the non-rod-shaped molecule approach [3.150]. Its moderate nonlinear optical coefficients are probably connected to the moderate first-order hyperpolarizability of the molecules themselves.

MC-PTS is an organic salt [3.151–153]. The crystallinity as well as the thermal and mechanical stabilities of salts are highly superior to those of the neutral dipolar molecules because of the strong ionic interaction (see also above).

D $\lambda_c \geq 550$ nm

All materials discussed here are orange to red in color. DR1, one of the standard materials used in poled polymers, is included in Table 3.3 for comparison, although it crystallizes centrosymmetrically [3.36]. A comparison of the first-order hyperpolarizabilities shows that huge macroscopic nonlinearities could be expected if single acentric crystals of a material like DR1 could be fabricated.

DCNP is an early example of a highly electro-optic crystal that is due to the good alignment of the chromophores [3.154]. In addition, it has a high melting point important for applications.

DAST is a successful example of the use of strong coulombic interactions [3.174]. It combines large nonlinear optical and electro-optic coefficients and easy crystal growth (crystals with a size of $1.8 \times 1.8 \times 0.4$ cm^3 were obtained) [3.155–157]. Phase-matched frequency-doubling at telecommunication wavelengths was experimentally investigated and interesting conditions for parametric interactions were predicted [3.157]. The electro-optic coefficients were determined up to a modulation frequency of 1 GHz, demonstrating the almost pure electronic origin of the large electro-optic effects [3.15]. In addition, photorefractive effects in the near-infrared were observed and intensively investigated [3.201].

DMSM (SPCD) is another organic salt with an optimized alignment of the chromophores for electro-optics [3.158–159]. Unfortunately, no large size crystals could be obtained up to now.

Mero-2-MDB, Mero-2-DBA and Mero-2-DAP are three examples of co-crystals (see the supramolecular synthetic approach above) [3.160–162]. As can be seen from Table 3.3 the crystalline structure can be tuned by a slight modification of the guest molecule. Interestingly, dependent on the growth conditions, two distinctively colored bulk crystals – phase I and phase II – were obtained for Mero-2-MDB [3.161]. The single crystal X-ray structural analysis proved that the two differently colored crystals possess an identical molecular packing but with absorption cut-offs, λ_c, that differ by about 65 nm. Both phases were characterized with regard to their linear optical, nonlinear optical and electro-optic properties (Table 3.3). The nature of the observed discrepancy between the two nominally structurally identical materials might be due to the different location of the proton that constitutes the short hydrogen bond; however, this still needs further confirmation. As the position of the hydrogen-bonded proton varies, it induces a different environment for the merocyanine dye. Therefore the nonlinearities of this dye in a crystal lattice are expected to vary as its nonlinear optical response in different solvents.

The last four molecules in Table 3.3 are all hydrazone derivatives, successful examples of the non-rod-shaped molecule approach. DANPH and MT-TNPH both have optimized angles for efficient parametric interactions [3.163–164]. In MTTNPH the importance of a two-dimensional charge-transfer could be nicely shown [3.164]. Whereas MTTNPH can be grown quite easily, the growth of large size DANPH crystals still imposes some problems. If these problems can be solved, DANPH is an extremely attractive candidate for nonlinear optical applications. NTMPH is an example of extreme polymorphism. No fewer than three crystalline structures based on the same molecule were found [3.165]. These three phases exhibit very different linear optical properties indicating important intermolecular interactions. 3,4-DHNPH is a newly developed hydrazone derivative interesting for electro-optic applications [3.166]. If the promise given by the large molecular hyperpolarizabilities holds, huge electro-optic effects can be expected from the crystal.

What can be learned from this extensive table? First, as already mentioned above, an increase in the nonlinear optical and/or electro-optic coefficients usually goes along with a shift of the cut-off wavelength towards longer wavelengths. Second, the limits of the nonlinear optical and electro-optic coefficients which are in the range of several thousand pm/V [3.133] is still far from being reached (see also Figs. 2.13a,b). It will be a topic of further research whether this is due to chromophores that are still not optimized or whether other reasons such as intermolecular interactions hinder these limits from being reached. Third, the values of the nonlinearities already achieved are nevertheless considerably larger than those of inorganic materials. It will be important in the future to demonstrate the potential of organic crystals by fabricating actual devices and to test their thermal, mechanical and (photo)chemical stability (see also Sect. 3.6).

3.4.3   Poled Polymers

Polymers are an important class of nonlinear optical materials as they combine the nonlinear optical properties of conjugated π-electron systems with the feasibility of creating new materials with appropriate optical and structural properties. The incorporation of nonlinear optical molecules in polymers is comparatively easy and can be done in different ways. The simplest one is the mixing of the active molecules in a polymer matrix forming a guest–host system. Alternatives are the covalent linking of the molecules to a polymer backbone in the form of a side-chain, their incorporation in the main-chain or their cross-linking between two polymer chains. Figure 3.13 depicts schematicaly these four types of nonlinear optical polymers and Table 3.4 gives a comparison of their advantages and disadvantages.

To show a second-order nonlinearity a material has to be noncentrosymmetric. In a polymer the molecules are randomly oriented leading to a cen-

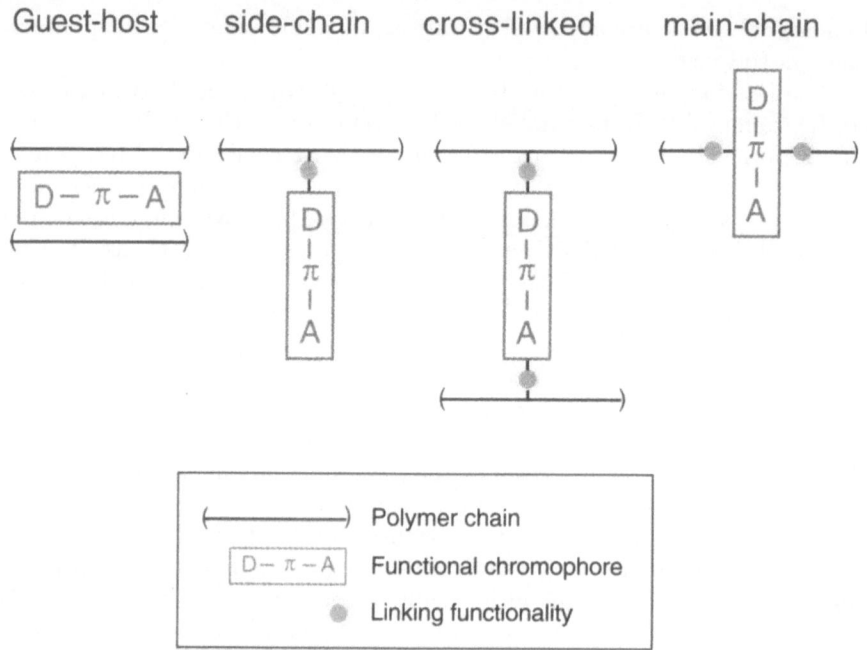

Fig. 3.13. Types of nonlinear optical polymers

trosymmetric structure. The symmetry can be broken by aligning the molecules in the direction of an applied strong electric field. If the polymeric system is brought to a glassy state by raising the temperature while the electric field is still applied, the opposing internal molecular forces decrease and the desired dipolar alignment is secured. The temperature at which the polymer goes from the solid state to the glassy state is called the glass transition temperature, T_g. Note that in order to be poled the nonlinear optical chromophores in the polymer have to have a permanent dipole moment. The two most common methods of chromophore alignment are the corona and electrode poling.

To be successfully used in nonlinear optical applications, poled polymers need to meet the following requirements:

- high electro-optic coefficient r_{33} ($> 35\,\text{pm/V}$) at $1.3\,\mu\text{m}$ and $1.5\,\mu\text{m}$
- no orientational relaxation at $80°C$ over a few years
- no orientational relaxation at $250°C$ over a short time
- no physical/chemical degradation up to $350°C$
- low optical loss ($< 2\,\text{dB/cm}$)
- compatibility with different substrates
- good thin-film processability
- broadband transparency
- low fabrication costs

Table 3.4. Advantages and disadvantages of the different types of polymers for nonlinear optics

Polymer type	Advantages	Disadvantages
Guest–host	• unlimited selection of desired nonlinear optical guests and polymer hosts • easy thin-film processing • inexpensive mass production	• decay of nonlinear optical activity due to orientational relaxation • low nonlinear optical activity due to limited solubility of nonlinear optical molecules in polymer matrix • scattering losses due to inhomogeneity • sublimation of nonlinear optical molecules at elevated temperatures
Side-chain	• high concentration of nonlinear optical molecules • tailoring of nonlinear optical properties via chemical modifications • increased orientational stability • low scattering losses	
Main-chain	• high concentration of nonlinear optical molecules • tailoring of nonlinear optical properties via chemical modifications • increased orientational stability • low scattering losses	• molecules difficult to orient to externally applied field • can have lowered solubility
Cross-linked	• tailoring of nonlinear optical properties via chemical modification • high orientational stability	• increased scattering losses • poor solubility

The comparison of polymer systems should always be done according to the application considered. For second-harmonic generation applications the nonlinear optical coefficient (d_{33}) has to be maximized at the fundamental wavelength. At the same time, the wavelength of maximum absorption (λ_{eg}) has to be kept far away from the wavelength of the second-harmonic light. For electro-optic applications the electro-optic coefficient (r_{33}) has to be maximized at the telecommunication wavelengths of 1.3 and 1.5 µm. For both applications the glass transition temperature (T_g) has to be high to prevent orientational relaxation. The chromophore concentration in the polymer, the method used for poling the material, and the strength of the applied electric field should also be considered when comparing different systems.

In Table 3.5 a selection of polymer systems for nonlinear optics is presented. The numbers in bold refer to the polymer entry number. The first 10 entries correspond to guest–host systems. Although not suitable for reliable applications due to their fast relaxation, they are broadly used for two reasons: either for checking the poling behavior of new molecules in standard polymers (usually polymethylmethacrylate, PMMA **1-5** or polycarbonate, PC **6**) or for checking the poling behavior of new polymer backbones using standard chromophores (e.g. thiophenes **8**, Disperse Red 1 (DR1) **9**). Lately, thermally stable, high T_g polyimides (PIQ-2200 **10**, Ultradel 4212 **7**, U-100 **9**) and polyquinolines (PQ-100 **8**) have often been used as hosts. Very large electro-optic coefficients (r_{33}) up to 38 and 45 pm/V at 1.3 μm have been obtained by mixing high $\mu\beta$ molecules like polyenes and thiophenes with polycarbonate **6** and polyquinoline **8** hosts, respectively.

The most common polymers for side-chain systems are the polyimides, as they have high glass transition temperatures leading to high thermal and temporal stabilities. Except system **16** all the other polyimides of Table 3.5 (**11**, **25**, **33–36**) use azo chromophores as nonlinear optical active material. Very good performances have been obtained for systems **33** and **34** where electro-optic coefficients of $r_{33} = 25\,\text{pm/V}$ and $r_{33} = 20\,\text{pm/V}$ at 1.3 μm, respectively, and good temporal stabilities at elevated temperatures have been reported. Polymer **34** (AM3-0095.11) developed by SANDOZ has been used for the fabrication of a polarization-independent integrated electro-optic phase modulator [3.202]. The maleimide copolymer **26** bearing a Disperse Red 1 analogue chromophore and synthesized using a protecting group during the processing, has both a substantially high glass transition temperature ($T_g = 255°C$) and a very large electro-optic coefficient ($r_{33} = 30\,\text{pm/V}$ at 1.3 μm).

The second group of polymers often used as a backbone are the methacrylates **13**, **17**, **22**, **29**, **31**, **32**. Although they have generally low glass transition temperatures, large electro-optic coefficients can be obtained either with thiophene ($r_{33} = 23\,\text{pm/V}$ at 1.52 μm [13]) or with azo ($r_{33} = 26\,\text{pm/V}$ at 1.32 μm **29**) chromophores. Polymer **29** (PMMA-NAT) has been widely used for the fabrication of phase modulators [3.203], Mach–Zehnder switches [3.204], and digital optical switches [3.205]. Polymer **32** (3RDCVXY) bearing a diazo chromophore, has been used for the fabrication of electro-optic channel waveguide modulators [3.206, 207]. A polymethylmethacrylate-based side-chain polymer with the nonlinear optical molecule 4-dimethylamino-4'-nitro-stilbene (DANS) has been used for the fabrication of electro-optic modulators [3.208, 209] and switches [3.210]. Cross-linking of methacrylate-based polymers has been achieved at low temperatures (150–250°C) which made possible the use of chromophores that are stable at these temperatures and led to improved nonlinearities (**41**, **44**).

Polyamides have been also investigated by several research groups. The side-chain systems, although exhibiting high glass transition temperatures,

Table 3.5. Selection of polymer systems for electro-optics and nonlinear optics

No.	Type	Polymer backbone	Chromophore	Chromophore concentration (wt.%)	λ_{eg} (nm)	T_g (°C)	d_{33} (pm/V) (λ (μm))	r_{33} (pm/V) (λ (μm))	Poling method (poling field)	Ref.
1	GH	PMMA	Disperse Red 1 (DR1)	2.74[a]	—	—	2.5[b] (1.580)	2.5 (0.633)	EP 37 V/μm	[3.215]
2	GH	PMMA		10	—	—	84[b] (1.356)	—	—	[3.216]
3	GH	PMMA		30	—	—	—	21 (1.06)	CP (9 kV)	[3.217]

Table 3.5. (Continued)

No.	Type	Polymer backbone	Chromophore	Chromophore concentration (wt.%)	λ_{eg} (nm)	T_g (°C)	d_{33} (pm/V) (λ (μm))	r_{33} (pm/V) (λ (μm))	Poling method (poling field)	Ref.
4	GH	PMMA		10	719	—	—	5 (1.3)	EP (50 V/μm)	[3.218]
5	GH	PMMA		45	—	—	—	30 (1.06)	CP	[3.219]
6	GH	Polycarbonate		10	750	120–140	—	38 (1.3)	EP (100 V/μm)	[3.220]

#		Polymer	Structure							Ref.
7	GH	Polyimide Ultradel 4212		20	—	—	—	10–12 (0.83)	EP (100 V/μm)	[3.57]
8	GH	Polyquinoline PQ100		20	—	—	180	45 (1.3)	EP (80 V/μm)	[3.221]
9	GH	U-100	Disperse Red 1 (DR1)	35	—	60 (1.064)	193	—	CP	[3.222]
10	GH	Polyamic acid PIQ-2200		12	700	≈ 220	—	10.8 (1.52)	EP (100 V/μm)	[3.223]

Table 3.5. (Continued)

No. Type	Polymer backbone	Chromophore	Chromophore concentration (wt.%)	λ_{eg} (nm)	T_g (°C)	d_{33} (pm/V) (λ (μm))	r_{33} (pm/V) (λ (μm))	Poling method (poling field)	Ref.
11 SC			57	496	210	—	13 (1.3)	EP (300 V/μm)	[3.224]
12 SC			—	—	—	50 (1.064)	—	CP (8 kV)	[3.225]
13 SC			35	—	≈115	—	23 (1.52)	EP (50 V/μm)	[3.226]
14 SC			27	—	175	—	21 (0.63) / 9 (0.83)	EP (100 V/μm)	[3.227]

15 SC		29	—	175	—	29 (0.63) 16 (0.83)	EP (100 V/μm)	[3.227]
16 SC		32	—	224	—	15 (0.83)	EP (50 V/μm)	[3.228]
17 SC		14.8[c]	—	135	—	42[b] (1.064)	CP (150 V/μm)	[3.229]
18 SC		—	306	165	2.9 (1.064)	—	CP (−10 kV)	[3.230]

R_1 :

R_2 : $-SO_2(CF_2)_8F$

Table 3.5. (Continued)

No. Type	Polymer backbone	Chromophore	Chromophore concentration (wt.%)	λ_{eg} (nm)	T_g (°C)	d_{33} (pm/V) (λ (μm))	r_{33} (pm/V) (λ (μm))	Poling method (poling field)	Ref.
19 SC			44.6	≈ 380	230	46[b] (1.064)	6.1 (0.63) 4.3 (0.78)	CP (200 V/μm)	[3.231]
20 SC			51	≈ 380	209	61[b] (1.064)	8 (0.63) 4 (0.78)	CP (200 V/μm)	[3.231]
21 SC	MB-TDI		12.3[d]	430	159	42 (1.064)	—	CP (6.3 kV)	[3.232]

22	SC		35	435	175	10 (1.06)	35.2 (0.63) 10.6 (0.83)	CP (60 V/μm)	[3.233]
23	SC	PMMA	68	480	121	56^c (1.06)	—	CP (4.6 kV)	[3.234]
24	SC		—	470	32-35	40 (1.064)	—	CP (8 kV)	[3.235]
25	SC		—	440	230	115^b (1.06)	—	CP	[3.236]

Table 3.5. (Continued)

No. Type	Polymer backbone	Chromophore	Chromophore concentration (wt.%)	λ_{eg} (nm)	T_g (°C)	d_{33} (pm/V) (λ (μm))	r_{33} (pm/V) (λ (μm))	Poling method (poling field)	Ref.
26 SC			28	—	255	—	30 (1.318)	CP (9 kV)	[3.237]
27 SC			—	458	164	—	4.6 (0.83) / 3.3 (1.3)	EP (60 V/μm)	[3.238]
28 SC			24	—	—	51 (1.06)	—	CP (6–9 kV)	[3.239]
29 SC	PMMA - NAT		43	420	121	—	26 (1.32)	EP (70 V/μm)	[3.203]

3 Second-Order Nonlinear Optical Organic Materials 243

No.		Structure								Poling	Ref.
30	SC	(azo dicyanovinyl chromophore; backbone: CH_2-$\overset{OH}{C}$-CH_2-O—; CH_3-C-CH_3 bisphenol; R—N—P)	—		600	140	66[b] (1.54)	—		CP	[3.240]
31	SC	(azo dicyanovinyl; H_5C_2 / HOH_2C_2 amine; PMMA: CH_2-$\overset{CH_3}{C}$, $\overset{O}{C}$-OCH_3)	8[a]		—	127	21[b] (1.58)	18 (0.799)		CP	[3.241]
32	SC	(azo cyanovinyl chromophore; CH_3 / CH_3 substituents; 3RDCVXY backbone CH_2-$\overset{CH_3}{C}$-$[CH_2$-$\overset{CH_3}{C}$-$COOCH_3]_m$ CH_2-$\overset{CH_3}{C}$-$\overset{O}{C}$-O-R)	23[c]		—	< 140	420[b] (1.064)	40 (0.63)		EP (200 V/µm)	[3.242]
33	SC	(NO_2 azo chromophore; P—N—CH_3; polyimide backbone with F_3C-$\overset{CF_3}{C}$, O—R)	32.3		477	235	146[b] (1.064)	25 (1.3)		CP (6 kV)	[3.243]

Table 3.5. (Continued)

No.	Type	Polymer backbone	Chromophore	Chromophore concentration (wt.%)	λ_{eg} (nm)	T_g (°C)	d_{33} (pm/V) (λ (μm))	r_{33} (pm/V) (λ (μm))	Poling method (poling field)	Ref.
34	SC			56	490	135	34 (1.54)	20 (1.313)	CP/EP (8 kV, 150 V/μm)	[3.244]
35	SC			—	480	—	—	10.8 (1.3)	EP (170 V/μm)	[3.245]
36	SC			62	—	172	—	12.5 (1.541)	EP (100 V/μm)	[3.246]
37	SC			23	728	—	—	11 (1.3)	EP (50 V/μm)	[3.247]

No.							Ref.
38 SC	14.4	443	205	—	1.7 (1.305)	EP (50 V/μm)	[3.248]
39 MC	—	—	113	10 (1.064)	—	CP	[3.249]
40 MC	67.7	471	125	40[b] (1.54)	16 (1.3)	CP	[3.250]
41 CL	36[c]	435	119	28 (1.064)	57 (0.632)	EP	[3.251]

Table 3.5. (Continued)

No. Type	Polymer backbone	Chromophore	Chromophore concentration (wt.%)	λ_{eg} (nm)	T_g (°C)	d_{33} (pm/V) (λ (μm))	r_{33} (pm/V) (λ (μm))	Poling method (poling field)	Ref.
42 CL			63	397	110	42 (1.064)	6.5 (0.53)	CP	[3.252]
43 CL	PUR- DR19		32	475	—	72 (1.064)	39 (0.633); 15 (0.8)	EP/CP	[3.253]
44 CL			32	437	—	60^b (1.064)	—	CP (10 kV)	[3.254]

45 SG	35	432	—	27[b] (1.064)	—	CP (9 kV)	[3.255]
46 SG	43	≈ 460	—	36 (1.064)	—	CP (5.6 kV)	[3.256]
47 SG	40	—	—	72 (1.064)	—	CP	[3.257]

(Silica host)

a ($N \times 10^{20}/\text{cm}^3$)
b Calibration reference not reported
c Mol%
d $N(\times 10^{20}\ \text{mol}/\text{cm}^3)$
e Tensor element not specified

have relatively poor nonlinear optical performance (**18, 27**). Main-chain poly-amides, on the other hand, **40**, have remarkable orientational stability and large nonlinearities.

Side-chain polyurethanes **21, 23, 24** have increased rigidity but show poor nonlinear optical performance whereas the cross-linked ones (**43**) are quite stable and efficient. Polymer **43** (PUR-DR19) has been used for the fabrication of packaged Mach–Zehnder modulators [3.211] and high frequency phase modulators [3.212–214]. Polyquinolines **14** and polyesters **38** can have high glass transition temperatures and very good temporal stabilities. Side-chain **30** and cross-linked **42** epoxies have been also investigated but without promising results. On the contrary, some sol–gel polymer systems **45–47** with Dispersed Red 1 analogues as nonlinear optical active chromophores seem to be very promising indicating large nonlinearities, excellent optical quality and acceptable thermal stability. A comparison of various polymer systems for nonlinear optical and electro-optical applications is given in Table 3.6.

3.4.4   Inorganic Dielectrics and Semiconductors

A list of inorganic materials is given for comparison (Table 3.7, see also Table 6.1). A general conclusion is that organic crystals have larger nonlinearities. However, inorganic materials are further developed and are superior with regard to mechanical, thermal and photochemical stability. Especially for the generation of blue or UV light inorganic crystals (LBO, BBO) are advantageous due to their better photochemical stability. LBO, BBO and KTP are used in commercial optic parametric oscillators. $KNbO_3$ is a very interesting material for phase-matched parametric interactions in the bulk and in waveguides (see Chap. 6 for a thorough discussion). The first blue lasers based on phase-matched frequency doubling in $KNbO_3$ are already commercially available (Rainbow Photonics, Zurich, Switzerland). $LiNbO_3$ is also of great interest for various applications involving optics. Due to a large d_{33}, but only moderate off-diagonal elements, quasi-phase-matching is used for efficient frequency conversion (Chap. 6). In addition, $LiNbO_3$ waveguides are currently the standard device used in electro-optic modulators for telecommunication (see also Sect. 3.5.3). ADP and KDP are materials already developed quite a few years ago. Besides quartz, KDP is often used as a well-known reference material in nonlinear optical measurements. The semiconductors GaAs and CdTe are added for comparison. Both materials have large nonlinear optical coefficients but have not yet been used in nonlinear optical commercial devices to our knowledge.

Table 3.6. Comparison of polymer systems for nonlinear optical and electro-optic applications. Type of polymer: GH = guest–host, SC = side-chain, CL = cross-linked. T_g is the glass transition temperature, r_{33} is the electro-optic coefficient and d_{33} is the nonlinear optical coefficient. In brackets the measurement wavelength is given. No. is the number of the corresponding entry in Table 3.5

Polymer	Type	No.	T_g (°C)	r_{33} (pm/V) (@ μm)	d_{33} (pm/V) (@ μm)	Remarks
Polycarbonate	GH	6	120–140	38 (1.3)		+ Easy to process
Polymethyl-methacrylate	GH	1		3 (0.63)	3 (1.58)	+ Easy to process
		2			84 (1.35)	+ Low dielectric
		3		21 (1.06)		constant
		4		5 (1.3)		+ Used for
		5		30 (1.06)		applications
						+ Broadband transparency
						- Low T_g
	SC	13	115	23 (1.52)		
		17	135		42 (1.06)	
		22	175	11 (0.83)	10 (1.06)	
		29	121	26 (1.32)		
		31	127	18 (0.8)	21 (1.58)	
		32	< 140	40 (0.63)	420 (1.06)	
	CL	41	119	57 (0.63)	28 (1.06)	
		44			60 (1.06)	
Polyimide	GH	7		11 (0.83)		+ High T_g
		9	193		60 (1.06)	+ Used for
		10	220	11 (1.52)		applications
	SC	11	210	13 (1.3)		
		16	224	15 (0.83)		
		25	230		115 (1.06)	
		33	235	25 (1.3)	146 (1.06)	
		34	135	20 (1.3)	34 (1.54)	
		35		11 (1.3)		
		36	172	13 (1.54)		
Polyquinoline	GH	8	180	45 (1.3)		+ High T_g
	SC	14	175	9 (0.83)		
Polyamide	SC	18	165		3 (1.06)	+ High T_g
		27	164	3 (1.3)		
	MC	40	125	16 (1.3)	40 (1.54)	
Polyurethane	SC	21	159		42 (1.06)	+ Used for
		23	121		56 (1.06)	applications
		24	32–35		40 (1.06)	
	CL	43		15 (0.8)	72 (1.06)	
Polyester	SC	38	205	2 (1.3)		+ High T_g
Epoxy	SC	30	140		66 (1.54)	- Low T_g
						- Low non-linearities
	CL	42	110	7 (0.53)	42 (1.06)	
Maleimide copolymer	SC	26	255	30 (1.3)		+ High T_g
						+ High nonlinearity
Sol–gel		45			27 (1.06)	+ Good optical
		46			36 (1.06)	quality
		47			72 (1.06)	

Table 3.7. Selected inorganic crystals and semiconductors that are used as frequency doublers in optic parametric oscillators/amplifiers or as electro-optic modulators. The electro-optic coefficients r and low-frequency dielectric constants ε are always given at frequencies above the acoustic phonon resonances, that is as clamped values. λ_c is the cut-off wavelength, d is the nonlinear optical coefficient and n is the refractive index. PM means that phase-matched second-harmonic generation is possible

Material	Point group	λ_c of bulk (nm)	d, r (pm/V), n, ε	Ref.
			Inorganic dielectrics	
α-quartz	32	150	$d_{111}(632.8) = 0.31 \pm 0.01$, $d_{111}(1064) = 0.30 \pm 0.02$ $n_e(632.8) = 1.5426$, $n_e(632.8) = 1.5517$	[3.258, 259]
$KNbO_3$	$mm2$	400	$d_{333}(1064) = 22.3$, $d_{322}(1064) = 13.1$, $d_{311}(1064) = 9.4$, $d_{232}(1064) = 13.5$, $d_{131}(1064) = 9.4$ $r_{333}(632.8) = 34.4$, $r_{223}(632.8) = 7.1$, $r_{113}(632.8) = 20.1$, $r_{232}(632.8) = 360$, $r_{131}(632.8) = 27.8$ $\varepsilon_{11} = 37$, $\varepsilon_{22} = 780$, $\varepsilon_{33} = 24$ $n_1(632.8) = 2.2801$, $n_2(632.8) = 2.3296$, $n_3(632.8) = 2.1687$ PM	[3.260]
$LiNbO_3$	$3m$	400	$d_{333}(1064) = 25.2$, $d_{311}(1064) = 4.6$ $r_{333}(632.8) = 28.8$, $r_{113}(632.8) = 7.7$ $\varepsilon_{33} = 28$ $n_o(632.8) = 2.2859$, $n_e(632.8) = 2.1960$ PM	[3.67, 3.261–262]
KTP	$mm2$	350	$d_{333}(1064) = 14.6$, $d_{322}(1064) = 2.2$, $d_{311}(1064) = 3.7$, $d_{232}(1064) = 1.9$, $d_{131}(1064) = 3.7$ $d_{\mathrm{eff}}(1064) = 3.3$ $n_1(632.8) = 1.7635$, $n_2(632.8) = 1.7730$, $n_3(632.8) = 1.8634$ PM	[3.258, 261]
LBO (LiB_3O_5)	$mm2$	160	$d_{333}(1064) = 0.04$, $d_{322}(1064) = 0.67$, $d_{232}(1064) = 0.85$, $d_{311}(1064) = d_{131}(1064) = 0.85$ $n_1(632.8) = 1.5742$, $n_2(632.8) = 1.6014$, $n_3(632.8) = 1.6163$ PM	[3.258, 261]

BBO (BaB$_2$O$_4$)	$3m$	198	$d_{211}(1064) = d_{112}(1064) = 2.3$, $d_{22}(1064) = 2.3$ $d_{\text{eff}}(1064) = 2.0$ (type I PM) $n_{\text{o}}(632.8) = 1.668$, $n_{\text{e}}(632.8) = 1.5506$ PM	[3.258, 263]
ADP	$\bar{4}2m$	184	$d_{312}(632.8) = 0.55 \pm 0.02$, $d_{312}(1064) = 0.45 \pm 0.03$ $n_{\text{o}}(632.8) = 1.5220$, $n_{\text{e}}(632.8) = 1.4773$ PM	[3.258, 259]
KDP	$\bar{4}2m$	177	$d_{36}(1064) = 0.39$ $n_{\text{o}}(632.8) = 1.5074$, $n_{\text{e}}(632.8) = 1.4669$ PM	[3.263]

Semiconductors

GaAs	$\bar{4}3m$	–	$d_{132}(1064) = 170$, $d_{132}(1533) = 119$, $r_{231}(1020) = 1.2$ $\varepsilon = 13.2$ $n(1020) = 3.5$, $n(1064) = 3.4785$, $n(1533) = 3.3800$	[3.4, 3.261]
CdTe	$\bar{4}3m$	–	$d_{132}(1064) = 109$, $d_{132}(1548) = 73$, $r_{231}(10600) = 5.3$ $\varepsilon = 9.4$ $n(10600) = 2.67$, $n(1064) = 2.8180$	[3.4, 261]

3.5   Applications

3.5.1   Optical Frequency Conversion

Optical frequency conversion is achieved either in bulk single crystals or in waveguides. We present one example for each type. Depending on the application, a single pass of the incident beam through the device or a resonator is used e.g. in the case of parametric oscillation. Optical parametric oscillation is of particular importance because it allows us to turn a single-frequency laser (with frequency ω_3) into a broadly tunable laser system by adjusting the phase-matching condition (3.7) using e.g. angle or temperature tuning. In this device the nonlinear crystal is placed in a resonator with two mirrors that are transparent to the input beam (frequency ω_3) and usually either the signal (frequency ω_1) or idler beam (frequency ω_2), and highly reflective for the remaining frequency. This configuration allows multipass nonlinear interaction in the nonlinear material and therefore larger conversion efficiencies.

Organic crystals show promise for optic parametric oscillation. The first optic parametric oscillator using an organic crystal was demonstrated in 1983 in urea [3.184]. Tunable radiation in the visible and the infrared with conversion efficiencies up to 20% were demonstrated. Two years later single pass parametric emission was achieved for the first time in an organic crystal [3.264]. Several experimental configurations for parametric interactions were tested in single crystals of POM. Later on, record low oscillation thresholds of the order of $0.5\,\mathrm{MW/cm^2}$ could be experimentally demonstrated in the highly nonlinear crystal NPP with a tuning range of 1 μm to 1.5 μm [3.96]. Very recently, near-infrared optic parametric oscillation was observed for the first time in an organomineral crystal [3.190]. The experiments gave an oscillation threshold of $6\,\mathrm{MW/cm^2}$ and a total conversion efficiency of 8%. Highly efficient optic parametric generation was also achieved in the highly nonlinear crystal AANP [3.265].

The recent development of new organic crystals with even larger second-order nonlinearities promises even further progress. Among interesting materials are the two hydrazone derivatives DANPH and MTTNPH and the organic crystal DAST. Although no experimental verification of parametric fluorescence has been obtained so far, exciting phase-matching possibilities exist for all three of the above materials. Figure 3.14 shows an example for the case of MTTNPH [3.164].

Another exciting application based on frequency-conversion is the cascading of second-order nonlinearities discussed in detail in Chap. 2.

Work on frequency conversion using poled polymers concentrates on second-harmonic generation. This field has very recently been thoroughly reviewed [3.266]. Besides efficient new light sources, one of the major applications is in the area of cascaded frequency doubling and difference frequency generation for all-optical switching purposes (see Chap. 2). Five different phase-matching techniques are used for the efficient generation of new wavelengths (see also

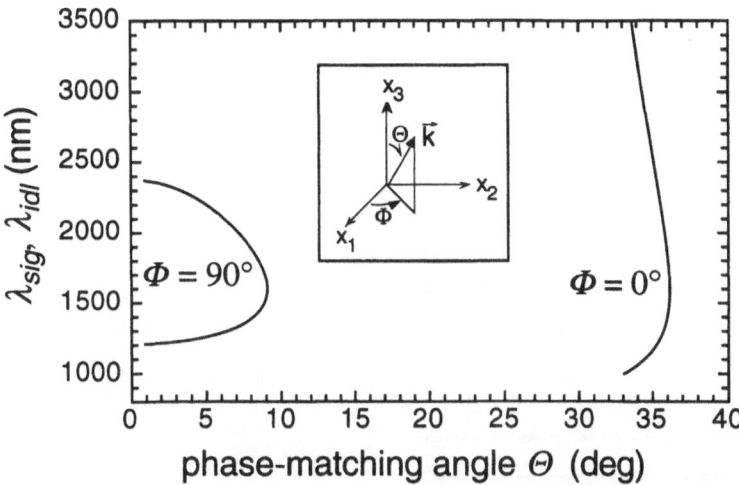

Fig. 3.14. Phase-matching curves for sum frequency generation or optical parametric oscillation of type I in a MTTNPH crystal. Tuning curves for one pump wavelength, $\lambda_3 = 800\,\mathrm{nm}$ (Ti:Sapphire laser), were calculated from the Sellmeier coefficients. $\Phi = 90°$: polarization of pump beam in bc-plane. $\Phi = 0°$: polarization of pump beam in ac-plane

Sect. 6.1.3.and Figs. 6.3–6.8): modal dispersion phase-matching, quasi-phase-matching, anomalous dispersion phase matching, birefringent phase matching, and Cerenkov-type phase matching (Fig. 3.15). Birefringent phase matching has not yet been demonstrated in polymer waveguides.

In contrast to phase matching in the bulk with the phase-matching condition

$$n^{\omega} = n^{2\omega} \tag{3.37}$$

we have

$$N^{\omega} = N^{2\omega} \tag{3.38}$$

in waveguides, where N is the effective refractive index of a waveguide mode (see also Chap. 6).

A typical polymeric channel waveguide structure for second-harmonic generation is shown in Fig. 3.16.

Much progress in poled polymers has been achieved over the last few years. Jäger and Otomo provided a nice graph (Fig. 3.17) of the figure of merit for guided wave second-harmonic generation reported in poled polymers [3.266].

Table 3.8 lists examples of polymeric frequency doublers together with a few inorganic waveguides (for a detailed general discussion on guided-wave frequency conversion we refer to Chap. 6). Listed are normalized conversion efficiencies: η' (in units of %/W) to take into account the input power and η (in units of %/(Wcm^2)) to further account for the interaction length in

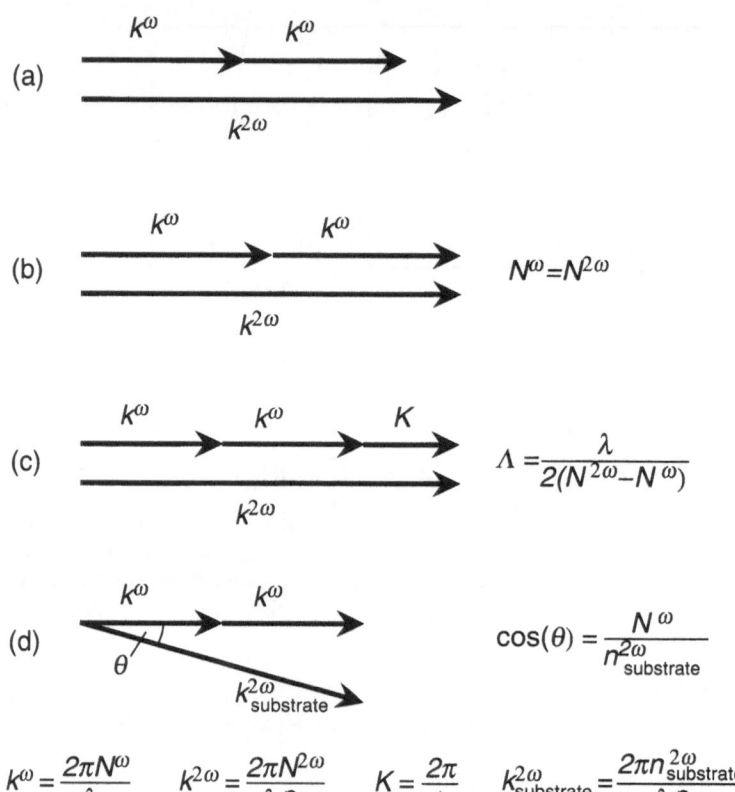

$$k^\omega = \frac{2\pi N^\omega}{\lambda} \qquad k^{2\omega} = \frac{2\pi N^{2\omega}}{\lambda/2} \qquad K = \frac{2\pi}{\Lambda} \qquad k^{2\omega}_{substrate} = \frac{2\pi n^{2\omega}_{substrate}}{\lambda/2}$$

Fig. 3.15. Phase-matching diagram and condition for (**a**) phase-mismatch, (**b**) modal dispersion, anomalous dispersion, or birefringent phase-matching, (**c**) quasi-phase-matching, and (**d**) Cerenkov-type phase matching (after [3.266]). N is the effective waveguide refractive index, θ is the grating spacing for quasi-phase matching and θ is the angle at which the second harmonic is emitted into the substrate

the waveguide. Most progress with an increase of three orders of magnitude in conversion efficiency was achieved with modal dispersion phase matching. Multilayer structures for an optimization of the overlap integral are easily fabricated and conversion efficiencies η close to the best inorganic waveguides were achieved.

Quasi-phase matching is the most versatile phase-matching technique. However, the polymer technology is not mature enough yet for significant conversion efficiencies.

The largest figures of merit so far were achieved with anomalous dispersion phase matching. However, up to now the short interaction lengths (due to large absorption effects) have led to rather small conversion efficiencies, inhibiting any applications.

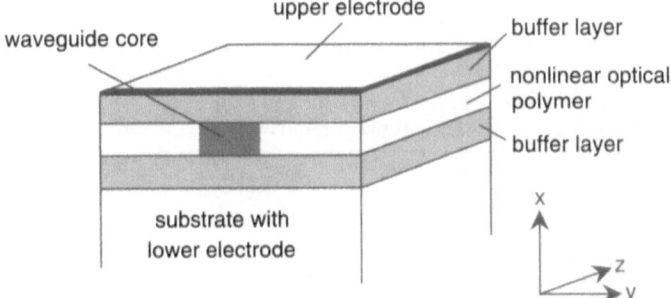

Fig. 3.16. Polymeric channel waveguide geometry for second-harmonic generation. The polar order is achieved by parallel plate poling. The buffer layers ensure that there is no waveguide loss due to the metal electrodes. Most often bleaching is used to confine the laser light along the y-direction

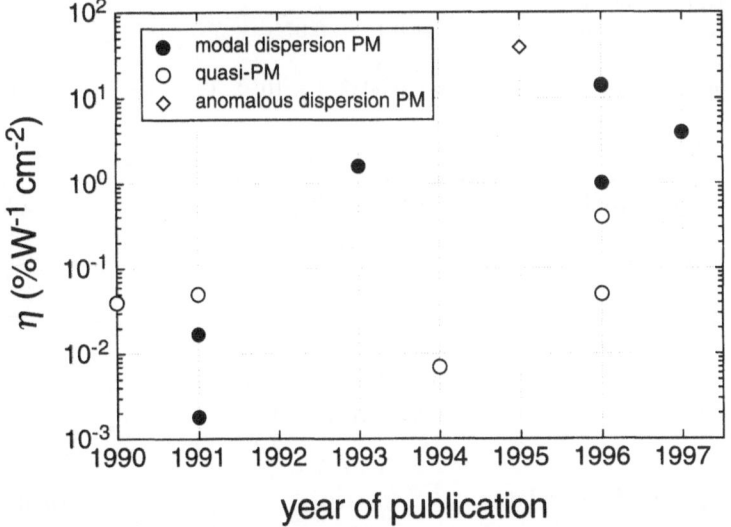

Fig. 3.17. Figure of merit for guided-wave second-harmonic generation in poled polymers vs. year of publication. The cases of modal dispersion phase matching, quasi phase matching and anomalous dispersion phase matching are shown (adapted from [3.266])

3.5.2   Short-Pulse Laser Applications

An important issue in the development and application of short pulse laser systems is the determination of the laser pulse width. The standard technique

Table 3.8. Examples of poled polymers used for phase-matched frequency doubling with $\eta \geq 1\%/(\mathrm{W\,cm^2})$. For comparison a few inorganic waveguides are also given

Waveguide structure	Type of phase matching	λ (nm)	L (cm)	η' (%/W)	η (%/(W cm²))	Ref.
Polymers						
Channel waveguide (TM$_{00}$ → TM$_{20}$)	Modal dispersion	1535	0.2	0.04	1	[3.267]
Channel waveguide (TM$_{00}$ → TM$_{10}$)	Modal dispersion	1510	0.7	0.7	1.3	[3.268]
Channel waveguide (TM$_{00}$ → TM$_{10}$)	Modal dispersion	1510	0.1	0.14	14	[3.268]
Slab waveguide (TM$_0$ → TM$_1$)	Modal dispersion	900	0.5	0.4	1.6	[3.269]
Channel waveguide (TM$_{00}$ → TM$_{10}$)	Modal dispersion	1610	0.1	0.04	4.0	[3.270]
Channel waveguide (TM$_{00}$ → TM$_{20}$)	Modal dispersion	1540	0.1	0.07	7.0	[3.270]
Slab waveguide	Anomalous dispersion	815	0.0032	4×10^{-4}	39	[3.271]
Inorganic materials						
Channel waveguide (KNbO$_3$	Birefringence	871	0.7	10	20	Chap. 6
Channel waveguide (KTP)	Birefringence	1018	0.82	2.9	4.3	[3.272]

that is used is autocorrelation [3.273] (Fig. 3.18). The autocorrelation function $G(\tau_s)$ is the convolution of a function $f(t)$ with itself:

$$G(\tau_s) = \int_{-\infty}^{+\infty} f(t)f(t - \tau_s)\,dt \ . \tag{3.39}$$

$G(\tau_s)$ gives information on the function $f(t)$ in a time frame determined by the time τ_s. This is illustrated in Fig. 3.18.

In an autocorrelation experiment a laser pulse is split into two pulses, which are then time-delayed with respect to each other. The pulses are focused on a noncentrosymmetric material and the generated second-harmonic intensity is measured as a function of the delay between the pulses. Most people use a noncollinear geometry in which a second-harmonic signal is detected only if the pulses overlap spatially (Fig. 3.19).

For laser pulses typically below 100 fs pulse broadening of the second harmonic occurs due to group-velocity dispersion in the materials [3.274]. The

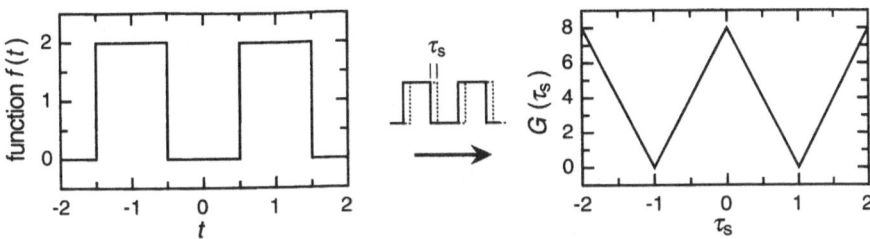

Fig. 3.18. Autocorrelation function $G(\tau_s)$ of a square wave function

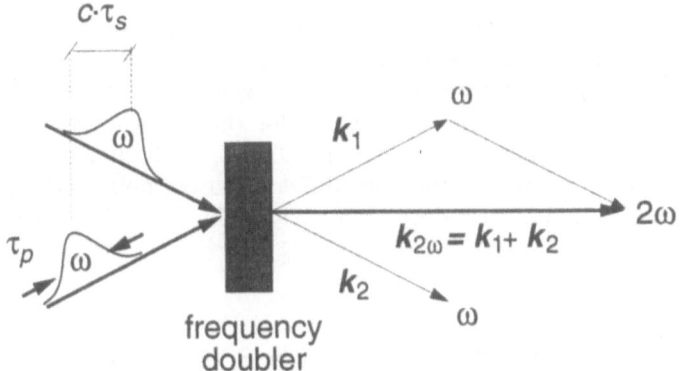

Fig. 3.19. Principle of noncollinear phase matching for determination of the auto-correlation of two laser pulses

pulse broadening of the second harmonic induced by group velocity mismatch between the fundamental and the second harmonic is given by [3.275]:

$$\Delta\tau_{\text{SHG}} = \left| \frac{L}{c} \left\{ \lambda_1 \left. \frac{dn}{d\lambda} \right|_{\lambda_1} - \lambda_2 \left. \frac{dn}{d\lambda} \right|_{\lambda_2} + n(\lambda_2) - n(\lambda_1) \right\} \right|, \qquad (3.40)$$

where L is the path length of the nonlinear material and c is the velocity of light *in vacuo*.

This effect requires short optical path lengths, typically below $100\,\mu\text{m}$ for $100\,\text{fs}$ laser pulses (depending on the dipersion of the material and on the wavelength used). Shorter pulses consequently require shorter samples. The cutting and polishing of crystals to thicknesses below $100\,\mu\text{m}$ or even down to about $10\,\mu\text{m}$ (for $10\,\text{fs}$ pulses) is very costly or sometimes even impossible. Therefore poled polymers are an ideal candidate for the autocorrelation of very short laser pulses. Since the nonlinear optical coefficients in poled polymers can be extremely large, the short interaction length (film thickness is smaller than the coherence length $l_c = \lambda/4(n^{2\omega} - n^\omega)$) is no drawback.

In the case of absorption at the second harmonic (especially severe in the case of blue or UV generation) the intensity at the second harmonic is decreased by [3.275]

$$F\left(\alpha\left(\lambda_2\right), \Delta k\right) = \exp\left[-\alpha\left(\lambda_2\right) L\right] \times \left|\operatorname{sinhc}\left(\frac{\left(\alpha\left(\lambda_2\right) + \mathrm{i}\Delta k\right) L}{2}\right)\right|^2 \quad (3.41)$$

with

$$\Delta k = \frac{4\pi\cos\theta}{\lambda_1}\left[n\left(\lambda_2\right) - n\left(\lambda_1\right)\right], \quad (3.42)$$

where θ is the angle between the two incident beams in the medium and $\operatorname{sinhc}(x) = \sinh(x)/x$. Besides autocorrelation thin polymeric films are also suitable for UV light generation by second harmonic generation [3.275].

The only minor drawback of poled polymers for the applications mentioned is the fact that the films are poled perpendicular to the film surface which means that the projection of the optical field onto the polar polymer axis is not optimized, leading to reduced effective nonlinear optical coefficients

$$d_{\mathrm{eff}} = \left|3d_{31}\cos^2\theta + d_{33}\sin^2\theta\right|. \quad (3.43)$$

It is possible to pole polymer films in-plane [3.276, 277]. However, in this case the electrode distance is only of the order of 20–50 µm, requiring strong focusing and a perfect alignment of the interacting beams. An interesting solution could be Langmuir–Blodgett films, in which an in-plane alignment of the molecular dipoles has been achieved [3.278].

Recently, the potential of poled polymer films for amplitude and phase measurements of fs pulses using frequency-resolved optical gating (FROG) was demonstrated [3.275].

Also, thin crystalline films of MNA were used to frequency double all components of a broadband light source in a non-phase-matched configuration, again showing the potential of organic films for the processing and measurement of fs pulses [3.110].

3.5.3 Polymer-Based Electro-Optic Modulators

The application of nonlinear optical polymers which has turned out to be the most important one in the last few years is their use as electro-optic waveguide modulators. Electro-optic modulators are needed in telecommunication applications to transmit information through an optical fiber at very high bit rates. In order to reach bit rates of gigabits or terabits per second it is necessary to develop modulators having a bandwidth of 100 GHz or more. Owing to their intrinsically high dielectric constants this task is not yet possible to fulfill with inorganic materials like for example lithium niobate (LiNbO$_3$), the material commercially used for electro-optic modulators up to 20 GHz.

Even though it was demonstrated by Nogouchi et al. in 1994 [3.279] that it is possible to reach 75 GHz by choosing special modulation electrode arrangements to reduce the effective dielectric constant at microwave velocities, the cost-effective limits seem to be reached.

In contrast to expensive LiNbO$_3$ waveguide technology, electro-optic polymer modulators are very cost-effective and have made huge progress during the last few years mainly due to the work performed at the University of California Los Angeles, where lately an electro-optic phase modulator having an electrical bandwidth of 113 GHz [3.212] was reported. Such high bandwidths are, from the material point of view, easily possible because the electro-optic effect in organic materials is mainly electronic in origin yielding subpicosecond response times and low dielectric constants.

In this present section we will discuss the speed limits of high-bandwidth electro-optic modulators and how such high-speed modulation can be measured. Further on we are going to discuss in detail the principle of the most commonly used Mach–Zehnder modulator concept but also a relatively new approach, so-called in-line fiber modulators. These basically modulate the light transmitted through a side-polished fiber and therefore suffer almost no insertion and waveguide losses – a big advantage compared to Mach–Zehnder modulators.

A   Speed Limits

One of the big problems arising in the context of electro-optic amplitude modulation in waveguides is the bandwidth: the wavelength of the modulating electric field becomes shorter than the modulator length at high frequencies. In this case the modulation of the optical beam has to be achieved with traveling microwaves. This sets limits to the modulation bandwidth since the 3 dB optical bandwidth of the modulator (frequency at which the power in the optical side bands is reduced by one-half) is given by (see e.g. [3.280])

$$\Delta f_{3\,\text{dB}} = \frac{1.4\,c}{\pi\,|n_\text{o} - n_\text{m}|\,L}\,, \tag{3.44}$$

where n_o and n_m $(= \sqrt{\varepsilon_m})$ are the refractive indices at optical and microwave frequencies, respectively, c is the speed of light, and L is the waveguide length.

Equation (3.44) indicates that in order to have efficient modulation the optical and electrical waves must travel at the same speed along the waveguide, or in other words, the refractive indices must match. Inorganic materials like LiNbO$_3$ show a strong dispersion of the refractive index between optical and microwave frequencies, thus limiting the maximum bandwidth. As mentioned above, despite this problem it was possible to achieve 75 GHz even with LiNbO$_3$ by choosing special electrode arrangements and therefore reducing the effective index for the microwaves, but realizing this concept, from a technological viewpoint, would be rather expensive.

In contrast to inorganic materials, organics show a very weak dispersion of the index of refraction, yielding an intrinsically high modulation bandwidth (see Chap. 2). Several groups have reported on polymer-based traveling wave modulators having bandwidths above 40 GHz using different polymers and standard traveling wave microstrip electrodes. The bandwidth-limiting factors in polymers are so far only electrical properties such as the connections and losses of the striplines, and not material properties of the polymers.

Further details on how to design electrodes and waveguides for high-speed electro-optical modulation will be given in Sect. 3.5.3.C.

B   Detection of Fast Electro-Optic Effects

Until a few years ago, detection of the electro-optical modulation response was in principle no problem since high-speed photodetectors commercially available ranged up to 60 GHz for visible light and up to 45 GHz for the near-infrared. For higher bandwidths people had to think about appropriate measurement techniques to probe such ultrafast response times. They had to find a way to reliably down-convert the signal to a frequency range that can be measured with the available detectors.

Optical heterodyne detection using two lasers provides an efficient way to fulfill this task and it was demonstrated up to 26.5 GHz by Tan et al. [3.281] and further improved by Wang et al. up to 110 GHz [3.212, 282] beyond the limits of commercial photodetectors. Furthermore such a system allows us to detect not only amplitude modulation but also phase modulation.

A typical optical heterodyne detection setup for a phase modulator is shown in Fig. 3.20. The setup uses two lasers having the same wavelength, whose lasing frequencies can be adjusted by changing crystal temperatures and cavity lengths, and whose output power is controlled externally by a combination of polarizing beam-splitters and half-wave plates. The beam of

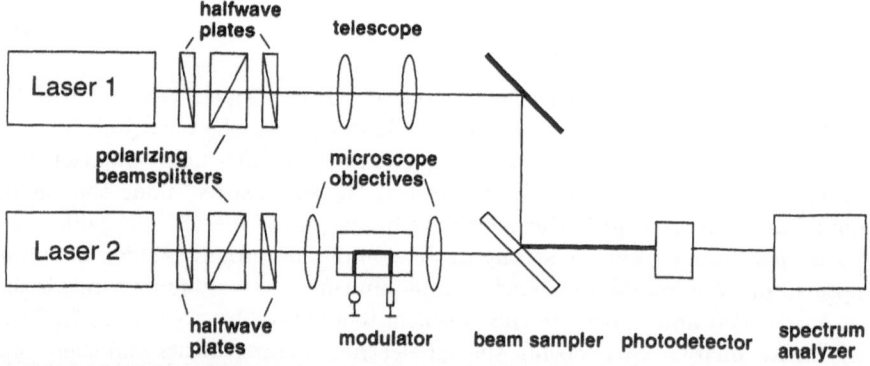

Fig. 3.20. Experimental setup for optical heterodyne detection of millimeter-wave electro-optic modulation

the first laser is coupled to the device under test and the beam of the second laser, whose divergence is controlled by a telescope, is aligned collinearly with the modulated beam using a beam sampler.

If a modulation voltage is applied the excited photocurrent in the detector can be written as

$$i_D \propto \sqrt{I_1 I_2} \left\{ J_0(\phi_1) \cos(\Delta\omega\, t + \Delta\phi) - J_1(\phi_1) \sin\left[(\Delta\omega + \Omega)\, t + \Delta\phi\right] \right.$$
$$\left. - J_1(\phi_1) \sin\left[(\Delta\omega - \Omega)\, t + \Delta\phi\right] \right\}\,, \tag{3.45}$$

where I_1 and I_2 are the intensities of the lasers, $\Delta\phi$ is the phase difference of the lasers, J_0, J_1 are zeroth- and first-order Bessel functions, $\Delta\omega$ is the frequency offset of the two lasers, and Ω is the millimeter wave modulation frequency. ϕ_1 is the phase modulation amplitude given as

$$\phi_1 = k\, \Delta n = \frac{\pi\, n^3 r_{\text{eff}} L}{\lambda\, h} V\,, \tag{3.46}$$

where n is the refractive index, $k = 2\pi/\lambda$ is the wave vector, L is the interaction length, h is the electrode separation (i.e. in the case of polymers the film thickness), r_{eff} is the effective electro-optic coefficient and V is the magnitude of the applied modulation voltage.

The last term in (3.45) represents the down-converted optical heterodyne signal. Its frequency $(\Delta\omega - \Omega)$ can easily be detected using commercial photodetectors if the two lasers are detuned by an amount $\Delta\Omega$ which is about the same as the applied modulation frequency Ω to the device under test.

For small driving voltages, where the phase shift is much smaller than π, we get

$$i_D(\Delta\omega - \Omega) \propto J_1(\phi_1) \cong \frac{\phi_1}{2} \propto V\,, \tag{3.47}$$

i.e. a linear dependence of the excited heterodyne current on the microwave driving voltage.

C Mach–Zehnder Interferometer

Most work in the field of electro-optic modulators is performed using so called Mach–Zehnder interferometric structures. A typical polymeric Mach–Zehnder interferometeric electro-optic modulator is shown in Fig. 3.21a. The incoming light is split into two arms using a y-branch. In one arm the phase of the light can be changed via the electro-optic effect by applying an electric field. If the induced phase change is π, the overlap of the two branches vanishes after the second y-branch. This constructive or destructive behavior of a Mach–Zehnder interferometer output is illustrated in Fig. 3.21b.

The voltage necessary to induce a phase change of π, the so called half-wave voltage V_π, is an important parameter for electro-optic device applications. It is given by [3.283]

$$V_\pi = \frac{\lambda}{(n^3 r)_{\text{eff}}} \frac{d}{L} = v_\pi \frac{d}{L} \tag{3.48}$$

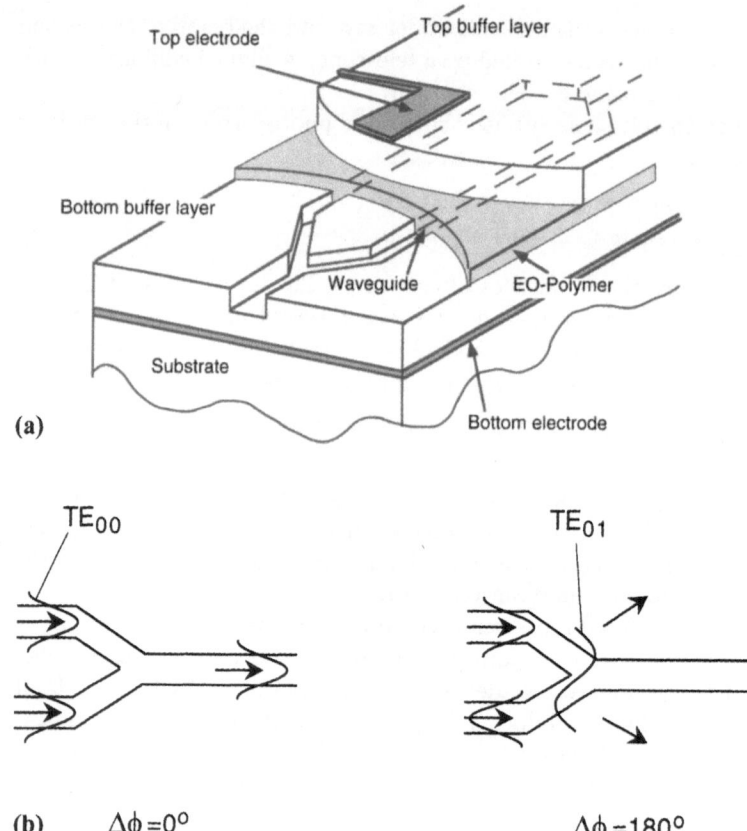

(a)

TE$_{00}$ TE$_{01}$

(b) $\Delta\phi = 0°$ $\Delta\phi = 180°$

Fig. 3.21. (a) Typical structure of a single-arm polymer-based Mach–Zehnder interferometric modulator. (b) Amplitude modulation through electro-optically-induced interference in a Mach–Zehnder interferometer

with $(n^3 r)_{\text{eff}}$ being the electro-optic figure of merit, a material parameter indicating the strength of the electro-optic effect for a certain material according to (3.13). λ is the operating wavelength. The geometry factor d/L indicates that a small electrode distance d and large electro-optic interaction lengths L are important for good device performance. To get a purely material-dependent parameter sometimes the reduced half-wave voltage $v_\pi = \lambda/(n^3 r)_{\text{eff}}$ is given.

Electro-optic Mach–Zehnder interferometric modulators have been commercially available for a few years based on the inorganic crystal LiNbO$_3$. For the above-mentioned reasons of higher bandwidth and lower costs, much effort has been put into the development of such modulators based on electro-optic polymers by many research groups and industry. Below we will describe how such devices are fabricated and operated and what are the major challenges that have to be faced.

Waveguide and Device Fabrication. The different polymeric materials used to date for electro-optic and nonlinear optic applications have been given in Sect. 3.4.3. The most commonly used are aminonitrostilbene based chromophores incorporated as a side-chain into a polyimide-, polyurethane- or polyquinoline backbone (see Table 3.5).

Fabricating a Mach–Zehnder interferometric waveguide structure requires the electro-optically active guiding layer to be sandwiched between two buffer or cladding layers to prevent optical losses occurring on metal surfaces. These cladding layers have to have a lower index of refraction for the light to be confined inside the active layer. The choice of buffer polymers is usually mainly limited by solvent compatibility to the active layer, i.e. taking care not to partially dissolve the polymer film during spinning of the next layer. Furthermore, the buffer polymers also have to show low losses since the guided light field partially extends into the buffers. To find the appropriate buffer polymers is one of the main problems when creating multilayer structures.

The most commonly used substrate is silicon because of its good cleavability necessary to create good endfaces for pigtailing of fiber ends or for endfire coupling. Before spinning the first buffer layer the lower electrode has to be deposited and eventually to be structured.

In order to create the horizontal channel for the light confinement in the active region there are two different approaches, both of which have been demonstrated to yield successfully operating modulators above 40 GHz: photobleaching and ridges produced using reactive ion etching (RIE). Both structuring methods have several advantages which will briefly be discussed.

Photobleaching was the method of choice for Teng [3.284], who demonstrated the first modulator having a 40 GHz performance. In the bleaching process the samples are irradiated with an intense, usually UV, light source, i.e. xenon or mercury lamp, through an optical mask. If the light is intense and close enough to the absorption peak of the chromophores the uncovered regions undergo a photochemical transition, thereby lowering the refractive index and creating the channel for the light to be guided. This method limits the number of fabrication steps compared to the etching techniques described later. The structuring processes can be controlled by varying exposure time and light intensity.

The structuring technique used by Chen et al. [3.212] who recently demonstrated an electro-optic phase modulator operating at 113 GHz is reactive ion etching. In contrast to photobleaching, RIE results in more sharply defined structures, but requires more structuring steps since application of photoresists is necessary. A big advantage of the RIE technique is the better controllability of the structuring process, there is also the possibility of creating vertical tapers [3.285]. Vertical tapers play an important role when trying to reduce the losses occurring due to fiber pigtailing, responsible for the major contribution to the overall losses of polymer-based optic devices, since the active layer is usually thinner than the fiber core diameter. The thickness of the active and buffer layers on the other hand must be small to achieve low switching voltages. When using RIE the upper buffer is spun after the ridge is made. This is not necessary in the case of photobleaching, where the bleaching can occur after the upper layer is already deposited.

Before depositing operating electrodes on the device, the electro-optic polymer has to be poled in order to show a strong electro-optic effect. The two most important poling techniques, electrode and corona poling, are described in Sect. 3.4.3. In device fabrication electrode poling is commonly used, since it can be controlled more easily and gives more reproducible results. Usually the poling electrodes are larger than the final operating electrodes and have to be removed before adding the electrodes for high-frequency modulation, as described in the next paragraph.

Electro-optic modulators operating at high frequencies, i.e. 1 GHz and above, have to have a traveling wave electrode design, since the microwave wavelengths become smaller than the modulator lengths (see Sect. 3.5.3.A). In order to match the characteristic impedance Z_0 of the device to the modulating source, the traveling wave electrode must have a Z_0 of 50 Ω. To achieve this, one has to design thin microstrip lines. Their impedance can be calculated taking into account capacitive and inductive resistances that depend mainly on the ratio between widths of the microstrip line and the distance to the ground electrode.

A thorough analysis of microstrip lines for microwave integrated circuits is given in the book of Gupta et al. [3.286]. It is possible to make a static approximation for the quasi-TEM fundamental mode of a microstrip line if the radio frequency is below f_g:

$$f_g \approx \frac{0.07c}{(w + 2h)\sqrt{\varepsilon + 1}}, \tag{3.49}$$

where w and h are the width of the microstrip and electrode spacing, respectively, and ε the dielectric constant of the underlying material. Inserting typical values for microstrip lines for w and h of the order of several micrometers and dielectric constants around 4, one sees that the approximation is valid up to several hundred gigahertz.

The closed forms for Z_0 and ε_{eff} are given by:

$$Z_0 = \frac{120\pi}{\sqrt{\varepsilon_{\text{eff}}}} \frac{1}{2\pi} \left\{ F(1.0) \ln \left(\frac{8}{R_{\text{we}}} + 0.25 R_{\text{we}} \right) \right.$$

$$\left. + (1 - F(1.0))(R_{\text{we}} + 1.393 + 0.667 \ln(R_{\text{we}} + 1.444))^{-1} \right\} \tag{3.50}$$

$$\varepsilon_{\mathrm{eff}} = \frac{\varepsilon+1}{2} + \frac{\varepsilon-1}{2}\left(1+\frac{12}{R_{\mathrm{w}}}\right)^{-1/2}$$

$$+ 0.04F(1.0)(1-R_{\mathrm{w}})^2 - \frac{(\varepsilon-1)R_{\mathrm{s}}}{4.6\sqrt{R_{\mathrm{w}}}}, \tag{3.51}$$

where

$$R_{\mathrm{w}} = w/h \quad R_{\mathrm{s}} = s/h\ ,$$

$$R_{\mathrm{we}} = R_{\mathrm{w}} + 0.4R_{\mathrm{s}}\left[1 + \ln(2/R_{\mathrm{s}}) + F(1/2\pi)\ln(2\pi R_{\mathrm{w}})\right]\ ,$$

and

$$F(a) = \frac{1}{2}\left(1 - \frac{R_{\mathrm{w}} - a}{|R_{\mathrm{w}} - a|}\right)$$

is a step function of R_{w}. s is the thickness of the microstrip.

Figure 3.22 illustrates the dependence of characteristic impedance and effective dielectric constant on the ratio of microstrip width and electrode distance for a dielectric constant $\varepsilon = 3.6$ and $R_{\mathrm{s}} = 0.3$. It can be seen that for these values a ratio of $w/h = 2.0$ yields $Z_0 = 50\,\Omega$.

Microstrip electrodes are usually fabricated using electroplating techniques. Hereby a thin metallic film is first deposited by thermal evaporation. Then the sample is masked with photoresist, which is patterned using conventional photolithographic processes. After patterning the photoresist, the unmasked part of the sample is covered with metal using electroplating. Then, photoresist and residual metal are etched away, with only the microstrip electrode remaining. An important task to be considered when designing the microstrip electrodes is the trade-off between conduction losses

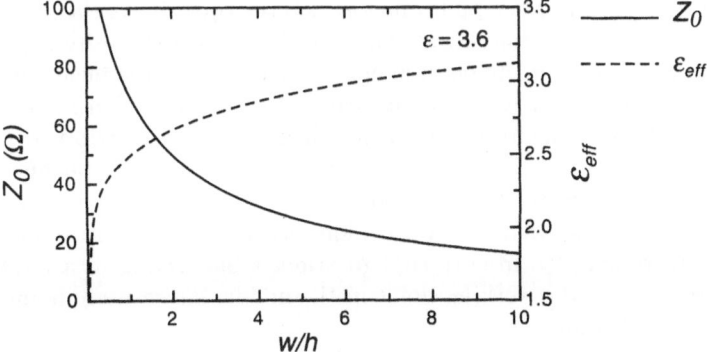

Fig. 3.22. Characteristic impedance and effective dielectric constant versus ratio of microstrip width and electrode distance

and the half-wave voltage V_π. In order to keep the operating voltage as low as possible, the electrode distance should be small. On the other hand, a small electrode distance results in a narrow microstrip line having higher conduction losses and therefore increasing the rf power necessary for efficient modulation (see e.g. [3.287]).

Another critical part of designing electrodes for high-speed electro-optic modulators is the coupling from the microwave connectors to the microstrip line. Usually the connectors have a launching pin a few hundred micrometers in size that is contacted to a rectangular contact pad. A linear taper from the contact pad to the microstrip line will result in low microwave loss in contrast to direct connection of the launching pin to the microstrip which would result in very high losses due to severe size mismatch.

Last, but not least, a very important topic in the fabrication of a waveguide electro-optic modulator is the pigtailing of fiber ends, unavoidable for field application. Several approaches have been undertaken to reduce this main optical loss factor. The first step is the preparation of the device endfaces. Whereas this is no problem in $LiNbO_3$, since crystals can be easily polished yielding good endfaces, in polymers polishing or cleaving is usually very difficult due to the relatively soft character of most polymers. Nevertheless, Teng cleaved the endfaces of his demonstration high-speed modulator, giving acceptable quality. Shi et al. [3.211] developed a method to cut their samples with a dicing blade and successfully polish the endfaces afterwards. The fiber ends were in both cases bonded to the polymer samples and fixed using standard epoxy. To achieve reproducible alignment of fiber ends, several groups have developed V-groove etching techniques in silicon wafers developed for semiconductor integrated optics [3.288, 289].

Electro-Optic Modulation. For Mach–Zehnder modulators two configurations for modulating the passing light are possible: single-arm or push–pull modulation. Schematic drawings for single-arm and push–pull modulators are sketched in Fig. 3.23. In the push–pull configuration the two arms of the modulators must be poled in opposite directions in order to have opposite signs of the electro-optic coefficient. As pointed out by Teng [3.287], the push–pull design for high frequencies does not necessarily lead to a reduction for the half-wave voltage V_π by a factor of two, since the ratio of microstrip width and electrode distance has to be much smaller to receive an overall impedance of $50\,\Omega$, because each arm has to have $100\,\Omega$ (see also Fig. 3.22). This may result in a reduction of the overlap of waveguide channel and microstrip line, therefore again increasing V_π. In both configurations a bias voltage is always necessary in order to set the working point in the middle between minimum and maximum transmission.

Teng used a single arm configuration for his P2ANS/MMA50/50-based 40 GHz modulator. The operating voltage V_π was rather low being 6 V at 1.3 micrometers. The electro-optic coefficient r_{33} of the active polymer was 21 pm/V. Data on the stability of the nonlinearity was not given. In sev-

(a)

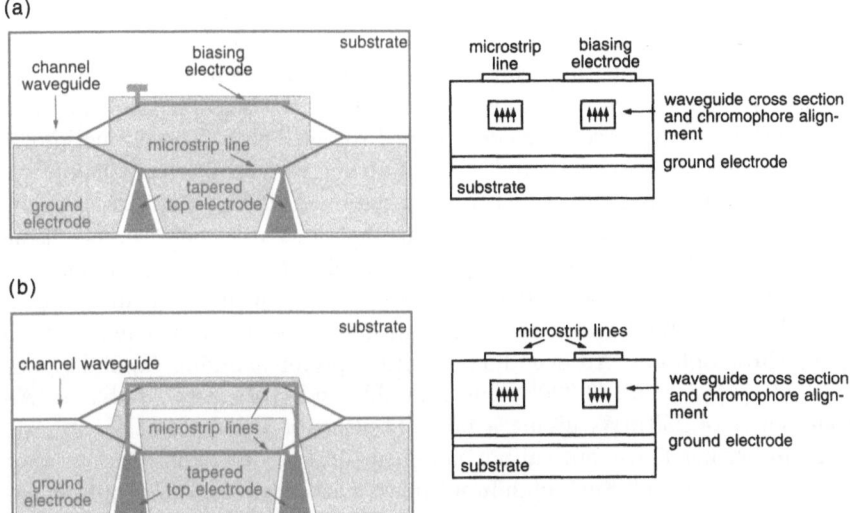

(b)

Fig. 3.23. Schematic diagram and cross-section for (**a**) single-arm and (**b**) push–pull Mach–Zehnder electro-optic modulators

eral publications (e.g. [3.211]) it was pointed out that the aminonitrostilbene chromophores used suffer serious degradation due to two-photon absorption. Information about the overall losses of the modulator has not been provided. The extinction ratio η, which is defined as

$$\eta = 10\log\frac{P_{\max}}{P_{\min}} , \tag{3.52}$$

where $P_{\max}$ and $P_{\min}$ are the maximum and minimum transmitted powers, respectively, was found to be 20 dB, i.e. 100:1 for his modulator.

Lee et al. [3.209] used a similar DANS/PMMA based side-chain polymer to report on a single-arm Mach–Zehnder modulator with the highest extinction ratio, 31 dB, to date. The r_{33} value was 13 pm/V and the operating voltage V_π of the device was 9 V. The 3 dB electrical bandwidth was around 2 GHz. The structuring method to define the channel waveguide was photobleaching, similar to Teng's modulator. The overall fiber to fiber losses, containing losses from input and output coupling to the fiber, waveguide and y-branch losses, turned out to be 13.2 dB. For comparison, the typical losses of commercially available LiNbO$_3$ modulators are around 4 dB.

Shi et al. [3.211] give a very nice overview of their push–pull high-speed polymer electro-optic modulator. Fabrication, properties and operating and stability issues concerning the modulator are thoroughly discussed. The polymer of choice was a thermally cross-linkable polyurethane-disperse red 19 side-chain polymer. The thermal stability of the second-order nonlinearity was tested using temperature dependent *in situ* second-harmonic generation experiments. A short term stability up to 110°C and a long term stability at 90°C have been observed. As already discussed in the paragraph about waveguide fabrication, Shi et al. used RIE to perform the structuring. As the modulating design they chose a push–pull arrangement, resulting in a V_π of 11 V. The device was tested up to 60 GHz modulation frequency and the overall insertion losses were determined to be 13.5 dB. The extinction ratio was typically between 15 and 20 dB. Furthermore they tested the optical power handling capability, which is, as mentioned above, a serious problem in DANS polymers. The tests indicated no degradation in performance using input powers of 10 mW at the operating wavelength of 1.32 micrometer. A high input power test with 150 mW resulted in evident photochemical degradation but still about two orders of magnitude slower than in polymers with DANS chromophores. An even higher optical power-handling capability was achieved by Shi et al. [3.290] using a double end cross-linked polymer. An input power of 250 mW, giving a peak intensity of about 0.9 MW/cm^2, resulted in neither linear optical nor nonlinear optical degradation, even after one week of exposure. Such high input powers are neccessary for applications like externally modulated community access televison (CATV) using analog fiber optic transmitters.

For completeness, it should be noted that high-speed Mach–Zehnder electro-optic modulators have also been built based on the semiconductors GaAs and AlGaAs. Spickermann et al. [3.291] reported on such a modulator having a 3 dB electrical bandwidth larger than 40 GHz. Phase velocity matching was achieved using coplanar electrodes with a special slow wave design. The half-wave voltage was 14 V and the extinction ratio 20 dB. Data on optical losses was not provided.

The results on electro-optic Mach–Zehnder modulators mentioned above are summarized in Table 3.9.

Table 3.9. Material and device parameters for several Mach–Zehnder electro-optic modulators. Data are given for $\lambda = 1.3\,\mu$m

Material	Bandwidth (GHz)	V_π (V)	η (dB)	Losses (dB)	Ref.
P2ANS/MMA50/50	40	6	20	–	[3.284]
DANS/PMMA	2	9	31	13.2	[3.209]
PU-DR19	60	11	20	13.5	[3.211]
LiNbO$_3$	75	5	20	6	[3.279]
GaAs/AlGaAs	> 40	14	20	–	[3.291]

D   Inline Fiber Modulator

General Remarks. Most approaches to realizing an electro-optic modulator, based on LiNbO$_3$ or polymers, use Mach–Zehnder interferometers or similar structures, requiring the light to pass through the electro-optically active medium. For this reason fiber pigtailing and losses of at least a few dB due to coupling are unavoidable. In polymers additional losses of about 1 dB/cm occur in the waveguide caused by waveguide absorption. One approach where these losses are drastically reduced was taken in 1980 by Bergh et al. [3.292], who reported a single-mode fiber optic directional coupler based on so called fiber half-couplers and proposed the use of such half-couplers to build fiber inline polarizers [3.293]. This approach can also be exploited to build very low loss electro-optic modulators especially suited for analog modulation, as will be described below.

A fiber half-coupler consists of a single-mode fiber epoxied into the arcing groove of a glass block. The block and the fiber are both polished down close to the fiber core, which allows the light passing through the fiber to couple evanescently to a suitable, e.g. electro-optically active, overlay or waveguide. The coupling condition is given by the eigenvalue equation [3.294]:

$$2\delta(n_o^2 - n_{ef}^2)^{1/2} = m\lambda_0 + \frac{2\lambda_0}{m} \arctan \zeta \left(\frac{n_{eo}^2 - n_c^2}{n_o^2 - n_{eo}^2} \right)^{1/2}, \qquad (3.53)$$

where δ is the overlay thickness, n_o is the overlay refractive index, n_{eo} is the mode effective index, n_{ef} is the effective index of the fiber, n_c is the index of the fiber cladding, λ_0 is the input wavelength, m is the order of the excited mode in the overlay, and ζ is a polarization-dependent constant: $\zeta = 1$ for TE and $\zeta = (n_o/n_c)^2$ for TM polarized light. The coupling occurs only at specific wavelengths where the transmission through the half-coupler decreases strongly, resulting in sharp dips in a wavelength scan of the transmitted light. The application of an electric field across the overlay induces a refractive index change via the electro-optic effect and therefore a change of the coupling condition, resulting in an amplitude modulation of the light transmitted through the coupler block. An inline fiber modulator based on a fiber half-coupler using an electro-optic polymer overlay is sketched in Fig. 3.24. The operating principle described above is illustrated in Fig. 3.25.

In this section we will discuss the fabrication and use of such a fiber half-coupler to build a low-loss, high-speed electro-optic modulator and discuss the additional advantages and challenges. At the end we will give a brief overview

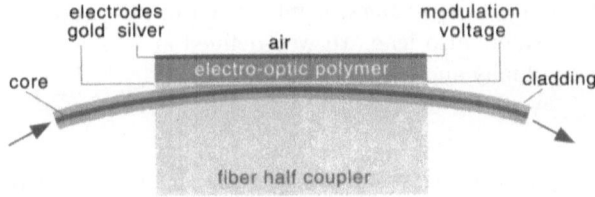

Fig. 3.24. Polymer-based inline fiber modulator using a fiber half-coupler

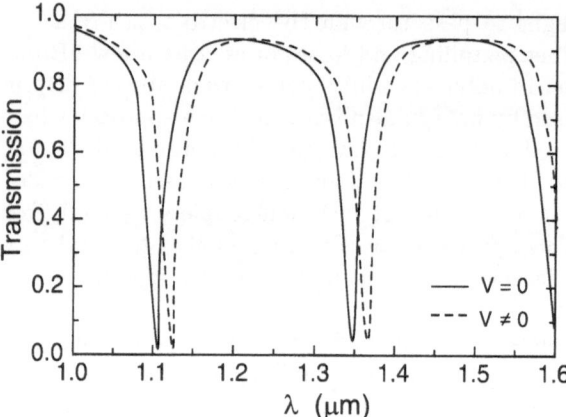

Fig. 3.25. Transmission spectrum of an inline fiber modulator with and without applied voltage

of other applications of fiber half-couplers that have become important in the last few years, such as inline polarizers and fiber optic bandpass filters.

Fabrication of Half-Couplers. Even though fiber half-coupler blocks are commercially available it is of interest to describe briefly the fabrication of these, since usually only standard couplers are available. As already mentioned, such couplers consist of single-mode fibers epoxied into the arcing grooves of a glass block. The crucial part of the fabrication is to polish block and fiber in such a way that only a thin layer of fiber cladding remains. A nondestructive method of measuring the remaining cladding thickness very accurately has been published by Digonnet et al. [3.295], which is basically a calibration of the throughput loss of the half-coupler when putting an appropriate refractive index liquid versus cladding thickness. This method has been further improved by Das et al. [3.296] who gave a closed quantitative form to derive the residual thickness from fiber parameters and loss calibration curves. Usually fiber half-couplers are fabricated using glass blocks. Another possible method, which would lead to reproducible well-defined structures and compatibility with current semiconductor technology would be the use of V-grooves in silicon. V-grooves are made by anisotropic etching of silicon wafers, a method widely used to align fibers in integrated optics. Stacking several layers of V-grooved wafers also leads to well-defined arcs, where the interaction lengths can individually and rather easily be chosen by the lengths of the upper stack.

Electro-Optic Modulation. Several overlays have been used to build an electro-optic modulator with a fiber half-coupler. A first demonstration was based on nematic liquid crystal overlays [3.297]. In 1991 Johnstone et al. [3.298] discussed in more detail the use of $LiNbO_3$ to perform electro-optic inline

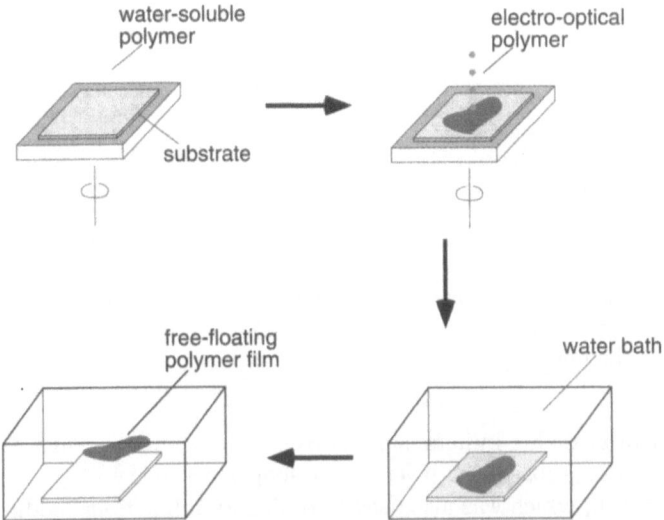

Fig. 3.26. Principle of preparing free-floating polymer films

modulation, and polymers were used for the first time in 1991 by Wilkinson et al. [3.299]. These demonstrations showed the working principle and possible use of such devices for electro-optic modulation, but driving voltages, extinction ratios, and bandwidths were far from being competitive with high-speed electro-optic modulators. However, recently inline fiber modulators having bandwidths of several GHz have been reported [3.300–302], showing promising performance and good potential for commercialization. Whereas Creaney et al. again used $LiNbO_3$, Hamilton et al. used side-chain electro-optic polymers, as will be described in more detail below.

In reference [3.302] two methods of designing and fabricating electro-optic inline fiber modulators are described, based on poly(disperse red 1 methacrylate-co-methyl methacrylate), P(DR1-MMA). A first approach uses a free-floating technique which is illustrated in Fig. 3.26. This technique was first used by Khanarian et al. [3.303] to produce stacked polymer films for quasi-phase-matched second-harmonic generation. The polymer is spun onto ITO after a water soluble polymer, e.g. polyacrylic acid, has been deposited. After the sample is corona poled the film is lifted from the ITO substrate in water, where the polyacrylic acid dissolves. The free-floating polymer films can then be transferred onto the half-coupler, where a thin transparent gold electrode has been deposited previously. The main advantage of this method is that the whole overlay can be spun and poled independently, resulting in several identical overlays being placed on different half-couplers. The upper electrode is applied using vapor deposition before lifting off the film or after the transfer onto the half-coupler.

It should be noted that this method could also be used to produce multilayers for Mach–Zehnder modulators (see Sect. 3.5.3.C). This method in-

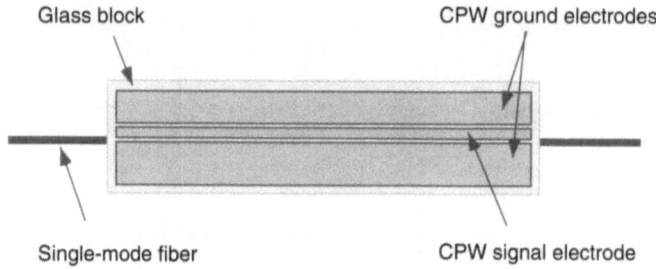

Glass block   CPW ground electrodes

Single-mode fiber   CPW signal electrode

Fig. 3.27. Fiber half-coupler (*top view*) with coplanar wave (CPW) electrodes for high-speed electro-optic modulation. Note that one of the two CPW electrode gaps must be placed above the fiber core

creases the number of available buffers, since solvent compatibility, which is important when spinning subsequent layers, is no longer necessary.

In a second approach, which was also used to design an inline fiber modulator operating at 18 GHz [3.301], the polymer is spun directly onto the half-coupler, after a coplanar traveling wave electrode has already been placed. The electrode is structured in a way that a 10 micrometer wide gap is placed over the fiber core (see Fig. 3.27) before spinning the polymer which fills the gap between the electrodes. The coplanar wave electrodes are then applied to perform in-plane electrode poling. To exploit the maximum electro-optic coefficient r_{33} for modulation it is favorable to use TE polarized light in this design, whereas in the design chosen in the first approach TM polarized light is more suitable. An advantage of the second geometry is also that the thickness and therefore the operating wavelength of the device can be accurately trimmed using reactive ion etching.

Note that coplanar electrodes are similar to microstrip lines suitable for high-speed traveling wave electrodes, since they can be designed in such a way as to yield a characteristic impedance of $Z_0 = 50\,\Omega$. Detailed analyses of coplanar wave and coplanar strip electrodes, which have been used by Creaney et al. [3.300], can also be found in [3.286].

What are the main advantages of inline fiber modulators compared to Mach–Zehnder structures? One important factor is certainly the extremely low losses of less than 1 dB compared with more than 5 dB of Mach–Zehnder structures as discussed at the beginning of Sect. 3.5.3.D. Furthermore, it is possible to exploit the fact that the light does not have to be guided in the polymer. For this reason the device can be operated even in the polymer's absorption band, where the electro-optic coefficients are resonance-enhanced, therefore reducing the half-wave voltage of the device. Another way to reduce the half-wave voltage of inline fiber modulators is to increase the interaction length, since the change in the coupling condition and therefore in transmission depends on the interaction length. In commercially available fiber half-couplers the interaction distance is given as approximately one millimeter, which could be increased using custom made couplers with optimized interaction length. However, as pointed out by Hamiliton et al. [3.302] the interaction length cannot be arbitrarily long because the slope of the dip is reduced with larger interaction lengths.

The main advantage of the inline fiber modulators compared with Mach–Zehnder modulators lies in their very large linear dynamic range, which makes them especially suitable for analog applications. One example of analog applications where a large linear dynamic range is of crucial importance is remote antenna controlling in radar systems. This demand cannot be met with Mach–Zehnder modulators where modulation over a large dynamic range introduces nonlinearities into the modulated signal giving distortions that might rise above the noise level [3.302].

In contrast to the excellent performance of inline fiber modulators for analog modulation their usefulness for digital modulation has not yet been demonstrated. The reason for this is the limited extinction ratio. Whereas for reliable digital operation an extinction of 20 dB, i.e. 99%, is required, the maximum extinction ratio of inline fiber modulators merely exceeds 10 dB at the moment as can be seen in the paper of Hamiliton et al. A way out of this dilemma might be the approach that Creaney et al. [3.300] used for their LiNbO$_3$-based inline fiber modulator. Whereas in conventional inline fiber devices the half coupler is polished down close to the fiber core as mentioned above, Creaney et al. polished in such a way that a small part of the fiber core is exposed. Therefore most of the light will usually not be transmitted through the fiber and only at certain wavelengths will steep peaks occur. This method of polishing into the core first introduced by McCallion et al. [3.304], who reported an inline fiber optic bandpass filter based on a half-coupler.

Finally it should be noted that industrial research is being performed to build inline fiber modulators exploiting the very high electro-optic coefficients of organic crystals, e.g. DAST (see Sect. 3.4.2), which could be used in a device similar to that reported in [3.300] (see [3.305]). The main advantage of organic crystals compared to LiNbO$_3$ is the strong resonance enhancement in the visible. The r_{11} value of DAST at 720 nm is 92 pm/V [3.156], three times the value of LiNbO$_3$, and the figure of merit $n^3 r$ is almost five times larger (see also Table 3.9).

Further Applications. The concept of inline components for single-mode optical fiber technology has given rise to a large field of applications besides electro-optic modulators, ranging from couplers, filters, polarizers and switches to sensors. As mentioned above the first proposed applications have been passive directional couplers [3.292] and polarizers [3.293]. These components, directional couplers even with manually tunable coupling ratios, are commercially available (e.g. [3.306]). Inline polarizers consist basically of metal-coated half-couplers (see e.g. [3.307]), where the TM polarization is strongly absorbed due to plasmon excitation on the metal surface. A recent publication of Lee et al. [3.308] describes the fabrication of an inline fiber polarizer using a birefringent polymer overlay. Further applications that have already been demonstrated are a 2×2 electro-optic switch using LiNbO$_3$ or zinc-sulfide as the active interlay between two half-couplers [3.309], an electric field sensor [3.310] and a fiber-optic bandpass filter using refractive index oils and LiNbO$_3$ [3.304]. Additionally, the possibility of all-optical switching with a liquid crystal overlay on a fiber half-coupler has been demonstrated recently [3.311].

3.5.4   Electro-Optic Sampling

Electro-optic sampling is a proven technique for making noninvasive time-domain measurements of high-bandwidth electrical circuits [3.312–314]. This is important to improve the high-frequency signal propagation on transmission lines. The basic principle is as follows: an electro-optic material is brought in close contact with the integrated circuit. As the electrical signal propagates the leaking field will induce a refractive index change in the electro-optic material which can be detected (Fig. 3.28). GaAs integrated circuits can be directly sampled using the electro-optic effect of the substrate by converting the electrical signals to amplitude modulation of the optical probe beam [3.314].

Organic materials are of great interest for optical sampling due to the large bandwidth arising from the almost purely electronic origin of the large nonlinearity and the low dielectric constants (see also Chap. 2). Low dielectric constants are of the utmost importance, since otherwise probe tips disrupt the circuit operation when brought into close contact with most circuits. As an example, LiTaO$_3$, often used for electro-optic sampling, has a high-frequency dielectric constant of about 43, compared with $\varepsilon = 5.2$ for DAST. Table 3.10 lists a few examples of materials interesting for electro-optic sampling.

If done carefully, it is possible to press organic crystals (or polymer films) on a circuit under test without causing any damage to the probe tip or

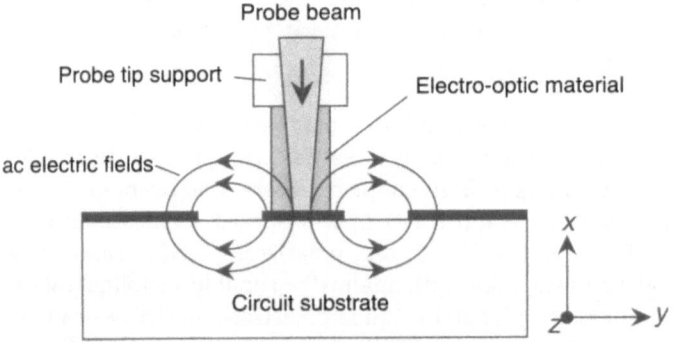

Fig. 3.28. Schematic of electro-optic sampling geometry for a coplanar waveguide configuration. For the best sensivity the largest electro-optic coefficient should be r_{22}

Table 3.10. Examples of organic and inorganic crystals and one polymer for electro-optic sampling. λ is the operating wavelength, $n^3 r$ is the electro-optic figure of merit, and ε is the high-frequency dielectric constant

Material	λ (nm)	$n^3 r$ (pm/V)	ε	Ref.
GaAs	1020	51	13.2	[3.4]
KNbO$_3$	632.8	351	24	[3.4]
LiNbO$_3$	632.8	305	28	[3.4]
LiTaO$_3$	550	315	43	[3.315]
COANP	632.8	67	2.6	[3.4]
DAST	720	1471	5.2	[3.4]
Alternating styrene-maleic-anhydrid copolymer	1300	89	3.6	[3.4]

the circuit. One good material for electro-optic sampling is DAST. A flat electro-optic frequency response from 1 to 40 GHz has been experimentally determined and the first sampling experiments have already been carried out [3.316].

Very recent work has demonstrated another possibility with the use of electric field-induced second-harmonic generation (see Sect. 3.4.1.A) [3.317]. In this detection scheme the second-order hyperpolarizability $\gamma(-2\omega, \omega, \omega, \Omega)$ (and third-order susceptibility $\chi^{(3)}(-2\omega, \omega, \omega, \Omega)$) is used. Ultrafast electrical pulses propagating on thin-film microstrip lines consisting of a benzocyclobutene polymer as dielectric layer on a silicon substrate were characterized. In contrast to classical electro-optic sampling, this is a totally contact-less field detection since the polymer serves as a dielectric layer and as the nonlinear optical probe. The experimental set-up is illustrated in Fig. 3.29. For these time-domain EFISH experiments a 100 fs Ti:Sapphire with a repetition rate of 75.6 MHz was used. A pump-probe scheme was employed where the probe beam excites the electric signal by a photoconductive switch and the probe beam is used to detect the propagating electric field at a distance from

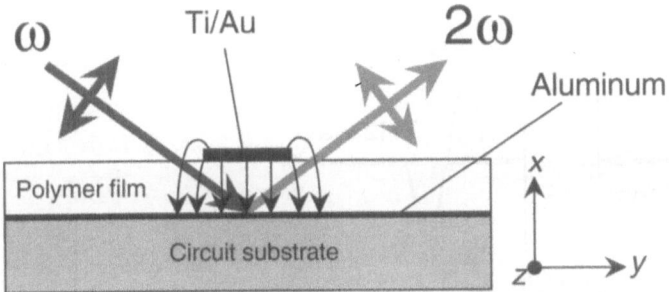

Fig. 3.29. Electric-field-induced second-harmonic generation in a thin film microstrip of a reported linewidth with an upper frequency of 4 THz

the photoconductive switch. For the maximum signal both the fundamental and the second-harmonic beams are p-polarized. Important polymer material parameters for efficient field detection using Ti:Sapphire lasers are a low dielectric constant, a large third-order susceptibility $\chi^{(3)}$, low absorption at around 400 nm and 800 nm, and a typical film thickness of 10 μm.

3.5.5   THz Generation

The development of coherent and tunable THz radiation is of great interest due to the abundance of excitations in molecular systems and condensed media, and because these frequencies can bridge the gap between optical waves and microwaves for high-frequency telecommunication. THz wave generation and detection has therefore attracted wide attention from both fundamental and applied points of view. Among the various applications THz time-domain spectroscopy should be mentioned in particular [3.318]. This is a new technique with capabilities far beyond linear far-infrared spectroscopy. It allows the generation and detection of THz waveforms propagating in a sample under investigation. A detailed analysis of the experimental results yields the frequency-dependent dielectric properties of the sample [3.318].

The first pioneering work in the area of THz generation was carried out using nonlinear difference-frequency mixing between two laser sources [3.319, 320]. More recent methods have allowed the generation of widely tunable THz waves based on laser light scattering from a polariton mode [3.321]. Other techniques include the use of photoconductivity and optical rectification [3.322–324].

Optical rectification is the generation of a quasi-static polarization in a noncentrosymmetric material due to an intense laser beam [3.325]. In the usual case this polarization does not radiate. However, if subpicosecond laser pulses are used, the polarization in the material becomes time-dependent and can radiate electromagnetic waves (Fig. 3.30). More accurately this effect can be described by a difference frequency generation process: when short laser pulses with a broad frequency spectrum are incident on a nonlinear material any two frequency components ω_1 and ω_2 will induce a polarization

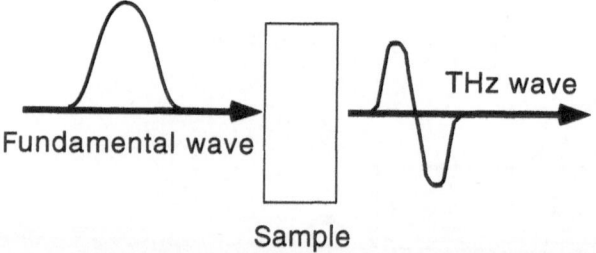

Fig. 3.30. Schematic setup of THz generation using optical rectification (see text for details). The short laser pulse ideally generates one cycle of the THz wave

and radiate electromagnetic waves at the frequency $\omega_1 - \omega_2$. This radiation contains frequencies from 0 to several THz. In the past organic materials as well as commonly used semiconductors have also been investigated.

Research on THz generation based on optical rectification using organic materials has been going on for some years now, concentrating on the organic salt DAST [3.326–329] and on poled polymers [3.330]. The ratios of the relevant tensor elements of the nonlinear optical coeffcients of DAST could be determined [3.329] and it was shown that mosaics of these crystals considerably enhance the intensity of the generated THz signals [3.328]. However, there are quite a few important points that have not been addressed so far. On the one hand, phase-matching possibilities for larger conversion efficiencies were mostly neglected. On the other hand, oscillatory structures visible in the generated THz waveforms of DAST could not be identified.

In the future one can expect some exciting results using organic crystals for the generation of THz waves based on optical rectification due to the large electro-optic coefficients combined with low dielectric constants as well as from the investigation of the excitation processes in molecular solids in this frequency range.

3.5.6   Thermo-Optic Switches

Extensive research in the field of guided wave optical components based on polymers was mainly done on *electro-optic* polymers for many years, as described in the previous sections. The use of polymers to build *thermo-optic* switches was recognized only a few years ago by Diemeer et al. [3.331], who demonstrated for the first time a thermo-optic waveguide switch based on polymers. Since then such polymer-based switches have developed very rapidly and are already commercially available. They are becoming more and more basic constituents of today's fiber optic networks, expanding continously due to the increasing demands on telecommunications network capacity. In this section we will give a brief introduction to the working principle of thermo-optic switching and present some data on the state of the art of such switches.

Figure 3.31 illustrates a thermo-optic switch based on a *y*-branch. Application of a current to the upper electrode resistively heats the electrode and the temperature of the waveguide below rises. This increase in temperature induces a reduction of the effective refractive index seen by the waveguide mode propagating in this arm of the waveguide via the thermo-optic effect. This effect is very large in polymers. The temperature coefficient is $dn/dT = -3.3 \times 10^{-4}\,°C^{-1}$ in polyurethane, for example, compared with LiNbO$_3$ having a temperature coefficient of only $dn/dT = +4 \times 10^{-5}\,°C^{-1}$ [3.331]. The index change in polymers results almost exclusively from density changes, as shown by Cariou et al. [3.332], whereas in LiNbO$_3$, a change in the UV absorption band also contributes to a large extent. The change in the effective refractive index causes the waveguide not to act as a power splitter

278   Ch. Bosshard et al.

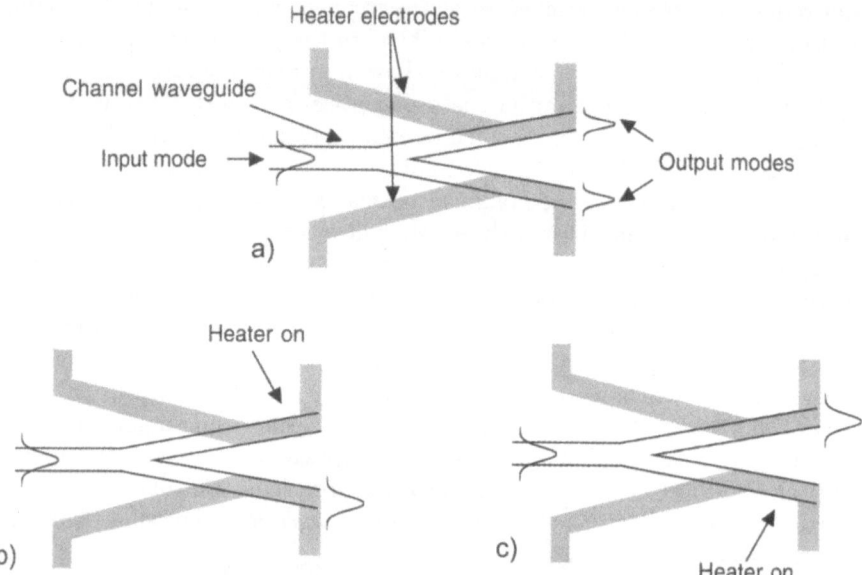

Fig. 3.31. Principle of thermo-optic y-branch switch: (**a**) no electrical power applied to y-branch; (**b**) electrical power applied to y-branch arm 1; (**c**) electrical power applied to y-branch arm 2

that equally splits the optical power into both arms of the y-branch due to its symmetry. Rather, it forces the mode to be propagated only in the lower channel of the y-branch, since the effective index in the heated arm is too small to allow the mode to propagate. Thus all the light is switched to the waveguide opposite the activated electrode. The main characteristics of the switch-like cross-talk, drive voltage, and extinction ratio, are optimized via the design parameters of the y-branch and the heating electrodes.

The typical response curve of a thermo-optic switch is shown in Fig. 3.32. One clearly sees that the extinction ratio is above 20 dB, as is required for operating as a digital switch, when the device is driven above 50 mW of electrical power. A main advantage of thermo-optic switches, compared to electro-optic switches for example, becomes obvious. The device does not have to be controlled via an external feedback loop, since the switch retains its digital character when operated above a certain threshhold. Furthermore, thermo-optic switches show a polarization dependence below 0.5 dB, so they are basically polarization-independent. The device can operate at both important telecommunications wavelengths, 1310 nm and 1550 nm. Additionally, thermo-optic switches can be easily integrated on a large scale and even fully packaged 8 × 8 switching matrices have already been demonstrated [3.333].

Table 3.11 summarizes the data on a commercially available polymer-based thermo-optic switch [3.334].

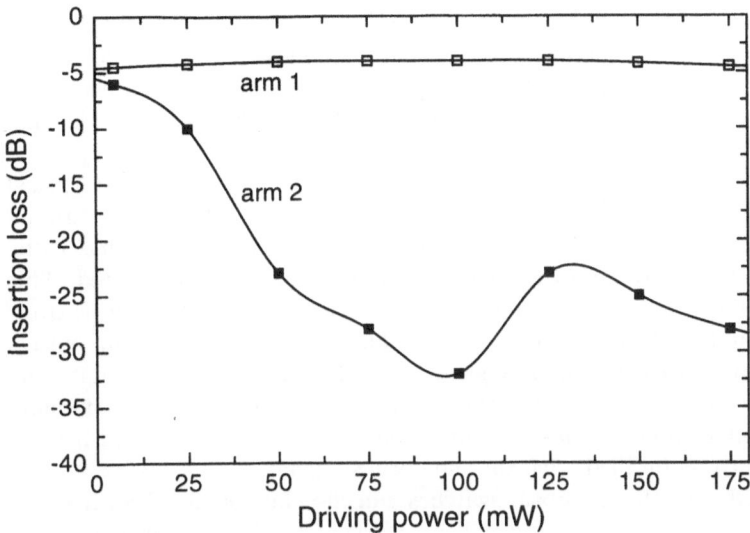

Fig. 3.32. Typical switching response of 1×2 thermo-optic switch when electrical power is applied to arm 2

Table 3.11. Performance characteristics of a commercially available thermo-optic switch [3.334]

Operating wavelengths	1310 nm and 1550 nm
Insertion loss	< 5 dB
Isolation	> 30 dB
Polarization sensitivity	< 0.5 dB
Drive Power	60 mW
Switching time	2 ms

Besides the design as y-branches, thermo-optic switches are also realized as directional couplers showing in principle the same characteristics [3.335, 336]. Furthermore, a polymeric 1×3 switch has recently been demonstrated [3.337], allowing further optimization of large-scale integration.

It should be noted that although thermo-optic switches show characteristics like low operating voltages, low losses, and polarization independence, the thermo-optic effect cannot be used for high-speed electro-optic modulation. As already described in Sect. 3.5.3 these applications require response times of nano- or even picoseconds, which cannot be achieved with the thermo-optic effect, which exhibits response times of about one millisecond.

Nevertheless there are two main fields of application where thermo-optic polymer switches can already be used today [3.338]. The first is so-called protection switching. One problem in fiber optic transmitting applications over large distances is the breaking of a fiber due to, for example, digging activities. For this reason usually the whole transmission line is duplicated, including transmitters and receivers. Such a duplication of electrical components is rather expensive. Thermo-optic switches allow the transmission line to be switched in less than two milliseconds without needing additional transmitters and receivers.

A second field is optical cross-connection applications. Even in optical networks, cross-connections in network nodes are achieved using electronic-based digital cross-connects. This means that all the information has to be converted into electronic form, processed and optically retransmitted. This optical–electrical–optical conversion creates a bottleneck which can be overcome by switching in the optical domain. This means that incoming optical channels can be thermo-optically switched and cross-connected to outgoing channels of the network node without conversion to the electrical domain. The required response times for cross-connection switching are also sufficiently long to be fulfilled with thermo-optic switches.

Polymer-based thermo-optic switches can also be combined with wavelength divison multiplexing and will play a key role in the ever-expanding optical networks in the near future.

3.6   Stability of Nonlinear and Electro-Optic Materials and Their Properties

We concentrate here on the discussion of two stability issues: the optical damage threshold in single crystals and orientational relaxation of poled polymers. There are other important issues, such as light-induced degradation, that are not treated here.

3.6.1   Optical Damage Threshold

The optical damage threshold is one of the most important criteria for frequency-conversion applications. Table 3.12 lists examples of organic and inorganic crystals in which damage threshold experiments were carried out. The first observation is that inorganic crystals are clearly superior with respect to the onset of optical damage. Second, as expected, damage intensities clearly depend on the laser pulse duration: the shorter the pulses the higher the damage intensities. This shows that optical damage effects are fluence dependent, not intensity dependent. In the following we will describe an example with an inorganic and an organic crystal in more detail.

In the 500 ps to 700 ps pulse range a total of eight $KNbO_3$ samples were investigated in [3.339]. The samples were irradiated with well-defined fluences and polarization. After irradiation the samples were inspected with microscopic techniques (bright and dark field Nomarski). The threshold fluence was defined as the average of the highest nondamaging and lowest damaging level.

Table 3.12. Optical damage thresholds of organic and inorganic crystals

Material	λ (nm)	Pulse duration	Repetition rate (Hz)	Damage threshold (GW/cm^2 or J/cm^2)	Ref.
			Organic crystals		
MBANP	1064	15 ns	4	$0.35\,GW/cm^2$	[3.200]
			0.5	$> 0.40\,GW/cm^2$	
		32 ps	1	$10\,GW/cm^2$	
	532	12 ns	0.1	$0.10\,GW/cm^2$	
		23.5 ps	1	$10\,GW/cm^2$	
NPLO	1064	8 ns	–	$6\,GW/cm^2$	[3.145]
DAN	1064	15 ns	30	$0.08\,GW/cm^2$	[3.117]
		100 ps	–	$5\,GW/cm^2$	
			Inorganic crystals		
$KNbO_3$	1064	100 ps	20	$> 100\,GW/cm^2$, $> 10\,J/cm^2$	[3.340]
	1054	700 ps	single shot	$26.4\,J/cm^2$	[3.339]
	527	500 ps	single shot	$4.1\,J/cm^2$	
BBO	1064	10 ns	–	$5\,GW/cm^2$	[3.341]
		1.3 ns	–	$10\,GW/cm^2$	
	532	10 ns	–	$1\,GW/cm^2$	
		250 ps	–	$7\,GW/cm^2$	
LiB_3O_5	1053	1.3 ns	–	$18.9\,GW/cm^2$	[3.341]
				$24.6\,GW/cm^2$	

Damage was assumed if a change in the surface appearance was detected using a magnification of 200. Among other things, initial free carrier generation induced by defect or multiphoton absorption and followed by avalanche amplification was believed to contribute to the damage mechanism. The damage threshold was found to depend on the polarization of the incident beam with a maximum damage threshold of $26.4\,J/cm^2$ for single-shot operation.

In the organic crystal MBANP transmission experiments were performed with several samples and for different laser pulse durations [3.200]. The transmitted energy was monitored as a function of the incident laser fluence. At a certain fluence level the transmission usually diverged from the linear regime. If this decrease in transmission was permanent the optical damage was defined for the fluence at which the nonlinear transmission change occurred. The damage was attributed to thermal heating in connection with low thermal diffusion in the MBANP crystals. In agreement with this finding, the damage threshold is much higher for ps pulses than for ns pulses and also dependent on the pulse repetition rate.

Applications of organic crystals for efficient parametric interactions therefore seem to lie in the ps to fs pulse duration range, where high intensities with corresponding low energies per pulse are feasible.

3.6.2   Orientational Relaxation of Poled Polymers

Understanding relaxational processes in nonlinear optical polymeric materials
is of critical importance in order to evaluate the long-term stability of poled
polymers that are in development for potential electro-optic applications. The
actual usefulness of these materials relies on sufficient stability of the poling-
induced order of the nonlinear optical chromophores within these polymers.
An essential requirement for stabilizing polymeric nonlinear optical materials
is the formation of a glassy state at relatively high temperatures. Amorphous
polymers, among other glasses, typically show evidence of a phase transition
from a liquid-like to a glassy state when cooled from high temperatures. The
temperature at which this transition occurs is known as the glass transition
temperature, T_g. The physical origin of the glass transition is primarily associated with the cooperative motions of large-scale molecular segments of the
polymer. The actual experimental observation of a glass transition is most
easily probed by measuring enthalpic changes in the polymer as a function of
temperature via differential scanning calorimetry (DSC). Several phenomenological theories describe the primary aspects of the glass transition, at least
as they are experimentally observed.

Relaxation processes in nonlinear optically active modified polyimide polymers with sidechain azo chromophores having glass transition temperatures
in the range $140°C < T_g < 170°C$ have been studied by differential scanning
calorimetry, dielectric relaxation, and second-harmonic generation experiments (Fig. 3.33; [3.342]). In these latter experiments the second-harmonic

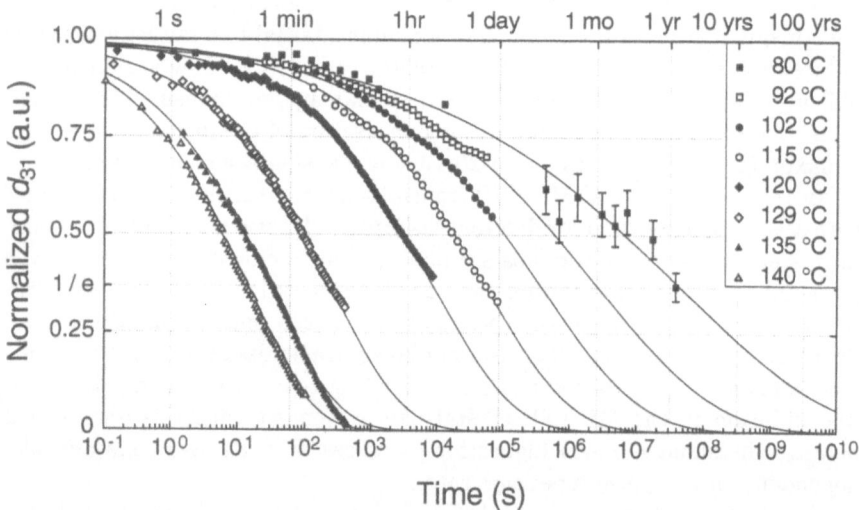

Fig. 3.33. Normalized second-harmonic generation relaxation for a polyimide side-
chain polymer ($T_g = 137°C$) at different temperatures below the glass transition
[3.342]. Solid lines are fits to the Kohlrausch–Williams–Watts (KWW) function

signal of different poled polymers stored at temperatures between 80°C and 140°C was measured at various time intervals [3.342]. A useful functional representation for describing the observed non-exponentiality of structural relaxations in polymeric glasses in the time domain is the stretched exponential or Kohlrausch–Williams–Watts (KWW) function [3.343]

$$\phi(t) = \exp\left[-(t/\tau)^b\right] \tag{3.54}$$

with $0 < b \le 1$ and τ the relaxation time. The theoretical curves in Fig. 3.33 are based on this function. The issue of nonlinearity of structural relaxations was addressed by Tool by introducing the concept of fictive temperature, T_f, which describes the influence of the actual structure of a glass on the relaxation time τ.

Using an assumption of thermorheological simplicity, Narayanaswamy has given an explicit procedure, which will be denoted the TN (Tool–Narayanaswamy) procedure, for the quantitative calculation of the fictive temperature [3.344]. Further details for incorporating nonexponentiality in this procedure have been provided by Moynihan et al., by making use of the KWW function in describing the enthalpic relaxation and recovery process [3.345]. Using this procedure, the value of T_f can be calculated from the previous thermal history by using the following expression:

$$T_f = T_0 + \int_{T_0}^{T} dT' \left\{ 1 - \exp\left[-\left(\int_{T'}^{T} \frac{dT''}{q\tau} \right)^b \right] \right\}, \tag{3.55}$$

where $q = dT/dt$ is the heating/cooling rate and T_0 is an arbitrary reference temperature above the glass transition temperature.

The structural relaxation time of polymeric liquids, as interpreted by DiMarzio and Gibbs involves the cooperative rearrangement of a number of molecular segments of the polymer molecule [3.346]. Adams and Gibbs subsequently obtained an expression for the structural relaxation time in terms of the configurational entropy [3.347]. Hodge has shown that the following expression can be derived for the relaxation time associated with the cooperative chain rearrangements [3.348]:

$$\tau = A \exp\left(\frac{B}{T(1 - T_2/T_f)} \right), \tag{3.56}$$

where $B \cdot R$ is an activation energy with R the gas constant, T is the temperature, and A is a time parameter. We assume that the temperature T_2 in this formulation is equivalent to the "Kauzmann temperature", T_K, which is described by the zero point of the configurational entropy.

Extended DSC (Differential Scanning Calorimetry) measurements of the polyimide NLO polymers indicate the following correlation between the parameters A, B, and T_2 [3.349]:

$$A = \tau_g \exp\left(\frac{-B}{T_g - T_2} \right) = \tau_g \exp\left(-2.303 C_1^g\right) \tag{3.57}$$

with τ_g the relaxation time at T_g, typically in the range of 10–100 s. The parameter C_1^g is one of the Williams–Landel–Ferry (WLF) parameters with a "universal" value of about 17 [3.350].

Above the glass transition the polymer system is in thermal equilibrium, and therefore T_f equals T. Together with the replacement of A in (3.56), this yields the functional form of the Vogel–Fulcher (VF) equation for the normalized relaxation time, $\tau_n = \tau / \tau_g$:

$$\ln \tau_n = -\frac{B}{T_g - T_2} + \frac{B}{T - T_2}. \tag{3.58}$$

Below the glass transition T_f reaches a final limiting value in simple cooling processes, i.e. without sub-T_g annealing. Assuming the final fictive temperature to be T_g, we obtain the following temperature dependence of the normalized relaxation time:

$$\ln \tau_n \approx \frac{B}{T_g - T_2} \frac{T_g - T}{T} = 2.303 \, C_1^g \frac{T_g - T}{T}. \tag{3.59}$$

This result indicates that the primary or α-relaxation processes should have an Arrhenius temperature dependence below the glass transition. In addition, from the "universality" of the (WLF) parameter, C_1^g, one might expect normalized relaxation times to scale with the scaling parameter $(T_g - T)/T$, at least in the same class of glass-forming liquids.

The differential scanning calorimetry, dielectric relaxation and second-harmonic generation experiments revealed important information on stability

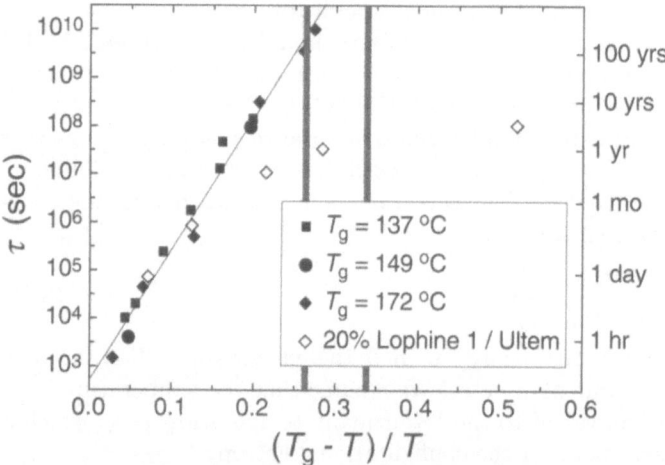

Fig. 3.34. Temperature scaling of normalized dielectric loss and second-harmonic relaxation times with respect to the scaling variable $(T_g - T)/T$ of three side-chain and one guest–host polymer (20% Lophine 1/Ultem). It is cleary seen that the nonlinear optical moiety has to be coupled to the polymer main chain for increased stability. The two thick vertical lines are related to examples discussed in the text

[3.342, 349]. We have shown that it is possible to model the relaxational behavior of nonlinear optical chromophores both above and below the glass transition over more than 15 orders of magnitude in time using the Tool–Narayanaswamy procedure incorporating the appropriate WLF parameters for the nonlinear optical polymers [3.349]. This leads to a scaling prediction for relaxation times in the glassy state with the scaling parameter $(T_g - T)/T$ (Fig. 3.34).

The relaxation time therefore depends on only one parameter: T_g. Figure 3.34 and the underlying theory now allow us to define the requirements of T_g for a certain required stability. As an example, for a polymer with $T_g = 172°C$ the relaxation time $(1/e)$ at an operating temperature of 80°C is about 100 years. With a T_g of 200°C this relaxation time dramatically increases. Annealing of polymers at elevated temperatures further increases the orientational stability [3.342]. Finally, it should be noted that the relaxation results of the polyimide side-chain polymers could be slightly altered for different polymer systems.

3.7   Concluding Remarks and Outlook

We have presented an overview of organic materials for second-order nonlinear optics. In the case of single crystals the largest potential lies in the generation of new wavelengths (including the generation of THz waves) and in fast electro-optic modulation. Much progress has been made in the last few years, ranging from large increases in the macroscopic nonlinearity to the growth of large size crystals sufficient for applications. A recent approach towards the fabrication of optical thin films is the self-assembly of organic molecules during molecular beam deposition, especially because it allows the growth of noncentrosymmetric structures for second-order nonlinear optics [3.351]. In Japan, much research is directed towards the generation of blue light using frequency doubling in organic crystals [3.88]. In this case a reduction of the molecular nonlinearity has to be taken into account in order to fulfill the transparency requirements. Others, such as our group, concentrate on applications based on parametric interactions in the near-infrared and on electro-optic effects. We have discussed the potential of organic crystals and have outlined the challenges still to be solved to have a commercial product. Several companies are currently trying to commercialize organic nonlinear optical crystals for electro-optic devices (MOEC [3.305]). Rainbow Photonics, a spin-off company of our group at ETH, also plans to offer nonlinear optical organic crystals in the near future [3.352].

Poled polymers are of great interest for the characterization of ultrashort pulses (autocorrelation, frequency-resolved optical gating) as well as for the efficient generation of UV ultrashort pulses. There is also great potential in the area of optical signal processing. Thermo-optic switches are already commercially available [3.334]. All these applications depend on the still possible improvements of the material properties (thermal and optical stability, electro-optic response) and the convincing demonstration thereof.

The commercialization of organic materials currently depends on the successful establishment of niche applications where they outperform traditional materials or at least offer similar performance at lower cost. Such a model device appears to be an electro-optic modulator, where poled polymers offer, besides high electro-optic coefficients, a very high frequency response in excess of 100 GHz. The low fabrication costs also expected have prompted research and development towards polymeric modulators in several companies (e.g. TACAN Corp., MOEC, Siemens). The advent of the first commercial products will promote the acceptance of organic materials as reliable and high-quality optical materials and will also stimulate further technological development. This process is also aided by the progress organics have made in other optical fields, such as linear optical polymers in photonics [3.353] and the recent advances in organic LEDs [3.354]. The possible multi-functionality is after all one of the exciting advantages that organic materials have to offer.

References

3.1 J. Zyss, *Molecular Nonlinear Optics: Materials, Physics, Devices*, (Academic Press, Boston, 1994).

3.2 J. Zyss, J.-F. Nicoud, Current Opinions in Solid State & Materials Science **1**, 533 (1996).

3.3 Ch. Bosshard, K. Sutter, Ph. Prêtre, J. Hulliger, M. Flörsheimer, P. Kaatz, P. Günter, *Organic Nonlinear Optical Materials* (Gordon & Breach Science Publishers, Amsterdam, 1995).

3.4 Ch. Bosshard, P. Günter, in *Nonlinear Optics of Organic Molecular and Polymeric Materials*, edited by S. Miyata and H.S. Nalwa (CRC Press., Inc, 1997), p. 391.

3.5 Ch. Bosshard, M.S. Wong, F. Pan, R. Spreiter, S. Follonier, U. Meier, P. Günter, in *Electrical and Related Properties of Organic Solids*, edited by R.W. Munn, A. Miniewicz, and B. Kuchta, NATO ASI Series 3 (Kluwer Academic Publishers, 1997), p. 279.

3.6 D. Burland, Optical Nonlinearities in Chemistry, in *Chemical Reviews* (American Chemical Society, Washington, D.C., 1994), Vol. 94, p. 1.

3.7 H. S. Nalwa, T. Watanabe, S. Miyata, in *Nonlinear Optics of Organic Molecular and Polymeric Materials*, edited by S. Miyata and H. S. Nalwa, (CRC Press., Inc, 1997), p. 89.

3.8 R. J. Twieg, C. W. Dirk, in *Organic Thin Films for Waveguiding Nonlinear Optics*, edited by F. Kajzar and J.D. Swalen, Advances in Nonlinear Optics Vol. 3 (Gordon and Breach Science Publishers, Amsterdam, 1996).

3.9 A. Yariv, *Quantum Electronics* (John Wiley and Sons, New York, 1975).

3.10 F. Brehat, B. Wyncke, Journal of Physics **B 22**, 1891 (1989).

3.11 B. Wyncke, F. Brehat, Journal of Physics **B 22**, 363 (1989).

3.12 S. H. Wemple, M. DiDomenico, Jr., J. Appl. Phys. **40**, 720 (1969).

3.13 J. Zyss, J. L. Oudar, Phys. Rev. **A 26**, 2028 (1982).

3.14 P. Günter, in *Proceedings of Electro-optics/Laser International '76*, edited by H. G. Jerrard, IPC Science and Technology (1976), p. 121.
3.15 R. Spreiter, R. Bosshard, F. Pan, P. Günter, Opt. Lett. **22**, 564 (1997).
3.16 Ch. Bosshard, M. Flörsheimer, M. Küpfer, P. Günter, Opt. Commun. **85**, 247 (1991).
3.17 Ch. Bosshard, G. Knöpfle, P. Prêtre, P. Günter, J. Appl. Phys. **71**, 1594 (1992).
3.18 M. S. Paley, J. M. Harris, H. Looser, J. C. Baumert, G. C. Bjorklund, D. Jundt, R. J. Twieg, J. Org. Chem. **54**, 3774 (1989).
3.19 S. K. Kurtz, T. T. Perry, J. Appl. Phys. **39**, 3798 (1968).
3.20 R. C. Eckardt, H. Masuda, Y. X. Fan, R. L. Byer, IEEE J. Quant. Electron. **26**, 922 (1990).
3.21 D. A. Kleinman, Phys. Rev. **126**, 1977 (1962).
3.22 K. D. Singer, A. F. Garito, J. Chem. Phys. **75**, 3572 (1981).
3.23 R. W. Terhune, P. D. Maker, C. M. Savage, Phys. Rev. Lett. **14**, 681 (1965).
3.24 S. Kielich, Acta Phys. Polon. **25**, 85 (1964).
3.25 S. Kielich, Bull. Acad. Polon. Sci., Ser. Math. Astron. Phys. **12**, 53 (1964).
3.26 S. Kielich, IEEE J. Quantum Electron. **QE-4**, 744 (1968).
3.27 S. Kielich, J. R. Lalanne, F. B. Martin, Phys. Rev. Lett. **26**, 1295 (1971).
3.28 K. Clays, A. Persoons, Phys. Rev. Lett. **66**, 2980 (1991).
3.29 J. Zyss, T. C. Van, Ch. Dhenaut, I. Ledoux, Chem. Phys. **177**, 281 (1993).
3.30 S. Stadler, R. Dietrich, G. Bourhill, Ch. Bräuchle, Opt. Lett. **21**, 251 (1996).
3.31 C. C. Teng, A. F. Garito, Phys. Rev. **B 28**, 6766 (1983).
3.32 L. T. Cheng, W. Tam, S. H. Stevenson, G. R. Meredith, G. Rikken, S. Marder, J. Phys. Chem. **95**, 10631 (1991a).
3.33 R. D. Miller, in *Organic Thin Films for Waveguiding Nonlinear Optics*, edited by F. Kajzar and J. D. Swalen, Advances in Nonlinear Optics Vol. 3 (Gordon & Breach Publishers, Amsterdam, 1996), p. 329.
3.34 R. A. Huijts, G. L. J. Hesselink, Chem. Phys. Lett. **156**, 209 (1989).
3.35 K. D. Singer, J. E. Sohn, L. A. King, H. M. Gordon, H. E. Katz, C. W. Dirk, J. Opt. Soc. Am. **B 6**, 1339 (1989).
3.36 M. S. Wong, U. Meier, F. Pan, Ch. Bosshard, V. Gramlich, P. Günter, Adv. Mater. **8**, 416 (1996).
3.37 C. Moylan, R. Miller, R. Twieg, V. Lee, in *Polymers for Second-Order Nonlinear Optics*, edited by G. Lindsay and K. Singer, Vol. 601 (American Chemical Society, Washington, DC, 1995), p. 66.
3.38 G. Knöpfle, Ch. Bosshard, R. Schlesser, P. Günter, IEEE J. Quantum. Electron. **30**, 1303 (1994).
3.39 M. Barzoukas, D. Josse, J. Zyss, P. Gordon, J. O. Morley, Chem. Phys. **139**, 359 (1989).
3.40 C. Moylan, C. Walsh, Nonlinear Optics **6**, 113 (1993).
3.41 L. T. Cheng, W. Tam, S. R. Marder, A. E. Stiegman, G. Rikken, C. W. Spangler, J. Phys. Chem. **95**, 10643 (1991).
3.42 P. Nguyen, G. Lesley, T. B. Marder, I. Ledoux, J. Zyss, Chem. Mater. **9**, 406 (1997).
3.43 S. Ermer, S. M. Lovejoy, D. S. Leung, H. Warren, C. R. Moylan, R. J. Twieg, Chem. Mater. **9**, 1437 (1997).
3.44 V. P. Rao, A. K-Y. Jen, K. Y. Wong, K. J. Drost, J. Chem. Soc., Chem. Commun. **16**, 1118 (1993).

288 Ch. Bosshard et al.

3.45 A. K. Y. Jen, Y. M. Cai, P. V. Bedworth, S. R. Marder, Adv. Mater. **9**, 132 (1997).

3.46 C. Cai, I. Liakatas, M.-S. Wong, M. Bösch, Ch. Bosshard, P. Günter, S. Concilio, N. Tirelli, and N.W. Suter, Org. Lett. **1**, 1847 (1999).

3.47 C. F. Shu, W. J. Tsai, J. Y. Chen, A. K. Y. Jen, Y. Zhang, T. A. Chen, Chem. Commun. **19**, 2279 (1996).

3.48 C. C. Hsu, C. F. Shu, T. H. Huang, C. H. Wang, J. L. Lin, Y. K. Wang, Y. L. Zang, Chem. Phys. Lett. **274**, 466 (1997).

3.49 F. Steybe, F. Effenberger, S. Beckmann, P. Kramer, C. Glania, R. Wortmann, Chem. Phys. **219**, 317 (1997).

3.50 V. P. Rao, K. Y. Wong, A. K. Y. Jen, K. J. Drost, Chem. Mater. **6**, 2210 (1994).

3.51 I. Liakatas, M. S. Wong, Ch. Bosshard, M. Ehrensperger, P. Günter, Ferroelectrics **202**, 299 (1997).

3.52 M. Ahlheim, M. Barzoukas, P. V. Bedworth, M. Blanchard-Desce, A. Fort, Z. Y. Hu, S. R. Marder, J. W. Perry, C. Runser, M. Staehelin, B. Zysset, Science **271**, 335 (1996).

3.53 M. Blanchard-Desce, V. Alain, P. V. Bedworth, S. R. Marder, A. Fort, C. Runser, M. Barzoukas, S. Lebus, R. Wortmann, Chem. Eur. J. **3**, 1091 (1997).

3.54 S. Marder, L.-T. Cheng, B. G. Tiemann, A. C. Friedli, M. Blanchard-Desce, J. W. Perry, J. Sindhoj, Science **263**, 511 (1994).

3.55 I. Ledoux, J. Zyss, A. Jutand, C. Amatore, Chem. Phys. **150**, 117 (1991).

3.56 S. Ermer, D. Girton, D. Leung, S. Lovejoy, J. Valley, T. Van Eck, L. Cheng, Nonlinear Optics **9**, 259 (1995).

3.57 V. P. Rao, A. K. Y. Jen, Y. M. Cai, Chem. Commun. **10**, 1237 (1996).

3.58 O. K. Kim, A. Fort, M. Barzoukas, J. M. Lehn, in *Proceedings of Organic Thin Films for Photonics Applications*, OSA technical digest series, edited by Optical Society of America (1997), p. 201.

3.59 U. Behrens, H. Brussaard, U. Hagenau, J. Heck, E. Hendrickx, J. Kornich, J. G. M. Vanderlinden, A. Persoons, A. L. Spek, N. Veldman, B. Voss, H. Wong, Chem. Eur. J. **2**, 98 (1996).

3.60 J. D. Bewsher, G. R. Mitchell, Opt. Eng. **36**, 167 (1997).

3.61 S. M. Lecours, H. W. Guan, S. G. Dimagno, C. H. Wang, M. J. Therien, J. Am. Chem. Soc. **118**, 1497 (1996).

3.62 M. S. Wong, Ch. Bosshard, F. Pan, P. Günter, Adv. Mater. **8**, 677 (1996).

3.63 H. E. Katz, K. D. Singer, J. E. Sohn, C. W. Dirk, L. A. King, H. M. Gordon, J. Am. Chem. Soc., 6561 (1987).

3.64 Ch. Bosshard, M. Küpfer, in *Oriented molecular systems*, edited by F. Kajzar and J. D. Swalen, Advances in Nonlinear Optics (Gordon & Breach Science Publishers, Amsterdam, 1996), p. 163.

3.65 G. D. Boyd, H. Kaspar, J. H. McFee, IEEE J. Quantum Electron. **QE-7**, 563 (1971).

3.66 J. Jerphagnon, S. K. Kurtz, J. Appl. Phys. **41**, 1667 (1970).

3.67 M. M. Choy, R. L. Byer, Phys. Rev. **B 14**, 1693 (1976).

3.68 A. Kitamoto, T. Kondo, I. Shoji, R. Ito, Opt. Rev. **2**, 280 (1995).

3.69 C. C. Teng, T. H. Man, Appl. Phys. Lett. **56**, 1734 (1990).

3.70 Y. Levy, M. Dumont, E. Chastaing, P. Robin, P. A. Chollet, G. Gadret, F. Kajzar, Mol. Cryst. Liq. Cryst. Sci. Technol.-Sec. B: Nonlinear Optics **4**, 1 (1993).

3.71 E. Kretschmann, Z. Phys. **241**, 313 (1971).

3.72 P. Prêtre, L.-M. Wu, R. A. Hill, A. Knoesen, J. Opt. Soc. Am. **B 15**, 379 (1998).

3.73 A. R. Johnston, Appl. Phys. Lett. **7**, 195 (1965).

3.74 J. L. Oudar, D. S. Chemla, J. Chem. Phys. **66**, 2664 (1977).

3.75 R. W. McQuaid, Appl. Opt. **2**, 310 (1963).

3.76 R. W. Lee, J. Opt. Soc. Am. **59**, 1574 (1969).

3.77 C. Cassidy, J. M. Halbout, W. Donaldson, C. L. Tang, Opt. Commun. **29**, 243 (1979).

3.78 J. A. Morell, A. C. Albrecht, K. H. Levin, C. L. Tang, J. Chem. Phys. **71**, 5063 (1979).

3.79 J. M. Halbout, S. Blit, W. Donaldson, C. L. Tang, IEEE J. Quantum Electron. **QE-15**, 1176 (1979).

3.80 K. Betzler, H. Hesse, P. Loose, J. Mol. Struct. **47**, 393 (1978).

3.81 I. Ledoux, J. Zyss, Chem. Phys. **73**, 203 (1982).

3.82 M. Kitazawa, R. Higuchi, M. Takahashi, T. Wada, H. Sasabe, Appl. Phys. Lett. **64**, 2477 (1994).

3.83 H. O. Marcy, M. J. Rosker, L. F. Warren, P. H. Cunningham, C. A. Thomas, L. A. DeLoach, S. P. Velsko, C. A. Ebbers, J.-H. Liao, M. G. Kanatzidis, Opt. Lett. **20**, 252 (1995).

3.84 N. Zhang, D. R. Yuan, X. T. Tao, D. Xu, Z. S. Shao, M. H. Jiang, M. G. Liu, Opt. Commun. **99**, 247 (1993).

3.85 X. T. Tao, D. R. Yuan, N. Zhang, M. H. Jiang, Z. S. Shao, Appl. Phys. Lett. **60**, 1415 (1992).

3.86 H. Yamamoto, S. Funato, T. Sugiyama, I. Jung, T. Kinoshita, K. Sasaki, J. Opt. Soc. Am. **B 14**, 1099 (1997).

3.87 T. Kinoshita, S. Horinouchi, K. Sasaki, H. Okamoto, N. Tanaka, T. Fukaya, M. Goto, J. Opt. Soc. Am. **B 11**, 986 (1994).

3.88 M. Sagawa, H. Kagawa, A. Kakuta, M. Kaji, M. Saeki, Y. Namba, Appl. Phys. Lett. **66**, 547 (1995).

3.89 M. Sagawa, H. Kagawa, M. Kaji, A. Kakuta, Appl. Phys. Lett. **63**, 1877 (1993).

3.90 H. Kagawa, M. Sagawa, A. Kakuta, M. Kaji, H. Nakayama, K. Ishii, J. Cryst. Growth **139**, 309 (1994).

3.91 Z. Li, B. Wu, G. Su, G. Huang, Appl. Phys. Lett. **70**, 562 (1997).

3.92 S. Genbo, G. Shouwu, F. Pan, H. Youping, L. Zhengdong, J. Phys. D: Appl. Phys. **26**, B236 (1993).

3.93 G. Puccetti, A. Perigaud, J. Badan, I. Ledoux, J. Zyss, J. Opt. Soc. Am. **B 10**, 733 (1993).

3.94 R. Masse, J. Zyss, Mol. Engin. **1**, 141 (1991).

3.95 Z. Kotler, R. Hierle, D. Josse, J. Zyss, J. Opt. Soc. Am. **B 9**, 534 (1992).

3.96 S. Khodia, D. Josse, I. D. W. Samuel, J. Zyss, Appl. Phys. Lett. **67**, 3841 (1995).

3.97 J. Zyss, I. Ledoux, M. Bertault, E. Toupet, Chem. Phys. **150**, 125 (1991).

3.98 H. Endoh, M. Kawaharada, E. Hasegawa, Appl. Phys. Lett. **68**, 293 (1996).

3.99 G. J. Zhang, T. Kinoshita, K. Sasaki, Y. Goto, M. Nakayama, Appl. Phys. Lett. **57**, 221 (1990).

3.100 Y. Kitaoka, T. Sasaki, S. Nakai, Y. Goto, M. Nakayama, Appl. Phys. Lett. **56**, 2074 (1990).

3.101 Y. Kitaoka, T. Sasaki, S. Nakai, Y. Goto, Appl. Phys. Lett. **59**, 19 (1991).

3.102 J. Kawamata, K. Inoue, T. Inabe, Appl. Phys. Lett. **66**, 3102 (1995).

3.103 C. H. Grossman, T. Wada, S. Yamada, H. Sasabe, A. F. Garito, in *Springer Proceedings in Physics, Nonlinear Optics of Organics and Semiconductors*, edited by T. Kobayashi, Vol. 36 (Springer-Verlag, Berlin, Heidelberg, 1989), p. 214.

3.104 C. H. Grossman, S. Schulhofer-Wohl, E. R. Thoen, Appl. Phys. Lett. **70**, 283 (1997).

3.105 A. Harada, Y. Okazaki, K. Kamiyama, S. Umegaki, Appl. Phys. Lett. **59**, 1535 (1991).

3.106 B. F. Levine, C. G. Bethea, C. D. Thurmond, R. T. Lynch, J. L. Bernstein, J. Appl. Phys. **50**, 2523 (1979).

3.107 G. F. Lipscomb, A. F. Garito, R. S. Narang, J. Chem. Phys. **75**, 1509 (1981).

3.108 R. Morita, N. Ogasawara, S. Umegaki, R. Ito, Jap. J. Appl. Phys. **26**, L1711 (1987).

3.109 E. S. S. Ho, K. Iizuko, A. P. Freundhofer, C. K. L. Wah, J. Lightwave Tech. **9**, 101 (1991).

3.110 J. O. White, D. Hulin, M. Joffre, A. Migus, A. Antonetti, E. Toussaere, R. Hierle, J. Zyss, Appl. Phys. Lett. **64**, 264 (1994).

3.111 S. M. Rao, K. X. He, R. B. Lal, R. A. Evans, B. H. Loo, J. M. Chang, R. M. Metzger, W. J. Lee, A. S. Shields, B. G. Penn, D. O. Frazier, J. Mater. Science **30**, 179 (1995).

3.112 M. Stähelin, D. M. Burland, J. E. Rice, Chem. Phys. Lett. **191**, 245 (1992).

3.113 M. Kato, M. Kigichi, N. Sugita, Y. Taniguchi, J. Phys. Chem. **B 101**, 8856 (1997).

3.114 K. Sutter, Ch. Bosshard, M. Ehrensperger, P. Günter, R. J. Twieg, IEEE J. Quantum Electron. **24**, 2362 (1988).

3.115 A. Nahata, K. A. Horn, J. T. Yardley, IEEE J. Quantum Electron. **26**, 1521 (1990).

3.116 P. Kerkoc, M. Zgonik, Ch. Bosshard, K. Sutter, P. Günter, Appl. Phys. Lett. **54**, 2062 (1989).

3.117 P. Kerkoc, M. Zgonik, Ch. Bosshard, K. Sutter, P. Günter, J. Opt. Soc. Am. **B 7**, 313 (1990).

3.118 A. Carenco, J. Jerphagnon, A. Perigaud, J. Chem Phys. **66**, 3806 (1977).

3.119 J. G. Bergman, G. R. Crane, J. Chem. Phys. **66**, 3803 (1977).

3.120 J. L. Stevenson, J. Phys. D: Appl. Phys. **6**, (1973).

3.121 J. Zyss, D. S. Chemla, J. F. Nicoud, J. Chem. Phys. **74**, 4800 (1981).

3.122 M. Sigelle, R. Hierle, J. Appl. Phys. **52**, 4199 (1981).

3.123 R. Hierle, J. Badan, J. Zyss, J. Cryst. Growth **69**, 545 (1984).

3.124 J. L. Oudar, R. Hierle, J. Appl. Phys. **48**, 2699 (1977).

3.125 I. Ledoux, C. Lepers, A. Périgaud, J. Badan, J. Zyss, Opt. Commun. **80**, 149 (1990).

3.126 M. Barzoukas, P. Fremaux, D. Josse, F. Kajzar, J. Zyss, in *Proceedings of Mat. Res. Soc. Symp. Proc.*, edited by A. J. Heeger et al., Materials Research Society (1988), p. 171.

3.127 M. Barzoukas, D. Josse, P. Fremaux, J. Zyss, J. F. Nicoud, J. O. Morley, J. Opt. Soc. Am. **B 4**, 977 (1987).

3.128 R. Morita, P. V. Vidakovic, Appl. Phys. Lett. **61**, 2854 (1992).

3.129 T. Tomono, L. S. Pu, T. Kinoshita, K. Sasaki, S. Umegaki, J. Phys. D: Appl. Phys. **26**, B217 (1993).

3.130 H. Minemoto, Y. Ozaki, N. Sonoda, T. Sasaki, J. Appl. Phys. **76**, 3975 (1994).
3.131 H. Minemoto, Y. Ozaki, K. Wakiti, T. Sasaki, Jap. J. Appl. Phys. **34**, 497 (1995).
3.132 H. Minemoto, Y. Ozaki, N. Sonoda, T. Sasaki, Appl. Phys. Lett. **63**, 3565 (1993).
3.133 Ch. Bosshard, K. Sutter, R. Schlesser, P. Günter, J. Opt. Soc. Am. **B10**, 867 (1993).
3.134 K. Sutter, Ch. Bosshard, W. S. Wang, G. Surmely, P. Günter, Appl. Phys. Lett. **53**, 1779 (1988).
3.135 K. Sutter, Ch. Bosshard, L. Baraldi, P. Günter, in *Proceedings of International Conference on Materials for Nonlinear and Electro-Optics*, edited by M.-H. Lyons, IOP Publishing Ltd (1989).
3.136 Ch. Bosshard, K. Sutter, P. Günter, J. Opt. Soc. Am. **B 6**, 721 (1989).
3.137 P. Günter, Ch. Bosshard, K. Sutter, H. Arend, G. Chapuis, R. J. Twieg, D. Dobrowolski, Appl. Phys. Lett. **50**, 486 (1987).
3.138 R. T. Bailey, F. R. Cruickshank, S. M. G. Guthrie, B. J. McArdle, H. Morrison, D. Pugh, E. A. Shepherd, J. N. Sherwood, C. S. Yoon, Opt. Commun. **65**, 229 (1988).
3.139 R. T. Bailey, G. H. Bourhill, F. R. Cruickshank, D. Pugh, J. N. Sherwood, G. S. Simpson, K. B. R. Varma, J. Appl. Phys. **75**, 489 (1994).
3.140 T. Kondo, R. Morita, N. Ogasawara, S. Umegaki, R. Ito, Jap. J. Appl. Phys. **28**, 1622 (1989).
3.141 T. Kondo, F. Akase, M. Kumagai, R. Ito, Opt. Rev. **2**, 128 (1995).
3.142 S. Tomaru, S. Matsumoto, T. Kurihara, H. Suzuki, N. Coba, T. Kaino, Appl. Phys. Lett. **58**, 2583 (1991).
3.143 A. Yokoo, S. Tomaru, S. Yokohama, H. Itoh, T. Kaino, J. Cryst. Growth **156**, 279 (1995).
3.144 K. Sutter, G. Knöpfle, N. Saupper, J. Hulliger, P. Günter, W. Petter, J. Opt. Soc. Am. **B 8**, 1483–1490 (1991).
3.145 T. Ukachi, T. Shigemoto, H. Komatsu, T. Sugiyama, J. Opt. Soc. Am. **B 10**, 1372 (1993).
3.146 G. S. Chen, J. K. Wilbur, C. L. Barnes, R. Glaser, J. Chem. Soc. Perkin Trans. **2**, 2311 (1995).
3.147 J. D. Bierlein, L. K. Cheng, Y. Wang, W. Tam, Appl. Phys. Lett. **56 (5)**, 423 (1990).
3.148 R. T. Bailey, G. H. Bourhill, F. R. Cruickshank, D. Pugh, J. N. Sherwood, G. S. Simpson, K. B. R. Varma, J. Appl. Phys. **71**, 2012 (1992).
3.149 G. Bourhill, F. Cruickshank, D. Pugh, G. Simpson, J. Sherwood, R. Bailey, in *Proceedings of Nonlinear Optical Properties of Organic Materials V*, SPIE **1775**, edited by D. J. Williams, (1992), p. 224.
3.150 M. Kitazawa, R. Higuchi, M. Takahashi, T. Wada, H. Sasabe, J. Phys. Chem. **99**, 14787 (1995).
3.151 S. Okada, A. Masaki, H. Matsuda, H. Nakanishi, M. Koto, R. Muramatsu, M. Otsuka, Jap. J. Appl. Phys. **29**, 1112 (1990).
3.152 S. Okada, A. Masaki, H. Matsuda, H. Nakanishi, T. Koike, T. Ohmi, N. Yoshikawa, Proc. SPIE **1337**, 178 (1990).
3.153 X. M. Duan, S. Okada, H. Nakanishi, A. Watanabe, M. Matsuda, K. Clays, A. Persoons, H. Matsuda, in *Proceedings of Organic, Metallo-Organic, and Polymeric Materials for Nonlinear Optical Applications*, SPIE **2143**, edited by S. R. Marder and J. W. Perry (1994), p. 41.

3.154 S. Allen, T. D. McLean, P. F. Gordon, B. D. Bothwell, M. B. Hursthouse, S. A. Karaulov, J. Appl. Phys. **64**, 2583 (1988).

3.155 G. Knöpfle, R. Schlesser, R. Ducret, P. Günter, Nonlinear Optics **9**, 143 (1995).

3.156 F. Pan, G. Knöpfle, Ch. Bosshard, S. Follonier, R. Spreiter, M. S. Wong, P. Günter, Appl. Phys. Lett. **69**, 13 (1996).

3.157 U. Meier, M. Bösch, Ch. Bosshard, F. Pan, P. Günter, J. Appl. Phys. **83**, 3486 (1998).

3.158 T. Yoshimura, J. Appl. Phys. **62**, 2028 (1987).

3.159 T. Yoshimura,Y. Kubota, in *Nonlinear Optics of Organics and Semiconductors*, edited by T. Kobayashi, Springer Proceedings in Physics Vol. 36 (Springer-Verlag, Berlin, Heidelberg, 1989), p. 222.

3.160 M. S. Wong, F. Pan, V. Gramlich, Ch. Bosshard, P. Günter, Adv. Mater. **9**, 554 (1997).

3.161 M. S. Wong, F. Pan, M. Bösch, R. Spreiter, Ch. Bosshard, P. Günter, V. Gramlich, J. Opt. Soc. Am. **B 15**, 426 (1998).

3.162 F. Pan, M. S. Wong, V. Gramlich, Ch. Bosshard, P. Günter, J. Am. Chem. Soc. **118**, 6315 (1996).

3.163 S. Follonier, Ch. Bosshard, G. Knöpfle, U. Meier, C. Serbutoviez, F. Pan, P. Günter, J. Opt. Soc. Am. **B 14**, 593 (1997).

3.164 F. Pan, M. S. Wong, M. Bösch, Ch. Bosshard, U. Meier, Ch. Bosshard, Appl. Phys. Lett. **71**, 2064 (1997).

3.165 F. Pan, Ch. Bosshard, M. S. Wong, C. Serbutoviez, K. Schenk, V. Gramlich, P. Günter, Chem. Mater. **9**, 1328 (1997).

3.166 I. Liakatas, M. S. Wong, V. Gramlich, Ch. Bosshard, P. Günter, Adv. Mater. **10**, 777 (1998).

3.167 J. C. MacDonald, G. M. Whitesides, Chem. Rev. **94**, 2383 (1994).

3.168 M. S. Wong, Ch. Bosshard, P. Günter, Adv. Mater. **9**, 837 (1997).

3.169 A. I. Kitaigorodskii, *Molecular Crystals and Molecules* (Academic Press, New York, 1973).

3.170 T. Tsunekawa, T. Gotoh, M. Iwamoto, Chem. Phys. Lett. **166**, 353 (1990).

3.171 T. Tsunekawa, T. Gotoh, H. Mataki, T. Kondoh, S. Fukada, M. Iwamoto, Proc. SPIE **1337**, 272 (1990).

3.172 W. Tam, B. Guerin, J. C. Calabrese, S. H. Stevenson, Chem. Phys. Lett. **154**, 93 (1989).

3.173 G. R. Meredith, in *Nonlinear Optical Properties of Organic and Polymeric Materials*, edited by D. J. Williams, ACS Symposium Series Vol. 233 (Washington, DC, 1983), p. 27.

3.174 S. R. Marder, J. W. Perry, W. P. Schaeffer, Science **245**, 626 (1989).

3.175 S. R. Marder, J. W. Perry, B. G. Tiemann, R. E. Marsh, W. P. Schaefer, Chem. Mater. **2**, 685 (1990).

3.176 F. Pan, M. S. Wong, Ch. Bosshard, P. Günter, Adv. Mater. **8**, 592 (1996).

3.177 Ch. Serbutoviez, Ch. Bosshard, G. Knöpfle, P. Wyss, P. Prêtre, P. Günter, K. Schenk, E. Solari, G. Chapuis, Chem. Mater. **7**, 1198 (1995).

3.178 M. C. Etter, G. M. Frankenbach, Chem. Mater. **1**, 10 (1989).

3.179 J. Zyss, I. Ledoux, Nonlinear optics in multipolar media: theory and experiments, in *Chem. Rev.* (1994), Vol. 94, p. 77.

3.180 R. Hoss, O. König, V. Kramer-Hoss, U. Berger, P. Rogin, J. Hulliger, Angew. Chemie **35**, 1664 (1996).

3.181 J. Hulliger, O. König, R. Hoss, Adv. Mater. **7**, 719 (1995).

3.182 Y. Le Fur, M. Bagieu-Beucher, R. Masse, J.-F. Nicoud, J.-P. Levy, Chem. Mater. **8**, 68 (1996).

3.183 D. Bäuerle, K. Betzler, H. Hesse, S. Kapphan, P. Loose, Phys. Stat. Sol. (a) **42**, K119 (1977).

3.184 W. R. Donaldson, C. L. Tang, Appl. Phys. Lett. **44**, 25 (1984).

3.185 K. Kato, IEEE J. Quantum Electron. **QE-16**, 810 (1980).

3.186 M. Kitazawa, M. Takahashi, M. Matsuoka, J. Cryst. Growth **141**, 425 (1994).

3.187 H. Yamamoto, T. Sugiyama, I. Jung, T. Kinoshita, K. Sasaki, J. Opt. Soc. Am. **B 14**, 1831 (1997).

3.188 J. G. Bergman, G. R. Crane, B. F. Levine, C. G. Bethea, Appl. Phys. Lett. **20**, 21 (1972).

3.189 S. Khodia, D. Josse, J. Zyss, Appl. Phys. Lett. **67**, 3081 (1995).

3.190 S. Khodia, D. Josse, J. Zyss, J. Opt. Soc. Am. **B 15**, 751 (1998).

3.191 C. K. L. Wah, K. Iizuka, A. P. Freundorfer, Appl. Phys. Lett. **63**, 3110 (1993).

3.192 M. Kiguchi, M. Kato, N. Kumegawa, Y. Tanigichi, J. Appl. Phys. **75**, 4332 (1994).

3.193 P. Kerkoc, Ch. Bosshard, H. Arend, P. Günter, Appl. Phys. Lett. **54**, 487 (1989).

3.194 S. Ducharme, W. P. Risk, W. E. Moerner, V. Y. Lee, R. J. Twieg, G. C. Bjorklund, Appl. Phys. Lett. **57**, 537 (1990).

3.195 D. Y. Kim, W. E. Torruellas, J. Kang, Ch. Bosshard, G. I. Stegeman, P. Vidakovic, J. Zyss, W. E. Moerner, R. Twieg, G. Bjorklund, Opt. Lett. **19**, 868 (1994).

3.196 J. Sapriel, R. Hierle, J. Zyss, M. Boissier, Appl. Phys. Lett. **55**, 2594 (1989).

3.197 D. Josse, S. X. Dou, J. Zyss, P. Andreazza, A. Périgaud, Appl. Phys. Lett. **61**, 121 (1992).

3.198 S. X. Dou, D. Josse, J. Zyss, J. Opt. Soc. Am. **B 10**, 1708 (1993).

3.199 B. Y. Shekunov, E. E. A. Shepherd, J. N. Sherwood, G. S. Simpson, J. Phys. Chem. **99**, 7130 (1995).

3.200 S. Nitti, H. M. Tan, G. P. Banfi, V. Degiorgio, R. T. Bailey, F. Cruickshank, D. Pugh, E. A. Shepherd, J. N. Sherwood, G. S. Simpson, J. Phys. D: Appl. Phys. **26**, B225 (1993).

3.201 S. Follonier, Ch. Bosshard, F. Pan, P. Günter, Opt. Lett. **21**, 1655 (1996).

3.202 A. Bräuer, T. Gase, L. Erdmann, P. Dannberg, W. Karthe, SPIE **2042**, 438 (1994).

3.203 J. Thackara, M. Jurich, J. Swalen, J. Opt. Soc. Am. **B 11**, 835 (1994).

3.204 J. I. Thackara, J. C. Chon, G. C. Bjorklund, W. Volksen, D. M. Burland, Appl. Phys. Lett. **67**, 3874 (1995).

3.205 S. S. Lee, S. W. Ahn, S. Y. Shin, Opt. Commun. **138**, 298 (1997).

3.206 Y. Shuto, S. Tomaru, M. Hikita, M. Amano, IEEE J. Quant. Electr. **31**, 1451 (1995).

3.207 M. Hikita, Y. Shuto, M. Amano, R. Yoshimura, S. Tomaru, H. Kozawaguchi, Appl. Phys. Lett. **63**, 1161 (1993).

3.208 M. C. Oh, W. Y. Hwang, H. M. Lee, S. G. Han, Y. H. Won, IEEE Phot. Techn. Lett. **9**, 1232 (1997).

3.209 M. H. Lee, H. J. Lee, S. G. Han, H. Y. Kim, K. H. Kim, Y. H. Won, S. Y. Kang, Thin Solid Films **303**, 287 (1997).

3.210 W. Y. Hwang, M. C. Oh, H. M. Lee, H. Park, J. J. Kim, IEEE Phot. Techn. Lett. **9**, 761 (1997).

3.211 Y. Shi, W. Wang, J. H. Bechtel, A. Chen, S. Garner, S. Kalluri, W. H. Steier, D. Chen, H. R. Fetterman, L. R. Dalton, L. Yu, IEEE J. Select. Top. Quantum Electron. **2**, 289 (1996).

3.212 D. Chen, H. R. Fetterman, A. Chen, W. H. Steier, L. R. Dalton, W. Wang, Y. Shi, Appl. Phys. Lett. **70**, 3335 (1997).

3.213 W. Wang, D. Chen, H. R. Fetterman, Y. Shi, W. H. Steier, L. R. Dalton, IEEE Phot. Techn. Lett. **7**, 638 (1995).

3.214 W. Wang, D. Chen, H. R. Fetterman, Y. Shi, W. H. Steier, L. R. Dalton, Appl. Phys. Lett. **65**, 929 (1994).

3.215 K. D. Singer, M. G. Kuzyk, J. E. Sohn, J. Opt. Soc. Am. **B 4**, 968 (1987).

3.216 H. E. Katz, C. W. Dirk, M. L. Schilling, K. D. Singer, J. E. Sohn, in *Nonlinear optical properties of polymers*, edited by A. J. Heeger, J. Orenstein, and D. R. Ulrich, Vol. 109 (Materials Research Society, Pittsburgh, Pennsylvania, 1988), p. 127.

3.217 S. S. Sun, C. Zhang, L. R. Dalton, S. M. Garner, A. Chen, W. H. Steier, Chem. Mater. **8**, 2539 (1996).

3.218 Y. Kubo, T. Takaba, S. Aramaki, Chem. Lett. **3**, 255 (1997).

3.219 J. Chen, J. Zhu, A. Harper, F. Wang, M. He, S. Mao, L. Dalton, Polymer Preprints **38**, 215 (1997).

3.220 M. Stähelin, B. Zysset, M. Ahlheim, S. R. Marder, P. V. Bedworth, C. Runser, M. Barzoukas, A. Fort, J. Opt. Soc. Am. **B 13**, 2401 (1996).

3.221 Y. M. Cai, A. K. Y. Jen, Appl. Phys. Lett. **67**, 299 (1995).

3.222 M. Ozawa, M. Nakanishi, H. Nakayama, O. Sugihara, N. Okamoto, K. Hirota, in *Poled Polymers and Their Applications to SHG and EO Devices*, edited by S. Miyata and H. Sasabe, Advances in Nonlinear Optics Vol. 4 (Gordon and Breach Science Publishers, Amsterdam, 1997), p. 223.

3.223 K. Y. Wong, A. K.-Y. Jen, J. Appl. Phys. **75**, 3308 (1994).

3.224 C. Moylan, R. Twieg, V. Lee, R. Miller, W. Volksen, J. Thackara, C. Walsh, in *Proceedings of Nonlinear Optical Properties of Organic Materials VII*, SPIE **2285**, edited by G. Möhlmann (1994), p. 17.

3.225 N. Tsutsumi, O. Matsumoto, W. Sakai, T. Kiyotsukuri, Appl. Phys. Lett. **67**, 2272 (1995).

3.226 K. J. Drost, V. P. Rao, A. K.-Y. Jen, J. Chem. Soc., Chem. Commun. **17**, 369 (1994).

3.227 T. Chen, A. Jen, Y. Cai, Chem. Mater. **8**, 607 (1996).

3.228 A. Jen, V. Rao, J. Chandrasekhar, in *Polymers for Second-Order Nonlinear Optics*, edited by G. Lindsay and K. Singer, Vol. 601 (American Chemical Society, Washington, DC, 1995), p. 147.

3.229 C. Samyn, G. Claes, M. Vanbeylen, A. Dewachter, A. Persoons, Macromol. Symp. **102**, 145 (1996).

3.230 N. Nemoto, F. Miyata, Y. Nagase, J. Abe, M. Hasegawa, Y. Shirai, Chem. Mater. **8**, 1527 (1996).

3.231 D. Yu, A. Gharavi, L. Yu, Appl. Phys. Lett. **66**, 1050 (1995a).

3.232 C. Branger, M. Lequan, R. M. Lequan, M. Large, F. Kajzar, Chem. Phys. Lett. **272**, 265 (1997).

3.233 J. Y. Chang, T. J. Kim, M. J. Han, D. H. Choi, N. Kim, Polymer **38**, 4651 (1997).

3.234 K. J. Moon, H. K. Shim, K. S. Lee, J. Zieba, P. N. Prasad, Macromolecules **29**, 861 (1996).

3.235 N. Tsutsumi, O. Matsumoto, W. Sakai, T. Kiyotsukuri, Macromolecules **29**, 592 (1996).

3.236 D. Yu, Z. Peng, A. Gharavi, L. Yu, in *Polymers for Second-Order Nonlinear Optics*, edited by G. Lindsay and K. Singer, Vol. 601 (American Chemical Society, Washington, DC, 1995), p. 172.

3.237 M. Dörr, R. Zentel, M. Sprave, J. Vydra, M. Eich, Adv. Mater. **9**, 225 (1997).

3.238 Y. W. Kim, J. I. Jin, M. Y. Jin, K. Y. Choi, J. J. Kim, T. Zyung, Polymer **38**, 2269 (1997).

3.239 Z. Y. Liang, Z. X. Yang, S. J. Sun, B. Wu, L. R. Dalton, S. M. Garner, S. Kalluri, A. T. Chen, W. H. Steier, Chem. Mater. **8**, 2681 (1996).

3.240 X. G. Wang, J. Kumar, S. K. Tripathy, L. Li, J. I. Chen, S. Marturunkakul, Macromolecules **30**, 219 (1997).

3.241 K. D. Singer, M. G. Kuzyk, W. R. Holland, J. E. Sohn, S. J. Lalama, Appl. Phys. Lett. **53**, 1800 (1988).

3.242 Y. Shuto, M. Amano, T. Kaino, IEEE Trans. Photon. Techn. Lett. **3**, 1003 (1991).

3.243 D. Yu, A. Gharavi, L. P. Yu, Macromolecules **29**, 6139 (1996).

3.244 P. Prêtre, P. Kaatz, A. Bohren, P. Günter, B. Zysset, M. Ahlheim, M. Stähelin, F. Lehr, Macromolecules **27**, 5476 (1994).

3.245 W. Sotoyama, S. Tatsuura, T. Yoshimura, Appl. Phys. Lett. **64**, 2197 (1994).

3.246 M. Eich, H. Beisinghoff, B. Knödler, M. Ohl, M. Sprave, J. Vydra, M. Eckl, P. Strohriegl, M. Dörr, R. Zentel, M. Ahlheim, M. Stähelin, B. Zysset, J. Liang, R. Levenson, J. Zyss, in *Proceedings of Nonlinear Optical Properties of Organic Materials VII*, SPIE **2285**, edited by G. Möhlmann (1994), p. 104.

3.247 S. Aramaki, Y. Okamoto, T. Murayama, Y. Kubo, in *Proceedings of Nonlinear Optical Properties of Organic Materials VII*, SPIE **2285**, edited by G. Möhlmann (1994), p. 58.

3.248 A. Nahata, C. Wu, C. Knapp, V. Lu, J. Shan, J. Yardley, Appl. Phys. Lett. **64**, 3371 (1994).

3.249 Y. D. Zhang, L. M. Wang, T. Wada, H. Sasabe, Macromol. Chem. Phys. **197**, 1877 (1996).

3.250 C. Weder, B. Glomm, P. Neuenschwander, U. Suter, P. Prêtre, P. Kaatz, P. Günter, in *Poled Polymers and Their Applications to SHG and EO Devices*, edited by S. Miyata and H. Sasabe, Advances in Nonlinear Optics Vol. 4 (Gordon & Breach Science Publishers, Amsterdam, 1997), p. 63.

3.251 N. Kim, D. Choi, S. Park, in *Poled Polymers and Their Applications to SHG and EO Devices*, edited by S. Miyata and H. Sasabe, Advances in Nonlinear Optics Vol. 4 (Gordon & Breach Science Publishers, Amsterdam, 1997b), p. 47.

3.252 D. Jungbauer, B. Reck, R. Twieg, D. Yoon, C. Willson, J. Swalen, Appl. Phys. Lett. **56**, 2610 (1990).

3.253 Y. Shi, W. Steier, M. Chen, L. Yu, L. Dalton, Appl. Phys. Lett. **60**, 2577 (1992).

3.254 Y. Shi, P. M. Ranon, W. H. Steier, C. Xu, B. Wu, L. R. Dalton, Appl. Phys. Lett. **63**, 2168 (1993).

3.255 S. Kalluri, Y. Shi, W. Steier, Z. Yang, C. Xu, B. Wu, L. Dalton, Appl. Phys. Lett. **65**, 2651 (1994).

3.256 D. Riehl, F. Chaput, Y. Lévy, J. Boilot, F. Kajzar, P. Chollet, Chem. Phys. Lett. **245**, 36 (1995).

3.257 H. Hayashi, H. Nakayama, O. Sugihara, N. Okamoto, Opt. Lett. **20**, 2264 (1995).

3.258 V. G. Dmitriev, G. G. Gurzadyan, D. N. Nikogosyan, *Handbook of Nonlinear Optics Crystals*, (Springer-Verlag, Berlin, 1991).

3.259 A. Mito, K. Hagimoto, C. Takahashi, Nonlinear Optics **13**, 3 (1995).

3.260 Ch. Bosshard, unpublished results (1998).

3.261 I. Shoji, T. Kondo, A. Kitamoto, M. Shirane, R. Ito, J. Opt. Soc. Am. **B 14**, 2268 (1997).

3.262 R. S. Weis, T. K. Gaylord, Appl. Phys. Lett. **A 37**, 191 (1985).

3.263 D. A. Roberts, IEEE J. Quantum Electron. **28**, 2057 (1992).

3.264 J. Zyss, I. Ledoux, R. B. Hierle, R. K. Raj, J. L. Oudar, IEEE J. Quantum Electron. **21**, 1286 (1985).

3.265 A. Yokoo, S. Yokohama, H. Kobayashi, T. Kaino, J. Opt. Soc. Am. **B 15**, 432 (1998).

3.266 M. Jäger, A. Otomo, in *Electrical and Optical Polymer Systems*, edited by D. Wise et al. (Marcel Dekker, Inc., 1998), p. 495.

3.267 M. Jäger, G. I. Stegeman, G. R. Möhlmann, M. C. Flipse, M. B. J. Diemeer, Electron. Lett. **32**, 2009 (1996).

3.268 M. Jäger, G. I. Stegeman, M. C. Flipse, M. Diemeer, G. Möhlmann, Appl. Phys. Lett. **69**, 4139 (1996).

3.269 G. L. J. A. Rikken, C. J. E. Seppen, E. G. J. Staring, A. H. J. Venhuizen, Appl. Phys. Lett. **62**, 2483 (1993).

3.270 W. Wirges, S. Yilmaz, W. Brinker, S. Bauer-Gogonea, S. Bauer, M. Jäger, G. I. Stegeman, M. Ahlheim, M. Stähelin, B. Zysset, F. Lehr, M. Diemeer, M. C. Flipse, Appl. Phys. Lett. **70**, 3347 (1997).

3.271 T. C. Kowalczyk, K. D. Singer, P. A. Cahill, Opt. Lett. **20**, 2273 (1995).

3.272 W. P. Risk, S. D. Lau, R. Fontana, L. Lane, C. Nadler, Appl. Phys. Lett. **63**, 1301 (1993).

3.273 E. P. Ippen, C. V. Shank, in *Ultrashort Light Pulses*, edited by S. L. Shapiro, Topics in Applied Physics Vol. 18 (Springer-Verlag, Berlin, 1977), p. 83.

3.274 W. H. Glenn, IEEE J. Quantum Electron. **QE-5**, 248 (1969).

3.275 D. R. Yankelevich, A. Dienes, A. Knoesen, R. W. Schoenlein, C. V. Shank, IEEE J. Quant. Elect. **28**, 2398 (1992).

3.276 A. Otomo, S. Mittler-Neher, Ch. Bosshard, G. I. Stegeman, W. H. G. Horsthuis, G. R. Möhlmann, Appl. Phys. Lett. **63**, 3405 (1993).

3.277 A. Otomo, G. I. Stegeman, W. H. G. Horsthuis, G. R. Möhlmann, Appl. Phys. Lett. **65**, 2389 (1994).

3.278 Ch. Bosshard, A. Otomo, G. I. Stegeman, M. Küpfer, M. Flörsheimer, P. Günter, Appl. Phys. Lett. **64**, 2076 (1994).

3.279 K. Noguchi, H. Miyazawa, O. Mitomi, Elect. Lett. **30**, 949 (1994).

3.280 E. Voges, in Electro-Optic and Photorefractive Materials, edited by P. Günter, Springer Proceedings in Physics Vol. 18 (Springer-Verlag, Berlin, 1987), p. 150.

3.281 T. S. Tan, R. L. Jungerman, S. S. Elliott, IEEE Transactions on Microwave Theory and Techniques **37**, 1217 (1989).

3.282 W. Wang, D. Chen, H. R. Fetterman, Y. Shi, W. H. Steier, L. R. Dalton, P. Md. Chow, Appl. Phys. Lett. **67**, 1806 (1995).

3.283 F. S. Chen, Proceedings of the IEEE **58**, 1440 (1970).

3.284 C. C. Teng, Appl. Phys. Lett. **60**, 1538 (1992).

3.285 A. Chen, V. Chuyanov, F. I. Marti Carrera, S. Garner, W. H. Steier, J. Chen, S. Sun, L. R. Dalton, Proc. SPIE **3005**, 65 (1997).

3.286 K. C. Gupta, R. Garg, I. Bahl, P. Bhartia, *Microstrip Lines and Slotlines* (Artech House, Boston, 1996).

3.287 C. C. Teng, in *Nonlinear Optics of Organic Molecular and Polymeric Materials*, edited by S. Miyata and H. S. Nalwa (CRC Press., Inc, 1997), p. 441.

3.288 M. Ziari, A. Chen, S. Kalliuri, W. H. Steier, Y. Shi, W. Wang, D. Chen, H. R. Fettermam, *Polyimides for Electrooptic Applications* (Washington, DC, USA, 1995).

3.289 S. Ermer, W. W. Anderson, T. E. Van Eck, D. G. Girton, S. M. Lovejoy, D. S. Leung, J. A. Marley, A. Harwit, Proc. SPIE **3006**, 397 (1997).

3.290 Y. Shi, W. Wang, L. Weiping, D. J. Olson, J. H. Bechtel, Appl. Phys. Lett. **70**, 1342 (1997).

3.291 R. Spickermann, S. R. Sakamoto, M. G. Peters, N. Dagli, Electr. Lett. **32**, 1095 (1996).

3.292 R. A. Bergh, G. Kotler, H. J. Shaw, Elect. Lett. **16**, 260 (1980).

3.293 R. A. Bergh, H. C. Lefevre, H. J. Shaw, Opt. Lett. **5**, 479 (1980).

3.294 C. A. Millar, M. C. Brierley, S. R. Mallinson, Opt. Lett. **12**, 284 (1987).

3.295 M. J. F. Digonnet, J. R. Feth, L. F. Stokes, H. J. Shaw, Opt. Lett. **10**, 463 (1985).

3.296 A. K. Das, M. A. Mondal, A. Mukherjee, A. K. Mandal, Opt. Lett. **19**, 384 (1994).

3.297 K. Liu, W. V. Sorin, H. J. Shaw, Opt. Lett. **11**, 180 (1986).

3.298 W. Johnstone, S. Murray, G. Thursby, M. Gill, A. McDonach, D. Moodie, B. Culshaw, Elect. Lett. **27**, 894 (1991).

3.299 M. Wilkinson, J. R. Hill, S. A. Cassidy, Electr. Lett. **27**, 979 (1991).

3.300 S. Creaney, W. Johnstone, K. McCallion, IEEE Photon. Tech. Lett. **8**, 355 (1996).

3.301 S. A. Hamilton, D. R. Yankelevich, A. Knoesen, R. A. Hill, RT. Weverka, G. C. Bjorklund, in *Proceedings of LEOS '97*, IEEE Lasers and Electro-Optics Society 1997 Annual Meeting (1997), p. 387.

3.302 S. A. Hamilton, D. R. Yankelevich, A. Knoesen, R. T. Weverka, R. A. Hill, G. C. Bjorklund, J. Opt. Soc. Am. **B 15**, 740 (1998).

3.303 G. Khanarian, M. A. Mortazavi, A. J. East, Appl. Phys. Lett. **63**, 1462 (1993).

3.304 K. McCallion, W. Johnstone, G. Fawcett, Opt. Lett. **19**, 542 (1994).

3.305 MOEC, Molecular OptoElectronics Corp. (MOEC), 877 25th Street, Watervliet, New York 12189, USA.

3.306 CIR, Canadian Instruments & Research LTD, 1155 Appleby Line Unit E8, Burlington, Ontario L7L 5H9, Canada.

3.307 K. Thyagarajan, S. Diggavi, A. K. Ghatak, W. Johnstone, G. Stewart, B. Culshaw, Opt. Lett. **15**, 1041 (1990).

3.308 S. G. Lee, J. P. Sokoloff, B. P. McGinnis, H. Sasabe, Opt. Lett. **22**, 606 (1997).

3.309 K. McCallion, W. Johnstone, G. Thursby, Elect. Lett. **28**, 410 (1992).

3.310 W. Johnstone, K. McCallion, D. Moodie, G. Thursby, G. Fawcett, M. S. Gill, IEE Proc. Science, Measurement and Technology **142**, 109 (1995).

3.311 M. Swillo, M. A. Karpierz, M. Sierakowski, I. S. Mauchline, K. McCallion, W. Johnstone, Optica Applicata **26**, 315 (1996).

3.312 B. H. Kolner, D. M. Bloom, IEEE J. Quantum Electron. **QE-22**, 79 (1986).

3.313 J. A. Valdamis, G. Mourou, IEEE J. Quantum Electron. **QE-22**, 69 (1986).

3.314 K. Weingarten, M. J. W. Rodwell, D. M. Bloom, IEEE J. Quantum Electron. **QE-24**, 198 (1988).

3.315 I. P. Kaminov, in *CRC Handbook of Laser Science and Technology, Vol. IV, Optical Materials, Part 2: Properties*, edited by M. J. Weber, Vol. IV (CRC Press, Inc., Boca Raton, Florida, 1984), p. 253.

3.316 J. I. Thackara, Optical sampling of organic electro-optic materials, PhD thesis, Stanford, 1992.

3.317 M. Nagel, C. Meyer, H.-M. Heiliger, T. Dekorsy, H. Kurz, R. Hey, K. Ploog, Appl. Phys. Lett. **72**, (1998).

3.318 M. C. Nuss, J. Orenstein, in *Millimeter and Submillimeter Wave Spectroscopy of Solids*, edited by G. Grüner (Springer-Verlag, Berlin, 1998), p. 7.

3.319 F. Zernike, P. R. Berman, Phys. Rev. Lett. **15**, 999 (1965).

3.320 R. Morris, Y. R. Shen, Phys. Rev. **A 15**, 1143 (1977).

3.321 H. Ito, K. Kawase, J. Shikata, IEICE Trans. Electron. **E81**, 264 (1998).

3.322 D. H. Auston, K. P. Cheung, P. R. Smith, Appl. Phys. Lett. **45**, 284 (1984).

3.323 B. Hu, X.-C. Zhang, D. H. Auston, Phys. Rev. Lett. **67**, 2709 (1991).

3.324 D. H. Auston, M. C. Nuss, IEEE J. Quantum Electron. **24**, 184 (1988).

3.325 M. Bass, P. A. Franken, J. F. Ward, G. Weinreich, Phys. Rev. Lett. **9**, 446 (1962).

3.326 X.-C. Zhang, X. F. Ma, Y. Jin, T.-M. Lu, E. P. Boden, P. D. Phelps, K. R. Stewart, C. P. Yakymyshyn, Appl. Phys. Lett. **61**, 3080 (1992).

3.327 T. J. Carrig, G. Rodriguez, T. S. Clement, A. J. Taylor, K. R. Stewart, Appl. Phys. Lett. **66**, 121 (1995).

3.328 T. J. Carrig, G. Rodriguez, T. S. Clement, K. R. Stewart, Appl. Phys. Lett. **66**, 10 (1995).

3.329 X. F. Ma, X.-C. Zhang, J. Opt. Soc. Am. **B 10**, 1175 (1993).

3.330 A. Nahata, D. H. Auston, C. Wu, J. T. Yardley, Appl. Phys. Lett. **67**, 1358 (1995).

3.331 M. B. J. Diemeer, J. J. Brons, E. S. Trommel, IEEE J. Lightwave Tech. **7**, 449 (1989).

3.332 J. M. Cariou, J. Dugas, L. Martin, P. Michel, Appl. Opt. **25**, 168 (1986).

3.333 A. Borreman, T. Hoekstra, M. Diemeer, H. Hoekstra, P. Lambeck, ECOC '96. 22nd European Conference on Optical Communication **5**, 59 (1996).

3.334 Akzo, Akzo Nobel Photonics Beambox™ Newsletter & Datasheet **4**, (1996).

3.335 N. Keil, H. H. Yao, C. Zawadzki, B. Strebel, Elect. Lett. **30**, 639 (1994).

3.336 N. Keil, H. H. Yao, C. Zawadzki, Elect. Lett. **32**, 655 (1996).

3.337 K. Propstra, T. Hoekstra, A. Borreman, M. Diemeer, T. Hoekstra, P. Lambeck, in *Proc. ECIO '97. 8th European Conference on Integrated Optics and Technical Exhibition* (Opt. Soc. America, Washington, DC, 1997), p. 618.

3.338 T. A. Tumolillo Jr, M. Donckers, W. H. G. Horsthuis, IEEE Comm. Mag. **35**, 124 (1997).

3.339 U. Ellenberger, R. Weber, J. E. Balmer, B. Zysset, D. Ellgehausen, G. J. Mizell, Appl. Opt. **31**, 7563 (1992).

3.340 I. Biaggio, P. Kerkoc, L.-S. Wu, P. Günter, B. Zysset, J. Opt. Soc. Am. **B 9**, 507 (1992).

3.341 Fujian Castech Crystals, Inc., P. O. Box 143 XiheFuzhou, Fujian 350002, PR China, technical information (1998).

3.342 P. Prêtre, U. Meier, U. Stalder, Ch. Bosshard, P. Günter, P. Kaatz, C. Weder, P. Neuenschwander, U. W. Suter, Macromolecules **31**, 1947 (1988).

3.343 G. Williams, D. C. Watts, S. B. Dev, A. M. North, Trans. Far. Soc. **67**, 1323 (1971).

3.344 O. S. Narayanaswamy, J. Am. Cer. Soc. **54**, 471 (1971).

3.345 C. T. Moynihan, S. N. Crichton, S. M. Opalka, J. Non-Cryst. Solids **131–133**, 420 (1991).

3.346 E. A. DiMarzio, J. H. Gibbs, J. Chem. Phys. **28**, 373 (1958).

3.347 G. Adam, J. H. Gibbs, J. Chem. Phys. **43**, 139 (1965).

3.348 I. M. Hodge, Macromolecules **19**, 936 (1986).

3.349 P. Kaatz, P. Prêtre, U. Meier, U. Stalder, Ch. Bosshard, P. Günter, B. Zysset, M. Stähelin, M. Ahlheim, F. Lehr, Macromolecules **29**, 1666 (1996).

3.350 J. D. Ferry, Viscoelastic Properties of Polymers, (J. Wiley & Sons, New York, 1980).

3.351 C. Cai, M. Bösch, Y. Tao, B. Müller, Z. Gan, A. Kündig, Ch. Bosshard, I. Liakatas, M. Jäger, P. Günter, J. Am. Chem. Soc. **120**, 8563 (1998).

3.352 Rainbow Photonics, Einsteinstrasse HPF-E7, CH-8093 Zurich, Switzerland.

3.353 L. A. Eldada, L. W. Shacklette, R. A. Norwood, J. T. Yardley, SPIE Critical Reviews of Optical Science and Technology **CR 68**, 207 (1997).

3.354 S. Nalwa and H. S. Miyata, *Organic Electroluminescent Materials and Devices*, (Gordon & Breach, Amsterdam, 1997).

4   The Photorefractive Effect in Inorganic and Organic Materials

G. Montemezzani, C. Medrano, M. Zgonik, and P. Günter

4.1   Photoinduced Changes of Optical Properties and Photorefractive Effect

Light-induced changes in optical properties such as refractive index and absorption coefficients are always related to material excitation, i.e. deviation from (thermal) equilibrium [4.1]. Examples of material excitations are an increase in the number of conducting free electrons, the phonon density or the number of electrons occupying excited states. As a result of spatially inhomogeneous material excitation, the optical properties become spatially modulated in the interference region of two intense coherent light waves and create gratings that may be permanent or dynamic, depending on the mechanism and the material. Permanent gratings hold for longer times, while dynamic or transient gratings disappear after the inducing laser source is switched off. The latter have been produced in a large variety of materials and are usually detected by diffraction of a probe beam or by self-diffraction of the light waves inducing the grating.

While any grating can be ultimately reported to be a spatial modulation of the material's refractive index and/or absorption constant, the nature of the grating can have different origins. For instance, upon light excitation an absorption process occurs populating excited electronic states in the electron-volt energy range of the material. As a consequence, inside the interference pattern a population density grating can be created. When this electronic excitation decays, lower- energy electronic, vibrational or other states may become populated, forming secondary gratings. When the excitation finally thermalizes it produces a thermal grating, accompanied by stress, strain and density variations, and in a mixture of different compounds it may even produce a concentration grating. All these excitations form optical amplitude and/or phase gratings, the former deriving from a modulation of the absorption constant, and the latter from a modulation of the refractive index. Within the above class, one could cite hundreds of different mechanisms for which light-matter interaction can produce optical gratings: for example, in semiconductors the absorption creates a spatial modulation of the conduction electron density; in fluids the spatial modulation comes from molecular orientation; in mixtures it comes from the concentration; and in photorefractive materials it comes from the space charges and their accompanying fields.

While the word "photorefraction" may literally describe all kinds of photo-induced changes of the refractive index of a material and therefore any photoinduced phase grating would belong to this category, it has become customary in the literature to consider only a smaller class of materials as being photorefractive. These materials possess two important properties; they are photoconductive and exhibit an electro-optic effect. Photoconductivity ensures charge transport, resulting in the creation of a space charge distribution under inhomogeneous illumination. The electro-optic effect translates the internal electric fields induced by the inhomogeneous space charges into a modulation of the material's refractive index. In recent literature, the concept of photorefraction has been slightly expanded to include also the effects observed in some polymeric compounds, where the refractive index change is governed by a field-assisted molecular reorientation. The photorefractive effect is distinguished from many other mechanisms leading to optical induced refractive index gratings also by the fact that it is an intrinsically nonlocal effect, in the sense that the maximum refractive index change does not need to occur at the same spatial locations where the light intensity is largest.

In the strict sense mentioned above, the photorefractive effect was first observed in the mid-1960s by Ashkin and co-workers [4.2]. They found that intense laser radiation focused on ferroelectric $LiNbO_3$ and $LiTaO_3$ crystals induced semipermanent index changes. This phenomenon was unwanted for their purposes; therefore they referred to it as "optical damage". However, the potential of this new effect for use in high-density optical storage of data was realized soon after by Chen and co-workers [4.3]. The effect later became known as the photorefractive effect and it is understood as a modulated refractive index change. Although the photorefractive effect was first discovered in inorganic ferroelectric materials, in the last few years highly polarizable and photoconductive organic crystals and polymeric materials with extended π-electron systems have presented themselves as possible candidates for photorefractive applications.

The photorefractive process, which culminates in the formation of the phase grating, is described by the mechanisms shown in the schematic diagram of Fig. 4.1. The three most important properties which a material must fulfill are shown in the outlined boxes: optical absorption, charge transport and electro-optic effect or field-assisted molecular reorientation. Optical absorption and charge transport together give rise to photoconductivity, while the electro-optic effect or molecular reorientation translates the internal electric field into refractive index changes. The mechanisms in the top loop indicated by dashed arrows are also necessary in most materials under low-intensity continuous illumination. Under these conditions, a large number of trapping sites allow the creation of a considerable space charge modulation amplitude, even though the number of mobile charges is small at any moment in time. Large photorefractive effects may be observed without the necessity of trapping by studying the initial response to intense short-pulsed light [4.4],

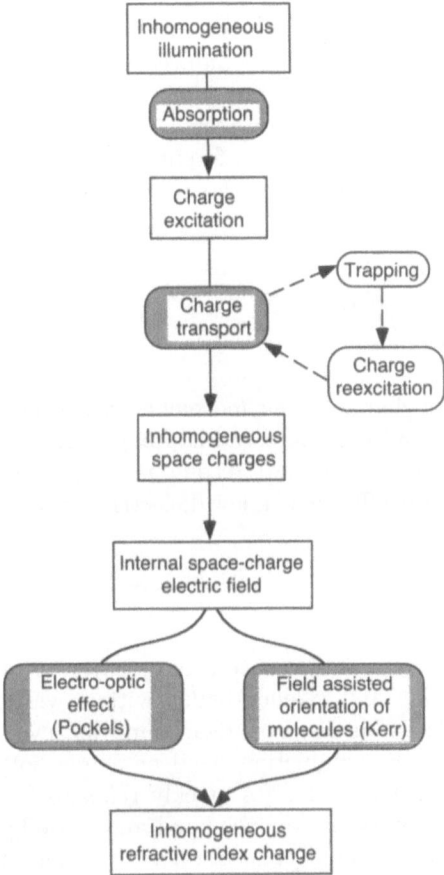

Fig. 4.1. The important mechanisms involved in the photorefractive effect and in field-assisted chromophore orientation in polymers

or, in some conditions, by using a continuous light with a wavelength short enough to produce interband photoexcitations [4.5] and thus create a large number of mobile electrons and holes.

This chapter concentrates on the discussion of the photorefractive effect as defined by the mechanisms shown in Fig. 4.1. The photorefractive properties and performances of inorganic and organic photorefractive materials are analyzed. It treats mainly inorganic and organic crystals, as well as photorefractive polymers. Some other materials which can also show a space charge-induced refractive index modulation, such as liquid crystals and ceramic compounds, are mentioned as well. First, the important aspects of the charge transport mechanisms which are active in different materials are discussed in Sect. 4.2. Section 4.3 deals with a simple band transport model for the overall photorefractive response of crystalline materials, as well as a model describing the formation of space charge fields in unordered polymers.

The electro-optic and molecular reorientation response is treated in Sect. 4.4. Measuring techniques to detect the photorefractively induced changes are presented in Sect. 4.5. A selection of a few applications and the potential for different materials is discussed in Sect. 4.6. Finally, Sect. 4.7 discusses the most important materials and compares them in terms of appropriate figures of merit. Some applications of photorefractive volume holograms memories for optical processing are discussed in Chap. 5.

4.2   Charge Transport in Inorganic and Organic Materials

This section considers the properties of charge carrier movement in photorefractive materials. We discuss briefly band transport and hopping transport of carriers. We also introduce the concept of geminate recombination, a process limiting the photoexcitation quantum efficiency in low dielectric constant organic crystals and polymers.

4.2.1   Band Transport

Single crystalline semiconductors or inorganic insulators have covalent or ionic bonds, leading to wide conduction and valence bands with a width often exceeding 1 eV. Charge photoexcitation from bandgap impurity levels is likely to generate a free electron in the conduction band or a free hole in the valence band (see also Fig. 4.5). These charges quickly relax to the bottom of the conduction band or the top of the valence band and move by a band transport mechanism or as large or small polarons. The direction of collective movement is driven by drift in an electric field, random diffusion, or the photogalvanic effect.

Drift currents obey Ohm's law

$$j_{\text{Drift}} = \sigma E , \tag{4.1}$$

where the conductivity σ is related to the densities $(n_{\text{e}}, n_{\text{h}})$ and mobilities $(\mu_{\text{e}}, \mu_{\text{h}})$ of mobile electrons and holes, respectively, as

$$\sigma = e\left(\mu_{\text{e}} n_{\text{e}} + \mu_{\text{h}} n_{\text{h}}\right) , \tag{4.2}$$

with e being the unit charge. For anisotropic mobilities the conductivity σ may be replaced by a second-rank tensor, i.e. the current density j_{Drift} and the electric field vector E may not be parallel in general. The number of photoexcited charges contributing to the photoconductivity is proportional to the crystal absorption constant α and the quantum efficiency. For instance, in the case of electrons

$$n_{\text{e}} = \frac{\alpha \phi_{\text{e}} \tau_{\text{e}}}{h\nu} I , \tag{4.3}$$

where I is the light intensity, ϕ_e is the quantum efficiency for generation of a free electron upon absorption of a photon with energy $h\nu$, and τ_e is the lifetime of the electron in the conduction band before recombination in a deep trap. In inorganic crystals the electron mobility ranges between a few $1000\,\mathrm{cm^2/(Vs)}$ in semiconductors like GaAs, InP or CdTe, few $\mathrm{cm^2/(Vs)}$ in materials such as $\mathrm{Bi_{12}SiO_{20}}$ where the conduction properties are described by large polarons [4.6] and a few tenths of $1\,\mathrm{cm^2/(Vs)}$ in materials like KNbO$_3$, BaTiO$_3$ where conduction is rather of the small polaron type. Very often, even lower values are measured in a cw experiment in oxide crystals. In such cases it must be assumed that one is not measuring the true band mobility, but a trap-limited mobility [4.7–8]. Photoexcited electrons are not continuously in the mobile state. They fall occasionally into a shallow trap level from which they can be thermally re-excited. The trap limited mobility for electrons μ_{eff} is related to the actual conduction band mobility μ_c by the expression

$$\mu_{\mathrm{eff}} = \mu_c \frac{\mu_t}{\mu_t + \mu_r} \, , \tag{4.4}$$

where τ_t is the average time spent by the electron in the mobile state between trapping events and τ_r is the average trapping time during which the electron cannot move. The shallow trap level may be viewed as equivalent to a conduction band which, because of imperfections in the lattice, has an irregularly shaped lower edge with dips acting as very shallow electron traps.

Diffusion currents j_{Diff} are present if the carrier density is not homogeneous in space, as is the case under the non-homogeneous illumination in a photorefractive experiment. With Einstein's relation for the diffusion coefficient one has

$$j_{\mathrm{Diff}} = \mu_e k_B T \nabla n_e + \mu_h k_B T \nabla n_h \, , \tag{4.5}$$

where k_B is the Boltzmann constant and T is the absolute temperature.

Photogalvanic currents can occur in noncentrosymmetric crystals and reflect the fact that carriers are emitted in a preferential direction due to asymmetric local potentials [4.9]. The photogalvanic current j_{PV} can be expressed in terms of the electric field vector $\mathcal{E}$ of the light wave through

$$j_{\mathrm{PV},i} = \beta_{ijk} \mathcal{E}_j \mathcal{E}_k^* \, . \tag{4.6}$$

The third rank tensor β_{ijk} depends on crystal absorption and is symmetric upon interchange of the last two indices. Alternatively the photogalvanic current may be expressed in terms of an average charge displacement distance $\boldsymbol{L}_{\mathrm{ph}}$ (also called the photovoltaic length) produced by photoexcitation of an impurity in a certain crystal host, i.e.

$$\boldsymbol{j}_{\mathrm{PV}} = \dot{n} e \boldsymbol{L}_{\mathrm{ph}} \, , \tag{4.7}$$

where $\dot{n}$ is the number of photoexcitations per unit time, proportional to the light intensity. Note that the vector $\boldsymbol{L}_{\mathrm{ph}}$ depends quadratically on the

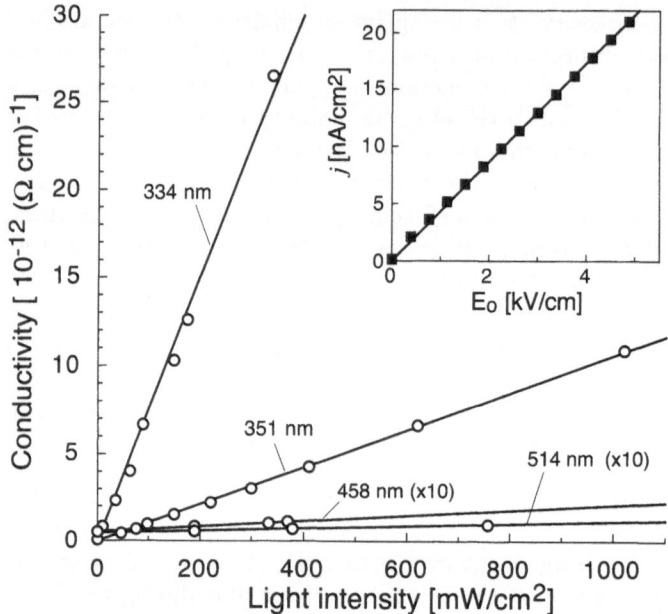

Fig. 4.2. Intensity dependence of the conductivity σ in photorefractive $Bi_4Ge_3O_{12}$ crystals for illumination at different wavelengths. *Inset*: current–field characteristics at $\lambda = 351\,nm$ for an illumination intensity of $I_0 = 0.41\,W/cm^2$. [4.142]

direction of the total electric field of the light wave $\mathcal{E}$ in a way consistent with (4.6). In doped $LiNbO_3$ crystals, which show a strong photogalvanic effect, the elements β_{311} and β_{333} of the photogalvanic tensor are of the order of 10^{-8}–$10^{-7}\,V^{-1}$ [4.10]. The photovoltaic length $|L_{ph}|$ may be of the order of a few tens of picometers to a few nanometers in typical oxides.

We should note at this point that in addition to electrons and holes, in some crystals nonphotoexcitable ionic charges can move by drifting in an internal electric field. These effects normally occur at elevated temperatures of approximately 100–200°C and are the basis for thermal hologram fixing in photorefractive materials [4.11]. Charge transport and photorefractive response with the presence of movable ions have been described in [4.12–14].

Figure 4.2 shows the conductivity σ of photorefractive $Bi_4Ge_3O_{12}$ crystals as a function of light intensity for four different wavelengths. The conductivity increases strongly at the ultraviolet wavelengths due to an increase in the absorption α and the quantum efficiency ϕ in (4.3). The inset shows an example of the Ohm's Law behavior of (4.1).

4.2.2 Hopping Transport

The band transport mechanism is inadequate for a complete description of charge transport in molecular crystals or polymers. The molecules of organic

Fig. 4.3. Hopping charge transport

crystals are usually held together by van der Waals forces or hydrogen bonds. These are weak interactions; therefore the electronic states of a molecule in a solid do not strongly differ from the states of the isolated molecule. This results in very narrow conduction bands with widths comparable to the thermal energy $k_B T$ at room temperature. In very pure molecular crystal, the room temperature carrier transport falls into an intermediate category between band-like movement and charge hopping (Fig. 4.3) between localized states [4.15]. With increasing amounts of impurities, which are necessary for good photorefractive performance, hopping transport becomes dominant and the mobilities decrease considerably with respect to $\mu \approx 1\,\mathrm{cm}^2/(\mathrm{Vs})$, which may be measured in pure samples at room temperature.

Disordered saturated polymers do not possess any true conduction or valence bands to allow band-like charge transport. The lack of any long-range order precludes the formation of extended electron states. In these materials charge movement is clearly of the hopping type [4.16]. Charges travel mainly by hopping through the side-chains or guest molecules. In general, most polymers are hole-conducting with carrier mobilities of the order of 10^{-9}–$10^{-6}\,\mathrm{cm}^2/(\mathrm{Vs})$, much lower than in organic or inorganic crystals. The hopping mobility increases with electric field, and hole mobilities approaching $10^{-3}\,\mathrm{cm}^2/(\mathrm{Vs})$ [4.17,18] and electron mobilities of the order of $10^{-5}\,\mathrm{cm}^2/(\mathrm{Vs})$ [4.19] have been measured in polymers for fields exceeding $50\,\mathrm{V/\mu m}$.

The trap-limited mobility mentioned in Sect. 4.2.1 in connection with inorganic crystals (4.4) may also be viewed as a manifestation of a hopping-like transport. In a certain sense, charges hop between localized shallow trap levels by drifting briefly in the conduction band. As shown by Feinberg et al. [4.20], for a photorefractive material charge hopping may be modeled by considering equidistant localized sites at different electric potentials ϕ_i and different local values of the light intensity I_i. Light-induced charge motion is expressed in terms of changes in the probability W_n that site n is occupied as a result of hopping to and from sites $m \neq n$:

$$
\begin{aligned}
\frac{\mathrm{d}W_n}{\mathrm{d}t} = &-\sum_m D_{mn} W_n I_n \exp\left(\frac{q\,(\phi_m - \phi_n)}{2k_B T}\right) \\
&+ \sum_m D_{nm} W_m I_m \exp\left(\frac{q\,(\phi_n - \phi_m)}{2k_B T}\right),
\end{aligned}
\tag{4.8}
$$

where q is the carrier charge and D_{mn} is the transition probability per unit time that a carrier in site m hops to site n. It has been shown [4.20,21] that upon appropriate choice of the transition probabilities D_{mn} a photorefractive model in which charge motion is described by (4.8) gives results equivalent to the conventional band model presented in Sect. 4.3.

4.2.3   Geminate Recombination

The quantum efficiency for carrier generation in polymers depends strongly
on the applied field. At low fields, geminate recombination of a carrier pair
is very likely to occur, while at higher fields the probability that one of the
carriers can escape increases. The strong influence of geminate recombination
in noncrystalline polymers can be attributed to two factors: the low dielectric
constant (of the order of 2.5 to 6) and the absence of clear conduction bands
as they are found in classic semiconductors.

The low dielectric constant leads to a larger capture radius r_C, which
is the distance at which the kinetic energy of a thermalized carrier is equal
to the attractive Coulomb potential energy of a stationary ion of opposite
charge. r_C can be calculated from the relation

$$r_C = \frac{e^2}{4\pi\varepsilon\varepsilon_0 k_B T} \,. \tag{4.9}$$

The capture radius is thus inversely proportional to the dielectric constant
of the enclosing medium, which is generally smaller in molecular compounds
($\varepsilon \approx 2\text{--}4$, corresponding to $r_C = 135\text{--}270\,\text{Å}$) than in semiconductors or oxides
($\varepsilon \approx 10\text{--}1000$, $r_C = 0.5\text{--}50\,\text{Å}$).

An external field can help the carrier to escape the attractive potential,
and thus increase the photogeneration quantum efficiency ϕ (Fig. 4.4). A
theoretical model explaining the field dependence of ϕ by means of geminate

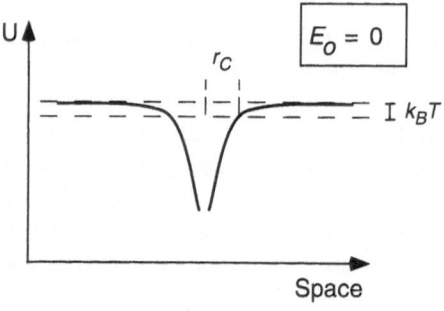

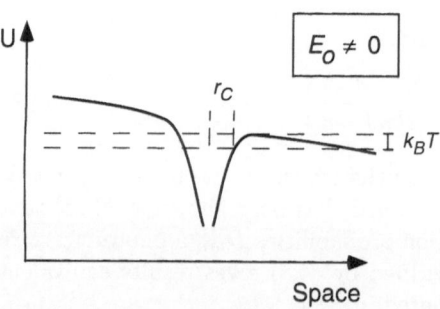

Fig. 4.4. The potential U created
by an ionic site after carrier generation with and without applied electric field. The capture radius r_C is
given by the distance at which a thermalized carrier has sufficient energy
to escape the potential. The applied
field leads to a decrease in r_C

recombination has been presented by Onsager [4.22]. The model assumes that absorption of a photon leads to the generation of a weakly bound hot hole–electron pair which subsequently thermalizes at a mutual separation distance r_0. This weakly bound pair can either recombine or separate into a free electron and hole. The ratio of the number of thermalized bound electron–hole pairs to the number of absorbed photons is the initial quantum yield ϕ_i, which is assumed to be independent on the applied field. The field-dependent quantum efficiency is then ϕ_i multiplied by the probability P of avoiding geminate recombination:

$$\phi\left(h\nu, r_0, E_0\right) = \phi_i\left(h\nu\right) P\left(r_0, E_0\right) . \tag{4.10}$$

The form for the function P is rather complicated. An explicit expression was presented by Mozumder [4.23]:

$$P\left(r_0, E_0\right) = 1 - \frac{1}{2\xi} \sum_{n=0}^{\infty} A_n\left(\eta\right) A_n\left(\xi\right) , \tag{4.11}$$

where

$$\xi = \frac{eE_0 r_0}{2k_B T} , \qquad \eta = \frac{r_C}{r_0} , \tag{4.12}$$

and the series $A_n(x)$ is given by

$$A_n\left(x\right) = 1 - e^{-x} \sum_{l=0}^{n} \frac{x^l}{l!} . \tag{4.13}$$

Photorefractive polymers may show initial quantum efficiencies ϕ_i of the order of 0.1–0.5 and thermalization distances r_0 of the order of a couple of nanometers [4.24]. The typical quantum efficiency ϕ may therefore increase from $\approx 10^{-5}$ at low fields (below $10\,\text{kV/cm}$) to $\approx 10^{-2}$ at high fields of $1000\,\text{kV/cm}$.

Geminate recombination is important not only in non-ordered polymers, but also in organic crystals. Due to the small dielectric constants, organic crystals also show a large value of the Coulomb radius r_C. However, the higher degree of order and the higher mobilities tend to produce larger thermalization distances r_0 than in polymers. The smaller influence of geminate recombination in organic crystals gives rise to quantum efficiencies in the 10^{-4}–10^{-2} region even at zero electric field [4.25]. Inorganic crystals are practically unaffected by geminate recombination, thanks to the large dielectric constants and high mobilities.

4.3 Model Descriptions of the Photorefractive Effect and Photoassisted Orientational Gratings

In this section we describe the simplest band model of the photorefractive effect, which has proved valuable to explain the experimental observations in most crystals. A slightly different model which takes into account some particularities of the charge generation and motion in polymers is presented in Sect. 4.3.2.

4.3.1   Band Model of the Photorefractive Effect

Crystalline photorefractive materials can be considered as wide bandgap semiconductors containing midgap impurity levels. The dynamics of charge redistribution can therefore be described by rate equations which are similar to the ones which are common in semiconductor physics. In the general case, there can be a number of deep and shallow defect levels active in the photorefractive process and numerous theoretical models have been analyzed in the literature; for a review see for instance [4.26].

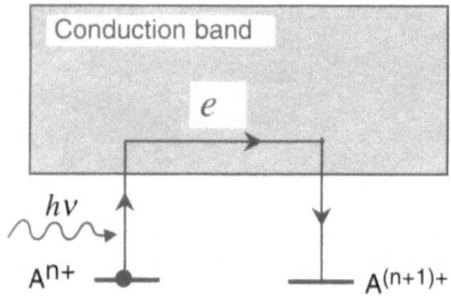

Fig. 4.5. Simple band scheme for a photorefractive crystal. A^{n+} indicates an impurity ion (e.g. Fe^{2+}, Cu^+, Cr^{3+})

Here we consider only the simplest model with the band scheme shown in Fig. 4.5. It is assumed that there is a single active impurity level between the two bands. The valence of these impurity centers can change between the states A^{n+} and $A^{[n+1]+}$ by excitation/retrapping of electrons to and from the conduction band. Charge transport occurs only in the conduction band by means of drift, diffusion or the photogalvanic effect. The equations governing charge redistribution and the creation of a space charge electric field are [4.27]

$$\frac{\partial N_D^+(r)}{\partial t} = (\chi w(r) + \beta)(N_D - N_D^+(r)) - \gamma n(r) N_D^+(r) \qquad (4.14)$$

$$\frac{\partial n(r)}{\partial t} = \frac{\partial N_D^+(r)}{\partial t} + \frac{1}{e}\nabla J(r) \qquad (4.15)$$

$$J(r) = e\mu n(r) E(r) + \mu k_B T \nabla n(r) + \chi w(r) \left(N_D - N_D^+(r)\right) e L_{ph} \quad (4.16)$$

$$\nabla E(r) = \frac{e}{\varepsilon_{eff}\varepsilon_0} \left(N_D^+(r) - n(r) - N_A\right) , \qquad (4.17)$$

where n is the free electron concentration in the conduction band, N_D is the total donor concentration, N_D^+ is the concentration of ionized donors, N_A is the concentration of ionized donors in the dark, J is the current density vector, E is the electric field vector, L_{ph} is the photogalvanic drift vector introduced in Sect. 4.2, $\chi \cong [2\hbar(N_D - N_A)]^{-1}$ is a normalization constant, β is the dark generation rate, γ is the recombination constant, μ is the electronic mobility, ε_0 is the permittivity of vacuum, ε_{eff} is the effective dielectric constant for the given photorefractive configuration, e is the absolute value of the elementary charge, k_B is the Boltzmann-constant,, T is the absolute temperature, and r is the position vector. Finally, the driving quantity in the above equations is the "usefully dissipated energy" $w(r)$, that is the optical energy which is locally dissipated for the generation of mobile charge carriers. It is defined as [4.28]

$$w(r) = \frac{1}{2}\varepsilon_0 \left[\mathcal{E}(r) \cdot \boldsymbol{\kappa} \cdot \mathcal{E}^*(r)\right] , \qquad (4.18)$$

where $\mathcal{E}(r)$ is the complex amplitude of the total optical electric field vector obtained by the coherent superposition of the illuminating optical waves. The second-rank tensor $\boldsymbol{\kappa}$ describes the anisotropy of the photoexcitation processes and is related to the imaginary part ε'' of the material dielectric tensor $\varepsilon = \varepsilon' + i\varepsilon''$ as

$$\kappa_{kl} = \phi_{kl} \left(\varepsilon''\right)_{kl} , \qquad (4.19)$$

where the quantities ϕ_{kl} describe the light polarization dependence of the quantum efficiency, that is, the probability that an absorbed photon of given polarization produces a photoexcited mobile carrier. In (4.19) no summing over equal indices is performed. In (4.14–17) we have chosen to introduce explicitly the "usefully" dissipated energy $w(r)$ instead of the light intensit $I(r)$ used by Kukhtarev et al. [4.27] because for materials with an anisotropic photoexcitation cross-section the modulation depths of $w(r)$ and $I(r)$ might differ considerably [4.28].

Equations (4.14–17) describe the formation of a nonhomogeneous charge distribution, and thus of a nonhomogeneous electric field as a result of a nonuniform photoexcitation of charge carriers. For the treatment in this section we consider specifically the interference of two coherent plane waves with the "usefully" dissipated energy given by ¡

$$w(r) = w_0 \left[1 + m \cos(K \cdot r)\right] , \qquad (4.20)$$

where K is the wave vector of the interference grating and m is the modulation index. In an electro-optic material the inhomogeneous electric field obtained from (4.17) modulates the optical indicatrix according to

$$\Delta \left(\frac{1}{\varepsilon'} \right)_{ij} (r) \equiv \Delta \left(\frac{1}{n^2} \right)_{ij} (r) = r_{ij}^{\text{eff}} \left| E(r) \right| , \tag{4.21}$$

where r^{eff} is a second-rank effective electro-optic tensor defined later in Sect. 4.4. Using (4.21) the spatially dependent second-rank real dielectric tensor at optical frequencies becomes

$$\varepsilon'(r) \equiv n^2(r) = \varepsilon'_u + \varepsilon'_u : \Delta \left(\frac{1}{\varepsilon'} \right) (r) : \varepsilon'_u , \tag{4.22}$$

where ε'_u is the optical dielectric tensor in absence of disturbances. The modulated squared refractive index (4.22) acts as a perturbation term in the wave equation

$$\nabla \times \nabla \times \mathcal{E}(r) + \frac{1}{c^2} \frac{\mathrm{d}^2 \left[\varepsilon'(r) \cdot \mathcal{E}(r) \right]}{\mathrm{d}t^2} = 0 . \tag{4.23}$$

where $\mathcal{E}(r)$ is the same complex amplitude of the optical waves appearing in (4.18). Therefore the wave equation (4.23) couples to the set (4.14–17) through the quantity $\mathcal{E}(r)$. Note that (4.23) is written under the assumption that no absorption gratings do exist. A more general expression valid also in the presence of strong absorption gratings in anisotropically absorbing materials can be found in [4.29].

A   Steady State Space Charge Field

The set of coupled equations (4.14–17) and (4.23) was solved by Kukhtarev et al. in some limiting cases [4.27, 4.30]. Here we concentrate only on the analytic solutions that are obtained in the case of small modulation ($m \ll 1$) and weak coupling. In this limit, the coupling between the waves given by (4.23) does not influence the magnitude of the space charge electric field $E(r)$. First, we assume no contributions from photogalvanic currents ($L_{\text{ph}} = 0$). In this case the steady state field distribution can be shown to be

$$E(r) - E_0 = \hat{k} \text{Re} \left[\frac{-imE_q(E_D - iE_0)}{E_q + E_D - iE_0} \exp(iK \cdot r) \right] , \tag{4.24}$$

which can be rewritten as

$$E(r) - E_0 = -m\hat{k} \left[\frac{E_q^2 E_0}{(E_q + E_D)^2 + E_0^2} \cos(K \cdot r) \right.$$
$$\left. + \frac{E_q E_D^2 + E_q E_o^2 + E_q^2 E_D}{(E_q + E_D)^2 + E_0^2} \sin(K \cdot r) \right] . \tag{4.25}$$

In (4.24) and (4.25) $\boldsymbol{E}_0$ is an externally applied electric field and $E_0 = \boldsymbol{E}_0 \cdot \hat{\boldsymbol{k}}$ is its projection in the direction of the unit vector along the grating $\hat{\boldsymbol{k}} = \boldsymbol{K}/|\boldsymbol{K}|$. The scalar quantities E_D and E_q have the dimension of an electric field and are called the diffusion and the trap-limiting field, respectively. They are defined as

$$E_D = \frac{|\boldsymbol{K}| \, k_B T}{e} \quad \text{and} \tag{4.26}$$

$$E_q = \frac{e}{\varepsilon_{\text{eff}}\varepsilon_0 \, |\boldsymbol{K}|} \frac{N_{Do}^+ \left(N_D - N_{Do}^+\right)}{N_D} \equiv \frac{e}{\varepsilon_{\text{eff}}\varepsilon_0 \, |\boldsymbol{K}|} N_{\text{eff}} \, , \tag{4.27}$$

where N_{Do}^+ is the number of ionized donors in the dark and N_{eff} is an effective trap concentration which become maximal if $N_{Do}^+ = N_D/2$. For the case of no applied field ($\boldsymbol{E}_0 = 0$) equation (4.25) simplifies to

$$\boldsymbol{E}(\boldsymbol{r}) = -m\hat{\boldsymbol{k}} \frac{E_q E_D}{E_q + E_D} \sin\left(\boldsymbol{K} \cdot \boldsymbol{r}\right) \, . \tag{4.28}$$

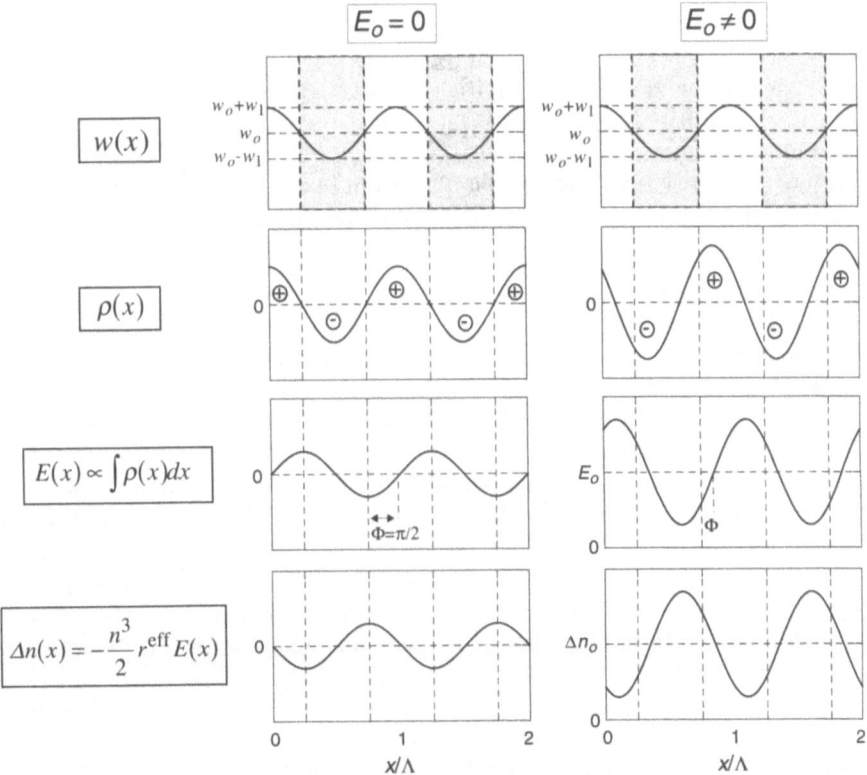

Fig. 4.6. Phase relationship between "usefully dissipated energy", space charge, space charge field and refractive index distributions

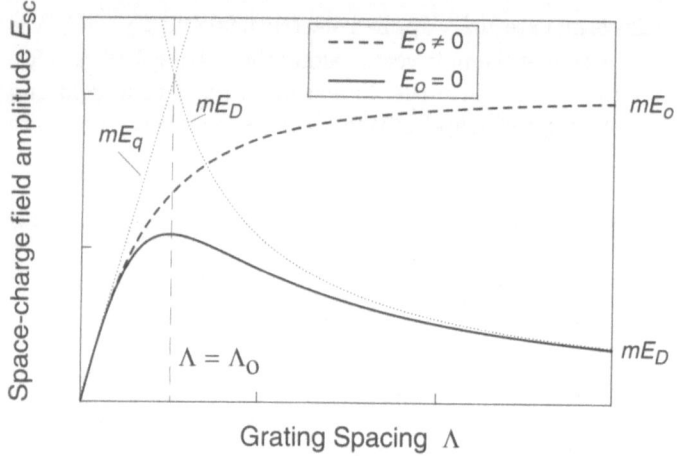

Fig. 4.7. Dependence of the space charge field amplitude on grating spacing

For this purely diffusive transport mechanism the electric-field grating is phase-shifted by $\pi/2$ with respect to the fringes defined by the rate of photoexcitation, as is seen by comparing (4.28) and (4.20). Figure 4.6 shows schematically the phase relationship between $w(\boldsymbol{r})$, the total space charge ρ, the space charge field and the refractive index change for the diffusion-only ($E_0 = 0$) and drift-assisted cases ($E_0 \neq 0$).

Figure 4.7 shows the space charge field amplitude as a function of the grating spacing $\Lambda = 2\pi/|\boldsymbol{K}|$, as obtained from (4.24) or (4.25). The maximum for the solid curve ($E_0 = 0$) is at the Debye grating spacing

$$\Lambda_0 = 2\pi \sqrt{\frac{\varepsilon_{\mathrm{eff}}\varepsilon_0 k_{\mathrm{B}} T}{e^2 N_{\mathrm{eff}}}} \,, \tag{4.29}$$

for which $E_{\mathrm{q}} = E_{\mathrm{D}}$. In typical inorganic crystals the Debye grating spacing ranges between $\approx 0.3\,\mu\mathrm{m}$ and $\approx 1.5\,\mu\mathrm{m}$.

In Fig. 4.8 we plot again the predictions of (4.24), this time with the external electric field E_0 as a variable. The unshifted ($\propto \cos \boldsymbol{K}\boldsymbol{r}$) and $\pi/2$ phase-shifted ($\propto \sin \boldsymbol{K}\boldsymbol{r}$) components of the space charge field amplitude are plotted separately together with the total amplitude E_{sc}. They are denoted in the graph as $\mathrm{Re}(E_{\mathrm{sc}})$ and $\mathrm{Im}(E_{\mathrm{sc}})$, in accordance with their complex phase in the amplitude term inside the bracket of (4.24). For the grating spacing considered here ($5\,\mu\mathrm{m}$), the unshifted component dominates at intermediate electric fields, but $\mathrm{Im}(E_{\mathrm{sc}})$ becomes dominant again when E_0 exceeds the value of the field E_{q}.

Finally, we give the steady state expressions for space charge fields in materials showing the photogalvanic effect ($\boldsymbol{L}_{\mathrm{ph}} \neq 0$). In this case it is important to consider explicitly the electrical boundary conditions. If the crystal surface is fully illuminated, the sample is connected to an external voltage supply

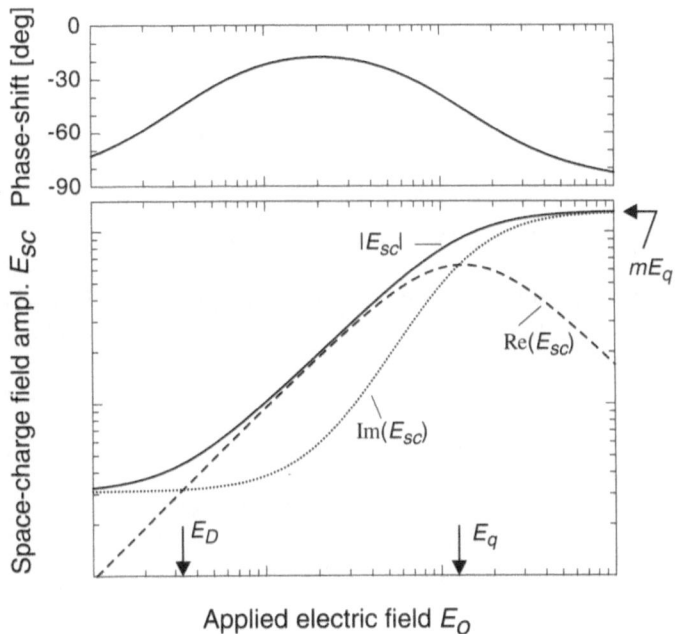

Fig. 4.8. Space charge field amplitude (*lower graph*) and refractive index grating phase shift (*upper graph*) as a function of the applied electric field ($\Lambda = 5\,\mu\mathrm{m}$)

(applied electric field E_0), and dark conductivity is negligible ($\beta \ll \chi w_0$), (4.24) transforms into

$$\boldsymbol{E}(\boldsymbol{r}) - \boldsymbol{E}_0 = \hat{\boldsymbol{k}}\,\mathrm{Re}\left[\frac{-imE_q\left[E_\mathrm{D} - \mathrm{i}\left(E_0 + E_\mathrm{pv}\right)\right]}{E_q + E_\mathrm{D} - \mathrm{i}\left(E_0 + \eta E_\mathrm{pv}\right)}\exp\left(\mathrm{i}\boldsymbol{K}\cdot\boldsymbol{r}\right)\right]\,, \qquad (4.30)$$

where $\boldsymbol{E}_\mathrm{pv} \equiv \boldsymbol{L}_\mathrm{ph}\gamma N_{D_0}^+/\mu$ is the photovoltaic field, that is the equivalent applied external field that would give rise to a photocurrent equal to the photogalvanic current, $E_\mathrm{pv} \equiv \boldsymbol{E}_\mathrm{pv}\cdot\hat{\boldsymbol{k}}$ is its projection along the grating spacing direction, and $\eta \equiv N_{D_0}^+/N_D$ is a factor that takes into account the reduction state of the crystal. Note that in many photorefractive materials, such as LiNbO$_3$, LiTaO$_3$ and KNbO$_3$, the longitudinal photogalvanic current is always antiparallel to the spontaneous polarization [4.31]; therefore E_pv has opposite sign to an electric field applied along the positive c-axis.

For short-circuited crystals in absence of external voltage (4.30) is still valid by inserting $E_0 = 0$. In contrast, for an open circuit the boundary conditions require a vanishing steady state current density across the sample; this is satisfied if an internal electric field $E_0 = -E_\mathrm{pv}$ builds up to compensate the photogalvanic current. Therefore from (4.30) the open-circuit steady state space charge field is

$$\mathbf{E}(\mathbf{r}) + \mathbf{E}_{\mathrm{pv}} = \hat{\mathbf{k}} \operatorname{Re} \left[\frac{-imE_{\mathrm{q}}E_{\mathrm{D}}}{E_{\mathrm{q}} + E_{\mathrm{D}} - iE_{\mathrm{pv}}(\eta - 1)} \exp(i\mathbf{K} \cdot \mathbf{r}) \right]$$

$$= -m\hat{\mathbf{k}} \left[\frac{E_{\mathrm{q}}E_{\mathrm{D}}E_{\mathrm{pv}}(1 - \eta)}{(E_{\mathrm{q}} + E_{\mathrm{D}})^2 + E_{\mathrm{pv}}^2(\eta - 1)^2} \cos(\mathbf{K} \cdot \mathbf{r}) \right. \tag{4.31}$$

$$\left. + \frac{E_{\mathrm{q}}E_{\mathrm{D}}^2 + E_{\mathrm{q}}^2 E_{\mathrm{D}}}{(E_{\mathrm{q}} + E_{\mathrm{D}})^2 + E_{\mathrm{pv}}^2(\eta - 1)^2} \sin(\mathbf{K} \cdot \mathbf{r}) \right] .$$

Figure 4.9 shows an example of the expected amplitude of the real ($\operatorname{Re}(E_{\mathrm{sc}})$) and imaginary components ($\operatorname{Im}(E_{\mathrm{sc}})$) of the steady state space charge field in

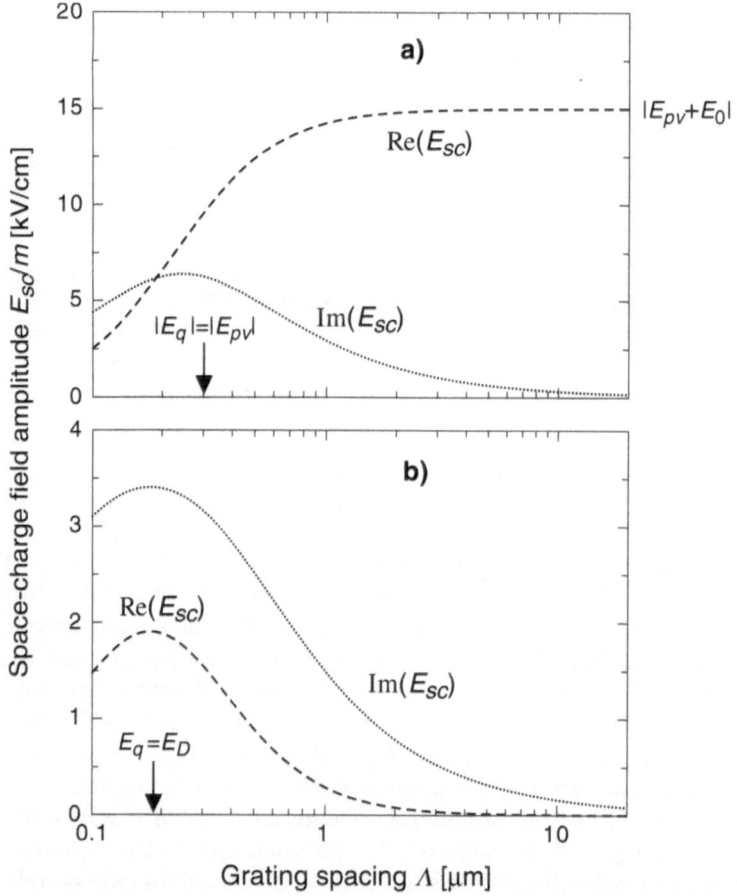

Fig. 4.9a,b. Real and imaginary component of the space charge field amplitude as a function of grating spacing in the presence of photogalvanic effect. (**a**) Short circuit (4.30), $E_0 = 5\,\mathrm{kV/cm}$; (**b**) open circuit (4.31). The additional parameters are: $E_{\mathrm{pv}} = -20\,\mathrm{kV/cm}$, $E_{\mathrm{D}}(\Lambda = 1\,\mu\mathrm{m}) = 1.6\,\mathrm{kV/cm}$, $E_{\mathrm{q}}(\Lambda = 1\,\mu\mathrm{m}) = 50\,\mathrm{kV/cm}$, $\eta = 0.5$

the case where the photogalvanic effect is strong. The short-circuit configuration usually gives rise to stronger fields with a dominating real component. In open circuit, in contrast, the field is generally smaller and the imaginary component, which is important for two-wave mixing energy coupling (Sect. 4.5.1), can exceed the value of the real one. However, it should be remarked that, due to electrode effects and surface conduction influence, in materials showing strong photogalvanic effects it is not always easy to experimentally define the electrical boundary conditions.

B   Space Charge Field Dynamics

The response time of the photorefractive material can be calculated from the set of equations (4.14–20) as well. The amplitude E_{sc} of the space charge field evolves exponentially in time as

$$E_{sc}(t) = E_{sc,sat}\left[1 - \exp\left(-t/\tau\right)\right] \tag{4.32}$$

during the build-up of the grating, and as

$$E_{sc}(t) = E_{sc}(t=0)\exp\left(-t/\tau\right) \tag{4.33}$$

during the grating decay, either under homogeneous illumination or in the dark. The time constant τ is in general complex, thus allowing for damped oscillations in the dynamics:

$$\text{Re}\,(\tau) = \tau_{die}\left[\frac{\left(K^2 + K_e^2\right)^2 + K^2 K_D^2}{K_e^2\left(K^2 + K_e^2\right)\left(1 + \frac{K^2}{K_0^2}\right) + \frac{K_e^2 K^2}{K_0^2}K_D^2}\right], \tag{4.34}$$

$$\text{Im}\,(\tau) = \tau_{die}\left[\frac{\left(K^2 + K_e^2\right)^2 + K^2 K_D^2}{K_e^2\left(1 - \frac{K_e^2}{K_0^2}\right)K K_D}\right]. \tag{4.35}$$

These equations are valid in absence of photogalvanic effects, the situation for photogalvanic materials is more involved because the electrical boundary conditions can change dynamically during the processes of hologram recording and erasure [4.32]. In (4.34) and (4.35), $K = |\boldsymbol{K}|$, $K_0 = 2\pi/\Lambda_0$,

$$K_e = \sqrt{\frac{e\gamma N_A}{\mu k_B T}} \tag{4.36}$$

is the inverse diffusion length, and

$$K_D = \frac{eE_0}{k_B T} \tag{4.37}$$

is the wavevector for which the applied field equals the diffusion field. The dielectric relaxation time τ_{die} is expressed by

$$\tau_{die} = \frac{\varepsilon_{eff}\varepsilon_0}{e\mu n_0}, \tag{4.38}$$

with the concentration of free electrons n_0 which depends on the average value w_0 of the "usefully" dissipated energy as

$$n_0 = \frac{(\chi w_0 + \beta)}{\gamma} \frac{(N_D - N_{Do}^+)}{N_{Do}^+} . \tag{4.39}$$

In materials with an isotropic absorption constant, in most cases w_0 is proportional to the average light intensity I_0. Therefore, with (4.34), (4.35) and (4.38) both $\mathrm{Re}(\tau)$ and $\mathrm{Im}(\tau)$ are expected to be inversely proportional to the light intensity. In the absence of applied fields and photogalvanic currents the time constant τ becomes real and the predicted dynamics is simply exponential, so (4.34) simplifies to

$$\tau = \tau_{\mathrm{die}} \left[1 + \frac{K^2}{K_e^2} \middle/ 1 + \frac{K^2}{K_0^2} \right] . \tag{4.40}$$

Figure 4.10 shows the dependence of the response time τ on grating spacing as given by (4.40) for a ratio $K_0/K_e = \sqrt{10}$. With no applied fields the correction terms to the dielectric time constants are only important for small grating spacings. Figure 4.11 shows $\tau(\Lambda)$ in unreduced $KNbO_3$ doped with 1000 at. ppm in the melt [4.33]. The solid curve is a fit with (4.40) for $K_e = 16.4\,\mu m^{-1}$ and $K_0 = 10.8\,\mu m^{-1}$, corresponding to a Debye length $\Lambda_0 \approx 0.6\,\mu m$. The discontinuity around $\Lambda = 0.25\,\mu m$ is due to the fact that the point at the shortest grating spacing is measured in a different geometry, and thus relates to a different dielectric constant, from the others.

Equation (4.40) and Figs. 4.10 and 4.11 indicate that the small grating spacing response can be considerably faster or slower than the dielectric relaxation time τ_{die}. It becomes faster if the Debye grating spacing Λ_0 is much larger than the diffusion length $1/K_e$. However, in this regime the

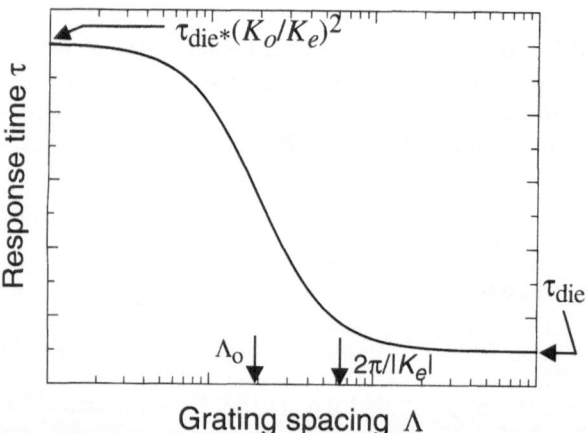

Fig. 4.10. Grating spacing dependence of the photorefractive response time in absence of an applied field

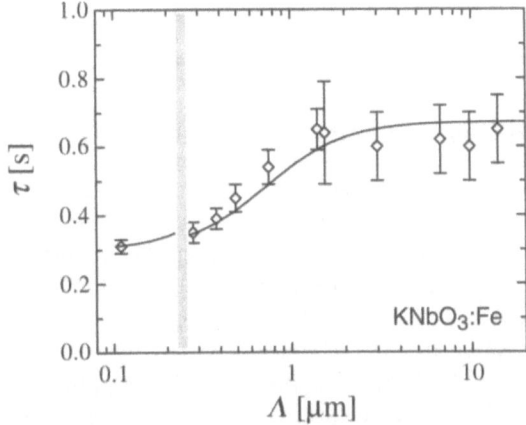

Fig. 4.11. Response time of unreduced $KNbO_3$:Fe (1000 at. ppm) as a function of grating spacing at a wavelength $\lambda = 488$ nm. Writing intensity is $1\,W/cm^2$

space charge field is normally not very big because $E_{sc} \approx mE_q \propto (\Lambda_o)^{-2}$. This limitation can be overcome considering that at a given light intensity and grating spacing the response time also changes as a function of the applied electric field following (4.34). In Fig. 4.12 we plot the real part of the response time for a grating spacing $\Lambda = 3\Lambda_0$, and $K_e/K_o = \sqrt{10}$. The response can be accelerated to a time constant of $\tau_{die}(K_0/K_e)^2$. Note that the point $K_D = K_e$ at which the response time begins to saturate corresponds to the condition where the drift length $L_{Drift} = E_0\mu\tau_R$ equals the diffusion length $L_{Diff} = (D\tau_R)^{1/2}$, where $D = \mu k_B T/e$ is the diffusion coefficient and $\tau_R = (\gamma N_A)^{-1}$ is the carrier recombination time. It is still true that the re-

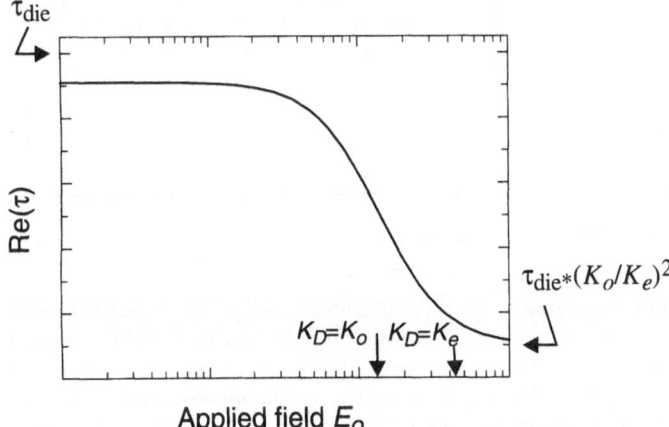

Fig. 4.12. Electric field dependence of the photorefractive response time for $\Lambda = 3\Lambda_0$

sponse time is never shorter than τ_{die} unless $K_{\mathrm{e}} > K_0$. The latter condition is equivalent to $L_{\mathrm{Drift}}(E_0 = E_{\mathrm{q}}) < \Lambda/2\pi$.

4.3.2   Model for the Space Charge Fields in Polymers

The mechanisms leading to photoinduced space charge fields in electro-optic polymers is similar to the one present in inorganic materials; however, some major differences do exist [4.34–37]. We start by indicating the important constituents of a photorefractive polymer. Figure 4.13 corresponds to the diagram presented in [4.34], which shows these constituents and their energy location. During the photorefractive process a hole is created by photoexcitation of a sensitizer molecule. This hole is transported by hopping between charge transport agent (CTA) molecules until it falls in a trap. The nonlinear optical chromophore (NLO) finally provides the necessary electro-optic response to the generated space charges, but does not participate in the charge transport. We notice in Fig. 4.13 that there is an ideal ordering of the optical absorption energies for the various building blocks discussed above. The absorption energy of the charge generator should be smaller than

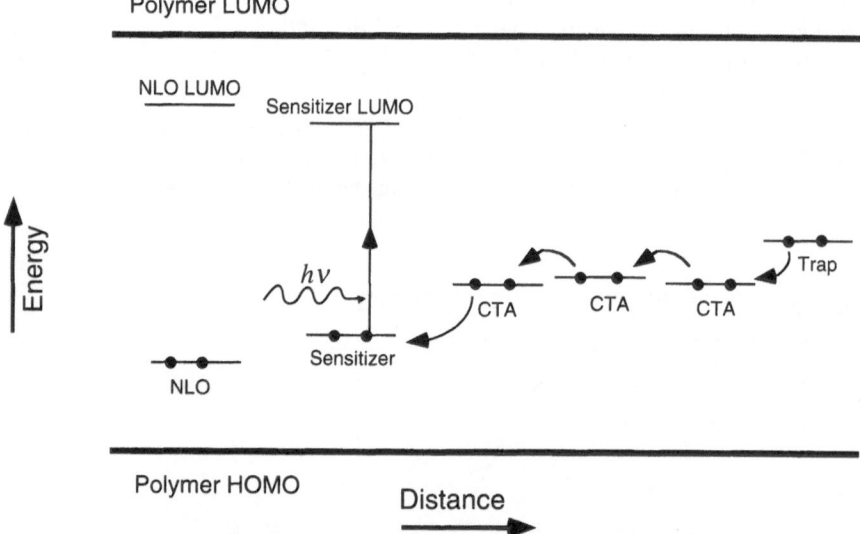

Fig. 4.13. The various components of a "photorefractive" polymer and the ideal location of the energy levels which ensures optimum hole transport and low overall absorption. The HOMO-LUMO optical transition energy of the sensitizer agent must be smaller than those of the nonlinear optical chromophores (NLO) and the host polymer. The charge transport agents (CTA) are not photoexcited; they form a network over which photogenerated holes can hop into trapping sites. Note that here holes hop to the right (redrawn after [4.34])

that of the polymer host, which corresponds to the energy difference between the lowest unoccupied molecular orbital (LUMO) and the highest occupied molecular orbital (HOMO) of the polymer. The charge generator absorption energy must also be smaller than that of the NLO molecules. This prevents unwanted absorption in channels not participating in the photorefractive effect. To allow the hopping transport of holes in the CTAs, it is important that their energy levels are higher than the levels for the sensitizer and NLO, which means that the electrochemical oxidation potential E_{ox} must be smaller for the CTA than for the other molecules [4.34].

Examples of the most commonly used polymer hosts, sensitizer, CTA and NLO molecules are given in Sect. 4.7. Here we discuss instead a model for the calculation of the space charge field in a polymer by means of rate equations, which has been developed in two papers by Schildkraut and co-workers [4.38–39]. The treatment follows closely the analysis of semiconductor-like crystals described in Sect. 4.3.1 and we discuss only the key differences from that analysis. Unlike in the original work, we use here dimensioned expressions for all physical quantities.

As mentioned in Sect. 4.2, in contrast to inorganics, in the case of polymers the quantum efficiency for charge generation and the charge carrier mobility are highly dependent on the electric field [4.40]. The quantum efficiency ϕ ($\propto \kappa$ in (4.14, 4.18)) is low at small fields due to geminate recombination, which is encouraged by the small dielectric constants of organics [4.41]. In Schildkraut's work an empirical dependence of the form

$$\phi(E) \propto s(E) \propto E^p; \quad s(E) = s_0(E/E_0)^p \tag{4.41}$$

is assumed for the hole-conducting polymers. E is the local electric field, p is a numeric exponent that must be determined experimentally, and E_0 is the value of the projection of the applied field in direction of the grating vector. Note that, since [4.38–39] assume that the photoexcitation is isotropic with respect to light polarization, the product χw in (4.14) can be replaced by sI, where s is a photoexcitation constant and follows (4.41), and I is the light intensity. The functional dependence (4.41) is in accordance with the behavior of some organics in the range of electric fields between 10 and 100 V/μm [4.42]. The hole mobility μ is modeled as

$$\mu(E) = \mu_0 \exp\left[C\left(\left(\frac{E}{E_0}\right)^{1/2} - 1 \right) \right], \tag{4.42}$$

where C is an experimentally determined constant and μ_0 is the value of the mobility at the field E_0. There are theoretical and experimental foundations for this kind of dependence [4.43–45]. Schildkraut et al. use the Einstein relation $D = \mu k_B T/e$ to relate diffusion coefficient and mobility. There is evidence that this relationship may not hold for some polymers under the condition of high electric field [4.46]. However, carrier diffusion is only important in regions of small electric field, where the above relation is always fulfilled.

For polymers it is assumed that there exist some generator molecules G (the sensitizer in Fig. 4.13) which can be photoionized to generate a hole; the constant s governing this process is proportional to the quantum efficiency, as described above. The inverse process of hole recombination is mediated by a recombination constant γ_R

$$G \underset{\gamma_R}{\overset{sI}{\rightleftarrows}} G^- + \text{hole} .\tag{4.43}$$

In addition to the generator centers there are some trapping centers T for holes. Trapping is mediated by a trapping constant γ_T, while thermal detrapping can occur through a detrapping rate r

$$T + \text{hole} \underset{r}{\overset{\gamma_T}{\rightleftarrows}} T^+ .\tag{4.44}$$

For recombination and trapping processes following Langevin theory [4.47] the constants γ_R and γ_T can be expressed in terms of the carrier mobilities [4.39], e.g.

$$\gamma_R(E) = \gamma_T(E) = \frac{e\mu(E)}{\varepsilon_{\text{eff}}\varepsilon_0} .\tag{4.45}$$

The Langevin model is appropriate to the case where a mobile carrier and the generation site are both charged so that a strong Coulomb attraction between them controls the recombination probability when they are within a sphere with radius equal to the Coulomb radius $r_C = e^2/4\pi\varepsilon_{\text{eff}}\varepsilon_0 k_B T$. Since the generator molecule G is usually initially uncharged, this assumption is reasonable for the recombination process (4.43). However, Langevin trapping may not be appropriate if the trapping centers T are initially uncharged in the process (4.44).

Equations (4.41–45) are used to build a set of rate equations similar to the one valid for crystalline materials (4.14–17). The space charge field distribution under a sinusoidal light illumination of the form (4.20) was found first using a numerical approach [4.39]. In a second paper [4.38] the equations were linearized to give the first Fourier harmonic of the space charge field grating in some limiting situations. Under the assumption of very deep hole traps untrapping is negligible ($r = 0$) and in the steady state all traps have been filled. The steady state space charge field amplitude is then

$$E_{\text{sc}} = A\frac{imE_q\left(E_D + i\,E_0\right)}{B_1 + iB_2}\tag{4.46}$$

and may be directly compared with the amplitude inside the square brackets on the right-hand side of (4.24). Under the further assumption that the average hole density p_0 is small compared with the total concentrations of generator (N_G) and trapping molecules (N_T), the quantities appearing in (4.46) are expressed as follows:

$$A = \frac{s_0 I_0 + \gamma_T N_T}{\gamma_T N_T}\tag{4.47}$$

is a dimensionless factor, E_D is the diffusion field defined in (4.26), and

$$E_q = \frac{e}{\varepsilon_{\text{eff}}\varepsilon_0 K} \frac{\gamma_T N_T (N_G - N_T)}{s_0 I_0 + (e\mu_0 N_G / \varepsilon_{\text{eff}}\varepsilon_0)} \equiv \frac{e}{\varepsilon_{\text{eff}}\varepsilon_0 K} N_{\text{eff}} \tag{4.48}$$

is a limiting field similar to (4.27) but with the effective number of photorefractive centers N_{eff} which depends on light intensity and on the mobility (due to Langevin recombination). The quantities B_1 and B_2 are expressed as

$$B_1 = E_D + (1 + Ap)E_q + \left(1 + \frac{1}{2}C\left(\frac{E_0}{E_{\text{ref}}}\right)^{1/2}\right) E_I \tag{4.49}$$

and

$$B_2 = E_0 + \frac{E_D E_q}{E_0}\left[\frac{1}{2}C\left(\frac{E_0}{E_{\text{ref}}}\right)^{1/2} - Ap\right], \tag{4.50}$$

where E_{ref} is a reference value of the applied field for which the constant C in (4.42) is determined. Finally, E_I is a field similar to E_q:

$$E_I = \frac{s_0 I_0}{s_0 I_0 + \gamma_T N_T} \frac{e}{\varepsilon_{\text{eff}}\varepsilon_0 K} (N_G - N_T) . \tag{4.51}$$

If the traps are initially charged one can assume a Langevin trapping process, thus substituting the constant γ_T with $e\mu_0 / \varepsilon_{\text{eff}}\varepsilon_0$ in (4.47, 4.48 and 4.50).

By analogy with Fig. 4.8, we plot in Fig. 4.14 the dependence of the real and imaginary components of the space charge field on the electric field. The curves are obtained using (4.41–51) with the parameters $\Lambda = 1\,\mu\text{m}$, $E_{\text{ref}} = 1\,\text{V}/\mu\text{m}$, $s(E_{\text{ref}}) = 1\,\text{cm}^2/\text{Ws}$, $p = 1$, $I_0 = 1\,\text{W/cm}^2$, $\mu(E_{\text{ref}}) = 10^{-7}\,\text{cm}^2/\text{Vs}$, $C = 3$, $\varepsilon_{\text{eff}} = 3$, $N_G = 5 \times 10^{15}\,\text{cm}^{-3}$, $N_T = 10^{16}\,\text{cm}^{-3}$, and $T = 295\,\text{K}$. Some of the details seen in Fig. 4.14 are due to the particular assumptions that have been made and do not all necessarily need to represent real physical phenomena. However, some qualitative conclusions may be drawn. The most important is that, as is the case for conventional photorefractivity in inorganic materials, the 90°-shifted component $\text{Im}(E_{\text{sc}})$ can become dominant if the applied field is large enough. In Sect. 4.5 it will be shown that the component $\text{Im}(E_{\text{sc}})$ is responsible for the transfer of energy between the beams in a two-wave mixing experiment. With few exceptions, photorefractive polymers are always used in a very high-field regime, with E_0 ranging between 10 and $100\,\text{V}/\mu\text{m}$. Experimentally, in polymer composites the phase shift between the light fringes and the refractive index grating can approach 90° for large applied fields (see for instance [4.48]), thus confirming the suggestions of Fig. 4.14. In contrast, for small fields the predictions derived with the above model have little meaning because the assumed empirical relations (4.41) and (4.42) may not hold.

Like the one discussed above, any model describing the formation of photorefractive gratings in polymers may need a large number of parameters and

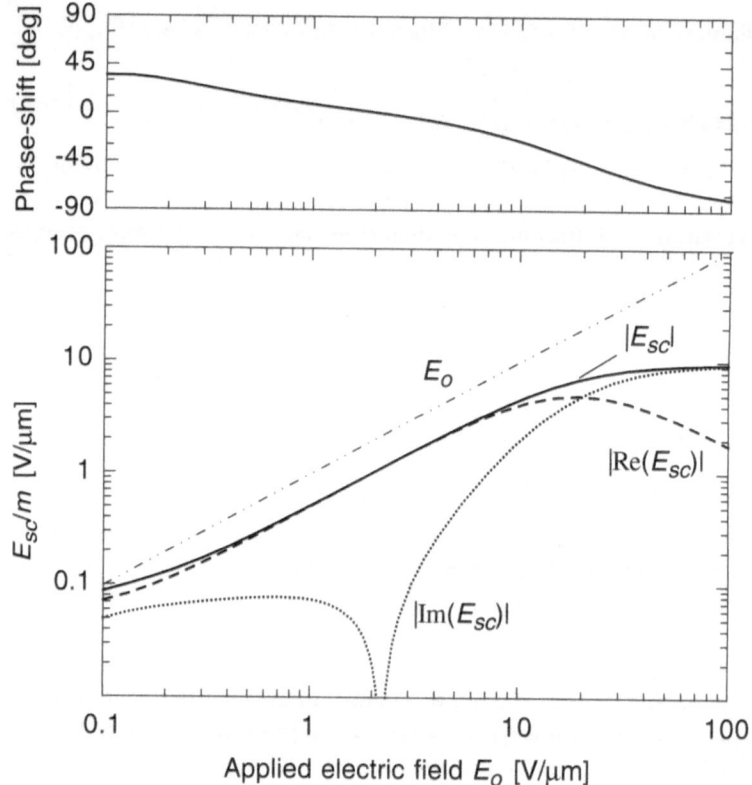

Fig. 4.14. Space charge field amplitude (*lower graph*) and refractive index grating phase shift (*upper graph*) as a function of the applied electric field as obtained using the photorefractive model for polymers described above. Note that the space charge field never exceeds the value of the applied electric field E_0. The parameters used are listed in the text

may be valid only in limiting situations. It is therefore not surprising that most researchers in polymeric photorefractive materials use the conventional band model presented in Sect. 4.3.1 to explain their results [4.36]. This might be considered as justified because the most relevant general features are common to the two situations, at least to a first approximation, as can be seen for instance by comparing Fig. 4.8 and Fig. 4.14 in the high-field limit. However, it is evident that more theoretical and experimental work is needed in order to establish a generally recognized theory for the charge redistribution in polymers.

To our knowledge, no detailed theoretical analysis of the grating dynamics has been presented to date in the case of polymers. As might be expected, by analogy with the case of inorganics expressed by (4.34), the grating response time is found to be of the order of the dielectric relaxation time (4.38). However, in organic materials the dielectric time itself is highly field-dependent

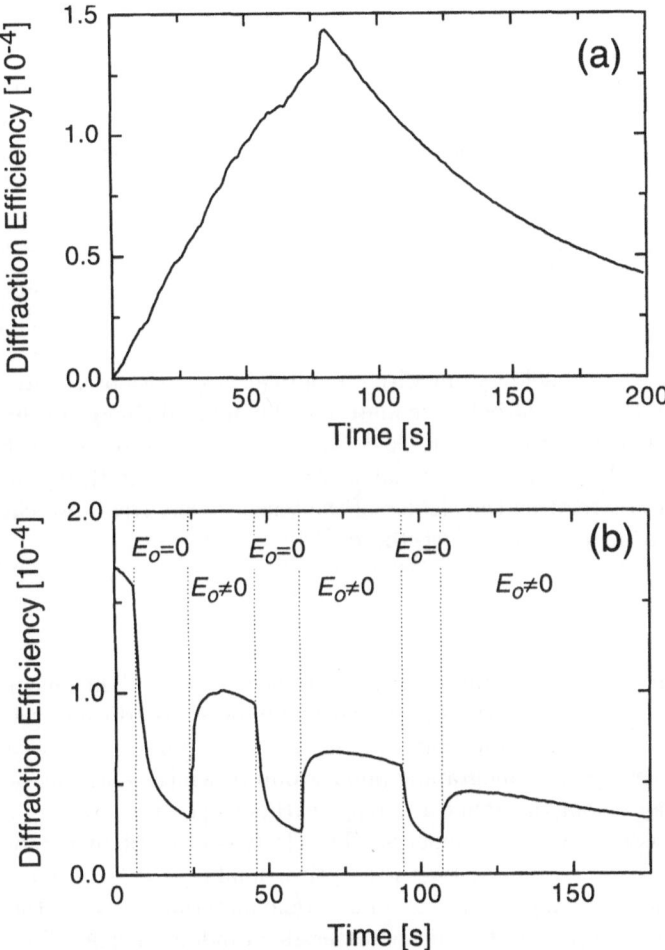

Fig. 4.15a,b. Dynamics of grating growth and decay in the photorefractive polymer bisA-NPDA:DEH at $\lambda = 676$ nm and with an applied field of 11.4 V/μm. (**a**) Grating recording ($t < 80$ s) and erasure under illumination with readout beam only ($t > 80$ s). (**b**) The applied field is switched off during grating decay at time $t = 20$ s and switched on again at $t = 47$ s. This process is repeated three times. The drop in diffraction efficiency in the absence of an external field is explained by chromophore orientation effects in this low glass temperature polymer (see Sect. 4.4) (redrawn after [4.34])

due to the strong dependencies of the quantum efficiencies, mobility and recombination cross-sections on the local electric field. Figure 4.15a shows an example of write/erasure curves in a photorefractive polymer [4.34]. Figure 4.15b shows the effect of switching on and off the applied field during decay in a polymer with low glass transition temperature. During the off-periods the diffraction efficiency drops due to the loss of the orientational quadratic electro-optic response, as discussed in Sect. 4.4. It recovers after switching the external field on again as long as inhomogeneously distributed space charges are still present. Note that by switching off the applied field during decay, the grating erasure process can be considerably slowed down because of a less efficient charge transport [4.39].

4.4   Electro-Optic Response

4.4.1   Pockels Effect

The presence of an electric field can modify the optical properties of a material. In noncentrosymmetric piezoelectric materials the induced change in the refractive index is linear in the electric field. This linear electro-optic effect (commonly known as the Pockels effect) may be expressed mathematically by a third rank tensor r_{ijk} relating the change of the tensor of the (real) optical indicatrix $(1/\varepsilon)_{ij}$ to the electric field vector $\boldsymbol{E} = (E_1, E_2, E_3)$ as

$$\Delta \left(\frac{1}{\varepsilon} \right)_{ij} \equiv \Delta \left(\frac{1}{n^2} \right)_{ij} = r_{ijk} E_k \,, \tag{4.52}$$

where the Einstein summation rule over equal indices is used. The linear electro-optic tensor r_{ijk} is symmetric in the first two indices. A reduced form r_{mk} (with $ij \rightarrow m$) is often used for easier tabulation. The symmetry-allowed tensor elements for every crystallographic point group to which inorganic or organic crystals may belong are found in many textbooks [4.49].

As already discussed in Chap. 3, noncrystalline electro-optic polymers rely on nonlinear optical chromophores for their optical nonlinearity. The non-poled polymers usually belong to the class of isotropic materials, because the orientation of the chromophores within the material is random in space. The contributions of the microscopic hyperpolarizabilities of individual molecules add up to give a vanishing macroscopic effect. To break this symmetry one has to apply an external electric field that aligns the nonlinear optical chromophores by coupling to their dipole moment. Contact electrodes [4.50] or corona poling [4.51] are the most commonly used poling techniques. The poling procedure imposes a cylindrical ∞m (Schönflies $C_{\infty v}$) point group symmetry on the polymer with the rotational symmetry axis ($z = x_3$ axis) parallel to the poling field. For this symmetry the electro-optic tensor has only three independent elements [4.52] and has the form

$$r_{ijk}(\text{poled polymer}) = \begin{pmatrix} 0 & 0 & r_{113} \\ 0 & 0 & r_{113} \\ 0 & 0 & r_{333} \\ 0 & r_{311} & 0 \\ r_{311} & 0 & 0 \\ 0 & 0 & 0 \end{pmatrix}. \tag{4.53}$$

The magnitude of the tensor elements depends on the degree of poling of the polymer compound. In organic materials the optical nonlinearity is usually of electronic nature and is generally characterized by the movement of electrons along a π-electron charge transfer axis within individual molecules. There are only negligible contributions from optical or acoustic phonons because of weak intermolecular forces. As a first approximation the elements of the electro-optic tensor are often estimated from the microscopic molecular hyperpolarizability using an oriented gas model discussed in Sect. 3.2 [4.53], which predicts a ratio $r_{333}/r_{113} = 3$. Another equality is brought about by Kleinman symmetry [4.54], $r_{113} = r_{311}$, stating that the DC polarizability and the polarizability at optical frequencies are the same. This equality may be well satisfied in polymers with a high glass temperature T_g, where the poled molecules are not able to reorient at room temperature. However, for low T_g compounds, molecules have been found to reorient easily in response to an electric field transversal to the poling field [4.55] (see Sect. 4.4.3). As this process is dissipative and not related solely to electron motion, but rather to the rearrangement of heavy molecules, it shows considerable dispersion [4.56–59]. Therefore, Kleinman symmetry is not valid at low frequencies for these polymer compounds.

4.4.2   Lattice Distortions and Electro-Optics

We have already mentioned that in organic materials the electro-optic response is predominantly of electronic nature. Often a shift in a π-electron cloud is responsible for the change in refractive index. Optical and acoustical phonons have only a small influence on the electro-optic nonlinearity, so that the electro-optic coefficients are essentially the same at low and high frequencies of the electric field. In inorganic crystals, in contrast, the phonon contributions are relevant. The value of the electro-optic coefficients may then depend not only on the frequency of the electric field but also on the mechanical state of the crystal. It has therefore become customary to measure the electro-optic response under two different mechanical boundary conditions, that is in an unclamped or in a clamped crystal. An unclamped crystal is either free or under constant stress conditions, whereas a clamped one is under constant strain conditions. The coefficients measured in the first case are denoted as r_{ijk}^T; those measured in clamped samples are denoted as r_{ijk}^S. Deformations of the crystal under the influence of a homogeneous electric field occur only in unclamped crystals, as shown schematically in Fig. 4.16b. Note that each volume element deforms in exactly the same way. Under these conditions the measured value of the unclamped electro-optic tensor element r_{ijk}^T is expressed by

$$r_{ijk}^T = r_{ijk}^S + p_{ijlm}^E d_{klm} \; , \tag{4.54}$$

where p_{ijlm}^E is the elasto-optic (Pockels) tensor [4.49] measured at constant electric field and d_{klm} is the piezoelectric tensor which describes the elastic

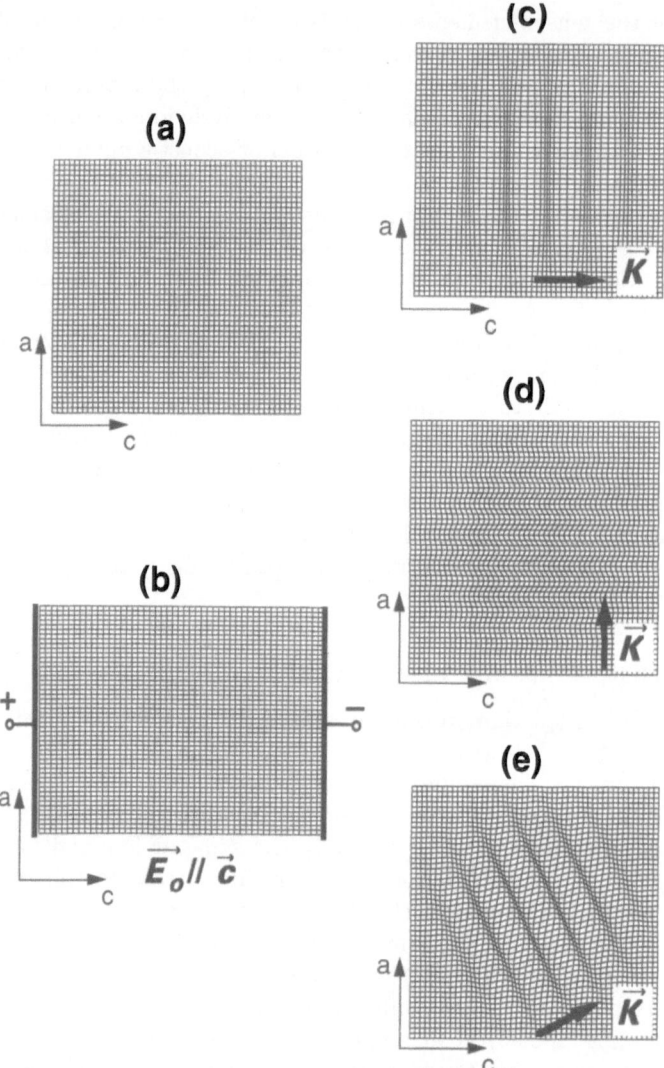

Fig. 4.16a–e. Schematic representation of elastic deformations of an ideal crystal belonging to the $4mm$ point group symmetry. (**a**) Undisturbed crystal; (**b**) homogeneous strain induced by a homogeneous electric field along the 4-fold c-axis; (**c**) periodic dilation/compression under the effect of a periodic space charge electric field with grating $\boldsymbol{K}$ vector $\parallel c$-axis; (**d**) shear deformations produced in the case of grating $\boldsymbol{K}$ vector $\parallel a$-axis; (**e**) same as (**c**) and (**d**) but for a $\boldsymbol{K}$ vector in the ac-plane: the deformations are a combination of dilation/compression and shear deformations. The space charge field in cases (**c**), (**d**) and (**e**) is assumed to affect only the central part of the crystal

strain generated by an electric field. The second term on the right-hand side contains all contributions of acoustic phonons to the electro-optic nonlinearity. Equation (4.54) indicates that the effect of crystal deformations on the electro-optic coefficients is given by a combination of the elasto-optic and the piezoelectric effect.

The relationship (4.54) holds only in the case where the electric field is homogeneous across the sample. In a photorefractive experiment, in contrast, the space charge field is in general of a periodic nature. Through the piezoelectric coupling the crystal deformation produced by the space charge electric field grating takes the form of a static plane wave, as shown schematically for three different geometries in Fig. 4.16c–e. The total refractive index change is then obtained by adding together the strain-free electro-optic contribution and the elasto-optic contribution, properly taking into account also the rotation of the axes of the optical indicatrix. This last roto-optic contribution is produced by shear deformations.

The above arguments, indicating the necessity of considering the correct elasto-optic contributions to the refractive index changes in photorefractive experiments have been pointed out in recent years [4.60–63]. To this aim, besides the refractive indices and the clamped electro-optic coefficients r_{ijk}^{S}, the piezoelectric stress tensor e_{ijk}, the elasto-optic (Pockels) tensor at constant electric field p_{ijlm}^{E} and the tensor of the elastic constants at constant electric field C_{ijkl}^{E} must be known. With this knowledge an effective second rank electro-optic tensor $r_{ij}^{\text{eff}}(\hat{k})$ can be calculated. It relates the change in the index ellipsoid with the absolute value of the local electric field $E(r)$, i.e.

$$\Delta\left(\frac{1}{n^2}\right)_{ij}(r) = r_{ij}^{\text{eff}}\,|E(r)| \tag{4.21}$$

under the assumption of a periodic field of the form

$$E(r) = \hat{k}E_{\text{sc}}\cos\left(K\cdot r\right) \tag{4.55}$$

with

$$\hat{k} \equiv K/|K|\,. \tag{4.56}$$

The elastic deformation in the form of a static wave also influences the low frequency dielectric constants of the crystal. Here again, in order to find the effective dielectric constant $\varepsilon_{\text{eff}}(\hat{k})$, in addition to the clamped dielectric constant ε_{ij}^{S}, the piezoelectric tensor e_{ijk}^{eff} and the elastic stiffness tensor C_{ijkl}^{E} are necessary.

Below, we show how to derive the expressions for $r_{ij}^{\text{eff}}(\hat{k})$ and $\varepsilon_{\text{eff}}(\hat{k})$. Let us consider a periodic electric field of the form (4.55). The change in the optical indicatrix can be rewritten in general as

$$\Delta\left(\frac{1}{n^2}\right)_{ij} = r_{ijk}^{S}E_k + p_{ijkl}^{\prime E}\frac{\partial u_k}{\partial x_l}\,, \tag{4.57}$$

where $u(r)$ is the displacement vector of the medium. Since piezoelectricity is a linear effect the static displacement wave has the same periodicity as the space charge field, that is

$$u(r) = u^A \sin(K \cdot r) \ . \tag{4.58}$$

Note however that the displacement amplitude vector u^A does not need to be parallel to the grating vector K, thus allowing not only compressive deformations (Fig. 4.16c) but also shear deformations (Fig. 4.16d) or combinations of the two (Fig. 4.16e). In (4.57) p'^E_{ijkl} is the modified elasto-optic tensor, which has no symmetry upon interchange of the last two indices [4.64]. It can be expressed as the sum of the conventional elasto-optic (Pockels) tensor p^E_{ijkl} [4.49] and the rotational part $p_{ij[kl]}$ [4.64] as

$$p'^E_{ijkl} = p^E_{ijkl} + p_{ij[kl]} \ , \tag{4.59}$$

where the second term is antisymmetric in the last two indices. In a coordinate system that coincides with the principal axes of the index ellipsoid this second term is equal to

$$p_{ij[kl]} = \frac{1}{2}\left[\left(\frac{1}{n^2}\right)_{ii} - \left(\frac{1}{n^2}\right)_{jj}\right](\delta_{il}\delta_{jk} - \delta_{ik}\delta_{jl}) \ . \tag{4.60}$$

The equation of motion for the displacement of a volume element under the action of the internal elastic forces is [4.49]

$$\frac{\partial T_{ij}}{\partial x_j} = \rho\frac{\partial^2 u_i}{\partial t^2} \ , \tag{4.61}$$

where ρ is the crystal density, and T_{ij} is the stress tensor, which can be expressed in terms of strain and electric field by

$$T_{ij} = C^E_{ijkl}S_{kl} - e_{kij}E_k \ . \tag{4.62}$$

C^E_{ijkl} and e_{ijk} have been defined above, and $S_{kl} \equiv 1/2\,(\partial u_k/\partial x_l + \partial u_l/\partial x_k)$ is the strain tensor. In equilibrium the right-hand side of (4.61) vanishes and the displacement vector amplitude u^A can be calculated inserting (4.58) and (4.62) into (4.61). The solution is [4.62]

$$u^A_k = \frac{E_{sc}}{|K|}A^{-1}_{ki}B_i \ , \tag{4.63}$$

where the matrices A_{ik} and B_i are calculated as

$$A_{ik} = C^E_{ijkl}\hat{k}_j\hat{k}_l \quad \text{and} \quad B_i = e_{kij}\hat{k}_k\hat{k}_j \ . \tag{4.64}$$

Inserting the displacement amplitude (4.63) into (4.57) one obtains the variations of the optical indicatrix which are expressed as

$$\Delta\left(\frac{1}{n^2}\right)_{ij} = \left(r^S_{ijk}\hat{k}_k + p'^E_{ijkl}\hat{k}_l A^{-1}_{km}B_m\right)E_{sc} \equiv r^{eff}_{ij}E_{sc} \ , \tag{4.65}$$

thus defining an effective second-rank electro-optic tensor r_{ij}^{eff}. As mentioned earlier, the dielectric constant $\varepsilon_{\text{eff}}(\hat{\boldsymbol{k}})$ wich relates the space charge amplitude ρ_{sc} and the electric field amplitude $E_{\text{sc}}(E_{\text{sc}} = \rho_{\text{sc}}/\varepsilon_{\text{eff}}\varepsilon_0|\boldsymbol{K}|)$ is also modified by the piezoelectric effect. In this case one starts from the dielectric response of a piezoelectric crystal given by

$$D_i = e_{ijk}S_{jk} + \varepsilon_0\varepsilon_{ij}^{\text{S}}E_j , \tag{4.66}$$

where $\boldsymbol{D} = (D_1, D_2, D_3)$ is the electric displacement vector and ε_0 is the permittivity of vacuum. Inserting the crystal deformations obtained from (4.63) and using the first Maxwell equation one obtains

$$\varepsilon_{\text{eff}} = \varepsilon_{ij}^{\text{S}}\hat{k}_i\hat{k}_j + \frac{B_kB_lA_{kl}^{-1}}{\varepsilon_0} . \tag{4.67}$$

This is the effective dielectric constant active in photorefractive experiments (where the electric field has a periodic form).

The full set of material constants necessary to calculate the expressions (4.65) and (4.67) has been determined for the crystals $KNbO_3$ [4.65] and $BaTiO_3$ [4.66]. Figure 4.17 shows an example of the theoretical and experimental dependence of the components r_{22}^{eff} and r_{33}^{eff} of the effective electro-optic tensor of $KNbO_3$ [4.67]. The experimental curves are for three differently doped crystals. They are extracted from photorefractive two-wave mixing measurements using a normalization constant to account for electron–hole competition effects in the charge transport. The nonmonotonic behavior of the effective electro-optic coefficient between $\beta = 0°$ and $\beta = 90°$ is a clear

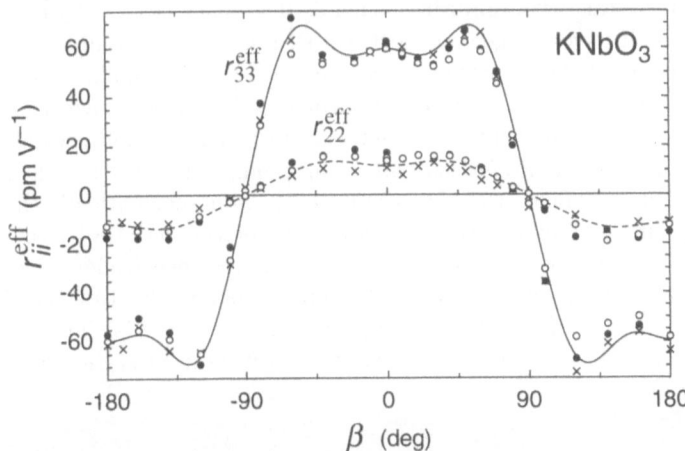

Fig. 4.17. Two components of the effective electro-optic tensor of $KNbO_3$ as determined from the two-beam coupling measurements with the grating vector in the bc-plane of the crystal. β is the angle between the grating vector $\boldsymbol{K}$ and the c-axis. ($\bullet$), (X), and (o) are for Fe, Cu and Rh doped samples, respectively. Solid and dashed lines are the calculated r_{33}^{eff} and r_{22}^{eff}, respectively (after data from [4.67])

manifestation of the piezoelectric contributions. It could also be shown that the peculiar photorefractive light fanning effects observed in BaTiO$_3$ are a direct consequence of these periodic crystal deformations [4.68].

4.4.3   Molecular Reorientation

In materials containing individual polar molecular entities, such as liquid crystals or dipolar nonlinear optical polymers, the polar molecules may reorient under the action of an electric field. If these molecules are strongly birefringent and are present in large quantities, this reorientation results in a large change in the refractive index of the material.

Reorientational effects upon application of an electric field are the basis for applications of liquid crystals in display technology [4.69]. The electric field vector $\mathcal{E}$ of the optical wave itself can induce a reorientation proportional to $\mathcal{E}^2$, which may be used to create optical gates [4.70] or to record phase gratings using the two-wave mixing configuration shown in Fig. 4.18 [4.71]. The two incident light waves are not symmetric with respect to the alignment of molecules, so that there exist components of the resultant light electric field vector both parallel and perpendicular to the average molecular director. In this optimum situation there is no threshold value of the electric field which would have to be exceeded in order to reorient the molecules. The configuration of Fig. 4.18 with oblique beams is also always employed for the study of photorefraction in polymers, where considerations of the symmetry of the electro-optic tensor impose this geometry with the grating wavevector K not normal to the poling direction.

Doping liquid crystals with impurities such as dye molecules or C_{60} may result in a photoconductive charge transport. Under inhomogeneous illumination static space charge fields may then be created by a process similar to the conventional photorefractive effect [4.72]. Recently, it has been shown that this space charge-induced molecular reorientation in nematic liquid crystals requires much smaller light intensities and is more efficient than the effect due to the optical electric field [4.73–81]. Figure 4.19 shows schematically the reorientation of nematic liquid crystal molecules under the action of forces produced by two oblique beams creating oblique interference fringes. Figure 4.19a shows the homeotropic cell (director axis perpendicular to the cell walls). The reorientation effect caused by the light electric field is shown in Fig. 4.19b, while reorientation by a phase-shifted light-induced static space

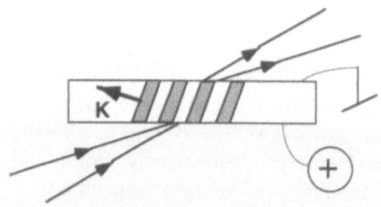

Fig. 4.18. Oblique geometry usually used to record phase gratings in nematic liquid crystal cells and polymers. The grating wave vector K of the light fringes is not perpendicular to the applied electric field

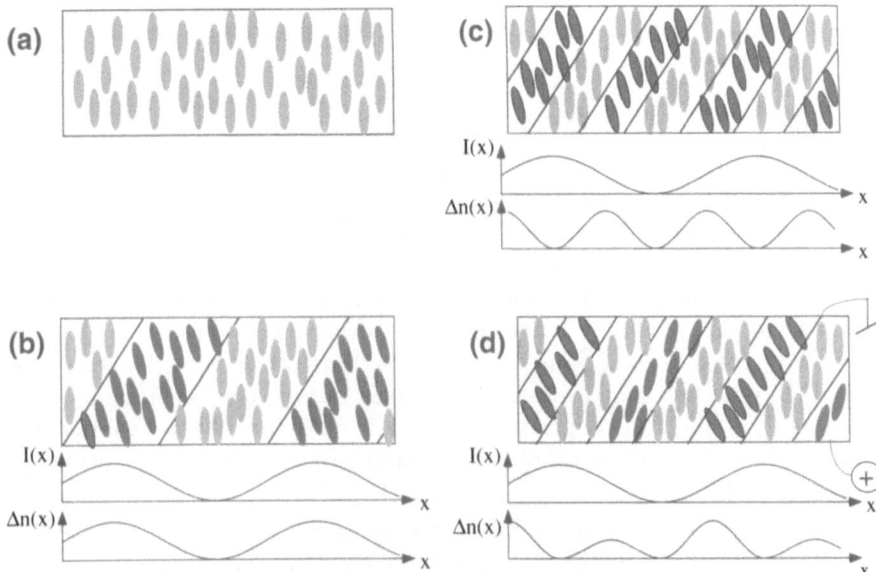

Fig. 4.19a–d. Orientational gratings in a homeotropic nematic liquid crystal cell. (a) cell configuration; (b) reorientation by the electric field vector of two optical waves creating oblique interference fringes; (c) reorientation by a static light-induced space charge electric field grating, no external electric field applied; (d) like (c), but with an externally applied field along the cell normal. The graphs for light intensity $I(x)$ and refractive index change $\Delta n(x)$ are given at the entrance face of the sample. The reorientation of nonlinear optical chromophores in polymers having low glass transition temperature is analogous to case (d)

charge field are shown in Figs. 4.19c and 4.19d for the cases without and with an external electric field applied normal to the cell walls, respectively. The induced changes in the optical properties are quadratic in the electric field; in case of Fig. 4.19c one has $\Delta n\left(x\right) \propto \left(E_{\mathrm{sc}} \cos\left(Kx\right)\right)^2 = E_{\mathrm{sc}}^2 \left(1 + \cos\left(2Kx\right)\right)/2$ and the spatial frequency of the refractive index change grating is twice the one of the light intensity. In contrast, applying the auxiliary external field induces also a linear response in the space charge field, because $\Delta n\left(x\right) \propto \left(E_0 + E_{\mathrm{sc}} \cos\left(Kx\right)\right)^2$ contains a term of the form $2E_0 E_{\mathrm{sc}} \cos(Kx)$. Therefore a component with the same spatial frequency as the light grating appears (case Fig. 4.19d). If the applied field largely exceeds the value of the internal space charge electric field E_{sc} the linear component dominates over the component with double spatial frequency.

Formally the reorientational effects in liquid crystals may be expressed in terms of quadratic electro-optic tensors $\mathrm{R}_{ijkl}^{\mathrm{DC}}$ and $\mathrm{R}_{ijkl}^{\omega}$ describing the response to quasi-static electric fields and optical electric fields, respectively. In analogy with (4.52) one can write

$$\Delta \left(\frac{1}{n^2} \right)_{ij} = \mathsf{R}^{\mathrm{DC}}_{ijkl} E_k E_l + \mathsf{R}^{\omega}_{ijkl} \mathcal{E}_k \mathcal{E}_l \,, \tag{4.68}$$

where $\boldsymbol{E} = \boldsymbol{E}_0 + \boldsymbol{E}_{\mathrm{sc}}$ is the total quasi-static electric field given by the sum of externally applied field and the space charge field, and $\mathcal{E}$ is the total electric field vector of the optical waves. The above quadratic electro-optic tensors are symmetric in the first two and last two indices. The second term on the right-hand side of (4.68) is proportional to the light intensity, so that the refractive index grating induced directly by the optical waves has the same periodicity as the light fringes, as discussed above. The coefficients of the tensors $\mathsf{R}^{\mathrm{DC}}_{ijkl}$ and $\mathsf{R}^{\omega}_{ijkl}$ depend on the elastic constants of the liquid crystal and on the static and optical dielectric anisotropy $\Delta\varepsilon$ and $\Delta\varepsilon_{\mathrm{op}}$, respectively. In general, $\mathsf{R}^{\mathrm{DC}}_{ijkl} \propto \Delta\varepsilon = \varepsilon_{||} - \varepsilon_{\perp}$ and $\mathsf{R}^{\omega}_{ijkl} \propto \Delta\varepsilon_{op} = n^2_{||} - n^2_{\perp}$ where the subscripts $||$ and $\perp$ indicate dielectric constants in direction parallel and perpendicular to the director axis, respectively. Most liquid crystal mixtures have a much larger static dielectric anisotropy than at optical frequencies, which explains the better holographic sensitivity obtained by making use of the photorefractive-like reorientational effect.

Electro-optic polymers with an elevated glass transition temperature may be designed for optimum stability, so that upon removal of the poling field their electro-optic and nonlinear optic response will remain stable for years [4.82] (see also Sect. 3.6). There is evidence that photorefractive polymers profit from exactly the opposite property. Low glass temperature (T_g) polymers with nonlinear chromophore guest molecules show very large photorefractively induced refractive index changes [4.83,84]. These changes are too large to be consistent with the independently determined values of the high-frequency electro-optic coefficients, which are measured with the modulating electric field parallel to the poling field (see Sect. 4.4.1). Under the assumption that the measured electro-optic coefficients are valid, the space charge field calculated from (4.52) would unrealistically exceed the applied field by a large amount. This dilemma can be resolved if one assumes that there are other contributions to the electro-optic effect, such as those of an overall molecule rotation. The reorienting birefringent chromophores then give a refractive index change in an analogous way, as depicted in Fig. 4.19d for the case of nematic liquid crystals. By analogy with Fig. 4.6, the photorefractive process in such low T_g polymers may be represented by the quantities plotted in Fig. 4.20.

Low T_g polymers are not true noncentrosymmetric materials and may be treated in the same way as centrosymmetric isotropic materials, for which the linear electro-optic tensor vanishes. Therefore, also in this case the refractive index change may be expressed by a quadratic electro-optic tensor R_{ijkl} using (4.68), where only the first term on the right-hand side is of practical

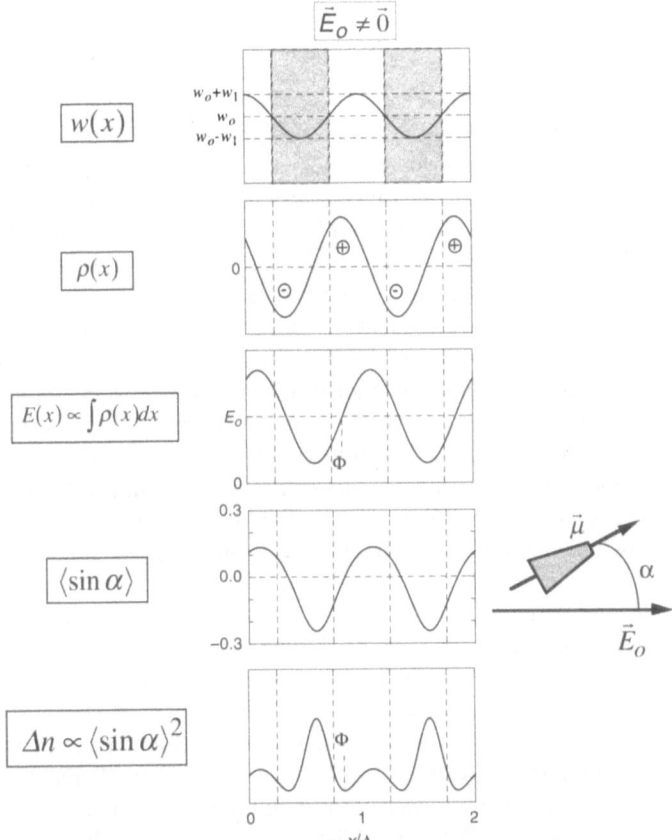

Fig. 4.20. Schematic diagram of "usefully dissipated energy", space charge distribution, space charge electric field, orientation direction and refractive index distribution along the direction x parallel to the grating vector $\boldsymbol{K}$ of photoassisted orientational gratings in polymers. E_0 is the projection of the external field along $\boldsymbol{K}$ and α is the angle between the field E_0 and the chromophore dipole moment

importance. For isotropic material the explicit form of R_{ijkl} in its reduced form is [4.52]

$$
R_{ijkl} \text{ (isotropic polymer)}
$$

$$
= \begin{pmatrix}
R_{1111} & R_{1122} & R_{1122} & 0 & 0 & 0 \\
R_{1122} & R_{1111} & R_{1122} & 0 & 0 & 0 \\
R_{1122} & R_{1122} & R_{1111} & 0 & 0 & 0 \\
0 & 0 & 0 & R_{1212} & 0 & 0 \\
0 & 0 & 0 & 0 & R_{1212} & 0 \\
0 & 0 & 0 & 0 & 0 & R_{1212}
\end{pmatrix}, \qquad (4.69)
$$

where

$$R_{1212} = \frac{1}{2} \left(R_{1111} - R_{1122} \right) . \tag{4.70}$$

The two independent elements of the above tensor might be determined using similar methods employed to measure the linear electro-optic effect. Considerable dispersion in the frequency Ω of the quasi-static electric field is expected because the contributions due to molecular reorientation will be present only at low enough Ω. At high frequencies the molecules are not able to follow the field, which has been confirmed by recent experiments [4.57–58, 4.85].

The treatment in terms of quadratic electro-optic effects avoids the necessity of using a local reference frame as proposed in an analysis of this so called "orientational enhancement of the photorefractive effect" which has been performed by Moerner et al. [4.55]. Two contributions to the overall refractive index change were considered. First, the spatially dependent birefringence was calculated in the local molecular frame using the vectorial sum of the externally applied field and the space charge electric field as a local poling field. This approach is possible because in such plasticized low T_g polymers the poling is dynamic and the molecules do not remain aligned upon removal of the external field. Second, since in polymers the high-frequency electro-optic effect is not independent on the poling field, the spatially dependent value of the electro-optic coefficients were calculated, again in a local frame with the 3-axis parallel to the local total electric field. Finally, the two contributions were transformed back to a laboratory frame with the 3-axis parallel to the sample normal to calculate the expected hologram diffraction efficiencies. The predictions of the above procedure were tested on the polymer compound composed of polyvinylcarbazole (PVK) host, 3-fluoro-4- (diethylamino)-(E)-β-nitrostyrene (FDEANST) as the nonlinear optical chromophore guest and 2,4,7-trinitro-9-fluorenone (TNF) as the optical sensitizer. Consideration of the anisotropy of the diffraction efficiencies for p- and s-polarized readout waves was in reasonable agreement with experiments. The enhancement in diffraction efficiency resulting from the reorientation effect was found to be of the order of 10 to 30 [4.55]. Such an enhancement is not observed in polymers having the nonlinear chromophores attached to the main polymer chain.

4.5   Measurement Techniques

This section describes two principal measuring techniques used to probe the strength of photorefractive gratings. Two-wave mixing probes only the component of the space charge field which is 90° out of phase with respect to the fringes defined by the distribution of dissipated energy. Bragg diffraction, in contrast, probes the total amplitude of the space charge field, however, with no information on its phase.

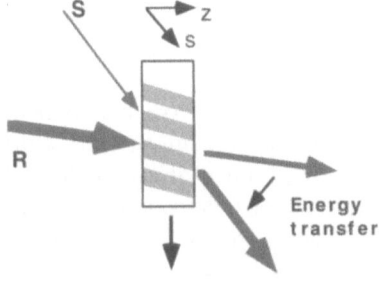

Fig. 4.21. Two-wave mixing configuration. The coordinate s corresponds to the direction of the Poynting vector of the signal beam S inside the crystal

4.5.1   Two-Wave Mixing

Light amplification by two-wave mixing is one of the key phenomena occurring in photorefractive materials. Figure 4.21 shows a general two-wave mixing configuration. The signal (S) and reference wave R interfere in the material. The nonlinear interaction gives rise to a transfer of energy from one wave to the other. The optical field in (4.23) is the superposition of the fields of the signal wave S and reference wave R

$$\mathcal{E}(\boldsymbol{r}) = \hat{e}^S S(\boldsymbol{r}) \exp\left(\mathrm{i}\boldsymbol{k}_S \cdot \boldsymbol{r}\right) + \hat{e}^R R(\boldsymbol{r}) \exp\left(\mathrm{i}\boldsymbol{k}_R \cdot \boldsymbol{r}\right) \ , \tag{4.71}$$

where $\hat{e}^S$ and $\hat{e}^R$ are the unit vectors along the direction of the electric field and $\boldsymbol{k}_R$, $\boldsymbol{k}_S$ are the propagation wavevectors. For the case of a weak signal ($S \ll R$) the amplification of the signal S in the direction of the coordinate s parallel to its direction of propagation (Poynting vector) is given by

$$\frac{\partial S(s)}{\partial s} = \frac{\Gamma}{2} S(s) \ . \tag{4.72}$$

For its intensity one has

$$\frac{\partial I_S(s)}{\partial s} = \Gamma I_S(s) \ , \tag{4.73}$$

and thus

$$I_S(s) = I_S(s = 0) \exp\left(\Gamma s\right) \ . \tag{4.74}$$

The exponential gain Γ can be calculated as follows [4.29]:

$$\Gamma = \frac{2\pi}{\lambda} n_R^2 n_S r_{\mathrm{eff}} \left(\hat{e}^R \cdot \hat{d}^R\right) \frac{\left(\hat{e}^S \cdot \boldsymbol{\kappa} \cdot \hat{e}^R\right)}{\left(\hat{e}^R \cdot \boldsymbol{\kappa} \cdot \hat{e}^R\right)} \mathrm{Im}\left(E_{\mathrm{sc}}\right) \ , \tag{4.75}$$

where n_R and n_S are the refractive indices seen by the reference and signal waves, respectively, r_{eff} is a scalar electro-optic coefficient obtained from

$$r_{\mathrm{eff}} = \hat{d}_i^S r_{ij}^{\mathrm{eff}} \hat{d}_j^R \ , \tag{4.76}$$

where r_{ij}^{eff} is the effective second-rank tensor of (4.65), $\hat{\boldsymbol{d}}^R$, $\hat{\boldsymbol{d}}^S$ are the unit vectors along the direction of the dielectric displacement (polarization) for the optical waves R and S, respectively, and the tensor $\boldsymbol{\kappa}$ describes the photoexcitation process and its anisotropy and was defined in (4.18–19). The quantity $\mathrm{Im}(E_{\mathrm{sc}})$ is the $\pi/2$ phase-shifted component of the space charge field amplitude (second term in the bracket of (4.25)); its electric field dependence has been shown in Fig. 4.8. Note that both r_{eff} and $\mathrm{Im}(E_{\mathrm{sc}})$ in (4.75) can be both positive or negative, depending on experimental geometry and material properties. The sign of $\mathrm{Im}(E_{\mathrm{sc}})$ is opposite for hole-conductive samples with respect to electron-conductive ones, and therefore the sign of the gain also switches.

It should be noticed that an anisotropy of the photoexcitation cross-section can influence the exponential gain in a dramatic way by means of the factor $(\hat{e}^S \cdot \boldsymbol{\kappa} \cdot \hat{e}^R / \hat{e}^R \cdot \boldsymbol{\kappa} \cdot \hat{e}^R)$ in (4.75). Figure 4.22 shows a contour plot of the exponential gain Γ expected for p-polarized signal and reference waves propagating in any directions in the ac-plane of BaTiO$_3$. Figure 4.22a is for an isotropic photoexcitation tensor $\boldsymbol{\kappa}$ and Fig. 4.22b for a tensor with an anisotropy $\kappa_{22}/\kappa_{33} = 10$. It is clearly seen that the gain landscape changes dramatically, the geometries for which the gain coefficient vanishes depend on this anisotropy. In the case where $\boldsymbol{\kappa}$ is anisotropic the maximum gain increases and more regions with non-negligible Γ appear throughout the two-dimensional parameter space (α_s, α_p) defined by the wave vector directions of the two waves (Fig. 4.22c). Note also that large gains become possible for geometries where signal and reference waves propagate in perpendicular directions. The strong influence of photoexcitation anisotropy on two-wave mixing has recently been confirmed experimentally using Ni-doped KNbO$_3$ for which $\kappa_{22}/\kappa_{33} = 3.4$ [4.28].

Besides its utility for a number of applications (Sect. 4.6), energy transfer by two-wave mixing is also often used to characterize the important parameter N_{eff}, the effective number of trapping centers contained in a material. Typically measurements are performed by varying the grating spacing while keeping the direction of the grating vector the same. In the example of BaTiO$_3$ one usually leaves the grating vector $\boldsymbol{K}$ along the crystal c-axis. For decreasing grating spacing, performing such an experiment corresponds to moving from the point $(\alpha_s = 90, \alpha_p = 90)$ towards the point $(\alpha_s = 0, \alpha_p = 180)$ in the plot of Fig. 4.22, which shows that the gain is influenced by an anisotropy of the tensor $\boldsymbol{\kappa}$ along this line as well. However, since in this kind of geometry the difference is quantitative rather than qualitative, this may lead to misinterpretations of the results in terms of electron–hole competition effects [4.86–87] or the analysis may deliver incorrect values of N_{eff}. Therefore, for the determination of the true value of the effective density of traps and of the correct amount of electron–hole competition, it is always advisable to perform experiments using s-polarized beams. In this case the unit electric field vectors $\hat{e}^S$ and $\hat{e}^R$ are parallel and most anisotropy terms in (4.75) lose their importance.

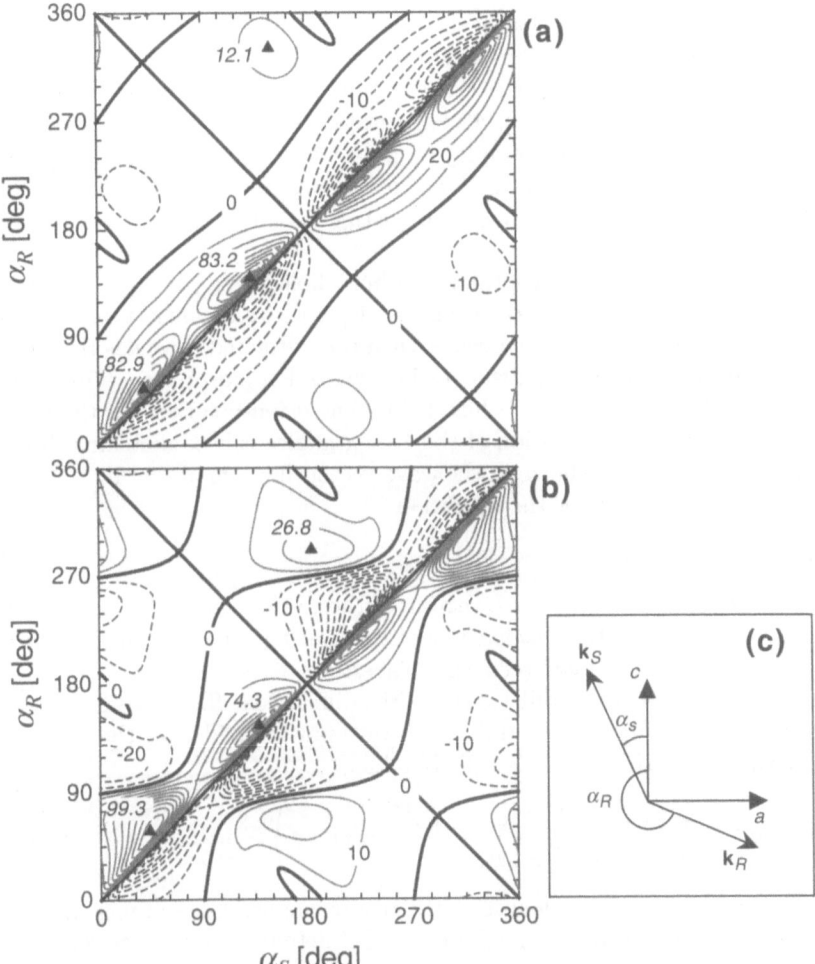

Fig. 4.22a–c. Contour plot of the two-wave mixing exponential gain Γ (in cm^{-1}) for p-polarized signal and reference waves propagating in the ac-plane of BaTiO$_3$. The wave vector propagation angles α_S and α_R are defined in (**c**) and are internal to the crystal. (**a**) No anisotropy of photoexcitation ($\kappa_{22}/\kappa_{33} = 1$); (**b**) anisotropic photoexcitation cross-section ($\kappa_{22}/\kappa_{33} = 10$). Peak gain values are denoted by triangles and dashed contours correspond to negative gains. The effective density of traps is assumed to be $N_{\mathrm{eff}} = 5 \times 10^{16}$ cm^{-3}. A particularly strong change in the landscape is observed along the lines $\alpha_R = \alpha_S \pm 90$, corresponding to geometries for which signal and reference waves are perpendicular to each other

4.5.2 Bragg Diffraction

In a photorefractive Bragg diffraction experiment a grating is recorded similarly as in the case of two-wave mixing by interference of two laser beams. A third beam (the readout beam) is then used to probe the grating. Under the appropriate conditions this beam generates a fourth scattered beam. In most cases a weak readout beam at a longer wavelength than the two recording beams is used. This approach ensures that the readout process does not influence the grating. However, a configuration with all the beams at the same wavelength is also often employed. In this case the readout beam counterpropagates one of the recording beams. Using the same wavelength for readout has the advantage of easier adjustment of the optimum incidence angles.

In birefringent materials containing volume holograms (see also Sect. 5.1) one distinguishes two possible situations; isotropic and anisotropic Bragg diffraction. Isotropic diffraction occurs when the diffracted wave has the same polarization (p or s) as the incident readout wave. For anisotropic diffraction, in contrast, the diffracted wave has a different eigen-polarization than the incident one. The determination of the Bragg angles (phase-matching direction) for the two cases is shown in the wave vector diagram of Fig. 4.23. Phase matching is achieved under the condition

$$\boldsymbol{k}_{\text{diff}} = \boldsymbol{k}_{\text{inc}} \pm \boldsymbol{K} ,\tag{4.77}$$

where $\boldsymbol{k}_{\text{inc}}$ and $\boldsymbol{k}_{\text{diff}}$ are the wave vectors of the readout and diffracted wave, respectively, and $\boldsymbol{K}$ is the wave vector of the grating produced by the interference of the two additional recording waves.

For a probe with two parallel surfaces the diffraction efficiency for a transmission grating, defined as the diffracted wave power divided by the incident wave power inside the crystal is calculated as

$$\eta = \sin^2 \left(\frac{\pi}{2\lambda} \frac{(n_{\text{inc}} n_{\text{diff}})^{3/2} (g_{\text{inc}} g_{\text{diff}})^{1/2}}{(\cos \theta_{\text{inc}} \cos \theta_{\text{diff}})^{1/2}} r_{\text{eff}} E_{\text{sc}} d \right) ,\tag{4.78}$$

where λ is the vacuum wavelength of the readout light, n_{inc} and n_{diff} are the refractive indices seen by the readout and diffracted wave, respectively, θ_{inc}

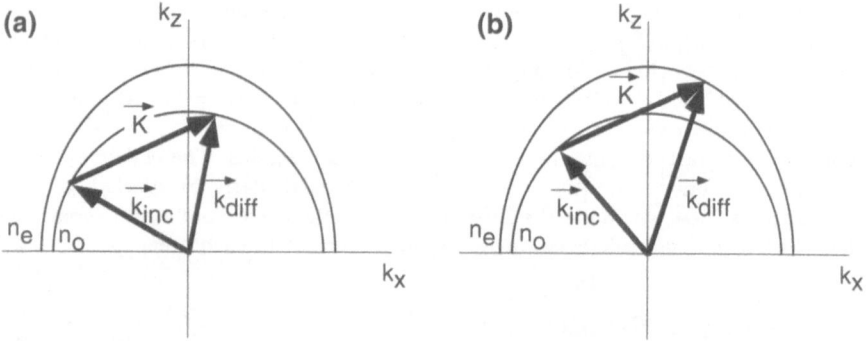

Fig. 4.23. Phase matching for (a) isotropic Bragg diffraction and (b) anisotropic Bragg diffraction in birefringent materials (n_0 and n_e: wave vector surfaces for ordinary and extraordinary waves)

and θ_{diff} are the angles between the sample normal and the Poynting vectors of the incident and diffracted wave, respectively, and d is the sample thickness. The quantities $g_{\mathrm{inc}} \equiv \hat{e}_{\mathrm{inc}} \cdot \hat{d}_{\mathrm{inc}}$ and $g_{\mathrm{diff}} \equiv \hat{e}_{\mathrm{diff}} \cdot \hat{d}_{\mathrm{diff}}$ are the cosines of the walk-off angles (angles between wavefront propagation direction and energy propagation direction) for the incident and diffracted wave, respectively. In optical isotropic materials one has always $g_{\mathrm{inc}} = g_{\mathrm{diff}} = 1$. The scalar electro-optic coefficient r_{eff} is calculated from (4.76) with the unit vectors $\hat{d}$ now corresponding to the polarization directions of the incident and diffracted wave. Equation (4.78) is thus valid for any arbitrary grating geometry and for both isotropic and anisotropic diffraction provided that the phase matching condition (4.77) is exactly satisfied. Effects of absorption are neglected in its derivation [4.29].

In the case of reflection gratings the projections of the incident and diffracted wave vectors on the surface normal have different sign. Instead as with (4.78), the diffraction efficiency is then calculated as

$$\eta = \tanh^2 \left(\frac{\pi}{2\lambda} \frac{(n_{\mathrm{inc}} n_{\mathrm{diff}})^{3/2} (g_{\mathrm{inc}} g_{\mathrm{diff}})^{1/2}}{(\cos \theta_{\mathrm{inc}} \cos \theta_{\mathrm{diff}})^{1/2}} r_{\mathrm{eff}} E_{\mathrm{sc}} d \right) , \tag{4.79}$$

which is again valid for both isotropic and anisotropic diffraction and neglecting absorption. Figure 4.24 compares the thickness dependencies of the diffraction efficiency for transmission (4.78) and reflection gratings (4.79). Due to backward diffraction, in reflection geometries the light penetrates less and less into deeper regions of the material; therefore in absorption-free samples the diffraction efficiency approaches asymptotically the value of 1.

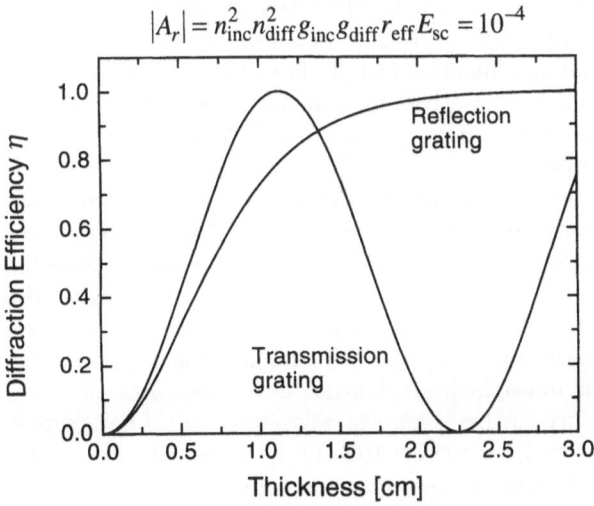

$$|A_r| = n_{\mathrm{inc}}^2 n_{\mathrm{diff}}^2 g_{\mathrm{inc}} g_{\mathrm{diff}} r_{\mathrm{eff}} E_{\mathrm{sc}} = 10^{-4}$$

Fig. 4.24. Diffraction efficiency vs. grating thickness for Bragg diffraction at thick transmission and reflection gratings (no absorption)

4.6   Applications

Applications of the photorefractive effect have been proposed in a number of areas. They include hologram storage and fixing [4.88], real-time holography [4.89], real-time image processing [4.90], edge enhancement [4.91], image amplification [4.92], phase conjugation [4.93], spatial light modulation [4.94] and all-optical associative memories [4.95]. Giving a complete list is beyond the scope of this work; the reader is referred to some excellent recent reviews [4.96–99]. Here we limit ourselves to some considerations on the potential of different photorefractive materials for a small selection of applications, classifying the latter into two main subareas, volume applications and applications which require only a relatively thin layer as the nonlinear holographic material.

4.6.1   Thick Volume Gratings

We consider in this category all those applications which profit from a large volume photorefractive hologram. *Holographic storage* evidently belongs here because, as discussed in more detail in Chap. 5, both the number of storable bits and the angular and wavelength selectivity of different data pages increase with increasing volume of the holographic medium [4.100]. Applications for *phase conjugation* [4.101] (Fig. 4.25) can also be considered as typical volume applications because most configurations for phase conjugation necessitate a complex three-dimensional grating structure in the material volume. Furthermore, the value of the gain–length product ΓL must be at least of the order of 1 to obtain reasonably efficient phase conjugation, which is achieved if the material thickness L is large enough. The latter argument applies also for a number of applications which are related in some way to *beam amplification*, such as weak signal amplification [4.102], laser beam cleanup [4.103], or applications involving photorefractive materials as processing elements in optical resonator systems [4.104–105]. All these applications require a considerable interaction volume.

Volume holographic storage is usually accomplished by interfering a signal wave carrying the image or bitmap to be stored with an encoding reference wave in the medium. Different data pages are encoded by changing the reference wave incident angle (angular multiplexing), by changing the recording wavelength (wavelength multiplexing), by imposing a phase code on the reference wave (phase multiplexing), or by changing the position in the recording medium (spatial multiplexing). Holographic storage is discussed in more detail in Chap. 5. Since the first demonstration of a digital holographic storage system by Heanue et al. in 1994 [4.106], intense research efforts have been accomplished towards the goal of achieving large capacity memories with fast readout rate. Inorganic photorefractive crystals have played a major role because they can offer a large interaction volume (often of the order of $1\,cm^3$ or more) and thus a large capacity. Due to their smaller achievable

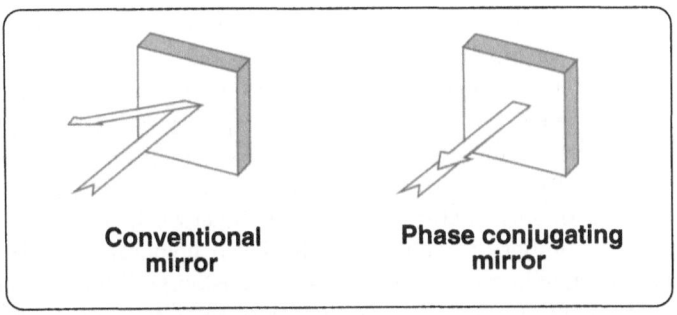

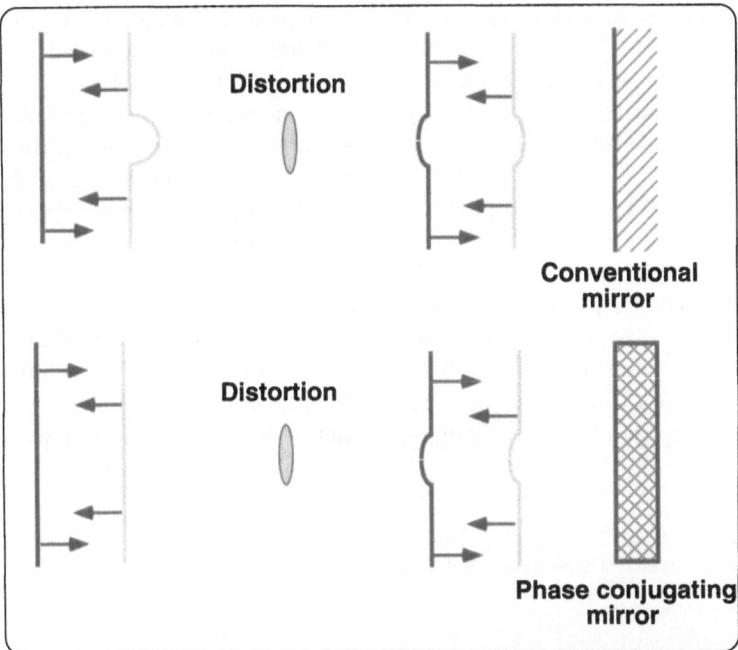

Fig. 4.25. The principle of a phase-conjugating mirror. In such a device reflected light exactly counterpropagates the incident wave and the wavefront emerging from the phase-conjugating mirror is identical to the incident wavefront. Unlike for conventional mirrors, distortions are compensated for after a double pass through the distorting medium

sample volume photorefractive polymers seem less promising than inorganic photorefractive crystals for holographic storage applications. To date, only relatively thin samples ($< 350\,\mu\text{m}$) have been fabricated. As discussed previously, photorefractive polymers perform best under the condition of large applied fields of the order of 50–$100\,\text{V}/\mu\text{m}$, so fabrication of a bulk sample of 1 cm thickness would require quite impractical voltages approaching 1 MV.

Furthermore, the maximum field that can be applied across a polymer film decreases substantially with increasing film thickness due to electrical break-down effects. The optimum thickness of polymeric samples seem to be of the order of 100–300 μm, much too small for a large capacity holographic storage system. Nevertheless, some attempts to overcome this problem by using stratified hologram structures [4.107] have been undertaken recently [4.108]. A schematic diagram of such a stacked structure is shown in Fig. 4.26. For maximization of the number of angularly multiplexed holograms the Bragg angle selectivity of each hologram should be as sharp as possible with possibly no side lobes in the angular selectivity. The width of the Bragg angle peak is proportional to Λ/D, where Λ is the grating period and $D = nd_A + (n-1)d_S$ is the total thickness of such a stratified device composed of n active layers of thickness d_A and $(n-1)$ spacers of thickness d_S [4.109]. Thus, as expected, the Bragg angle condition becomes sharper with increasing total thickness. Unfortunately the stratification of holographic layers also introduces side lobes in the angular response, which generate potential cross-talk problems during information readout [4.109–110]. The distance between the Bragg angle peak and the first sidelobe is proportional to $\Lambda/(d_A + d_S)$ and a small spacer thickness d_S contributes both to increase the side lobe distances and decrease their intensity. To reduce the secondary peak to less than 7% of the height of the Bragg matched peak requires the ratio of the active layer to spacer thickness to be greater than 15:1 [4.108]. It seems evident therefore that a device of the type shown in Fig. 4.26 must resemble closely a true volume holographic medium to become competitive in terms of storage capacity, i.e. it should be composed of a large number of active layers interspersed with

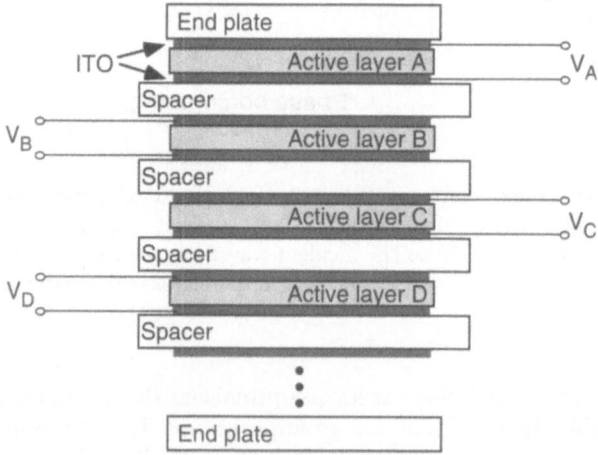

Fig. 4.26. Schematic diagram of a stratified volume holographic optical element. The active layers may be composed of photorefractive polymers. The photorefractive effect is activated in the individual layers by applying a voltage

very thin inactive buffer layers. Despite the limited material thickness, error-free digital holographic storage and reconstruction of a 64 kbit data page with photorefractive polymers based on polymethylmethacrylate [4.111] and polysiloxane [4.112–113] has been demonstrated, the latter compound giving better performance due to much better optical quality. However, the overall performance is still far from that obtainable with some inorganic materials.

As mentioned above, all the members of the second class of volume applications (phase conjugation, beam amplification, beam cleanup, etc.) require materials with a large gain–length product ΓL which should exceed the exponential losses αL. In photorefractive crystals with large electro-optic coefficients, such as $BaTiO_3$, $KNbO_3$ or SBN, high values of the gain coefficients Γ can be obtained in the diffusion regime, without the requirement of applying external electric fields. In polymers, net gain is usually achieved upon application of large electric fields of the order of $50\,V/\mu m$. Observation of gain–length products $\Gamma L \approx 10$ giving small-signal gains $\gamma_0 \exp(\Gamma L)$ exceeding 10^4 is not unusual in a number of inorganic crystals. Electro-optic ceramics can also show a net gain of the order of $\Gamma = 100\,cm^{-1}$ using moderate applied fields. In polymers the best gain coefficients are observed in materials which make use of the reorientational effect described in Sect. 4.4. To date, the maximum gains obtained in polymers are of the order of $\Gamma = 200\,cm^{-1}$ and are observed with fields of the order of 50–$100\,V/\mu m$. For instance, the polyvinylcarbazole (PVK) based compound presented by Meerholz et al. [4.84] had $\Gamma = 220\,cm^{-1}$ at $90\,V/\mu m$. The corresponding thickness of $105\,\mu m$ gives a gain–length product $\Gamma L \approx 2.3$ or small signal gain of $\gamma_0 \approx 10$. The gain–length product could be improved by using the stacked configuration of Fig. 4.26. Using a three-layer structure with total active thickness of $360\,\mu m$ Grunnet-Jepsen et al. [4.114] demonstrated a small signal gain $\gamma_0 = 500$ at $E_0 = 71\,V/\mu m$ corresponding to a gain–length product $\Gamma L \approx 6.2$. They employed a PVK-based compound and made used of moving fringes techniques [4.115] to enhance the gain. Whether such a stratified configuration will have any negative consequences on the phase conjugation performance or on the quality of an optically processed signal beam remains to be verified. We can conclude that inorganic crystals still have a certain edge over polymers for maximum light amplification; however, the performance improvement of the latter is fast. For instance, recent results demonstrating spontaneous oscillation and self-pumped phase conjugation using a polymer stack placed in an optical cavity appear very promising for the future [4.116]. In order to increase the nonlinear medium length in polymers, in alternative to the stacked configuration discussed above, a transverse waveguide-like geometry with the optical waves propagating in the plane of the $\approx 100\,\mu m$ thin polymer layer may be used [4.117]. This approach could increase the interaction distance L to the centimeter range. However, in such a geometry the space charge formation is no longer assisted by the large external electric field and the exponential gain coefficient Γ might become considerably smaller.

4.6.2 Thin Gratings

There are applications for which a large volume hologram with sharp angular selectivity is not a requirement or may be even a disadvantage. Optical pattern recognition systems based on joint Fourier-transform correlators (JTC) [4.118] or thin hologram Van der Lugt type correlators [4.119] are discussed in more detail in Chap. 5 and belong to this category because shift invariance is maintained only for thin gratings. Applications in the field of spatial light modulation or incoherent to coherent optical conversion [4.94,4.120–121] also require rather thin gratings in order to get sufficient optical resolution. Resolution issues may also point to thin gratings for applications aimed at imaging through turbid media [4.122–126].

Common to all the above image processing applications are three requirements: (1) the refractive index grating, though thin, should give rise to a high enough diffraction efficiency that can be safely detected; (2) the material resolution should be such as to allow the parallel processing of a large number of points; and (3), the processing should be as fast as possible. For instance, at a video rate of 25 Hz a resolution of 1000×1000 points leads to a data throughput of 25 Mbps, not enough to compete with the best computers. Clearly the cycle time has to be about two or three orders of magnitude smaller to ensure the competitiveness of an optical system.

For these optical applications, inorganic and organic photorefractive crystals have to compete also with semiconductor multiple quantum well (MQW) devices, where the photorefractive effect is mediated by resonant effects such as the Franz–Keldysh effect or the quantum confined Stark effect [4.127–128]. Local resonant effects in atomic vapors can also be used for the same kinds of application [4.129]. The first requirement of a large enough diffraction efficiency seems not to be an issue for inorganic crystals or the best polymers, even with sample thicknesses of only 10–$20\,\mu m$. Resolution is an issue for MQW devices and atomic vapors, rather than for inorganic and organic crystals or photorefractive polymers. The carrier mobilities in the latter materials are small enough, which prevents washing out of the space charge distribution even for grating spacings shorter than $1\,\mu m$. In contrast, due to fast carrier movement, the minimum feature size that can be used inside the nonlinear material exceeds $\approx 10\,\mu m$ for MQW and atomic vapors, giving roughly 100 times fewer bits per unit area. Optimized inorganic crystals may show a photorefractive response time of the order of $1\,ms$ at $1\,W/cm^2$ intensity level in the visible, even though faster processes have been observed. Due to the larger mobilities shorter response times (a few μs) can be measured in some bulk semiconductors or in semi-insulating MQW devices. Even faster response times (a few ns) are obtained with atomic vapors. Photorefractive organic crystals and polymers are rather slower; the fastest organic crystal has a response time of $0.5\,s$ at $I = 0.3\,W/cm^2$ intensity [4.130], while the fastest polymeric compound responds in around $5\,ms$ at $I = 1\,W/cm^2$ [4.131].

It appears that at the present stage of research, semiconductors, multiple quantum wells and atomic vapors have an advantage with respect to processing speed, while inorganic insulators and organic photorefractive materials guarantee better resolution and higher diffraction efficiencies. More material research is needed to increase the speed of polymeric materials for these applications. Inorganic insulator crystals may also need faster response times, which could be achieved by modifying the material properties using appropriate chemical reduction techniques. Alternatively, the speed may also be increased by employing a recording wavelength shorter than the bandgap which gives rise to efficient interband carrier photoexcitation and can guarantee response times of the order of $10\,\mu s$ [4.132]. With the latter approach the effective thickness of the photorefractive gratings can be adjusted externally by the light intensity level. Thin gratings are therefore obtained even in bulk samples without expensive crystal preparation.

4.6.3   Materials Requirements and Figures of Merit

As discussed above, a given material may have good or less good properties in view of a certain application. It is useful to establish some simple criteria to compare the potential of different materials [4.133]. In this section we deal with the definition of such criteria. A comparison of materials in terms of diffraction efficiencies, sensitivities and figures of merit is given in Sect. 4.7.

One first important criterion is the dynamic range, i.e. the maximum photoinduced refractive index change $\Delta n_{\max}$ achievable in a material. This quantity is of great importance for holographic storage and for many of the thick grating applications mentioned above. In many cases, rather than $\Delta n_{\max}$ it is the grating strength ($\Delta n_{\max} d/\lambda$) which is of practical importance (d: sample thickness). The total grating strength gives a better method of judging the achievable diffraction efficiencies and two-wave mixing gain–length products.

For many real-time dynamic applications it is not only important to achieve a large saturation refractive index change, but also to achieve it fast and at low costs in optical energy. In this case it is desirable to use a material with high sensitivity, i.e. large refractive index change per unit incident or absorbed energy. Four differently defined sensitivities have been used in the literature [4.92]:

$$S_{n1} = \frac{1}{\alpha} \frac{\partial (\Delta n)}{\partial W_0} , \tag{4.80}$$

$$S_{n2} = \frac{\partial (\Delta n)}{\partial W_0} , \tag{4.81}$$

$$S_{\eta 1} = \frac{1}{\alpha d} \frac{\partial \eta^{1/2}}{\partial W_0} , \tag{4.82}$$

$$S_{\eta 2} = \frac{1}{d} \frac{\partial \eta^{1/2}}{\partial W_0} , \tag{4.83}$$

where W_0 is the incident fluence of optical energy per unit area and Δn is the refractive index change induced by the photorefractive process. The derivatives in the above expressions are taken at the beginning of the recording process.

In order to predict the theoretical performance and the potential for applications of a given material starting from its physical parameters one can introduce some appropriate figures of merit. It has to be kept in mind that different experimental regimes may require a different figure of merit, so that materials may not be absolutely compared only in terms of a unique quantity. We discuss here a couple of figures of merit which are important in particular situations.

A   Figures of Merit for Refractive Index Change

A set of figures of merit describes the maximum refractive index change obtainable in a given experimental situation. We recall from Sect. 4.3 that the space charge field amplitude in (4.24) is proportional to the field E_q in two different regimes, at small fringe spacings Λ or at large fringe spacing in the case where the applied field E_0 exceeds E_q. In these cases the induced refractive index change is $\Delta n \approx 0.5(n^3 r_{\mathrm{eff}} E_q)$ and one can write a material figure of merit:

$$Q_1 \equiv \frac{n^3 r_{\mathrm{eff}}}{\varepsilon_{\mathrm{eff}}} N_{\mathrm{eff}} . \tag{4.84}$$

The effective number of traps N_{eff} depends strongly on the individual crystal sample and often is not considered within the expression for the figure of merit, giving the very common expression

$$Q_2 \equiv \frac{n^3 r_{\mathrm{eff}}}{\varepsilon_{\mathrm{eff}}} . \tag{4.85}$$

In many inorganic crystals the electro-optic coefficients are so high that the space charge field does not need to approach the value of the field E_q to obtain the desired large nonlinearities. The experiments are then performed at intermediate fringe spacings with only moderate applied fields, if so. The space charge field is proportional either to the applied field E_0 or to the diffusion field E_D in the case of no applied field. Both are independent of material parameters and the figure of merit becomes

$$Q_3 \equiv n^3 r_{\mathrm{eff}} . \tag{4.86}$$

Polymers are often operated in a regime where the figures of merit Q_1 and Q_2 are the limiting factors; inorganic crystals often rely on a regime where Q_3 is important.

B   Figures of Merit for Sensitivity

By analogy with the above figures of merit one can calculate expressions giving the combinations of material parameters for optimum sensitivity. Here we limit ourselves to expressions proportional to the photorefractive sensitivity S_{n1} defined above. In the limits $\Lambda \ll \Lambda_0, K_e^{-1}$ or $E_0 \gg E_q$ the sensitivity is proportional to $(\Delta n/\tau_{\mathrm{die}})(K_e/K_0)^2$ (see Sect. 4.3, (4.34, 4.40)) and a material figure of merit can be defined as

$$Q_4 \equiv \frac{n^3 r_{\mathrm{eff}}}{\varepsilon_{\mathrm{eff}}} \phi\lambda , \qquad (4.87)$$

where ϕ is the quantum efficiency and λ is the light wavelength. In contrast, in the regime of validity of (4.86) one has $S_{n1} \propto \Delta n/\tau_{\mathrm{die}}$ and the figure of merit is

$$Q_5 \equiv \frac{n^3 r_{\mathrm{eff}}}{\varepsilon_{\mathrm{eff}}} \phi\mu\tau_{\mathrm{R}}\lambda , \qquad (4.88)$$

where μ is the carrier mobility and τ_{R} is the carrier lifetime.

4.7   Materials

This section summarizes some of the properties of important photorefractive materials which have been investigated in the past. The materials may be partially compared in terms of some of the above defined figures of merit.

4.7.1   Photorefractive Materials

A   Inorganic Crystals

Since the early works with $LiNbO_3$ and $LiTaO_3$, photorefractive gratings have been observed in a number of other inorganic materials, including $BaTiO_3$ [4.134], $KNbO_3$ [4.135], $K(TaNb)O_3$ [4.135–136], $BaNaNb_5O_{15}$ [4.137], $Sr_xBa_{1-x}Nb_2O_6$ [4.138], $Bi_{12}(Si,Ge)O_{20}$ [4.139], $Bi_4Ti_3O_{12}$ [4.140], $Bi_4Ge_3O_{12}$ [4.141,142], $(K_xNa_{1-x})_{2z}(Sr_yBa_{1-y})_{1-z}Nb_2O_6$ [4.143], $Sn_2P_2S_6$ [4.144], $(Pb,La)(Zr,Ti)O_3$ ceramics [4.145], GaAs, InP, CdTe and other semiconductors [4.146–149], as well as in multiple quantum well semiconductor compounds [4.127–128].

In inorganic crystals the photorefractive effect is often induced by an intentional doping of the growing melt, though in some materials already nominally undoped samples contain a sufficient number of donors and traps. Reduction or oxidation treatments may also be used to optimize the performances of a material [4.150]. Doping and annealing treatments modify the properties of a material in many ways as a consequence of the creation of new electron levels in the bandgap and the related shift in the Fermi level energy. The most evident material change is the appearance of new absorption bands and thus new colors for the crystal samples. These absorption bands may facilitate photorefractive operation at a wavelength of interest. Other properties, such as the sign of the dominant charge carrier type or the effective carrier mobility may be influenced as well. It is not rare that an as-grown crystal with dominant hole conduction becomes electron conducting after a reduction treatment. In general the properties of every individual crystal sample have to be determined independently because of partially uncontrolled growing conditions. The production of crystal samples with standard predefined photorefractive properties remains one of the still unsolved challenges.

In oxygen-octahedra ferroelectrics of the form ABO_3 (Fig. 4.27), such as $KNbO_3$ or $BaTiO_3$, transition metal elements have been found to be useful dopants. These elements can enter either the A or B sites in the lattice. The valence of the dopants, which is usually different from that of the A or B metals, can be compensated by oxygen vacancies [4.151–152]. Iron is probably the most widely used dopant in this kind of materials, but other transition metals like Cu and Mn have proven to be important too. Copper, iron and manganese ions have been widely used also in the ilmenite crystal $LiNbO_3$. Popular dopants for photorefractive semiconductors are Cr or the intrinsic defect EL2 for GaAs, and Fe for InP. Recently it has been found that the near-infrared sensitivity of $BaTiO_3$ crystals can be highly improved by doping with Rh ions [4.153–154]. The same transition metal also improves the near-infrared performance in $KNbO_3$ [4.155].

Nowadays inorganic single crystals are still the most widely investigated photorefractive materials. The combination of large electro-optic coefficients with the possibility of growing large single crystalline samples makes these materials the most suitable candidates for many volume hologram applications. In addition, applications for real-time optical signal processing have proved very attractive for those materials which show fast response times. Among the approximately two dozen different photorefractive crystals we

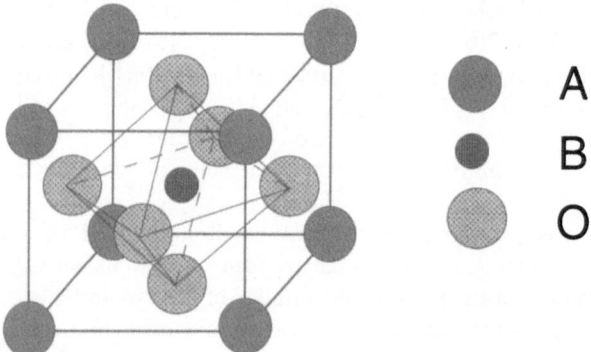

Fig. 4.27. Structure of ABO_3 oxygen-octahedra ferroelectrics with perovskite structure

review here the characteristics of $BaTiO_3$, $KNbO_3$, $LiNbO_3$, $Bi_{12}SiO_{20}$ and GaAs, which, in our opinion, cover the wide spectrum of requirements for different laboratory experiments as well as for numerous practical applications. Some properties of these crystals are compared with those of other classes of materials later in this chapter.

BaTiO$_3$ (barium titanate). Barium titanate is probably the most widely investigated photorefractive crystal. It belongs to the category of oxygen-octahedra ferroelectrics with perovskite structure. Large values of the spontaneous polarization at room temperature and the high packing density of oxygen octahedra give rise to very large electro-optic coefficients ($r_{232} \approx$ 1200 pm/V) and high nonlinear optical susceptibilities. $BaTiO_3$ shows large photorefractive gains ($\Gamma \approx 30\,cm^{-1}$) in the diffusion regime, so that high phase conjugate reflectivities are easily obtained without application of any external voltage. For this reason $BaTiO_3$ is the most popular material for laboratory experiments involving photorefractive phase conjugate mirrors or photorefractive resonators performed with Ar-ion lasers at visible wavelengths. Recently an increased near-infrared performance has been discovered in Rh-doped samples. In $BaTiO_3$ the photorefractive response is rather slow due to the combination of large dielectric constants and relatively small effective mobilities.

KNbO$_3$ (potassium niobate). Like $BaTiO_3$, $KNbO_3$ is an oxygen-octahedra ferroelectric and shows similarly large values of the electro-optic coefficients. However, the faster response of $KNbO_3$ gives rise to an increased sensitivity. With $Bi_{12}SiO_{20}$, $KNbO_3$ shows the best sensitivities in the blue/green spectral range. $KNbO_3$ is used in the diffusion regime for the kind of experiment mentioned above and is also one of the most promising candidates for real-time optical processing applications. Chemically reduced crystals show the fastest response and are often used in conjunction with an external applied field in the drift regime. For instance, electric field-assisted grating recording and optical phase conjugation with sub-ms response times and two-wave mixing gain of up to $16\,cm^{-1}$ at a moderate field of $20\,kV/cm$ have been demonstrated in reduced samples [4.92]. Alternatively, very large near-infrared sensitivity can be induced up to the $1.5\,\mu m$ telecommunication wavelength by ion implantation [4.156]. Unlike $BaTiO_3$, $KNbO_3$ is optically biaxial at room temperature. Undoped crystals can be easily phase matched for many nonlinear optical wavelength conversion experiments (see Chap. 6).

LiNbO$_3$ (lithium niobate). Lithium niobate was the first photorefractive crystal to be discovered [4.2]. It has ilmenite structure and, like $BaTiO_3$ and $KNbO_3$, it is ferroelectric at room temperature. The nonlinearity of $LiNbO_3$ is smaller than that of $BaTiO_3$ or $KNbO_3$ but is still among the best. In $LiNbO_3$ the charge transport is generally dominated by the bulk photogalvanic effect, therefore $LiNbO_3$ is not specially well suited for phase conjugation or other experiments requiring large two-wave mixing amplification.

The structure of LiNbO$_3$ can accommodate a large number of defects so that very large space charge fields, and thus large refractive index gratings can build up. The photoconductivity is small so that the photorefractive response time of LiNbO$_3$ is very slow, often exceeding several minutes. To date, this is the photorefractive crystal where the most efficient hologram fixing technique has been demonstrated. In our opinion, LiNbO$_3$ is the best candidate for holographic storage or other long-term applications based on photorefractive materials. Recent results indicate that crystals grown with stoichiometric instead of the more common congruent composition possess improved properties for dynamic holography.

Bi$_{12}$SiO$_{20}$ (bismuth sillenite). Bismuth sillenite (BSO) is cubic and does not need a poling procedure after growing. Its growing technique is rather simple. Due to its relatively small electro-optic coefficients, BSO is not particularly efficient in terms of refractive index changes at saturation. However, thanks to the efficient charge transport BSO is, together with KNbO$_3$, the most sensitive material in the visible. In order to increase its performance BSO is often used in conjunction with external electric fields. Its major drawback is its high natural optical activity, which prevents efficient use of thick samples. Usually one requires less than 90° rotation of the polarization, which limits the maximum thickness to about 2 mm for green light.

GaAs (gallium arsenide). In its semi-insulating form (undoped or Cr-doped) this widely known III–V semiconductor has proved one of the best photorefractive materials for the near-infrared spectral region (0.9–1.5 µm) of compact semiconductor lasers as well as the fundamental frequency of Nd:YAG lasers. GaAs has cubic structure and is not optically active. Therefore, unlike BSO, rather thick samples can be used in experiments. A principal advantage of GaAs and the other photorefractive semiconductors InP and CdTe is the very large carrier mobility, which leads to short response times. GaAs has two major drawbacks: small electro-optic coefficients and large dark conductivity. Due to the weak nonlinearity, auxiliary external electric fields are used to increase the magnitude of the photorefractive space charge fields. The large dark conductivity (dark relaxation time ≈ 0.1 ms) prevents the use of low light intensity levels. In addition to the bulk material, thin layered structures of GaAs/AlGaAs are used for real-time photorefractive experiments. These 2–3 µm thick multiple quantum well structures make use of resonant electro-optic effects near the exciton absorption wavelength [4.157].

B Electro-Optic Ceramics

Electro-optic ceramics form another category of photorefractive materials worth mentioning. Materials such as LiNbO$_3$ or BaTiO$_3$ can also be produced as ceramics, but limited optical quality has prevented their use in optics even for moderate thicknesses of a few tens of microns. However, there exist a group of ferroelectric ceramics which have a high transparency, exceeding

60%, and show interesting electro-optical effects [4.158]. These ceramics belong to the lead zirconate titanate family with perovskite structure and can be cheaply produced in large pieces. The lead ions are partially substituted with lanthanum ions, with a resulting chemical composition $(\text{Pb}_{1-x}\text{La}_x)(\text{Zr}_y, \text{Ti}_z)\text{O}_3$, abbreviated $\text{PLZT}_{x/y/z}$. Many applications of transparent ceramics for light shutters, optical switches, displays, spatial light modulators and holographic nonlinear optical devices have been explored over the last twenty years [4.159].

In PLZTs, the nature of the electro-optic effect can be influenced by adjusting the Zr/Ti ratio or the La content. At high La contents the compounds are paraelectric and exhibit the quadratic electro-optic effect (Kerr effect) described in Sect. 4.4. At low La contents the ceramics are ferroelectric and can be poled to exhibit the linear electro-optic (Pockels) effect.

PLZT ceramics have been found to exhibit the photorefractive effect. Most investigations have been performed using the ceramics in their paraelectric form [4.160–161]. A pseudo-linear electro-optic effect is achieved by mixing the modulated space charge field with a DC externally applied field, in a similar way to that discussed in Sect. 4.4 for photorefractive gratings recorded in liquid crystals or low T_g polymers. PLZT ceramics exhibit a slow response (typically of the order of a few minutes for $1\,\text{W/cm}^2$ intensities) and a relatively low photorefractive sensitivity, which can be only slightly increased by transition metal substitutions. However, the induced refractive index changes at saturation can be very large ($\Delta n \approx 10^{-3}$), two-beam coupling gain coefficients exceeding $\Gamma \approx 100\,\text{cm}^{-1}$ can be obtained for externally applied electric fields of the order of $10\,\text{kV/cm}$ [4.162]. Because of the high values of the dielectric constant (> 1000) the space charge field amplitude created in PLZT ceramics is generally limited by the field E_q, and thus by the effective trap density N_{eff} [4.163]. Consequently, the holographic resolution in PLZT is limited to $\approx 1000\,\text{lines/mm}$ because fringe spacings at least of the order of $1\,\mu\text{m}$ have to be used.

In the regime where the material exhibits only the quadratic electro-optic effect one can define a pseudo-linear electro-optic coefficient as $r = RE_0$, where R is the quadratic electro-optic coefficient and E_0 is the applied DC field. Therefore these materials offer the potential advantage that the holographic diffraction efficiency can be controlled during readout by the external field. Furthermore, in partially illuminated slow-responding materials like PLZTs the information on the magnitude of the electric field applied during information recording is kept for longer periods in the form of an internal screening field which is forming to counteract the applied field in the illuminated regions during the recording step [4.161]. During low-intensity readout the screening field adds to the applied field, which, combined with the above field-dependent nonlinearity, can also be used for the selective readout of digital information by means of the electric field [4.164].

C   Organic Crystals

A potential advantage of organic crystals for photorefractive applications is the possibility of altering the molecular structure for optimizing the electro-optic or nonlinear optical properties. Furthermore, organic materials have smaller dielectric constants than inorganic crystals. Therefore large photorefractive space charge fields can be created at short grating spacings (large field E_q in (4.27)) even with a crystal sample containing a relatively small number of traps. The electro-optic properties of some organic crystals have been discussed in Sect. 3.4.

To our knowledge, to date the photorefractive effect has been observed in only three different organic single crystals, known by the short names COANP:TCNQ, MNBA and DAST [4.130,4.165–166]. The first observation of the photorefractive effect in an organic material was reported in 1990 [4.165,4.167] in single crystals of 2-cyclooctylamino-5-nitropyridine (COANP) doped with 7,7,8,8-tetracyanoquinodimethane (TCNQ). The material in this study was chosen because of the large nonlinear optical and electro-optic effects. As a result of the doping with TCNQ the charge transport properties as well as the photorefractive effects were enhanced. The amplitudes of both absorption and phase gratings as well as their phase shifts with respect to the fringe pattern were investigated using an interferometric technique, which confirmed that the observed effect is based on a light-induced charge separation and a refractive index change mediated by the linear electro-optic effect. Photoabsorption or photochemical refractive index changes were excluded as mechanisms for the observed effects. The phase grating build-up in COANP:TCNQ was slow, with build-up times ranging between 30 and 50 min.

In an effort to find an organic material with better photorefractive response, single crystals of 4-nitrobenzylidene-3-acetamino-4-methoxyaniline (MNBA) [4.166] were investigated. MNBA belongs to the point group m and has a nearly optimized structure for electro-optic applications with all molecular charge transfer axes lying nearly parallel along the crystallographic a-axis. The largest electro-optic coefficient is $r_{111} = 50\,\mathrm{pm/V}$ at a wavelength of 532 nm [4.168]. High-quality single crystals can be grown from supersaturated solutions, and (010) faces can be used in optical experiments without prior polishing. Two-wave mixing experiments gave beam coupling gains as high as $2.2\,\mathrm{cm}^{-1}$ and response times between 5 and 300 s for writing intensities of $130\,\mathrm{mW/cm^2}$ at visible wavelengths. It was found that the dependence of the response time and two-beam coupling gain on fringe spacing is consistent with the predictions of the single carrier band model.

More recently the crystal 4-N, N-dymethylamino-4'-N'-methyl-stylbazolium toluene-p-sulfonate (DAST, Fig. 4.28) was established as the first organic crystal showing net photorefractive gain ($\Gamma = 2.2\,\mathrm{cm}^{-1}$, $\alpha = 1.8\,\mathrm{cm}^{-1}$ at $\lambda = 750\,\mathrm{nm}$) [4.130]. The response of this material was found to be relatively fast ($\tau = 500\,\mathrm{ms}$ at $0.3\,\mathrm{W/cm^2}$ intensity).

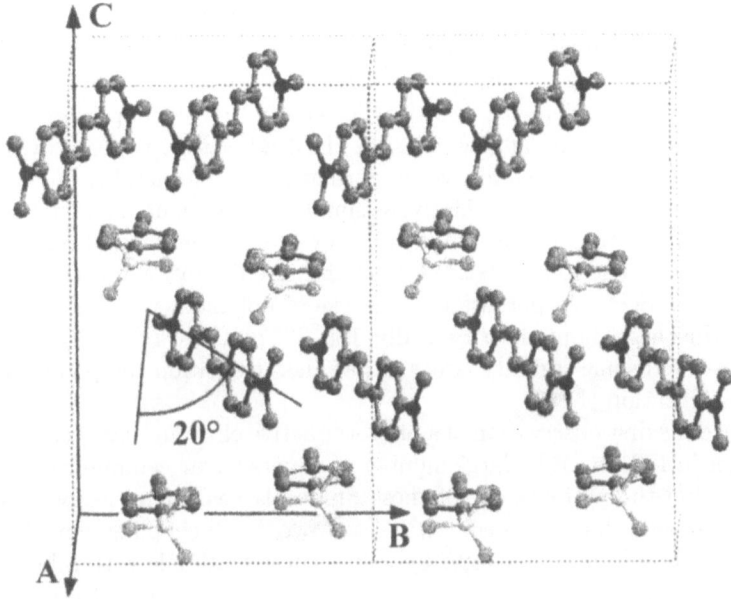

(a) **polar axis**

(b) Stilbazolium Chromophore

Tosylate

Fig. 4.28a,b. Crystal packing and chemical structure of the polar organic salt 4-N, N-dimethylamino-4'-N'-methyl-stylbazolium toluene-p-sulfonate (DAST)

Organic crystals potentially combine many of the advantages of polymers and inorganic crystals. Like polymers they have small dielectric constants and can be designed for optimum properties. Like inorganic crystals they can be grown in relatively large sizes and can show large electro-optic effects. It might be expected that future optimized organic crystals may enter in strong competition with the best inorganics for photorefractive applications.

4.7.2   Polymers and Liquid Crystals Showing Photorefractive and Photoassisted Orientational Gratings

A   Polymers

The mechanisms leading to photorefractivity in these classes of materials is similar to the one present in inorganic materials, but a few major differences do exist, as discussed in Sect. 4.3. There are two potential advantages of polymers with respect to inorganic crystals. First, the ingredients necessary for charge generation, charge transport, trapping and obtaining an electro-optic response can be introduced in the material quite easily by mixing appropriate agents in the polymer solution. One can design in this way a wide spectrum of materials with different properties. Second, but no less important, polymer films can be produced at low costs. A third advantage is given by the low dielectric constant, which allows the formation of considerably large electric space charge fields with a relatively small effective number of traps. This advantage, however, is important only in certain experimental regimes, and the large dielectric constants are not an absolute limitation for inorganics. As discussed in Sect. 4.6, potential drawbacks of polymers are their small maximum thickness, limited to typically 100–300 µm, the relatively slow time response, and some stability issues in low glass transition compounds related to crystallization [4.169].

Since the first observation of a photorefractive effect in a not-yet optimized polymer in 1991 [4.170] a large number of papers have been published on this subject and the performance of these materials has dramatically increased. The increase in diffraction efficiencies and two-beam coupling gains has been made possible to a large extent by using softer materials with a lower glass transition temperature [4.83–84]. In this regime a quadratic electro-optic effect related to reorientation of the nonlinear chromophore constituents dominates the material response to the space charge field grating (Sect. 4.4.3).

Essentially there are four classes of polymers used for photorefractive studies: (i) electro-optic polymers doped with sensitizer and charge transport molecules; (ii) photoconducting polymers doped with electro-optically active chromophores; (iii) inert host polymers used as a binder for all active components; and (iv) fully functionalized polymers having all the necessary components attached to the polymer backbone. The roles of all the key constituents of a photorefractive polymer are shown in the energy diagram of Fig. 4.13. The chemical structures of some of the most often used molecules in a photorefractive polymer compound are shown in Fig. 4.29.

Trapping centers represent the first necessary constituents for a photorefractive polymer, they are usually present in sufficient amounts in the material compound. The selection of trapping centers and the control of their concentration have received only a little attention up to the present [4.171–172]. In contrast, the other constituents need to be carefully chosen for material properties optimization. The second building block is a sensitizer which creates movable charge carriers upon photon absorption. In inorganics the charge generation process is guaranteed by the presence of impurity dopants, such as Fe, Cu, Co or Rh. In polymers the charge generator can be introduced by adding appropriate molecules to the polymer solution. 2,4,7-trinitro-9-fluorenone (TNF) and C_{60} fullerenes are often used for this purpose [4.84, 4.173]. Usually the sensitizing molecules are not attached directly to the polymer backbone. Their typical concentration is in the 0.1 wt.% region.

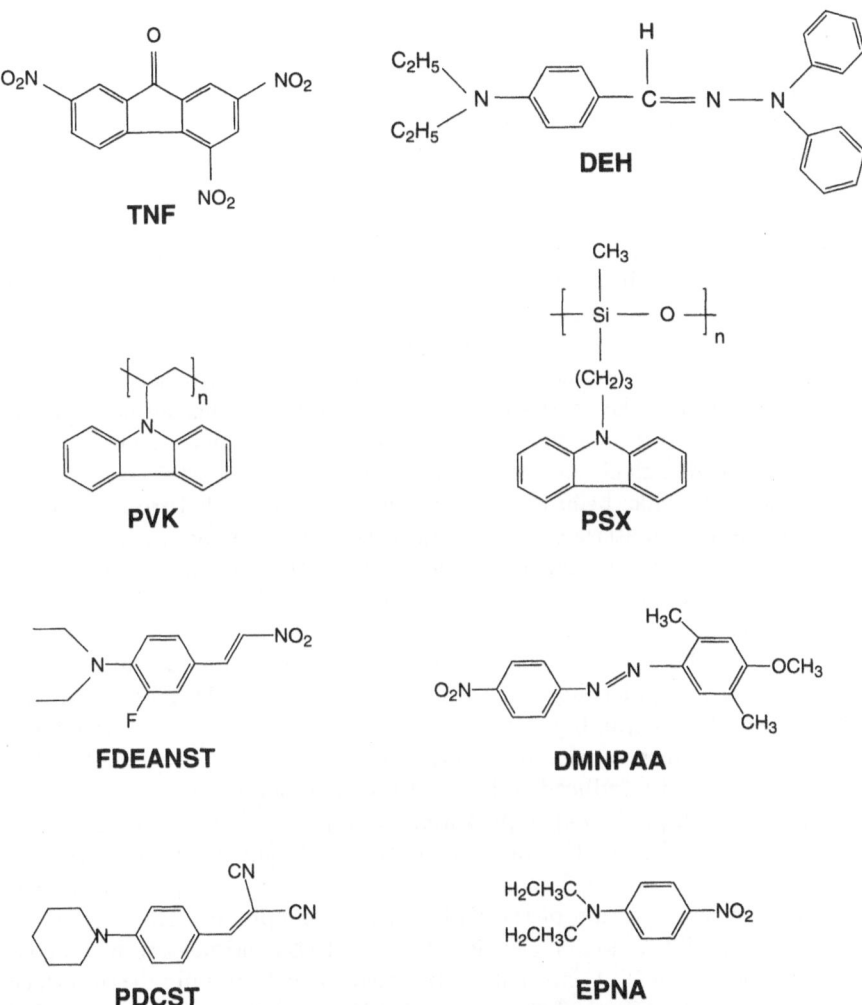

Fig. 4.29. Some constituents of photorefractive polymers. Sensitizer molecule 2,4,7-trinitro-9-fluorenone (TNF); charge transport agent (diethylamino)-benzaldehyde-diphenyl-hydrazone (DEH); photoconductive polymers poly(N-vinylcarbazole) (PVK) and carbazole substituted polysiloxane (PSX); nonlinear optical chromophores 3-fluoro-4-(diethylamino)-(E)-β-nitrostyrene (FDEANST), 2,5-dimethyl-4-(p-nitrophenylazo)anisole (DMNPAA), 4-piperidinobenzylidene-malononitrile (PDCST), and N,N-diethyl-substituted $para$-nitroaniline (EPNA)

The charge created by photoionization of the sensitizer moves by hopping through charge transport agents (CTA). Very often the molecule (diethylamino)benzaldehyde-diphenylhydrazone (DEH) is used for this purpose [4.34]. DEH molecules, as well as other CTAs are not attached to the polymer chain. Large concentrations of the order of 30–40 wt.% are necessary in order to increase the hopping probability. The alternative to doping with a CTA is to use a polymer host where charge transport can occur along the polymer chain itself. The most used photoconducting host is polyvinylcarbazole (PVK) [4.173], which constitutes the host in most of the best performing photorefractive polymers to date. Polysiloxane (PSX) has also proved to be a very attractive photoconducting polymer for photorefraction and very high two-wave mixing gains have already been obtained [4.174].

Finally, the nonlinear element which produces a change in the composite refractive index in response to an internal electric field is provided by doping with nonlinear optical chromophores (NLO). In some cases the NLO can also provide the necessary photosensitivity, so that doping with an additional sensitizer is not necessary. The NLO may be attached directly to the polymer chain like in the early experiment by Ducharme et al.[4.170] that used 4-nitro-1,2-phenylenediamine (NPDA) chromophores connected to the bisphenol-A-diglycidylether (bisA) chain. More recently, however, the best performance has been achieved using NLOs decoupled from the main polymer chain. In materials with low glass transition temperatures T_g these chromophores keep a sufficient rotational mobility to enhance the electro-optic response. Chromophores studied until now include 2,5-dimethyl-4-(p-nitrophenylazo)anisole (DMNPAA) [4.84], 3-fluoro-4-(diethylamino)-(E)-β-nitrostyrene (FDEANST) [4.83], N,N-diethyl-substituted $para$-nitroaniline (EPNA) [4.175], 4-piperidinobenzylidene-malononitrile (PDCST) [4.116] and many others. In general, high concentration doping with NLO chromophores is necessary to achieve optimum performance, however, high doping often has the consequence of a reduced stability of the compound.

The first high-performance photorefractive polymer was based on PVK: DMNPAA:ECZ:TNF in the ratio 33:50:16:1 wt.% [4.84]. N-ethylcarbazole (ECZ) is a plasticizer added in order to decrease the glass temperature T_g and facilitate the reorientation of the DMNPAA chromophores. This compound exhibits Bragg diffraction efficiencies of nearly 100% and net two-beam coupling gain of 207 cm^{-1} at external electric fields of 90 V/μm in a 105 μm thick sample. The performance of this compound has been approached or equaled later by other compounds based on PVK or PSX polymers [4.174, 4.176–178]. Recently, a compound based on the inert polymer poly(methyl methacrylate-co-tricyclodecylmethycrylate-co-N-cyclohexylmaleimide-co-benzyl methacrylate) (PTCB) and the NLO dye 2,N,N-dihexylamino-7-dicyanomethylidenyl-3,4,4,6,10-pentahydronaphtalene (DHADC-MNP) showed an impressive gain of $\Gamma = 225$ cm^{-1} for an applied field of 50 V/μm and the first maximum of diffraction efficiency at a field of 28 V/μm in 105 μm thick samples [4.179]. This compound has excellent thermal stability, and the chromophores play the triple function of charge generator, charge transport agent and electro-optic dye.

Many photorefractive polymers still show a relatively slow time response; however in the last few years quick progress has been achieved also in this respect. To our knowledge, to date the polymer compound with the fastest response is based on PVK:7-DCST:BBP:C_{60} in the ratio 49.5:35:15:0.5 wt.% [4.131], where 7-DCST is the nonlinear optical chromophore and BBP is a liquid plasticizer that shows better resistance against crystallization than ECZ. The response time for this material is 4 ms at the intensity of 1 W/cm^2 ($\lambda = 647$ nm) and with an applied field of 100 V/µm.

Besides the classes of photorefractive polymers mentioned above, there are two additional types of material that recently showed very encouraging results and merit consideration. These are organic glasses formed by glassy chromophores and organic oligomers that permit us to join all functionalities in a single component material. The potential of organic glasses for photorefraction has been proved by Lundquist et al. [4.180], who demonstrated a gain coefficient of 69 cm^{-1} at the comparatively low field of 40 V/µm in a compound containing nearly 90 wt.% of the glassy chromophore 2BNCM. The response time of this material is rather slow and could be decreased at least to the level of $\approx$ 80 s by adding about 10 wt.% of PMMA polymer to the glassy chromophore matrix. Organic oligomers [4.181–183] are very interesting because of their multifunctionality. In contrast to guest–host systems, phase separation and crystallization issues are not a major problem here. In [4.181] a maximum gain coefficient of 80 cm^{-1} and diffraction efficiency of 20% were demonstrated for a field of 31 V/cm in a 130 µm thick sample of a photoconductive electro-optic carbazole trimer doped with 0.06 wt.% TNF as the charge generator.

While the large majority of experiments using polymers is performed under the conditions of large external electric field applied longitudinal to the thin samples, few investigations have been reported also under zero-field conditions [4.184–185]. Yu et al. reported on a novel kind of conjugated photorefractive polymer where the backbone plays also the role of charge generator and charge transporter [4.185]. The synthesis of such fully functionalized materials is much more challenging than that of guest–host systems, but promises better long-term stability. In a compound where the nonlinear chromophores covalently bond to the backbone a gain coefficient of 5.7 cm^{-1} was measured at zero field in a 1.4 µm thin sample for a grating spacing of $\approx$ 1.25 µm [4.185]. While the grating phase shift and the spatial dispersion of the gain coefficient are consistent with the conventional photorefractive band model of Sect. 4.3.1, the value of the calculated refractive index modulation amplitude ($\Delta n = 3 \times 10^{-5}$) calculated assuming negligible dichroism is too large to be consistent with the measured electro-optic coefficient of 4.3 pm/V and a diffusion limited process. Some kind of enhancement of the electro-optic response might also be active in this case. An even larger gain coefficient of about 200 cm^{-1} at zero applied field has been observed in another polymer compound that contains an ionic tri(bispyridyl) ruthenium complex as the charge-generating species [4.186]. There are indications that some kind of internal electric field may enhance the zero external field charge transport in this kind of polymers [4.187]. More investigations are needed to clarify the origin of the refractive index gratings in these compounds.

B Liquid Crystals

It should be evident at this point that the best polymers for photorefraction profit from the strong chromophores reorientational effect that can be observed in compounds with low T_g. To optimize this effect, in the ultimate limit a compound would be composed predominantly of long rod-shaped molecules with large molecular birefringence that are able to reorient as easily in an electric field as they were in the liquid state. Nematic liquid crystals approximate this ultimate limit very well. As already mentioned in Sect. 4.4.3, the first investigations of photorefractive effects in doped nematic liquid crystals were performed in 1994 by Rudenko and Sukhov [4.72–73]. They used homeotropically aligned nematic liquid crystal 5CB doped with ≈ 0.6 moles/l rhodamine 6G and could demonstrate photoconductivity and a diffraction efficiency of few % in holographic experiments. After that the field has witnessed considerable increase in interest and the progress has been rapid [4.74–79, 4.81, 4.188]. Already in 1995 Wiederrecht et al. [4.75] were able to demonstrate transfer of 88% of the energy from the pump to an initially equally intense signal wave in beam coupling experiments performed in 37 μm thick samples. If translated to the two-wave mixing formalism valid for thick holograms (gratings in liquid crystals are of the thin type because of the use of very long grating spacings) such a coupling would correspond to an impressive gain coefficient of 640 cm^{-1}. The samples used in this work contained perylene and N, N'-di(n-octyl)-1,4,5,8-naphtalenediimide (NI) as electron donors and acceptors, respectively, allowing photoinduced electron transfer reactions to yield mobile ions and hence photoconductivity. An even larger equivalent two-wave mixing gain coefficient of $\Gamma \approx 2890$ cm^{-1} was found in recent work by Khoo et al. in pentylcyanobiphenyl liquid crystal doped with a small amount of C_{60} fullerene [4.78]. Recently, Miniewicz et al. demonstrated phase conjugation with a reflectivity of 2.6% using a 6.3 μm thick dye doped cell [4.189].

With respect to polymers, the main advantage of photorefractive liquid crystals is their low operation voltage. In general voltages of 1 to 5 V are sufficient to maximize the effect (as compared to 5 kV or more in polymers). The main disadvantage is the low intrinsic resolution limited to about twice the sample thickness [4.190]. This is because for shorter grating spacings the intermolecular elastic restoring forces are too weak to ensure the correct orientation modulation. Due to the low resolution, photorefractive liquid crystals must to be operated at rather large grating spacings, of the order of 50–100 μm, and diffraction is of the Raman–Nath type. In liquid crystals the minimum build-up and decay times of the refractive index modulation are governed by the orientational decay constant, which is proportional to γ/K, where γ and K are the effective rotational viscosity and elastic constant of the liquid crystal, respectively. Response times of the order of 1 ms have already been observed [4.77]. While there is general consensus that photorefraction in liquid crystals is connected with migration of ionic charges, there is not yet a full understanding of the details of the charge generation process and of the interaction between the electric space charge field and the liquid crystal molecules, which is the result of multiple physical effects.

A very interesting new approach that merges some of the advantages of polymers and liquid crystals has recently been proposed by Ono et al. and Golemme et al. by use of polymer-dispersed liquid crystals [4.191–192]. Small liquid crystals droplets with dimensions much smaller than the typical grating spacings are dispersed into a matrix formed by a photoconducting polymer. The polymer matrix serves to produce the space charge field and also plays the role of a solid spacer between the liquid crystal droplets, thus eliminating collective molecular relaxation effects. The electro-optic effect is fully due to reorientation within the liquid crystal droplets. This approach has the advantages of dramatically improving the resolution with respect to liquid crystals alone and reducing somewhat the magnitude of the necessary applied electric field with respect to photorefractive polymers due to an enhanced electro-optic effect. In the above mentioned works high refractive index change amplitudes of 3×10^{-3} and 2×10^{-3} were observed for electric fields of 40 and 22 V/μm, and grating spacings of 12 and 4.5 μm, respectively.

4.7.3   Wavelength Sensitivity

Ideally a photorefractive material would be sensitive over a wide spectral range, from the near-ultraviolet to the near-infrared. Unfortunately there is no single material that can operate equally well over all this wavelength region. Many of the most efficient photorefractive materials show an optimum response in the visible region.

In the near-infrared, photorefractive materials can take advantage of compact solid state lasers and may be operated at the important telecommunication wavelengths of 1.32 and 1.55 μm. This spectral region is presently occupied by semi-insulating semiconductor materials such as GaAs, InP and CdTe [4.193], which unfortunately possess relatively small nonlinearity. However, considerable efforts have been made recently to improve the infrared sensitivity of bulk oxide crystals such as $BaTiO_3$ and $KNbO_3$ by proper doping and processing [4.153–155, 4.194]. Short wavelengths in the near-ultraviolet are interesting in connection with excimer lasers and promise an increase in the achievable grating resolution and response speed. Up to now, photorefractive gratings recorded at near-ultraviolet wavelengths have been observed only in a few transparent materials [4.142, 4.195–198].

In Fig. 4.30 we show the wavelengths at which photorefractive gratings have been recorded in a selection of photorefractive materials. The short wavelength region from the near-ultraviolet to the blue–green spectral range is presently dominated by inorganic crystals. Organic crystals and photore-

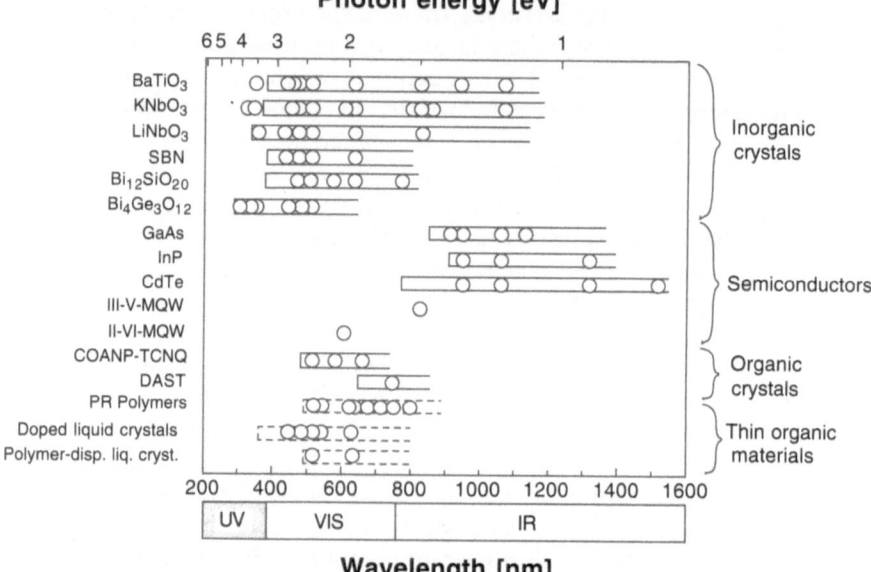

Fig. 4.30. Wavelengths at which photorefractive gratings have been observed in a selection of photorefractive materials. The horizontal bars indicate the transmission range of a material, with the band edge wavelength at their left end. Data for inorganic crystals are limited to bulk materials

fractive polymers occupy mainly the red to near-infrared border region, while semiconductors respond at infrared wavelengths larger than $\approx 800\,\mathrm{nm}$. Multiple quantum well photorefractive devices (MQW) [4.127] differ from the other materials by the fact that holographic gratings can be recorded and read out only in a few nanometer-wide spectral band corresponding to the excitonic resonances.

4.7.4 Comparison of Materials Properties

There are many ways in which the performance of different photorefractive materials may be compared. One approach is to consider the achievable diffraction efficiencies (or exponential gain) and the time needed to achieve them. The higher the diffraction efficiency and the shorter the response time, the better the material. The diagram of Fig. 4.31 shows the steady state diffraction efficiency and the inverse response time obtained in a selection of inorganic and organic crystals, photorefractive polymers and dye-doped liquid crystals. The diffraction efficiencies are scaled to a typical thickness d of actual samples, i.e. $d = 100\,\mu\mathrm{m}$ for polymers, $d = 20\,\mu\mathrm{m}$ for liquid crystalline cells and $d = 2\,\mathrm{mm}$ for inorganic and organic crystals. The response time is scaled to a light intensity of $1\,\mathrm{W/cm^2}$. The data for polymers are taken under an applied electric field of typically $500\text{--}1000\,\mathrm{kV/cm}$ and those

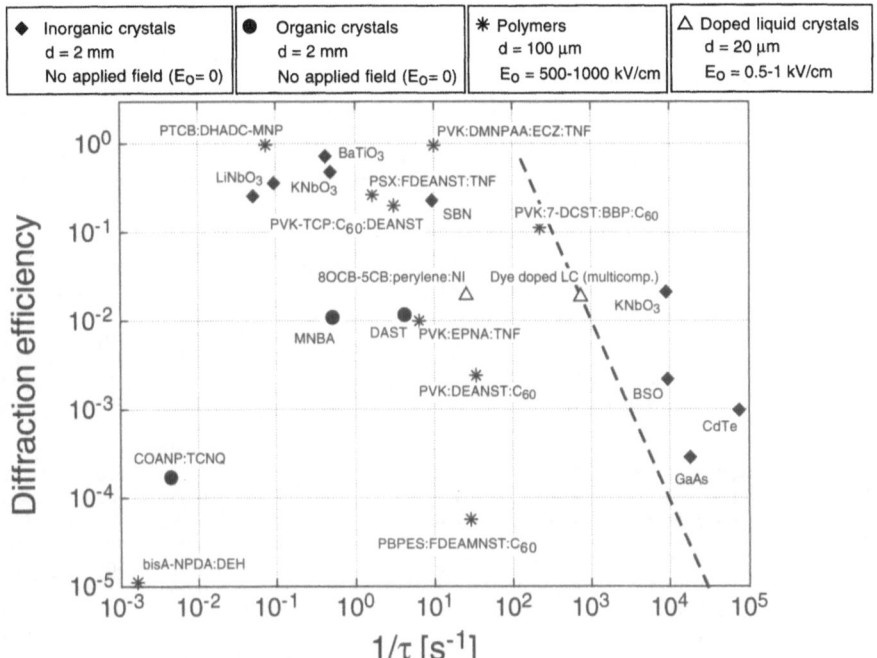

Fig. 4.31. Diffraction efficiency versus growth rate τ^{-1} for a selection of photorefractive crystals, polymers and doped liquid crystals. The values are scaled to $1\,\mathrm{W/cm^2}$ light intensity and to a typical thickness of $2\,\mathrm{mm}$ for crystals, $100\,\mathrm{\mu m}$ for polymer films and $20\,\mathrm{\mu m}$ for liquid crystal cells. The dashed line connects points of equal sensitivity

for liquid crystals under a field of $0.5\text{--}1\,\mathrm{kV/cm}$, while those for inorganic and organic crystals are under zero-field conditions. The dashed line in the diagram connects points of constant sensitivity. However, the same sensitivity for the different categories of materials is not on the same line because of the different values assumed for the sample thickness. The region of the best materials (the upper right corner in the diagram) is still occupied mainly by inorganic crystals; however, the best polymers and liquid crystals are approaching this region very fast. The best organic crystals have made it to the midfield region.

Another way of comparing the performance of different photorefractive materials is through the photorefractive sensitivities defined in Sect. 4.6.3. In Table 4.1 the sensitivities S_{n1} and S_{n2} are compared for a selection of experiments in organic and inorganic photorefractive materials. The refractive index change $\Delta n_{\max}$ achieved in the referenced works is also reported in Table 4.1. In many inorganic materials values of $\Delta n_{\max}$ considerably larger than those listed here have been observed. However, large refractive index changes are often obtained at the expense of a slower response and the sensitivities are

Table 4.1. Absorption constants, photorefractive sensitivities, refractive index changes and grating strengths for selected experiments reported in inorganic and organic photorefractive materials

Material	α (cm^{-1})	S_{n_1} (cm^3/kJ)	S_{n_2} (cm^2/kJ)	Δn_{max} [a]	$\frac{\Delta n_{max} d}{\lambda}$ [b]	E_0 (kV/cm)	Ref.
LiNbO$_3$	0.1–100	0.005–0.05	0.17^{c,d}	10^{-5}–10^{-3}	0.04–4	0	[4.199, 200]
KNbO$_3$	3.8	12	50	$\approx 10^{-5}$	0.04	7	[4.201]
			30	$\approx 10^{-5}$	0.04	0	[4.202]
	5500^c	0.4^c	2000^c	2×10^{-5}	0.08	2.2	[4.5]
BaTiO$_3$		1.3^e		5×10^{-5}	0.2	10	[4.203]
	2.2	3.4^f	7.6^f	1.5×10^{-4} ^{f}	0.6	0	[4.204]
SBN	1.0	0.4	0.4	1.5×10^{-4}	0.6	0	[4.205]
Bi$_{12}$SiO$_{20}$	2.3	72	167			6	[4.139]
	≈ 0.8	10	8	$\approx 10^{-6}$	0.004	0	[4.206]
GaAs	$\approx 1.2^g$	35^g	42^g	3×10^{-6}	0.006	0	[4.147]
CdTe	1.8^h	230^h	410^h	5×10^{-6}	0.007	0	[4.207]
GaAs/AlGaAs MQW	8000	2000^{i,j}	16×10^6 ^{i,j}	10^{-2} ^{i}	0.02^i	75	[4.208]
CdZnTe/ZnTe MQW		$< 700^{i,k}$	7×10^6 ^{i,k}	10^{-2} ^{i}	0.03^i	320	[4.128]
PLZT ceramic (9/65/35)	2–4	≈ 0.02	0.06	$\approx 10^{-3}$	≈ 1		[4.162]
COANP:TCNQ	5^l	7×10^{-7} ^{l}	4×10^{-6} ^{l}	2×10^{-6} ^{l}	0.006	0	[4.165]
MNBA	70	2×10^{-5}	0.001	10^{-5}	0.04	0	[4.166]
bisA-NPDA:DEH	10^k	1.4×10^{-5} ^{k}	1.4×10^{-4} ^{k}	2×10^{-6}	0.0003	114	[4.170]
PVK:DMNPAA:ECZ:TNF	13^p	0.8^{n,p}	$10^{n,p}$	5×10^{-3} ^{n,p}	0.7^n	900	[4.84]
PSX:FDEANST:TNF	32^l	0.07^{n,l}	2.1^{n,l}	1.4×10^{-3} ^{n,l}	0.29^n	770	[4.112]
PVK-7-DCST:BBP:C60	26^m	0.8^{m,n}	20^{m,n}	8×10^{-4} ^{m,n}	0.1^{m,n}	1000	[4.131]
Anthraquinone dye doped LC	462^q	16^q	7400^q	1.9×10^{-3} ^{q}	0.05^q	10	[4.77]

Data for $\lambda = 488$ nm or $\lambda = 514$ nm unless otherwise indicated.

[a] Representative for the cited experiments, higher values have been observed in most of the inorganic crystals in different experimental situations

[b] calculated for a typical thickness of 2 mm for crystals, 100 μm for photorefractive polymers, 500 μm for ceramics, and 2 μm for MQW devices

[c] $\lambda = 351$ nm

[d] $E_0 = 50$ kV/cm

[e] $\lambda = 458$ nm

[f] 45°-cut

[g] $\lambda = 1064$ nm

[h] $\lambda = 1523$ nm

[i] Quantum confined Stark effect

[j] $\lambda = 830$ nm

[k] $\lambda = 596$ nm

[l] $\lambda = 676$ nm

[m] $\lambda = 647$ nm

[n] Reorientationally enhanced;

[p] $\lambda = 675$ nm;

[q] $\lambda = 633$ nm

not increased. It can easily be shown that the maximum achievable sensitivity in a material is ultimately limited by the speed of charge production [4.209], explaining the fact that samples with moderate refractive index changes often show comparable or better sensitivities than crystal samples with stronger space charge gratings at saturation. In many cases rather than Δn_{max} it is the grating strength $(\Delta n_{\mathrm{max}} d / \lambda)$ which is of practical importance. The total grating strength gives a better method of judging the achievable diffraction efficiencies and two-wave mixing gain–length product. This figure is also given in Table 4.1 for typical sample thicknesses.

The performance of a material in a certain application may be related to material parameters by means of figures of merit, such as the quantities $Q_1 \ldots Q_5$ defined in Sect. 4.6.3. Large values of Q_1, Q_2 and Q_3 indicate materials where large refractive index changes can be obtained. Large values of Q_4, and Q_5 indicate materials with good holographic sensitivity. Table 4.2 lists the typical quantities $Q_1 \ldots Q_5$ for some photorefractive materials as well as a few other relevant parameters.

Owing to the small dielectric constants, some organic materials have a larger figure of merit Q_2 than the best inorganics; however, the latter are still better in terms of Q_1, and especially Q_3. The faster response of most inorganic crystals reflects also in higher values for the sensitivity figures of merit Q_4 and Q_5.

4.8   Conclusions

In the past several years the field of photorefraction has grown steadily, and today the intense activity in this field is reflected in a large volume of publications and presentations at international conferences. During the past decade major advances have been achieved in both the theoretical and experimental aspects of photorefractive materials. The classical photorefractive materials such as inorganic electro-optic crystals and transparent ceramics, are now being challenged by new materials, i.e. organic crystals, photorefractive polymers, dye-doped nematic liquid crystals and semiconductor multiple quantum well structures. The photorefractive effect has proved invaluable for demonstrating new techniques in industrial inspection, locking of laser diodes, image processing and holographic storage. At present, inorganic crystals are still the materials of choice for most realistic applications; however, we expect the rapid performance improvement of the new materials to lead to strong competition in the next several years. Ultimately it is not expected that a single class of materials will excel in all classes of applications. The selection of the materials which will reach the best commercial maturity will be based not only on pure performance characteristics, but also on other criteria, such as reliability, stability and cost. Research efforts must continue in known and new materials in order to optimize all these factors.

Table 4.2. Refractive index, electro-optic coefficients, dielectric constant, mobility and figures of merit $Q_1 \ldots Q_5$ for some inorganic and organic photorefractive materials

Material	λ (nm)	n	r_{eff} (pm/V)	ε_{eff}	μ (cm²/Vs)	$\phi\mu\tau_R$ (10^{-16} m²/V)	N_{eff} (10^{22} m⁻³)	Q_1 (V⁻¹ μm⁻²)	Q_2 (pm/V)	Q_3 (pm/V)	Q_4 (nm²/V)	Q_5 (10^{-32} m⁴/V²)
LiNbO₃	633	$n_e = 2.20$	$r_{333} = 31$	29	0.8	0.03–0.6	20	2.3	11.4	331		0.002–0.04
KNbO₃	514	$n_c = 2.21$	$r_{33}^{eff} = 59$	35	0.5	10–10^5	5	0.9	18.2	634	3.3	0.9–9000
		$n_b = 2.40$	$r_{23}^{eff} = 460$	989						6359	1.2	0.3–3000
BaTiO₃	514	$n_e = 2.42$	$r_{33}^{eff} = 90$	83	0.5	10–1000	5	0.8	15.4	1275	2.6	0.8–80
		$n_0 = 2.49$	$r_{23}^{eff} = 1326$	4350					4.7	20470	0.8	0.24–24
Bi₁₂SiO₂₀	633	2.54[a]	$r_{231} = 5.0$	55	3.4	10^5	2	0.03	1.5	82		940
GaAs	1064	3.50[b]	$r_{231} = 1.2$	13.2	$\mu_e = 8500$ $\mu_h = 400$	10^5	0.2	0.008	3.9	51		6900
CdTe	1550	2.74	$r_{231} = 6.8$[c]	9.4	$\mu_e = 1050$ $\mu_h = 5$	$\mu\tau \approx 10^7$	0.5	0.07	14.9	140		$< 10^6$
PLZT ceramic (10/65/35)[d]	514	≈ 2.5	200–800	4500		10	50	≈ 0.9	≈ 1.7	≈ 8000		≈ 0.1
COANP:TCNQ	633	$n_c = 1.65$	$r_{333} = 15$	2.6		0.05–0.2	0.1	0.03	25.9	67	2×10^{-4}	0.008–0.032
MNBA	514	$n_1 = 2.23$	$r_{111} = 62$	4.0		0.013	0.01	0.02	172	688	10^{-5}	0.01
bisA-NPDA:DEH[e]	647	1.63	0.33	2.9	small		0.2	0.001	0.5	1.4		
PVK:FDEANST: TNF[f]	647	≈ 1.7	≈ 1.8[g]	≈ 3	small				≈ 3[g]	≈ 9[g]		
PVK:DMNPAA: ECZ:TNF[h]	675	1.75	≈ 18[g]	≈ 3.1	small				≈ 32[g]	≈ 100[g]	< 0.2	

[a] $\lambda = 620$ nm;
[b] $\lambda = 1020$ nm;
[c] $\lambda = 3.39$ μm;
[d] $E_0 = 5$-10 kV/cm;
[e] $E_0 = 120$ kV/cm;
[f] $E_0 = 400$ kV/cm;
[g] estimated including the quadratic effect of chromophore reorientation;
[h] $E_0 = 900$ kV/cm.

References

4.1 H. J. Eichler, P. Günter, D. W. Pohl, *Laser-Induced Dynamic Gratings* (Springer Verlag, Berlin, 1986).

4.2 A. Ashkin, G. D. Boyd, J. M. Dziedzic, R. G. Smith, A. A. Ballman, J. J. Levinstein, K. Nassau, Appl. Phys. Lett. **25**, 233 (1966).

4.3 F. S. Chen, J. T. LaMacchia, D. B. Fraser, Appl. Phys. Lett. **13**, 223 (1968).

4.4 I. Biaggio, M. Zgonik, P. Günter, J. Opt. Soc. Am. **B 9**, 1480 (1992).

4.5 G. Montemezzani, P. Rogin, M. Zgonik, P. Günter, Phys. Rev. **B 49**, 2484 (1994).

4.6 I. Biaggio, R. W. Hellwarth, J. P. Partanen, Phys. Rev. Lett. **78**, 891 (1997).

4.7 J. Mort, D. M. Pai (eds.) *Photoconductivity and Related Phenomena* (Elsevier, Amsterdam, 1976).

4.8 J. P. Partanen, P. Nouchi, J. M. C. Jonathan, R. W. Hellwarth, Phys. Rev. **B 44**, 1487 (1991).

4.9 A. M. Glass, D. VonderLinde, T. J. Negran, Appl. Phys. Lett. **25**, 233 (1974).

4.10 R. Sommerfeldt, L. Holtmann, E. Krätzig, B. C. Grabmaier, Phys. Stat. Sol. (a) **106**, 89 (1988).

4.11 J. J. Amodei, D. L. Staebler, Appl. Phys. Lett. **18**, 540 (1971).

4.12 M. Carrascosa, F. Agulló-López, J. Opt. Soc. Am. **B 7**, 2317 (1990).

4.13 G. Montemezzani, M. Zgonik, P. Günter, J. Opt. Soc. Am. **B 10**, 171 (1993).

4.14 A. Yariv, S. S. Orlov, G. A. Rakuljic, J. Opt. Soc. Am. **B 13**, 2513 (1996).

4.15 M. Pope, C. E. Swenberg, *Electronic Processes in Organic Crystals* (Clarendon Press, Oxford, 1982).

4.16 D. Emin, in *Handbook of Conducting Polymers* (ed. T. A. Skotheim) (Dekker, New York, 1986), vol. 2, p. 915.

4.17 P. M. Borsenberger, E. H. Magin, J. J. Fitzgerald, J. Phys. Chem. **97**, 9213 (1993).

4.18 S. M. Silence, J. C. Scott, F. Hache, E. J. Ginsburg, P. K. Jenkner, R. D. Miller, R. J. Twieg, W. E. Moerner, J. Opt. Soc. Am. **B 10**, 2306 (1993).

4.19 H. Tokuhisa, M. Era, T. Tsutsui, S. Saito, Appl. Phys. Lett. **66**, 3433 (1995).

4.20 J. Feinberg, J. D. Heimen, A. R. Tanguay, R. W. Hellwarth, J. Appl. Phys. **51**, 1297 (1980).

4.21 R. A. Mullen, in *Photorefractive Materials and Their Applications I: Fundamental Phenomena* (eds. P. Günter, J. P. Huignard) (Springer-Verlag, Berlin, 1988), vol. 1, p. 167.

4.22 L. Onsager, Phys. Rev. **54**, 554 (1938).

4.23 A. Mozumder, J. Chem. Phys. **60**, 4300 (1974).

4.24 M. E. Orczyk, J. Zieba, P. N. Prasad, J. Phys. Chem. **98**, 8699 (1994).

4.25 J. H. Perlstein, P. M. Borsenberger, in *Extended Linear Chain Compounds* (ed. J. S. Miller) (Plenum Press, New York, 1982), vol. 2, p. 339.

4.26 K. Buse, Appl. Phys. B Lasers and Optics **64**, 273 (1997).

4.27 N. V. Kukhtarev, V. B. Markov, S. G. Odulov, M. S. Soskin, V. L. Vinetskii, Ferroelectrics **22**, 949 (1979).

4.28 G. Montemezzani, C. Medrano, P. Günter, M. Zgonik, Phys. Rev. Lett. **79**, 3403 (1997).

4.29 G. Montemezzani, M. Zgonik, Phys. Rev. **E 35**, 1035 (1997).

4.30 N. V. Kukhtarev, V. B. Markov, S. G. Odulov, M. S. Soskin, V. L. Vinetskii, Ferroelectrics **22**, 961 (1979).

4.31 B. I. Sturman, V. M. Fridkin, *The Photovoltaic and Photorefractive Effects in Noncentrosymmetric Materials* (Gordon & Breach, Philadelphia, 1992).

4.32 C. Gu, J. Hong, H. Y. Li, D. Psaltis, P. Yeh, J. Appl. Phys. **69**, 1167 (1991).

4.33 M. Ewart, Reduced $KNbO_3$ for photorefractive applications, PhD thesis, Diss. ETH No. 12484, Swiss Federal Institute of Technology (1998).

4.34 W. E. Moerner, S. M. Silence, Chem. Rev. **94**, 127 (1994).

4.35 S. M. Silence, D. M. Burland, W. E. Moerner, in *Photorefractive Effects and Materials* (ed. D. D. Nolte) (Kluwer Academic Publishers, Boston, 1995), p. 265.

4.36 S. Ducharme, Proc. SPIE **2526**, 144 (1995).

4.37 W. E. Moerner, A. Grunnetjepsen, C. L. Thompson, Ann. Rev. Mater. Sci. **27**, 585 (1997).

4.38 J. S. Schildkraut, Y. Cui, J. Appl. Phys. **72**, 5055 (1992).

4.39 J. S. Schildkraut, A. V. Buettner, J. Appl. Phys. **72**, 1888 (1992).

4.40 W. D. Gill, in *Photoconductivity and Related Phenomena* (ed. J. Mort, D. M. Pai) (Elsevier, Amsterdam, 1976), p. 303.

4.41 A. Twarowski, J. Appl. Phys. **65**, 2833 (1989).

4.42 C. L. Braun, J. Chem. Phys. **80**, 4157 (1984).

4.43 L. Pautmeier, R. Richert, H. Bässler, Synth. Mat. **37**, 271 (1990).

4.44 J. K. Mack, L. B. Schein, A. Peled, Phys. Rev. **B 39**, 7500 (1989).

4.45 P. Borsenberger, Phys. Stat. Sol. **B 173**, 671 (1992).

4.46 L. Pautmeier, R. Richert, H. Bässler, Phil. Mag. **B 63**, 587 (1991).

4.47 W. Helfrich, in *Physics and Chemistry of the Organic Solid State* (eds. D. Fox, M. M. Labes, A. Weissberger) (Wiley-Interscience, New York, 1967), p. 22.

4.48 M. Liphardt, S. Ducharme, J. Opt. Soc. Am. (Optical Physics) **15**, 2154 (1998).

4.49 J. F. Nye, *Physical Properties of Crystals* (Clarendon Press, Oxford, 1985).

4.50 M. Eich, B. Reck, D. Y. Yoon, C. G. Wilson, G. C. Bjorklund, J. Appl. Phys. **66**, 3241 (1989).

4.51 M. A. Mortazavi, A. Knoesen, S. T. Kowel, B. G. Higgins, A. Dienes, J. Opt. Soc. Am. **B 6**, 733 (1989).

4.52 Y. Sirotine, M. Chaskolskaia, *Fondements de la Physique des Cristaux* (Mir, Moscow, 1984).

4.53 C. Bosshard, K. Sutter, P. Prêtre, J. Hulliger, M. Flörsheimer, P. Kaatz, P. Günter, in *Organic Nonlinear Optical Materials* (eds. A. F. Garito, F. Kajzar), Advances in Nonlinear Optics (Gordon & Breach, Amsterdam, 1995).

4.54 D. A. Kleinmann, Phys. Rev. **126**, 1977 (1962).

4.55 W. E. Moerner, S. M. Silence, F. Hache, G. C. Bjorklund, J. Opt. Soc. Am. **B 11**, 320 (1994).

4.56 Sandalphon, B. Kippelen, K. Meerholz, N. Peyghambarian, Appl. Opt. **35**, 2346 (1996).

4.57 B. Swedek, N. Cheng, Y. P. Cui, J. Zieba, J. Winiarz, P. N. Prasad, J. Appl. Phys. **82**, 5923 (1997).

4.58 L. Mager, C. Melzer, M. Barzoukas, A. Fort, S. Mery, J. F. Nicoud, Appl. Phys. Lett. **71**, 2248 (1997).

4.59 J. D. Shakos, A. M. Cox, D. P. West, K. S. West, F. A. Wade, T. A. King, R. D. Blackburn, Opt. Commun. **150**, 230 (1998).

4.60 A. A. Izvanov, A. E. Mandel, N. D. Khatkov, S. M. Shandarov, Optoelectronics, Data Processing and Instrumentation **2**, 80 (1986).

4.61 S. I. Stepanov, S. M. Shandarov, N. D. Khatkov, Sov. Phys. Solid State **29**, 1754 (1987).

4.62 P. Günter, M. Zgonik, Opt. Lett. **16**, 1826 (1991).

4.63 G. Pauliat, M. Mathey, G. Roosen, J. Opt. Soc. Am. **B 8**, 1942 (1991).

4.64 D. F. Nelson, M. Lax, Phys. Rev. Lett. **24**, 1187 (1970).

4.65 M. Zgonik, R. Schlesser, I. Biaggio, E. Voit, J. Tscherry, P. Günter, J. Appl. Phys. **74**, 1287 (1993).

4.66 M. Zgonik, P. Bernasconi, M. Duelli, R. Schlesser, P. Günter, M. H. Garrett, D. Rytz, Y. Zhu, X. Wu, Phys. Rev. **B 50**, 5941 (1994).

4.67 M. Zgonik, K. Nakagawa, P. Günter, J. Opt. Soc. Am. **B 12**, 1416 (1995).

4.68 G. Montemezzani, A. A. Zozulya, L. Czaia, D. Z. Anderson, M. Zgonik, P. Günter, Phys. Rev. **A 52**, 1791 (1995).

4.69 P. G. deGennes, J. Prost, in *The Physics of Liquid Crystals* (eds. J. Birman, S. F. Edwards, C. H. L. Smith, M. Rees) (Clarendon Press, Oxford, 1993).

4.70 M. Ingold, P. Günter, M. Schadt, J. Opt. Soc. Am. **B 7**, 2380 (1990).

4.71 I. C. Khoo, IEEE J. Quantum Electron. **22**, 1268 (1986).

4.72 E. V. Rudenko, A. V. Sukhov, JETP Lett. **59**, 142 (1994).

4.73 E. V. Rudenko, A. V. Sukhov, JETP **78**, 875 (1994).

4.74 I. C. Khoo, H. Li, Y. Liang, Opt. Lett. **19**, 1723 (1994).

4.75 G. P. Wiederrecht, B. A. Yoon, M. R. Wasielewski, Science **270**, 1794 (1995).

4.76 S. Bartkiewicz, A. Miniewicz, Advanced Materials for Optics and Electronics **6**, 219 (1996).

4.77 S. Bartkiewicz, A. Januszko, A. Miniewicz, J. Parka: Pure Appl. Opt. **5**, 799 (1996).

4.78 I. C. Khoo, B. D. Guenther, M. V. Wood, P. Chen, M. Y. Shih, Opt. Lett. **22**, 1229 (1997).

4.79 G. P. Wiederrecht, B. A. Yoon, W. A. Svec, M. R. Wasielewski, J. Am. Chem. Soc. **119**, 3358 (1997).

4.80 G. Cipparone, A. Mazzulla, F. P. Nicoletta, L. Luchetti, F. Simoni, Opt. Commun. **150**, 297 (1998).

4.81 R. MacDonald, P. Meindl, G. Chilaya, D. Sikharulidze, Opt. Commun. **150**, 195 (1998).

4.82 P. Prêtre, P. Kaatz, A. Bohren, P. Günter, B. Zysset, M. Ahlheim, M. Stähelin, F. Lehr, Macromolecules **27**, 5476 (1994).

4.83 M. C. J. M. Donckers, S. M. Silence, C. A. Walsh, F. Hache, D. M. Burland, W. E. Moerner, R. J. Twieg, Opt. Lett. **18**, 1044 (1993).

4.84 K. Meerholz, B. L. Volodin, Sandalphon, B. Kippelen, N. Peyghambarian, Nature **371**, 497 (1994).

4.85 B. Kippelen, Sandalphon, K. Meerholz, N. Peyghambarian, Appl. Phys. Lett. **68**, 1748 (1996).

4.86 G. C. Valley, J. Appl. Phys. **59**, 3363 (1986).

4.87 F. P. Strohkendl, J. M. C. Jonathan, R. W. Hellwarth, Opt. Lett. **11**, 312 (1986).

4.88 D. L. Staebler, J. J. Amodei, Ferroelectrics **3**, 107 (1972).

4.89 J. P. Huignard, J. P. Herriau, Appl. Opt. **16**, 1807 (1977).

4.90 J. O. White, A. Yariv, Appl. Phys. Lett. **37**, 5 (1980).

4.91 J. P. Huignard, J. P. Herriau, Appl. Opt. **17**, 2671 (1978).

4.92 P. Günter, Phys. Rep. **93**, 199 (1982).

4.93 J. P. Huignard, J. P. Herriau, P. Aubourg, E. Spitz, Opt. Lett. **4**, 21 (1979).

4.94 A. A. Kamshilin, M. P. Petrov, Sov. Tech. Phys. Lett. **6**, 144 (1980).
4.95 Y. Owechko, IEEE J. Quantum Electron. **25**, 619 (1989).
4.96 P. Günter, J. P. Huignard, *Photorefractive Materials and Their Applications II: Applications* (Springer-Verlag, Berlin, 1989).
4.97 S. I. Stepanov, Rep. Progr. Phys. **57**, 39 (1994).
4.98 M. Gower, D. Proch (eds.) *Optical Phase Conjugation* (Springer-Verlag, Berlin, 1994).
4.99 L. Solymar, D. J. Webb, A. Grunnet-Jepsen, *The Physics and Applications of Photorefractive Materials* (Clarendon Press, Oxford, 1996).
4.100 P. J. vanHeerden, Appl. Opt. **2**, 393 (1963).
4.101 B. Y. Zel'dovich, N. F. Pilipetsky, V. V. Shkunov, *Principles of Phase Conjugation* (Springer-Verlag, Berlin, 1985).
4.102 F. Laeri, T. Tschudi, J. Albers, Opt. Commun. **47**, 387 (1983).
4.103 A. Takada, M. Cronin-Golomb, Opt. Lett. **20**, 1459 (1995).
4.104 D. Z. Anderson, Opt. Lett. **11**, 56 (1986).
4.105 G. Montemezzani, G. Zhou, D. Z. Anderson, Opt. Lett. **19**, 2012 (1994).
4.106 A. B. Vasil'ev, L. D. Kislovskii, S. V. Mednikov, S. F. Chernov, L. A. Shuvalov, Sov. Phys. Solid State **26**, 543 (1984).
4.107 R. V. Johnson, A. R. Tanguay, Opt. Lett. **13**, 189 (1988).
4.108 J. J. Stankus, S. M. Silence, W. E. Moerner, G. C. Bjorklund, Opt. Lett. **19**, 1480 (1994).
4.109 G. P. Nordin, R. V. Johnson, J. A. R. Tanguay, J. Opt. Soc. Am. **A 9**, 2206 (1992).
4.110 R. De Vre, L. Hesselink, J. Opt. Soc. Am. **A 13**, 285 (1996).
4.111 P. M. Lundquist, C. Poga, R. G. DeVoe, Y. Jia, W. E. Moerner, M. P. Bernal, H. Coufal, R. K. Grygier, J. A. Hoffnagle, C. M. Jefferson, R. M. Macfarlane, R. M. Shelby, G. T. Sincerbox, Opt. Lett. **21**, 890 (1996).
4.112 C. Poga, P. M. Lundquist, V. Lee, R. M. Shelby, R. J. Twieg, D. M. Burland, Appl. Phys. Lett. **69**, 1047 (1996).
4.113 D. M. Burland, R. G. Devoe, C. Geletneky, Y. Jia, V. Y. Lee, P. M. Lundquist, C. R. Moylan, C. Poga, R. J. Twieg, R. Wortmann, in Materials for Nonlinear Optics Val Thorens, France, 14–18 January 1996. Pure and Applied Optics **5**, 513 (1996).
4.114 A. Grunnet-Jepsen, C. L. Thompson, W. E. Moerner, Opt. Commun. **145**, 145 (1998).
4.115 S. I. Stepanov, V. V. Kukikov, M. P. Petrov, Optics Commun. **44**, 19 (1982).
4.116 A. Grunnet-Jepsen, C. L. Thompson, W. E. Moerner, Science **277**, 549 (1997).
4.117 L. Yu, W. Chan, Y. Chen, Z. Peng, Z. Bao, D. Yu, Proc. SPIE **2025**, 268 (1993).
4.118 C. S. Weaver, J. W. Goodman, Appl. Opt. **5**, 1248 (1966).
4.119 A. VanderLugt, IEEE Trans. Inf. Theory IT **10**, 139 (1964).
4.120 P. Amrhein, P. Günter, J. Opt. Soc. Am. **B 12**, 2387 (1990).
4.121 P. Bernasconi, G. Montemezzani, M. Wintermantel, I. Biaggio, P. Günter, Opt. Lett. **24**, 199 (1999).
4.122 S. C. W. Hyde, N. P. Barry, R. Jones, J. C. Dainty, P. M. W. French, M. B. Klein, B. A. Wechsler, Opt. Lett. **20**, 1331 (1995).
4.123 S. C. W. Hyde, R. Jones, N. P. Barry, J. C. Dainty, P. M. W. French, K. M. Kwolek, D. D. Nolte, M. R. Melloch, IEEE Journal Of Selected Topics In Quantum Electronics **2**, 965 (1996).

4.124 N. P. Barry, R. Jones, S. C. W. Hyde, J. C. Dainty, P. M. W. French, Electronics Letters **33**, 1732 (1997).
4.125 B. Kippelen, S. R. Marder, E. Hendrickx, J. L. Maldonado, G. Guillemet, B. L. Volodin, D. D. Steele, Y. Enami, Sandalphon, Y. J. Yao, J. F. Wang, H. Rockel, L. Erskine, N. Peyghambarian, Science **279**, 54 (1998).
4.126 D. D. Steele, B. L. Volodin, O. Savina, B. Kippelen, N. Peyghambarian, H. Rockel, S. R. Marder, Opt. Lett. **23**, 153 (1998).
4.127 D. D. Nolte, D. H. Olson, G. E. Doran, W. H. Knox, A. M. Glass, J. Opt. Soc. Am. **B 7**, 2217 (1990).
4.128 A. Partovi, A. M. Glass, D. H. Olson, G. J. Zydzik, K. T. Short, R. D. Feldmann, R. F. Austin, Opt. Lett. **17**, 655 (1992).
4.129 I. Biaggio, J. P. Partanen, B. Ai, R. J. Knize, R. W. Hellwarth, Nature **371**, 318 (1994).
4.130 S. Follonier, C. Bosshard, F. Pan, P. Günter, Opt. Lett. **21**, 1655 (1996).
4.131 D. Wright, M. A. Díaz-García, J. D. Casperson, M. DeClue, W. E. Moerner, R. J. Twieg, Appl. Phys. Lett. **73**, 1490 (1998).
4.132 G. Montemezzani, P. Rogin, M. Zgonik, P. Günter, Opt. Lett. **18**, 1144 (1993).
4.133 P. Delaye, J. M. C. Jonathan, G. Pauliat, G. Roosen, in Materials for Nonlinear Optics, Val Thorens, France, 14–18 January 1996. Pure and Applied Optics **5**, 541 (1996).
4.134 R. L. Townsend, J. T. LaMacchia, J. Appl. Phys. **41**, 5188 (1970).
4.135 P. Günter, U. Flückiger, J. P. Huignard, F. Micheron, Ferroelectrics **13**, 297 (1976).
4.136 B. Orlowski Krätzig, Opt. Commun. **35**, 45 (1980).
4.137 J. J. Amodei, D. L. Staebler, A. W. Stephens, Appl. Phys. Lett. **18**, 507 (1971).
4.138 J. B. Thaxter, Appl. Phys. Lett. **15**, 210 (1969).
4.139 M. Peltier, F. Micheron, J. Appl. Phys. **48**, 3683 (1977).
4.140 L. H. Lin, Proc. IEEE **57**, 252 (1969).
4.141 E. Moya, L. Contreras, C. Zaldo, J. Opt. Soc. Am. **B 5**, 1737 (1988).
4.142 G. Montemezzani, S. Pfändler, P. Günter, J. Opt. Soc. Am. **B 9**, 1110 (1992).
4.143 Y. H. Xu, Z. Q. Huang, W. L. Li, H. Wang, D. R. Zhu, H. C. Chen, Ferroelectrics **92**, 211 (1989).
4.144 A. A. Grabar, R. I. Muzhikash, A. D. Kostyuk, Y. M. Visochanskii, Sov. Phys. Solid State **33**, 1314 (1991).
4.145 F. Micheron, A. Hermosin, G. Bismuth, J. Nicolas, C. R. Acad. Sci. Paris **274**, 361 (1972).
4.146 A. M. Glass, A. M. Johnson, D. H. Olson, W. Simpson, A. A. Ballmann, Appl. Phys. Lett. **44**, 948 (1984).
4.147 M. B. Klein, Opt. Lett. **9**, 350 (1984).
4.148 J. Strait, A. M. Glass, Appl. Opt. **25**, 338 (1986).
4.149 J. Strait, A. M. Glass, J. Opt. Soc. Am. **B 3**, 342 (1986).
4.150 C. Medrano, E. Voit, P. Amrhein, P. Günter, J. Appl. Phys. **64**, 4668 (1988).
4.151 H. Kurz, E. Krätzig, W. Keune, H. Engelman, U. Gonser, B. Dichsler, A. Räuber, Appl. Phys. **12**, 355 (1977).
4.152 W. Keune, S. K. Date, I. Dézsi, U. Gonser, J. Appl. Phys. **46**, 3914 (1975).
4.153 G. W. Ross, P. Hribek, R. W. Eason, M. H. Garret, D. Rytz, Opt. Commun. **101**, 60 (1993).

4.154 B. A. Wechsler, M. B. Klein, C. C. Nelson, R. N. Schwartz, Opt. Lett. **19**, 536 (1994).

4.155 M. Ewart, R. Ryf, C. Medrano, H. Wuest, M. Zgonik, P. Günter, Opt. Lett. **22**, 781 (1997).

4.156 S. Brülisauer, D. Fluck, P. Günter, L. Beckers, C. Buchal, J. Opt. Soc. Am. **B 13**, 2544 (1996).

4.157 D. D. Nolte, R. M. Melloch, in *Photorefractive Effects and Materials* (ed. D. D. Nolte) (Kluwer Academic Publishers, Boston, 1995), p. 373.

4.158 G. H. Haertling, C. E. Land, J. Am. Cer. Soc. **54**, 1 (1971).

4.159 A. Sternberg, Ferroelectrics **131**, 13 (1992).

4.160 F. Micheron, C. Mayeux, A. Hermosin, J. Nicolas, J. Am. Ceram. Soc. **57**, 306 (1974).

4.161 J. M. Rouchon, F. Micheron, Czech. J. Phys. **B 25**, 575 (1975).

4.162 A. Krumins, Ferroelectrics **131**, 105 (1992).

4.163 A. Krumins, R. A. Rupp, K. Kerperin, Ferroelectrics **80**, 281 (1988).

4.164 F. Micheron, J. M. Rouchon, M. Vergnolle, Appl. Phys. Lett. **24**, 605 (1974).

4.165 K. Sutter, J. Hulliger, P. Günter, Solid State Commun. **74**, 867 (1990).

4.166 K. Sutter, J. Hulliger, R. Schlesser, P. Günter, Opt. Lett. **18**, 778 (1993).

4.167 K. Sutter, P. Günter, J. Opt. Soc. Am. **B 7**, 2274 (1990).

4.168 G. Knöpfle, C. Bosshard, R. Schlesser, P. Günter, IEEE J. Quantum Electron. **30**, 1303 (1994).

4.169 K. Meerholz, R. Bittner, Y. Denardin, C. Brauchle, E. Hendrickx, B. L. Volodin, B. Kippelen, N. Peyghambarian, Adv. Mat. **9**, 1043 (1997).

4.170 S. Ducharme, J. C. Scott, R. J. Twieg, W. E. Moerner, Phys. Rev. Lett. **66**, 1846 (1991).

4.171 G. G. Malliaras, V. V. Krasnikov, H. J. Bolink, G. Hadziioannou, Appl. Phys. Lett. **66**, 1038 (1995).

4.172 A. Grunnet-Jepsen, D. Wright, B. Smith, M. S. Bratcher, M. S. DeClue, J. S. Siegel, W. E. Moerner, Chem. Phys. Lett. **291**, 553 (1998).

4.173 Y. Zhang, Y. Cui, P. N. Prasad, Phys. Rev. **B 46**, 9900 (1992).

4.174 O. Zobel, M. Eckl, P. Strohriegl, D. Haarer, Adv. Mater. 7, 911 (1995).

4.175 G. G. Malliaras, V. V. Krasnikov, H. J. Bolink, G. Hadziioannou, Appl. Phys. Lett. **65**, 262 (1994).

4.176 A. M. Cox, R. D. Blackburn, D. P. West, T. A. King, F. A. Wade, D. A. Leigh, Appl. Phys. Lett. **68**, 2801 (1996).

4.177 A. Grunnet-Jepsen, C. L. Thompson, R. J. Twieg, W. E. Moerner, Appl. Phys. Lett. **70**, 1515 (1997).

4.178 F. Wang, Z. Chen, Q. Gong, Y. Chen, H. Chen, Solid State Commun. **106**, 299 (1998).

4.179 E. Hendrickx, J. Herlocker, J. L. Maldonado, S. R. Marder, B. Kippelen, A. Persoons, N. Peyghambarian, Appl. Phys. Lett. **72**, 1679 (1998).

4.180 P. M. Lundquist, R. Wortmann, C. Geletneky, T. R. J. Twieg, M. Jurich, V. Y. Lee, G. R. Moylan, D. M. Burland, Science **274**, 1182 (1996).

4.181 L. Wang, Y. Zhang, T. Wada, H. Sasabe, Appl. Phys. Lett. **69**, 728 (1996).

4.182 Y. D. Zhang, L. M. Wang, T. Wada, H. Sasabe, Appl. Phys. Lett. **70**, 2949 (1997).

4.183 T. Wada, Y. Zhang, H. Sasabe, RIKEN Review. No. 15, 334 (1997).

4.184 Y. M. Chen, Z. H. Peng, W. K. Chan, L. P. Yu, Appl. Phys. Lett. **64**, 1195 (1994).

4.185 L. Yu, Y. Chen, W. K. Chan, Z. Peng, Appl. Phys. Lett. **64**, 2489 (1994).

4.186 Z. H. Peng, A. R. Gharavi, L. P. Yu, Appl. Phys. Lett. **69**, 4002 (1996).

4.187 L. Yu, Y. M. Chen, W. K. Chan, J. Phys. Chem. **99**, 2797 (1995).

4.188 I. C. Khoo, IEEE J. Quantum Electr. **32**, 525 (1996).

4.189 A. Miniewicz, S. Bartkiewicz, J. Parka, Opt. Commun. **149**, 89 (1998).

4.190 I. C. Khoo, Opt. Lett. **20**, 2137 (1995).

4.191 H. Ono, N. Kawatsuki, Opt. Lett. **22**, 1144 (1997).

4.192 A. Golemme, B. L. Volodin, E. Kippelen, N. Peyghambarian, Opt. Lett. **22**, 1226 (1997).

4.193 A. M. Glass, J. Strait, in *Photorefractive Materials and Their Applications I* (eds. P. Günter, J.-P. Huignard) (Springer-Verlag, Berlin, 1988), p. 237.

4.194 Y. Ding, Z. G. Zhang, H. J. Eichler, D. Z. Shen, X. Y. Ma, J. Y. Chen, Opt. Lett. **20**, 686 (1995).

4.195 E. Krätzig, R. Orlowski, Appl. Phys. **15**, 133 (1978).

4.196 R. Jungen, G. Angelow, F. Laeri, C. Grabmaier, Appl. Phys. **A 55**, 101 (1992).

4.197 F. Laeri, R. Jungen, G. Angelow, U. Vietze, T. Engel, M. Wuertz, D. Hilgenberg, Appl. Phys. **B 61**, 351 (1995).

4.198 J. J. Xu, X. F. Yue, R. A. Rupp, Phys. Rev. **B 54**, 16618 (1996).

4.199 H. Kurz, Philips Tech. Rev. **37**, 109 (1977).

4.200 R. Orlowski, E. Krätzig, H. Kurz, Opt. Commun. **20**, 171 (1977).

4.201 P. Günter, F. Micheron, Ferroelectrics **18**, 27 (1978).

4.202 E. Voit, M. Z. Zha, P. Amrhein, P. Günter, Appl. Phys. Lett. **51**, 2079 (1987).

4.203 E. Krätzig, F. Welz, R. Orlowski, V. Doormann, M. Rosenkranz, Sol. State Commun. **34**, 817 (1980).

4.204 M. H. Garrett, J. Y. Chang, H. P. Jenssen, C. Warde, Opt. Lett. **17**, 103 (1992).

4.205 R. A. Vasquez, F. R. Vachss, R. R. Neurgaonkar, M. D. Ewbank, J. Opt. Soc. Am. **B 8**, 1932 (1991).

4.206 G. C. Valley, M. B. Klein, Opt. Engineering **22**, 704 (1983).

4.207 A. Partovi, J. Millerd, E. Garmire, M. Ziari, W. H. Steier, S. B. Trivedi, M. B. Klein, Appl. Phys. Lett. **57**, 846 (1990).

4.208 A. Partovi, A. M. Glass, D. H. Olson, G. J. Zydzik, H. M. O'Brien, T. H. Chiu, W. H. Knox, Appl. Phys. Lett. **62**, 464 (1993).

4.209 P. Yeh, Appl. Opt. **26**, 602 (1987).

5 Photorefractive Memories for Optical Processing

M. Duelli, G. Montemezzani, M. Zgonik, and P. Günter

The field of three-dimensional optical storage has lately witnessed a tremendous increase of interest from both the scientific and the industrial communities. Volume holographic storage was proposed by van Heerden [5.1] shortly after the introduction of the laser and interest in this technology has continued at varying degrees of intensity since. Holography potentially combines the advantages of high storage densities and ease of parallel data readout. The storage of multiple pages of information in the same volume and the parallel readout of many bits promise fast data access times and high data bandwidth. Furthermore, in holography the information is distributed throughout the storage medium and is therefore less sensitive to local material imperfections. If Fourier holograms are stored in holographic memories, the effects of imperfections in the medium are distributed throughout the page and are not linked to individual pixels; thus valuable bits of data are not lost, only the signal-to-noise ratio is diminished. This is in contrast to magnetic and conventional optical recording, where an individual bit of information is represented by a highly localized change in some physical property. This advantage may be less important in holographic digital data storage when additional redundancy of information storage and error correction can be easily implemented.

Optical data storage currently finds commercial viability mainly in two-dimensional low-cost data distribution on CD-ROM (optical disc) and in quasi-archival storage on removable disk read–write drives (magneto-optical disk). Both of these technologies were made possible by the development of low-cost diode lasers in the past twenty years. An example of a quasi-three-dimensional optical data storage is the new digital versatile disc (DVD) standard, where the improvement in storage capacity with respect to CD-ROMs is due to the smaller bit size (shorter wavelength), reduction of the storage area sacrified for tolerances, and the use of more than one layer as a storage plane. Volume holographic storage is notable for its complete absence from the data storage market, it has been incorporated only in a limited way into specialized commercial systems until now.

One of the reasons for the renewed interest in holographic data storage is the improved input–output (I/O) technologies, such as liquid crystal spatial light modulators and integrated detector arrays, that are capable of delivering high data bandwidth. The absence of these elements has put a strong brake on the development of this technology in the past. The next few years will tell us whether volumetric data storage will allow us to overcome the current I/O bottleneck in digital computers, where the processor speed far exceeds the ability of conventional rotating disk devices to import or export the data needed for manipulations.

Other than for high-speed retrieval of data in a holographic memory system, the high parallelism of optical beams is also very attractive for computationally intensive data processing tasks, such as pattern recognition, edge enhancement or spatial filtering, where vast amounts of data (a two-dimensional image is considered as a data page) have to be processed. Optical pattern recognition systems are being developed for many applications, such as identification of human pupils and partial or distorted fingerprints, robotic vision, or rapid searching of vast databases. Most of these applications make use of coherent light and the Fourier transform (FT) capability of lenses [5.2].

In particular, implementations of neural network models [5.3–4], which are based on the structure of the human brain, are quite well suited to utilizing the natural properties of light. These models of computation devices consist of many simple processing units or "neurons", which communicate with each other via interconnection weights. Neural networks are successful in solving ill-defined problems, e.g. problems which are partially random, ill-posed or combinatorially complex, so that an algorithm cannot be formulated. These problems typically involve processing vast amounts of information, preferably in parallel, and often result in answers which need not be of high numerical accuracy. Communication and parallelism is more important than the computing power of individual neurons. Ideally, the architecture of a neural network computer should reflect the highly parallel, associative and nonlinear analog nature of the neural network models. One approach to achieving such an architecture is to use optics for parallel communication and massive interconnection between a large number of processing units represented by planes of pixels having nonlinear optical properties.

A sophisticated aspect of analog optical computing that is based on neural network architectures is Associative Memory (AM) [5.3]. In contrast to usual computer memories, where a separate unique address points to each stored datum (location addressing), in associative memories the output data is recalled directly using the input data itself (content addressing), without the use of separate addresses. Data can flow through the system, exciting chains of associations until a decision is reached in a global and parallel manner. Associative memories also have error correction properties in that a complete undistorted set of data can be retrieved using a distorted or partial version of input data.

In this chapter we discuss holographic memory systems with particular emphasis on their applications to optical pattern recognition and associative memories. The problematics of volume digital data storage for computer memories is scratched only at the surface, the interested reader is reminded of some recent excellent reviews on this topic [5.5–9]. A stronger attention is given to implementations of optical processing and associative memory systems that use volume holograms recorded with the photorefractive effect, which is described in Chap. 4. For lack of space, optical associative memories based on matrix–vector multiplication (see for instance [5.10–13]) or more general optical neural network architectures [5.14] capable of learning will not be discussed here.

The structure of this chapter is as follows. The first part (Sects. 5.1–5.3) presents a general introduction to holographic data storage, optical pattern recognition and holographic associative memories, respectively. In the second part (Sects. 5.4–5.6) the same three topics are discussed by considering in detail implementations based on photorefractive materials. Section 5.1 reviews some recent advances in volumetric data storage, the storage capacity is discussed in relation to system architecture and multiplexing schemes. Section 5.2 focuses on optical pattern recognition systems and discusses specific features of architectures based on volume holograms. Section 5.3 gives a general discussion of linear and nonlinear holographic associative memories. Section 5.4 considers photorefractive materials as volume storage media. We discuss recording schemes, storage capacity and the possibilities of hologram fixing and nondestructive readout. Implementations of optical correlators using photorefractive crystals are discussed in Sect. 5.5, while photorefractive all-optical associative memories including a deeper discussion of a system developed in our laboratory are presented in the last section. For a detailed discussion of the photorefractive effect itself, the reader is referred to Chap. 4.

5.1 Volumetric Optical Data Storage

In this section we discuss the system architecture and the storage density limitations of holographic memories. This discussion concentrates on digital storage of information, i.e. each data page is composed of pixels which are either "on" or "off", as it is the most interesting case for very large capacity memories. However, most of the conclusions are valid also for the case of analog storage of information which is necessary to perform optical information processing, as described in Sect. 5.2.

5.1.1 Light Diffraction Volume Gratings

To analyze the storage capabilities of holographic volume media, we have to know the diffraction properties of volume refractive index or absorption gratings. Consider a thick holographic recording medium where a grating has been recorded by interference of two plane waves as it is shown in Fig. 5.1.

The grating with grating vector $K = k_1 - k_2$ can be read out using a probe beam I_i that may have the same or a different wavelength; the phase-matching conditions for the case of different wavelength are visualized in Fig. 5.1. For thick gratings one diffraction order can be observed (Bragg

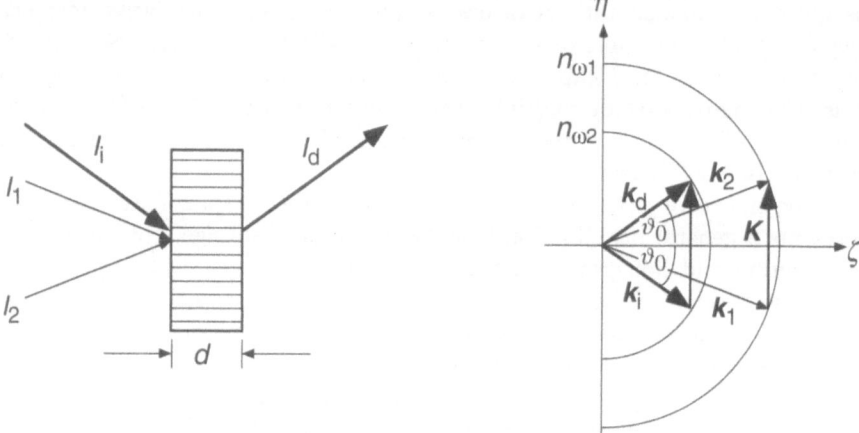

Fig. 5.1. Schematic of holographic grating recording and readout using a readout beam of longer wavelength than the recording beams. The right side shows the wave vector diagram for isotropic Bragg diffraction. I_1, I_2, I_i and I_d are the intensities of the two recording waves and the incident and diffracted readout waves, respectively. k_1, k_2, k_i and k_d are the corresponding wave vectors. K is the grating vector and ϑ_0 is the Bragg angle for readout at frequency ω_2 inside the crystal

diffraction regime), whereas for thin gratings the light is diffracted into many diffraction orders (Raman–Nath regime). The conditions to be fulfilled in order to have a thick hologram have been put forward by Gaylord and Moharam [5.15], they are

$$Q = \frac{2\pi\lambda d}{n\Lambda^2} > 1, \tag{5.1}$$

and

$$\rho = \frac{\lambda^2}{n\Lambda^2\sigma} \geq 10 \tag{5.2}$$

with λ the vacuum wavelength of the reading beam, d the interaction length (thickness of the medium), n the refractive index, and $\Lambda = 2\pi/|K|$ the grating spacing. The quantity σ is proportional to the strength (amplitude) of the grating modulation. For phase gratings one has $\sigma = \Delta n$, where Δn is the amplitude of the periodic refractive index modulation; for absorption gratings $\sigma = \Delta\alpha\lambda/2\pi$, where $\Delta\alpha$ is the amplitude of the absorption constant modulation. Figure 5.2 visualizes the conditions for a thick grating in the case of a phase grating only. The material thickness d as well as the refractive index modulation amplitude Δn are shown as a function of the grating spacing Λ for the conditions corresponding to $Q = 1$ and $\rho = 10$. In order to consider a grating as thick, the interaction length and the grating modulation amplitude mut both be in the gray areas. Note that with condition (5.2) the

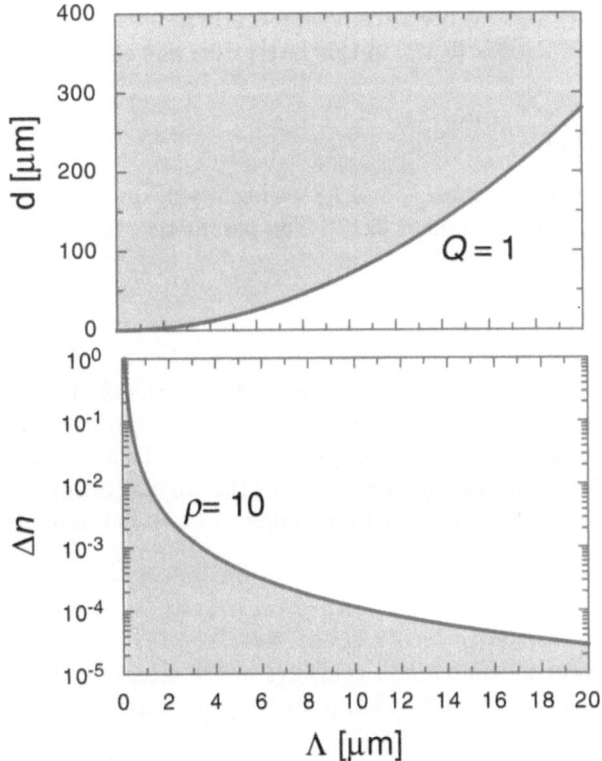

Fig. 5.2. Material thickness (interaction length) d and refractive index modulation amplitude Δn as a function of the grating spacing. The gray area give the regions where conditions (5.1) and (5.2) for a thick grating are fulfilled. Parameters: wavelength $\lambda = 500\,\mathrm{nm}$, refractive index $n = 2.2$

maximum amplitude Δn for which a grating can still be considered to be thick is approximately 10^{-4} for a grating spacing of $10\,\mu m$.

In thick gratings light is only diffracted if the angle of incidence is close to the Bragg angle. This angle is derived from the phase matching condition $k_\mathrm{d} - k_\mathrm{i} = K$. For the symmetric situation of Fig. 5.1 the value of the angle of incidence inside the crystal for which the Bragg-condition is exactly fulfilled is given by

$$\vartheta_0 = \arcsin\left(\frac{\lambda}{2n\Lambda}\right). \tag{5.3}$$

The most important property of a thick hologram is its diffraction efficiency η, that is the relative power being diffracted into the first (and single) Bragg order with respect to the total incident power in the material. A general form for η for a fixed transmission holograms with a sinusoidal refractive index modulation has been derived by Kogelnik [5.16] for isotropic materials using a coupled wave theory and has been extended recently to the anisotropic case by Montemezzani and Zgonik [5.17], in this latter case one obtains

$$\eta = \frac{I_d(d)}{I_i(0)} = \frac{\sin^2 \sqrt{\nu^2 + \xi^2}}{(1 + \xi^2/\nu^2)} e^{-2\alpha d}, \tag{5.4}$$

where it was assumed that both waves (I_d and I_i) are absorbed equally strong with an amplitude absorption constant α [5.17]. The parameter ν is proportional to the grating strength and is defined as

$$\nu = \frac{\pi \Delta n \, d}{\lambda \cos \theta}, \tag{5.5}$$

where the angle θ is obtained from $\cos \theta = (\cos \theta_i \cos \theta_d)^{1/2}$, that is from a geometrical average of the projection cosines of the Poynting vector directions of the incident (θ_i) and diffracted (θ_d) waves onto the direction of the medium surface normal ζ of Fig. 5.1. The parameter ξ describes the angular mismatch of the incident wave with respect to perfect Bragg angle of (5.3). It is defined as

$$\xi = \frac{\Delta k \, d}{2}, \tag{5.6}$$

where Δk is the wave vector mismatch and is always in the direction of the surface normal ζ. For the situation of Fig. 5.1, (5.6) is approximated by

$$\xi = \frac{\Delta \vartheta \, K \, d}{2}, \tag{5.7}$$

and is expressed in terms of the grating vector magnitude $K = |\boldsymbol{K}|$ and of the deviation $\Delta \vartheta = \vartheta - \vartheta_0$ of the internal angle for the wave vector $\boldsymbol{k}_i$ from the exact Bragg angle. In the case of perfect Bragg matching ($\xi = 0$), (5.4) reduces to the well-known formula

$$\eta = \sin^2 \left(\frac{\pi \Delta n \, d}{\lambda \cos \theta} \right) e^{-2\alpha d}. \tag{5.8}$$

It should be noted that (5.4) through (5.6) as well as (5.8) are valid in general transmission geometries and also include the case of anisotropic Bragg diffraction, provided that the refractive index change amplitude Δn is calculated in a consistent way for the studied beam interaction [5.17]. Equations (5.3) and (5.7), in contrast, are specific to the symmetric geometry shown in Fig. 5.1.

The coupled wave theory leading to the above equations for the diffraction efficiency of transmission phase gratings can also be applied to the case of absorption or mixed absorption and phase gratings, as well as to the case of reflection-type holograms. Expressions describing all these cases for isotropic and anisotropic materials can be found in [5.16–18].

5.1.2   Hologram Multiplexing Methods

Utilization of the full data storage capacity of volume recording media requires hologram multiplexing. Beside the possibility of storing different holograms at different locations in the volume (spatial multiplexing), three methods have been proposed and tested to store several pages of information in the same region of the holographic medium: angular multiplexing, wavelength multiplexing and phase multiplexing. In practice a combination of spatial multiplexing with one of the other multiplexing techniques seems to be ideal. Stacks of pages will be recorded in relatively small regions of the medium, because this approach allows stacks to be recorded and erased without affecting other regions of the storage volume. In the following, angular, phase and wavelength multiplexing will be discussed briefly.

A   Angular Multiplexing

For angular multiplexing each page of information is adressed by a particular angle at which the reference wave (usually a plane wave) illuminates the holographic material. In the recording step, object pages (having all the same average wave vector k_S) are recorded each using a slightly different wave vector direction k_R for the reference wave (see Fig. 5.3a) [5.1]. To read out a specific stored object, the hologram is illuminated with the reference beam at the angle which was used to create the hologram. The holographic systems discussed in the second part of this chapter are mostly based on this multiplexing technique.

The good angle selectivity of angular multiplexed holograms is a direct consequence of using thick gratings, the minimum angle separation between neighboring reference waves can be easily derived from (5.4). Let's consider

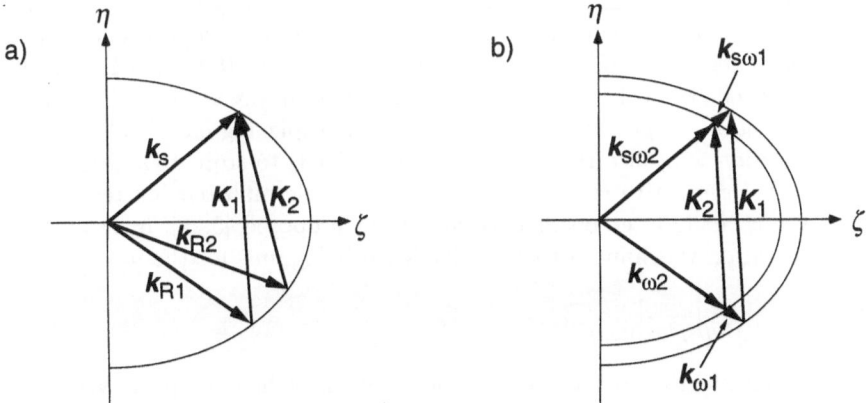

Fig. 5.3. Wave vector diagram of (a) angle multiplexing by deflecting the reference beams and (b) wavelength multiplexing

a hologram with a moderate grating strength ($\nu \ll 1$) as encountered if multiplexing a large number of pages. Then the diffraction efficiency falls to half of its peak value if the mismatch parameter is $\xi \cong 1.39$. With (5.7) and $K = 4\pi n \sin \vartheta_0 / \lambda$ one finds that for a symmetric geometry as in Fig. 5.1 this value of ξ corresponds to an angular mismatch of $\Delta \vartheta = 2.78\lambda/(4\pi n d \sin \vartheta_0)$ and is therefore inversely proportional to the grating thickness d. For an internal incidence angle $\vartheta_0 = 20°$, $\lambda = 0.5\,\mu\text{m}$, $n = 2.2$ and $d = 2\,\text{mm}$ one has $\Delta \vartheta \approx 0.004°$, so that approximately 1000 holograms can be recorded and independently read out by a total angular displacement of the order of $5°$ inside the medium. It should be noted that most implementations of digital holographic storage by angle multiplexing use an interaction angle of nearly $90°$ between the beams. This has the big advantage of a reduced scattering noise as well as a slightly better angle resolution.

Besides the above-described method of changing the reference beam angle of incidence, angular multiplexing can also be achieved by keeping a fixed crossing angle between the recording beams and rotating the storage medium between consecutive recordings. Each hologram in this orientation multiplexing is again characterized by a differently oriented main grating vector $\boldsymbol{K}$. The readout address of each hologram is given by the orientation of the probe beam relative to the sample. Peristrophic multiplexing [5.19], a method for which the hologram is rotated around the surface normal of the sample, is only one of the many possible variations. Although angular multiplexing requires accurate control of each reference angle for good reconstruction of the stored objects it is the most often used and investigated multiplexing technique.

The angular multiplexing technique alone was shown by Mok et al. to allow storage of 500 objects or 5000 edge enhanced objects (320×220 pixels) in a photorefractive $LiNbO_3$ crystal [5.20–21]. The reference beam direction was adjusted with an acousto-optic deflector in conjunction with a telescope. The holograms showed a diffraction efficiency of approximately 4×10^{-6}. Angular multiplexing together with spatial-multiplexing, i.e. the holograms are stored in different regions of the volume, has been used to store 750 (128×128 pixels, diffraction efficiency 1×10^{-4}) [5.22] and 10000 images [5.23] by Tao et al. and Burr et al., respectively. Recently, angle multiplexing has also been used in connection with electrical fixing (see Sect. 5.5.3) to store 1000 holograms in the photorefractive crystal SBN:75 [5.24]. Angle multiplexing techniques that use a single laser beam have also been proposed [5.25]; experimental demonstration was limited to 10 multiplexed holograms for the moment.

B Phase Multiplexing

Phase multiplexing uses a more complex reference than the plane wave-like beams employed in angle multiplexing. The reference wave may be thought as a linear combination of all possible reference beams used in the angular multiplexing scheme. The addresses of the stored images reside in the adjustable phases of different parts of the reference wave all overlapping with the signal wave in the holographic medium. The phase distortions of the reference beam can be random, as introduced by using a ground glass plate [5.26–27], or deterministic, as obtained by using a phase-only spatial light modulator. Multiple holograms may be superimposed without significant cross-talk by choosing a set of orthogonal functions for the phase pattern, i.e. $e^{i\varphi^j(x,y)}$, where $\varphi^j(x,y)$ describes the phase distribution of the jth reference wave which is related to the jth image. If the space x, y is divided into N discrete pixels, as obtained for instance by using a phase-only spatial light modulator, the orthogonality condition among the phase patterns is expressed by

$$\sum_{n=1}^{N} e^{i\varphi_n^j(x,y)} e^{i\varphi_n^m(x,y)} = 0, \quad \text{if } j \neq m, \tag{5.9}$$

where n is an index related to the position of a given pixel in the phase pattern. For readout the phase code corresponding to the desired object wave is used, and with efficient code selection only the desired object will significantly contribute to image reconstruction. The fixed alignment geometry allows faster access times than for angular multiplexing. However, the storage capacities of experimental implementations are rather low, partly due to the lack of high-precision phase-only spatial light modulators which are able to provide complex wavefront structures.

Denz et al. [5.28–29] presented a deterministic phase encoding method where N objects are stored with N pure and deterministic phase coded reference beams. Using this method the image beam simultaneously interferes with all N reference beams inside the storage medium. Each of these references is encoded in phase and they are angularly spaced by more than the Bragg angle. It has been shown that if N is a power of 2, N orthogonal phase addresses can be constructed by choosing the phases to be equal to 0 or π and using the Walsh–Hadamard transform [5.30]. This method of binary phase encoding has been experimentally verified by multiplexing 64 images (128×128 pixels) in a photorefractive $BaTiO_3$ crystal [5.31–32]. The access time to retrieve any of the stored images was 150 µs. By using a subsampling of phase codes, holograms stored by phase multiplexing can also be used for optical addition, substraction and inversion of images [5.33]. In an earlier experiment, sinusoidally phase-modulated reference beams [5.34] were also used for the storage of seven different holograms.

An alternative multiplexing technique, which may be seen as a a special case of phase multiplexing, has been proposed recently [5.35]. This shift multiplexing technique is based on a set of reference waves converging on a region of the recording medium. The medium is shifted by a few micrometers between two exposures. In the local frame of the medium the phase distribution of the reference rays is changed as a result of the shift, thus allowing discrimination of different hologram pages. Shift multiplexing may be implemented using various kinds of reference waves, including spherical waves [5.36] and highly non-collimated Gaussian wave. It offers the potential advantages of simplicity and better compatibility with today's spinning disk technologies. Multiplexing of a limited number of holograms with a required hologram-to-hologram shift of the order of 5 μm has been reported [5.35].

C   Wavelength Multiplexing

For wavelength multiplexing each hologram is recorded using a different laser wavelength. The direction of the reference and signal wave outside the medium remains constant [5.1]. With respect to angular multiplexing there is the advantage that no readjustment of angles is required, however there is a necessity for a tunable laser source. The wave-vector diagram for frequency multiplexing is shown in Fig. 5.3b for the case where the writing beams are symmetric with respect to the surface normal of the storage medium. A hologram written by interference of the reference wave with wave vector $\boldsymbol{k}_{\omega 1}$ and the object wave ($\boldsymbol{k}_{s,\omega 1}$) forms an index grating with grating vector $\boldsymbol{K}_1$. By illuminating the crystal with light from the same direction but of different frequency ω_2, no diffracted light results, because the vector $\boldsymbol{k}_{\omega 2}$ is not Bragg-matched.

Reflection gratings are often used in wavelength multiplexing because of their efficient use of k-space. The grating vectors are oriented nearly parallel to the reference beam and are of magnitude approximately twice that of the reference beam. Therefore, in this case the grating vectors may span a greater portion of k-space than grating vectors arising from oblique beams [5.37]. For reflection gratings recorded by counterpropagating beams the number of holograms N that can be distinguished in a medium of thickness d and within a wavelength range $\lambda - \Delta\lambda/2 \ldots \lambda + \Delta\lambda/2$ can be easily estimated. As for the case of transmission gratings, the diffraction efficiency of a weak reflection holograms falls to half its maximum value if the mismatch parameter of (5.6) is $\xi \approx 1.39$ [5.17], so that

$$N \cong \frac{2\pi n}{\lambda^2} \frac{d}{2.78} \Delta\lambda. \tag{5.10}$$

With this criterion, (5.10) tells us that about 400 holograms can be distinguished for a $d = 2\,\text{mm}$ thick medium with refractive index $n = 2.2$ and a total wavelength tuning of $\pm 5\,\text{nm}$ around a central wavelength of 500 nm. It must be remembered that, if broad bandwidth lasers are used, the maximum number of multiplexed pages might be limited by the laser bandwidth and not by the material thickness [5.38].

In a first experimental demonstration using a tunable solid state laser diode 26 images were multiplexed by tuning the wavelength over the range 670±6 nm [5.39]. More recently, a digital wavelength-multiplexed holographic data storage system based on the same tunable laser diode technology has been demonstrated [5.40]. A 60 kbyte data file split into 27 wavelength multiplexed pages was stored and retrieved. Instead of a counterpropagating (180°)

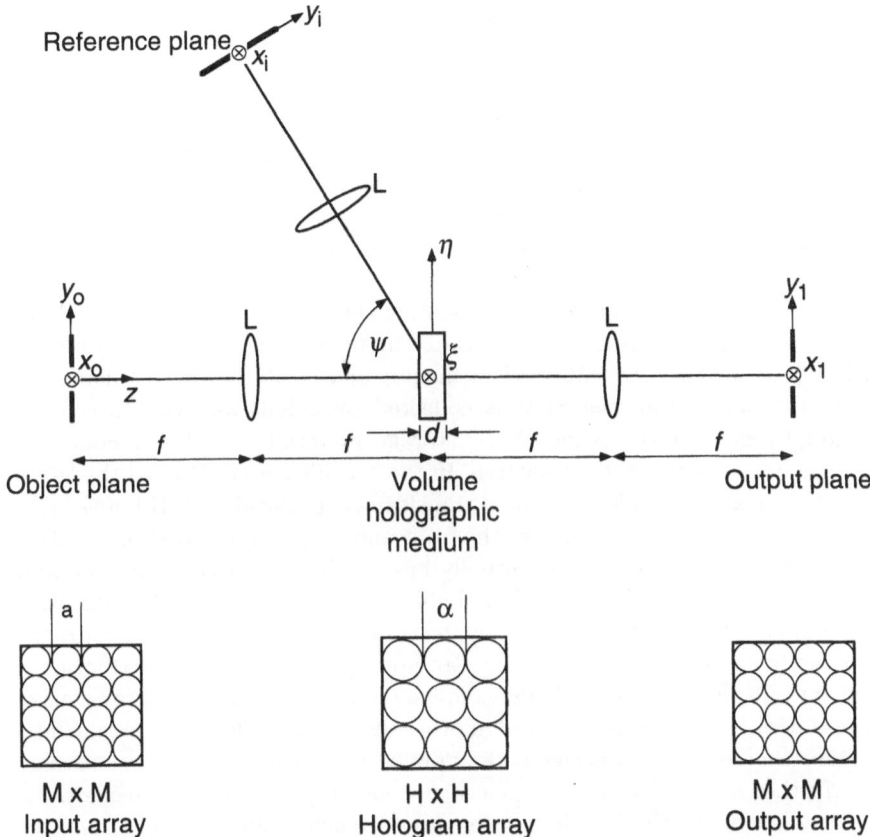

Fig. 5.4. Schematic diagram of the holographic storage system used to determine the storage capacity and storage density at a given signal-to-noise ratio. L: lenses; f: focal length of the lenses; a: pixel size; α: hologram size

configuration, this system uses a 90° angle the between reference and signal beams, which has the important advantage of reduced reflection noise.

5.1.3   System Architecture

Holographic storage systems using one of the multiplexing techniques described above may all be implemented with the system architecture depicted in Fig. 5.4. This schematic architecture is useful for estimating the limitations of storage capacity due to the optical system, as will be discussed in Sect. 5.1.4. The system consists of three distinct optical arms, the reference arm that provides the address of a data page, the object arm where the input information is injected during hologram recording, and the output arm that recovers the output page information upon readout.

In the case of angular multiplexing encoding corresponds to switching on one single pixel at the reference plane (x_i, y_i), which, for instance, may be achieved by placing a spatial light modulator (SLM) in that plane. The position of the bright pixel defines the reference beam direction after the transforming lens. For phase multiplexing, in contrast, a certain number of pixels with a well-defined phase relationship between each other is switched on at the reference plane and give the address of the recorded page. Finally, for wavelength multiplexing the wavelengths of both objects and reference waves are changed.

The electronic data stream representing the information to be stored (one data page) is transformed into a coherent optical signal by the use of an SLM (assumed to have $M \times M$ elements) placed at the object plane. The light diffracted from the SLM is collected by a lens and concentrated in a small region at the plane of the storage material as either a Fourier or Fresnel transform of the bit pattern. In the specific case of Fig. 5.4 the lens in the object arm provides a Fourier transform relationship [5.41] between the optical distributions $f(x_0, y_0)$ in the front and $g(\xi, \eta)$ in the back focal plane. Random phase shifters can eventually be used in conjunction with the SLM to reduce the dynamic range of the light distribution at the hologram plane.

Finally, the output arm provides the transformation of the light diffracted off the hologram upon readout into an appropriate intensity distribution at the output plane, where a charge-coupled device (CCD) sensor array consisting of $M \times M$ pixels can be placed in order to obtain an electronic signal of the reconstructed page. This signal can now be interfaced to the host processor for further manipulations using electronic signal processing. By using parallel readout of the CCD high transfer rates are achievable, possibly exceeding gigabits per second [5.6].

In general, if the holographic medium is large compared with the hologram size α, multiple holograms may be accommodated by spatially multiplexing $H \times H$ holograms across the surface. In the simplified scheme of Fig. 5.4 it was assumed for simplicity that the pixel distance (pitch) at the object arm SLM is equal to the pixel size a, and that, similarly, the hologram size α and the hologram distance are the same. As discussed above, for a volume holographic medium multiple images may be superimposed onto a stack at the location of a single element of the $H \times H$ hologram array. Storage in stacks of holograms is preferred due to the limited superposition of holograms and because it allows stacks of pages to be recorded and erased without affecting other regions of the storage medium.

5.1.4   Storage Capacity of Volume Media

The fundamental upper limit for storage capacity of volume media is determined from diffraction considerations [5.42]. However, besides diffraction, other effects, such as the optical quality of the holographic medium, the aperture of the optical system, lens aberrations and cross-talk noise, may limit the practically achievable storage capacity and density. Some of these aspects are discussed below.

A   Storage Capacity Limitation Due to Diffraction

The limiting storage capacity of volume media may be quickly estimated by extrapolating the results for two-dimensional media. Let us assume that the recording medium used for storage has an area A facing the recording and reconstructing beams. The area used for each hologram is $A/H^2 = \alpha^2$, as indicated in Fig. 5.4, and elementary considerations of diffraction show that the number of bits which can be stored in such a hologram is of the order of α^2/λ^2, where λ is the wavelength of the readout beam. Hence the maximum number of bits that can be stored if the full area of the medium is occupied by thin holograms is $\approx A/\lambda^2$. By making the hologram thick compared with the fringe spacing we add another dimension. Using angular multiplexing each hologram can be addressed by choosing the direction of incidence of the readout beam. Diffraction limits the number N of angularly multiplexed holograms to the order of d/λ, where d is the sample thickness. At each stack position the storage capacity in bits is the product of N and the storage capacity of thin holograms α/λ^2. The diffraction- limited storage capacity C of the full volume storage medium is therefore given by $C_{\max} \approx (\alpha/\lambda^2)(d/\lambda)H^2$, which can be rewritten as

$$C_{\max} \approx \frac{V}{\lambda^3} , \tag{5.11}$$

where $V = Ad$ is the volume of the storage sample. As expected, this result is similar to that expected from using a non-holographic sequential three-dimensional optical recording by "focusing a bit" to a localized volume of λ^3 (every cube with an edge of length λ acts as a separate storage cell). Indeed, it has been shown that, even after considering optical system limitations, holographic and layered-bit recording techniques will lead to similar storage capacity in volume media [5.43], with the holographic technique being slightly better if lenses with a F number larger than ≈ 1 are used.

The result (5.11) can also be derived by considering wavelength multiplexing [5.1] or by calculating the maximum number of orthogonal wave vectors that are available within the storage medium [5.9, 5.44–46]. For wavelengths in the visible region (5.11) predicts a tremendous storage density of 10^{13} bits per cm^3. However, as shown below, this number is too optimistic because of other less fundamental limitations that lead to lower, but more realistic, storage densities.

B   Storage Capacity Limitation Due to Optical System

We first consider the capacity limitations for a thin (two-dimensional) holographic storage medium on the basis of the limits of the optical system.

Several authors have considered this problem; specifically, three papers in the 1970s [5.47–49] reported conceptually similar analyses: they optimize the capacity of the holographic storage system with a proper selection of the input, output and hologram arrays and the apertures of the lenses in the system. Following these analyses the two-dimensional capacity C (total number of bits that can be stored in a given system) and areal storage density D (number of stored bits per area) can be expressed solely as a function of the parameters of a general optical system (Fig. 5.4) [5.47–48, 5.50]

$$C(\text{2-dim}) = \left(\frac{D_{\mathrm{L}}}{8Fn_{\mathrm{r}}\lambda}\right)^2 \tag{5.12}$$

$$D(\text{2-dim}) = \frac{1}{8}\left(\frac{1}{Fn_{\mathrm{r}}\lambda}\right)^2 \tag{5.13}$$

where F is the lens F-number (the focal length f divided by the diameter D_{L}) and the Raleigh criterion n_{r} is taken as a measure indicating when two bits can still be resolved. A Raleigh criterion of $n_{\mathrm{r}} = 1.22$ means that two bits can still be resolved if the centre of the Airy pattern generated by one bit falls on the first minimum of the Airy pattern of the neighboring bit. Equations (5.12) and (5.13) are derived under the assumption that the lens in the object arm of Fig. 5.4 has an "optimum diameter" D_{L} equal to the sum of the sizes of the input and hologram arrays [5.48]:

$$D_{\mathrm{L}} = \sqrt{2}Ma + \sqrt{2}H\alpha, \tag{5.14}$$

where the factor $\sqrt{2}$ accounts for measuring along the diagonals. It should be noted that the total storage capacity given by (5.12) can be eventually exceeded by mechanically shifting the holographic medium in the tranverse directions to allow for additional spatial multiplexing. In contrast the density given by (5.13) cannot be exceeded in thin planar gratings. Inserting $F = 1$, $n_{\mathrm{r}} = 1.22$ and $\lambda = 0.5\,\mu\mathrm{m}$, one obtains an area storage density of $D(\text{2-dim}) = 34\,\mathrm{Mbit/cm^2}$. This value is rather modest compared with the density of localized non-holographic two-dimensional storage devices such as CD-ROM or DVD. They present densities of the order of $(1/n_{\mathrm{r}}\lambda)^2$, corresponding to about $270\,\mathrm{Mbit/cm^2}$ if the same wavelength and Rayleigh criterion as above are assumed. This reduced density and the fact of having to use rather bulky systems are the price to be payed for parallel random access to the data.

The above analysis is valid for thin two-dimensional media. Generalization to the three-dimensional case is easy; the maximum storage density can be expressed as [5.44]

$$D_{\max}(\text{3-dim}) = \frac{1}{8}\left(\frac{1}{Fn_{\mathrm{r}}\lambda}\right)^2 \chi\frac{N_{\max}}{d} \tag{5.15}$$

where d is the material thickness and $N_{\max}$ is the maximum number of holograms that can be multiplexed at a given location. As will be seen below,

N_{max} is limited by cross-talk effects due to diffraction or other noise sources and depends on the minimum allowable signal-to-noise ratio (SNR) for detection. In (5.15) the detrimental effect of lens aberration that leads to further cross-talk among bits in the detector plane has been considered through the empirical quantity

$$\chi = \left(\frac{1}{1 + \alpha_L \sqrt{f/n_\mathrm{r}\lambda}} \right)^4 \tag{5.16}$$

first introduced by Akos et al. [5.51]. The constant α_L is of the order of $\alpha_L = 10^{-3}$ for conventional lenses, where it is assumed that, for a fixed Fresnel number, the aberrations are proportional to the focal length of the lens.

Recording of multiple hologram stacks may be accomplished in the easiest way by shifting the holographic medium. We can therefore put $H = 1$ and calculate explicitly the dependence of the storage density on the input and hologram sizes using (5.14) and (5.15)

$$D_{\mathrm{max}}(\text{3-dim}) = \frac{1}{4} \left(\frac{M}{\alpha} + \frac{1}{a} \right)^2 \chi \frac{N_{\mathrm{max}}}{d}. \tag{5.17}$$

Here we recall that a and α are the pixel and hologram sizes, respectively, which are approximately related by $\alpha = n_\mathrm{r}\lambda f/a$. Putting this in (5.17) shows that in general a certain compromise has to be made between hologram size and pixel size.

C   Interpage Cross-Talk and Storage Density

The last term in (5.15) and (5.17) contains implicitly the dependence of the storage density on the multiplexing method. To fully evaluate the density D_{max} the maximal number of multiplexed holograms N_{max} must be calculated. This number is limited by the condition that pixels have to be distinguishable in the output plane. Assuming that all other noise sources have been reduced to a minimum, the only fundamental noise source remaining is the cross-talk between the stored pages during readout (interpage cross-talk). For angular multiplexing, this cross-talk-limited storage capacity of volume holographic memories has been calculated by Ramberg [5.50] and Gu et al. [5.52]. Considering Fig. 5.4, it is found that the maximum number of objects can be stored when the angle between signal and reference beam is $\psi = 90°$ and the separation Δ between adjacent reference points on the y_i-coordinate in the reference plane is chosen as $\Delta = \lambda f/d$ [5.52]. For this optimum configuration the worst signal-to-noise ratio $(SNR)_{\mathrm{min}}$ in terms of intensity can be expressed as

$$(\mathrm{SNR})_{\mathrm{min}} \approx \frac{2d\,f}{\lambda Y\,N_{\mathrm{max}}}, \tag{5.18}$$

where $N_{\max}$ is the maximum number of angular multiplexed images, d is the medium thickness, f is the lens focal length, λ is the vacuum wavelength, and $Y = 2y_{1,\max}$ is the linear dimension of the output plane (object image). Equation (5.18) shows that the maximum number of image pages depends on the minimum SNR that the detection system can handle. Its derivation assumes that the paraxial approximation is valid and that the transverse dimensions of the medium are much larger than the spatial bandwidth of the objects, i.e. the whole Fourier spectrum (spatial frequency spectrum) of the input object can be recorded. For $f/Y = 1$ and assuming that an SNR equal to 1 can be tolerated one obtains from (5.18) $N_{\max} \approx 2d/\lambda$, in agreement with the estimate made at the beginning of this section.

Similar analysis of the interpage cross-talk noise can also be accomplished for the cases of wavelength and phase multiplexing. In the case of wavelength multiplexing the cross-talk noise is minimal for $\psi = 180°$ (antiparallel geometry) and the maximum noise appears for the middle hologram with frequency $\nu_m = \nu_0$, where ν_m is the frequency of the mth hologram and ν_0 is the middle frequency, in the frequency schedule [5.53]. Assuming that $\nu_0/\nu_{\max} \approx 1$ the signal-to-noise ratio for wavelength multiplexing is given by

$$(\text{SNR})_{\min} = \pi^2 \left(1 + \frac{Y^2}{8f^2}\right) \left(\sum_{m=1}^{P} \frac{1 - \cos\left(2\pi m Y^2/8f^2\right)}{m^2}\right), \qquad (5.19)$$

where $N = 2P + 1$ is again the number of stored holograms.

In Fig. 5.5 the SNR is shown as a function of number of stored holograms following (5.18) and (5.19) for angle and wavelength multiplexing, respectively. The latter has a slightly more favorable cross-talk-limited signal-to-noise ratio than angular multiplexing.

By examining a coupled wave theory in the Fourier regime Bashaw et al. [5.54] compared the cross-talk for orthogonal phase-encoded and random phase-encoded multiplexing with the cross-talk of angular multiplexing. For an ideal storage medium (with no background scatter, no grating vector dispersion and an infinite lateral extent), they find that orthogonal phase-encoded multiplexing and angular multiplexing have similar cross-talk characteristics. Orthogonal phase encoding exhibits an SNR which is better by a factor of two than angular multiplexing (5.18). This improvement occurs because the cross-talk is the page average, rather than the worst-page cross-talk of angular multiplexing. These results partially contradict [5.55], where the SNR is estimated by a numerical analysis. It is claimed that, at least until a capacity of about 500 holograms, the orthogonal Walsh–Hadamard binary phase encoding scheme [5.28] produces a much better SNR on reconstruction than angular encoding [5.55].

With the minimum affordable SNR set, the relationships (5.18) and (5.19) allow us to calculate the limiting storage density $D_{\max}(3\text{-dim})$ for a volume hologram by means of (5.15) or (5.17). Figure 5.6 shows this theoretical storage density for the most common case of angle multiplexing. Even after

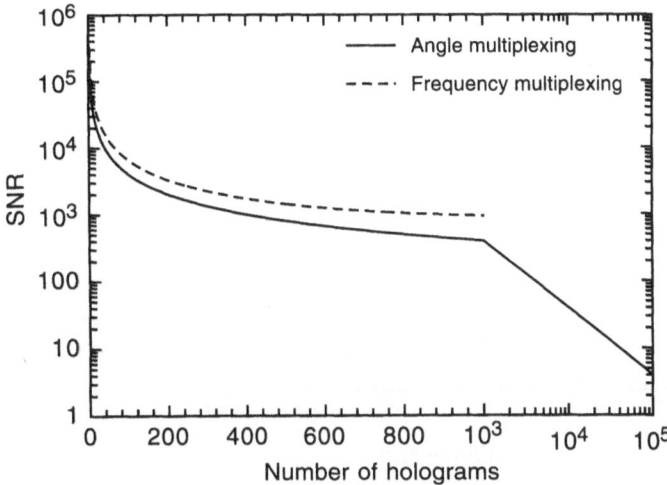

Fig. 5.5. Logarithm of the worst signal-to-noise ratio as a function of the number of stored holograms N for angle and frequency (wavelength) multiplexing. Notice that for $N > 1000$ a logarithmic scale has been used. The SNR for frequency multiplexing has not yet been calculated for large N. The parameters are material thickness $d = 1\,\text{cm}$, focal length $f = 30\,\text{cm}$, $\lambda = 500\,\text{nm}$ and output dimension $Y = 3\,\text{cm}$

consideration of the optical system limitations, if an SNR of unity can be afforded, a storage density of about $1\,\text{Tbit/cm}^3$ is theoretically achievable. This value is about one order of magnitude lower than the storage density $D_{\max} \approx \lambda^{-3}$ (5.11) expected from pure diffraction considerations. However, it should be noted that for such a low SNR a significant fraction of the stored bits might have to be used for error correction purposes. For SNR = 10 one expects a storage density of the order of $100\,\text{Gbit/cm}^3$ for typical array sizes of 1000×1000 pixels. If an SNR of 100 is required the storage density would decrease to the order of $10\,\text{Gbit/cm}^3$, which is in agreement with a statistical analysis by Yi et al. [5.56] that takes into account also intrapage interpixel cross-talk which arises from the limited spatial extend of the storage material. For both angular and wavelength multiplexing a storage density between 4 and $20\,\text{Gbit/cm}^3$ is predicted depending on the pixel size a for an SNR of 150 or a bit-error rate of 10^{-9} [5.56].

The decrease of the density for larger parallelism seen in Fig. 5.6 is due mainly to the effect of lens aberrations characterized by the constant α_L and is contained in the quantity χ in (5.16). More input pixels requires larger lenses with correspondingly longer focal lengths. Figure 5.6 also shows that the reduction in storage density with increasing parallelism is faster with coarser pixel size a. Thus at any chosen number of pixels, increasing the input–output pixel density also increases the storage density. To summarize,

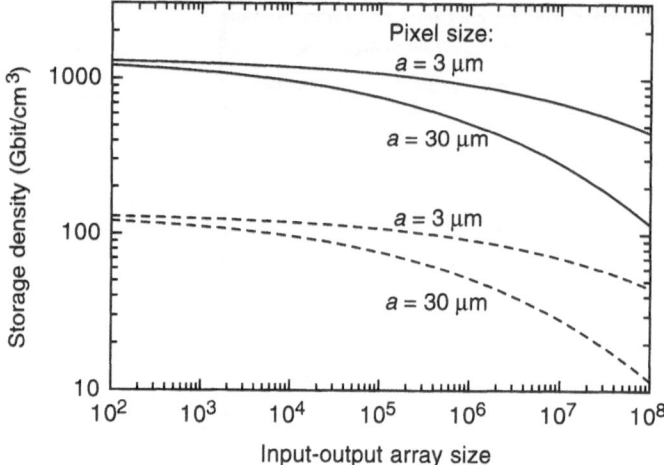

Fig. 5.6. Volume storage density (5.17) for angle multiplexing as a function of input–output parallelism with the input–output resolution as a parameter. The solid curves are for a signal-to-noise ratio SNR = 1, the dashed curves for SNR = 10. Parameters: lens F-number $F = 1$, $n_r = 1.22$, $\lambda = 500\,\text{nm}$, $d = 1\,\text{cm}$. The decrease of the density for larger parallelism is due mainly to the effect of lens aberrations characterized by the constant $\alpha_L = 10^{-3}$ in (5.16)

larger input dimensions reduce the storage density, as also do pixels that are too large.

Instead of storing Fourier holograms, image plane holograms can also be stored. It has been shown that the SNR is independent of the pixel location at the output plane for wavelength and angle multiplexing and that the worst-case signal-to-noise ratio is slightly better than the worst-case SNR for Fourier transformed holograms recorded with the same parameters [5.57–58].

D Storage Capacity Limitations Due to Material Quality

The cross-talk calculations discussed above show that the storage density of volume media is significantly lower than the ultimate limit given by diffraction. This is particularly true if large signal-to-noise ratios and low bit-error rates of the reconstructed information are required. However, it was assumed that interpage or interpixel cross-talk and lens aberrations are the only noise sources in the holographic recording system. In practice additional noise arises from imperfections of the storage material (such as scattering centers) and decreases the achievable storage density further. Despite the fact that their bad optical quality represents the principal limitation of many holographic media, unfortunately this noise source cannot be easily quantified. Bashaw et al. [5.54] use the quantity SNR_1 to express the ratio between signal and scattered noise if only one hologram is recorded. In photorefractive materials the diffraction intensity decreases as the square of the number of stored holograms N

(see Sect. 5.4.1) and so does the signal-to-noise ratio $\mathrm{SNR_{scatt.}} = \mathrm{SNR_1}/N^2$, which is due to scattering imperfections. As the cross-talk signal-to-noise ratio decreases only with the first power of N, in many material samples the scattering noise dominates over cross-talk noise for even a relatively small number of stored holograms.

For digital holographic storage, the construction of precise experimental testing beds for the characterization of the material quality is very important [5.59]. Furthermore, material research to produce low-noise media is still of crucial importance in order to reach the full potential of volumetric optical data storage [5.60].

5.2 Optical Pattern Recognition

Pattern recognition is one of the most natural signal processing tasks for which the parallelism of optics provides important advantages. The two common types of optical pattern recognition systems, the Joint Fourier Transform correlator and the VanderLugt-type correlator, will be discussed in this section. We will focus especially on the performance characteristics of these correlators when volume holograms instead of thin holograms are used. A more detailed discussion on using thin holograms in correlators and the possibilities of electronic pre- and postprocessing can be found for example in [5.61] and references therein.

5.2.1 Optical Correlators

A Joint Fourier Transform Correlator

In the Joint Fourier Transform correlator (JTC) which was first proposed by Weaver and Goodman [5.62], the object and scene patterns to be correlated are presented at the input plane simultaneously. Both are Fourier transformed by a single lens and the interference pattern of their Fourier transforms is recorded in a hologram placed in the Fourier plane. The subsequent interrogation of this hologram with a collimated beam yields, after Fourier transformation, the correlation products of the object and the scene patterns. The JTC is especially advantageous when two unknown signals have to be correlated in real time because of its simplicity. Provided real-time devices for the input as well as for recording are available, search routines can be performed at the input data rate.

Consider a system as sketched in Fig. 5.7. A transparency with a real transmittance $s_1(x_0, y_0 + h)$, which contains the object set, is positioned below the optical axis with its center at the point $(0, -h)$, and a second transparency with a real transmittance $s(x_0, y_0 - h)$ representing the input object is positioned above the optical axis with its center at the point $(0, h)$ at the front focal plane of a converging lens. The two transparencies are

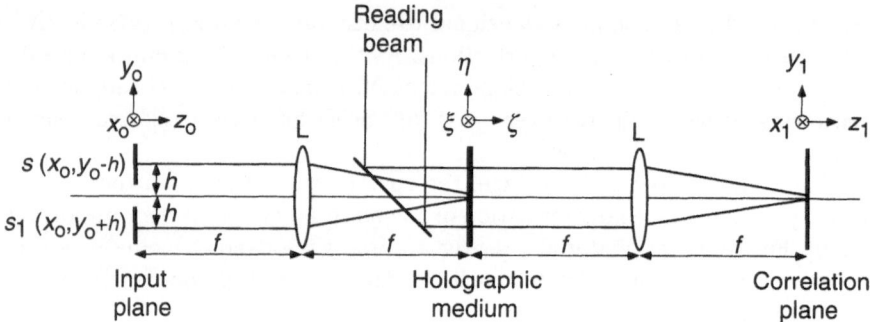

Fig. 5.7. Schematic setup of a single-axis joint Fourier transform correlator. L: lenses; f: focal length of the lenses

illuminated by a collimated beam from a laser and Fourier transformed by the lens. Then, in the back focal plane, the light amplitude is $\mathcal{F}\{s_1(x_0, y_0 + h) + s(x_0, y_0 - h)\}$, where $\mathcal{F}$ is the Fourier transform operator. The interference of the two input waves creates an intensity distribution at the holographic material that is proportional to the product of the two FTs of the inputs:

$$
\begin{aligned}
I(\xi, \eta) &= |\mathcal{F}\{s_1(x_0, y_0 + h) + s(x_0, y_0 - h)\}|^2 \\
&= |S_1(\xi, \eta)|^2 + |S(\xi, \eta)|^2 + S_1(\xi, \eta)S^*(\xi, \eta)\mathrm{e}^{-2\mathrm{i}\eta h} \\
&\quad + S_1^*(\xi, \eta)S(\xi, \eta)\mathrm{e}^{2\mathrm{i}\eta h}
\end{aligned}
\tag{5.20}
$$

where $S(\xi, \eta)$ and $S_1(\xi, \eta)$ are the Fourier transforms of $s(x_0, y_0 - h)$ and $s_1(x_0, y_0 + h)$, respectively, and * indicates complex conjugation.

This intensity distribution is recorded in a thin holographic medium. The correlation theorem states that the correlation of two functions is equivalent to the FT of the product of their complex field amplitude in the Fourier plane. Therefore an optical Fourier transformation of the amplitude of the recorded hologram is needed. Illumination of the hologram with a plane wave and Fourier transforming the transmitted amplitude using the second lens in Fig. 5.7 accomplishes this task. The output of the Joint Fourier Transform correlator can be found in the back focal plane of the second lens and is divided into three regions. In the first region around the coordinate $x_1 = y_1 = 0$ the sum of the autocorrelation of the two inputs is observed. The two side regions around the $x_1 = 0$, $y_1 = \pm 2h$ coordinates correspond to the last two terms of (5.20) and contain the cross-correlation of the object with the scene pattern. The observed intensity distribution is proportional to the absolute square of the amplitude correlation function of the two scenes

$$
\begin{aligned}
I(x_1, y_1 \pm 2h) &= |\mathcal{F}\{S_1(\xi, \eta) \cdot S^*(\xi, \eta)\}|^2 \\
&= |s_1(x_0, y_0 + h) \otimes s(x_0, y_0 - h)|^2 \\
&= \left|\iint s_1(x_0 \mp x_1, y_0 + h \mp y_1)s(x_0, y_0 - h)\mathrm{d}x_0\mathrm{d}y_0\right|^2,
\end{aligned}
\tag{5.21}
$$

Input plane   Correlation plane

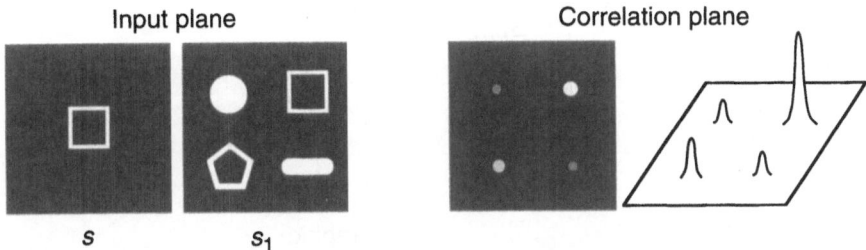

s s_1

Fig. 5.8. Input intensity distribution of a four-channel JTC consisting of the reference set of objects s_1 and the unknown input object s. The output plane centered at $(x_1 = 0, y_1 = 2h)$ contains the cross-correlation product of the two scenes

where $\otimes$ denotes the cross-correlation. Complex conjugates in the integral are omitted because the transmittances s and s_1 were assumed to be real.

In the case of two equivalent images the cross-correlation has a maximum at $(x_1 = 0, y_1 \pm 2h)$. In the case of an object set consisting of several images the cross-correlation shows maxima at positions corresponding to the relative shifts between the similar portions of images in the input plane. This property can be used to identify the position of a smaller image inside a larger image, or to find out if a certain object is present in a scene (multichannel JTC). If the input object s is similar to one of the objects contained in the reference set s_1, peaks of intensity will occur at the corresponding places (Fig. 5.8). The height of the peak is a measure of the degree of similarity of two objects. Thus the identification is performed by detecting the position and relative intensities of the correlation peaks in the output plane.

The JTC scheme described above is called a single-axis Joint Fourier Transform correlator because a single lens is used to perform the optical Fourier tranform of both object and reference image sets in front of the holographic medium. There exists a slightly modified technique (dual-axis JTC correlator) in which each signal is transformed by independent optics, greatly reducing the requirements on the Fourier transform lenses. For these practical reasons most experimental investigations with JTCs have been performed using the latter scheme. Although the dual-axis JTC correlator relieves the stringent requirements of the FT lens, it does introduce the potential problem that the two Fourier planes may have a relative tilt with respect to each other. It has been shown that this can lead to a degradation of the sharpness of the correlation peak [5.63].

JTC correlators have been demonstrated using photographic films in the Fourier plane or optical adressed spatial light modulators [5.64], as well as with real-time holographic recording media like thermoplastic plates [5.63], bacteriorhodopsin [5.65], atomic vapor [5.66], semi-insulating multiple quantum wells [5.67] and photorefractive crystals [5.68]. The latter are used as volume holograms which can significantly alter the performance of this kind of correlator, as we will see in Sect. 5.2.2.

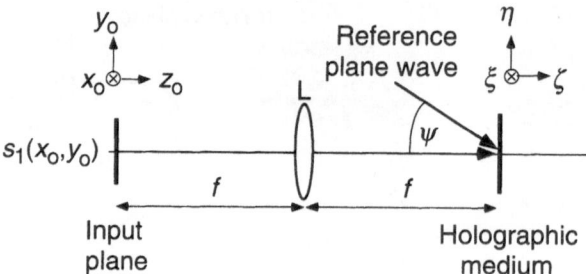

Fig. 5.9. Realization of a matched filter for the frequency plane correlator. L: lens; f: focal length of the lens

B   VanderLugt-Type Correlator

In the frequency plane correlator (FPC) or VanderLugt type correlator, the input pattern with a real transmittance $s(x_0, y_0)$ is compared with a fixed template. The template itself is recorded as a hologram. One method for the construction of the hologram was proposed by VanderLugt [5.69] and the correlator based on this method is called the VanderLugt correlator. The optical generation of the matched filter that contains the frequency spectrum of the object pattern $s_1(x_0, y_0)$ is shown in Fig. 5.9. The input device is located in the front focal plane of a lens and a holographic recording device (e.g. a photographic film) is placed in the back focal plane.

The amplitude and phase of the Fourier spectrum $S_1(\xi, \eta)$ of $s_1(x_0, y_0)$ are recorded by interference with a plane reference wave $R(\xi, \eta) = |R|e^{ic\eta}$ that is assumed, for simplicity, to propagate in the $\eta\zeta$-plane. If aberrations are neglected, the observed intensity pattern is given by

$$I(\xi, \eta) = |R(\xi, \eta) + S_1(\xi, \eta)|^2 = |R|^2 + |S_1(\xi, \eta)|^2 + R^*(\xi, \eta)S_1(\xi, \eta)$$
$$+R(\xi, \eta)S_1^*(\xi, \eta) . \tag{5.22}$$

The holographic medium is developed so that the transmission of the fixed recorded hologram is proportional to $I(\xi, \eta)$.

This filter is now reinserted in the optical system at the back focal plane of the first lens (Fig. 5.10). An adjustment precision of a few microns perpendicular to the optic axis and a few tens of microns along the optic axis must be fulfilled to achieve a proper result.

When a transparency $s(x_0, y_0)$ is present at the input plane and illuminated with a collimated beam, the intensity distribution at the output plane can then be divided into three regions. The first region which appears on the optical axis contains the autocorrelation beam. The observed intensity distribution at the second region, centered at $(x_1 = 0, y_1 = f \tan \psi)$ is proportional to the absolute square of the amplitude convolution function of the input and the stored scene. The third region appears with its center off the optical axis by an amount $(x_1 = 0, y_1 = -f \tan \psi)$. The intensity distribution

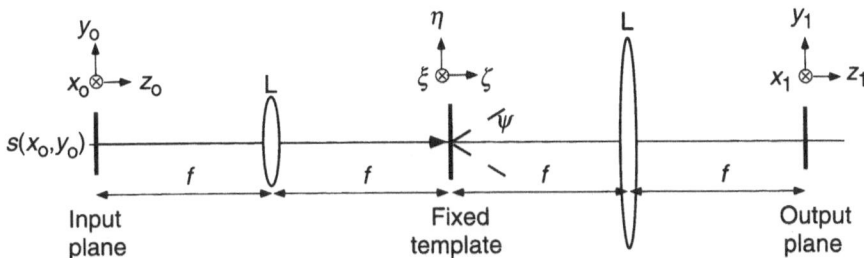

Fig. 5.10. $4f$ optical processing system for a frequency plane correlator. The fixed template is inserted at the back focal plane of the first lens

is proportional to the absolute square of the amplitude correlation function of the input object $s(x_0, y_0)$ and the stored object $s_1(x_0, y_0)$. As is the case in the JTC correlator, the detection of the correlation peak can be used either to identify the position of a smaller object inside a larger object or to find out if a certain object is present in a scene.

While conventional holographic pattern recognition systems can identify a target rapidly by matching it to the information stored either in a template (FPC) or a frame store (JTC), they suffer a basic limitation in that the target images must match the template or frame store image exactly, i.e. these types of correlators are not rotation or scale invariant with respect to the input object. Only shifts of the input object within the input plane are permitted, because a lateral displacement in the input plane will lead to an additional phase factor in the Fourier plane. In this case the correlation pattern will remain the same and will be displaced in the correlation plane. A significant amount of progress has been made in generalizing the pattern matching capabillity of the FPC using preprocessing operations (coordinate transforms, phase coding), optimized filtering (use of synthetic discriminant functions, circular harmonic filters, multiplexed filters, ...) and post-processing operations [5.61].

5.2.2   Optical Pattern Recognition Using Volume Holograms

The above description of optical correlators is valid for a thin nonlinear holographic material placed at the filter plane. If the medium is thick, the Fourier plane of the image scene is just one layer of the hologram, which has important consequences for the correlator's performance as described below.

A   Joint Fourier Transform Correlator with Volume Media

When using volume holograms like photorefractive crystals in optical correlators, the Bragg condition for diffraction plays an important role. Nicholson et al. [5.70] considered the Bragg condition in an optical joint transform correlator using photorefractive $Bi_{12}SiO_{20}$. Their simplified analysis is based on

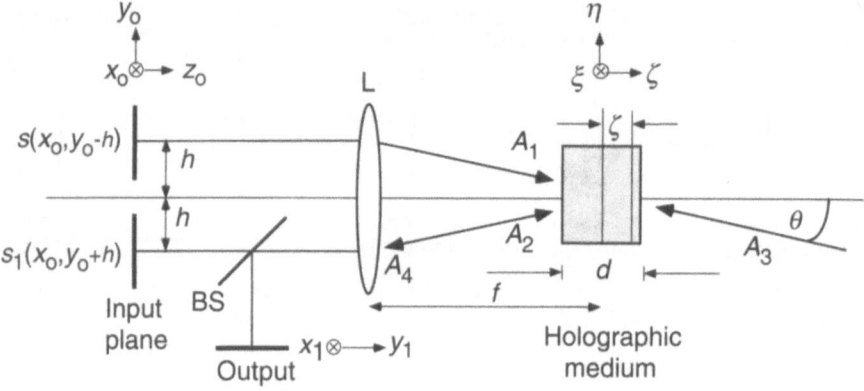

Fig. 5.11. Optical joint transform correlator using a thick real-time holographic recording medium. The output plane (x_1, y_1) can be observed after the beamsplitter BS

Kogelnik's results of diffraction from a volume hologram based on a plane wave grating approximation. Gheen et al. [5.71] presented a more detailed analysis, which takes into account the structure of the grating. The volume hologram is decomposed into a series of thin slices and the contributions from the composite thin holograms are added to get the final result. With this method the intensity distribution at the output plane is calculated for the system sketched in Fig. 5.11. It is assumed that the beams are not significantly affected through scattering or absorption as they pass through the material. Thus the result is only valid for small diffraction efficiencies. The x_0- and ξ-dimensions of the system are also ignored to simplify the analysis.

The two input signals to be correlated, $s(x_0, y_0 - h)$ and $s_1(x_0, y_0 + h)$ are Fourier transformed and illuminate the holographic medium leading to a phase grating proportional to the intensity distribution. This hologram is read by the third beam, a plane wave which satisfies the Bragg condition for the grating spacing written by the zero-order Fourier components of the two scenes. The diffracted output wave propagates backward through the lens and is observed at the output screen. This geometry is not unique: the readout beam might also differ in wavelength and enter the crystal from the left side at the appropriate Bragg angle.

The complex scattered amplitude at a distance ζ from the Fourier plane is calculated in a similar way as for thin holograms. Integrating these amplitudes along the crystal length d and propagating them to the output plane (x_1, y_1) one can show that the complex wave amplitude at that plane is

$$s(y_1) = C \int_{\text{Aperture}} s(y_0 - h) s_1(y_0 + h + y_1)$$

$$\text{sinc}\left(\frac{kd(y_0 - h)(y_1 - h)}{2nf^2}\right) dy_0, \tag{5.23}$$

where the integration is over the aperture of the input signals, the coordinate x_0 has been ignored, and the waves s and s_1 were again taken to be real functions. The constant C depends on material parameters, k is the wave number of the readout beam and n is the refractive index of the material.

The expression (5.23) is close to the desired correlation function (5.21). However, the Bragg effect has introduced an extra sinc term. The main effect of the sinc function is to limit the spatial extent of the input object $s(x_0, y_0 - h)$ so that $s_1(x_0, y_0 + h)$ is only being correlated with a portion of s. This effect varies over the output depending on the position with the most severe restriction for the y_1 values far from the center of the y_1-plane. The width w of the sinc function at the half-intensity level is given by $w = 0.9nf^2\lambda/(d(h - y_1))$ where λ is the vacuum wavelength of the readout wave. The output correlation is centered at $y_1 = -h$ when the coordinates y_0 and y_1 are chosen as mirror images with respect to BS.

As a general guide, the Bragg condition can be neglected if w is larger than the spatial extent of the input object $s(x_0, y_0 - h)$. The Bragg condition limits the spread of wave vectors that can be reconstructed, and thus the space–bandwidth product of the input scenes. For $n = 2.23$, $f = 300\,\text{mm}$, $\lambda = 500\,\text{nm}$ and a crystal thickness of $d = 1\,\text{mm}$ the aperture of the input scenes is limited to $y_{0,\text{max}} = 7.6\,\text{mm}$. The number of pixels that can be processed in parallel, i.e. the space–bandwidth product, is of the order of the ratio of the input dimension to the smallest feature area in the entrance plane of the nonlinear medium [5.72]. Diffraction limits the smallest feature size to approximately $\sqrt{\lambda d}$. For the above parameters the minimal feature size is about $22\,\mu\text{m}$, leading to a space–bandwidth product of $\approx 10^5$. Using a thin medium ($d = 1\,\mu\text{m}$) the minimum pixel size is around $1\,\mu\text{m}$, so that a space bandwidth product of $\approx 10^8$ can be reached for an input aperture of $10\,\text{mm}$.

Another serious problem is cross-talk between spatial components directed along the grating vector due to the non-vanishing crystal thickness (Fig. 5.12). For spatial components directed orthogonally to the plane of incidence no cross-talk occurs. A reduction of the recording angle φ and the crystal size d diminishes this problem.

As we have pointed out, several aspects must be considered when constructing a Joint Fourier Transform correlator using volume holograms. In order to avoid cross-talk effects the two beams have to be aligned almost parallel and Fourier transformed by a single FT lens. The resolution limit in the Fourier plane, which depends on the material thickness, determines the pixel size. Efficient reconstruction can only be realized for the writing beam components with incidence angles allowed by the Bragg acceptance, which determines the maximum input object size and leads to problems if wide fields of view are required from both input signals. These two aspects strongly limit the useful space–bandwidth product of such a device. On the other hand, utilization of a volume hologram does not lead to any advantage in such a system. Therefore we can conclude that holographic volume media

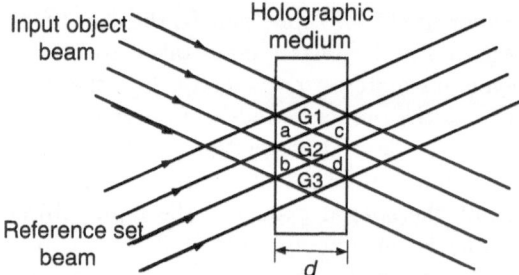

Fig. 5.12. Cross-talk problem. Spatial components along the grating vector lead to cross-talk. The regions G1–G3 are the desired gratings between the spatial components; a, b, c, d are the undesired cross-talk gratings

are not well suited for implementing a JTC [5.73]. Nevertheless several real-time correlators using volume holograms, especially photorefractive crystals, have been implemented due to the lack of fast, thin, real-time holographic recording media [5.68, 5.74–76].

B   *VanderLugt Correlator with Volume Media*

A calculation similar to that of Gheen et al. [5.71] for the joint transform correlator can also be performed for the VanderLugt-type correlator using the same assumptions. First the phase grating obtained during recording of the matched filter is calculated at a distinct plane in the holographic medium. The intensity distribution at the correlation plane is then obtained by coherently adding the contributions of each thin hologram. In order to simplify the analysis it is assumed that the normal to the front face of the storage medium is aligned parallel to the optical axis of the system.

The configuration used for recording of the holographic matched filter is shown in Fig. 5.13. While for thin media illumination of the recorded hologram with an object wave $s(x_0, y_0)$ produces three ouput beams (Sect. 5.2.1.B), for thick media the convolution beam cannot be observed because it is not Bragg-matched. Here we are interested in the correlation beam that propagates in the same direction as the reference beam that was used in the recording step. Neglecting any variations in the x_0- or ξ-dimension, the contribution δE of the hologram layer at the plane ζ to the diffracted amplitude in the correlation beam is expressed as [5.77]

$$\delta E(\eta, \zeta) = C_1 S_1^*(\eta, \zeta) RS(\eta, \zeta), \tag{5.24}$$

where C_1 is a complex constant proportional to the refractive index change amplitude in the hologram and S_1 is the complex amplitude of the optical field of the stored scene at the plane ζ given by

$$S_1(\eta, \zeta) = C \exp(ik\zeta) \int_{\text{Aperture}} s_1(y_0) \exp\left(-i\frac{ky_0^2\zeta}{2f^2}\right) \exp\left(-i\frac{k\eta y_0}{f}\right) dy_0, \tag{5.25}$$

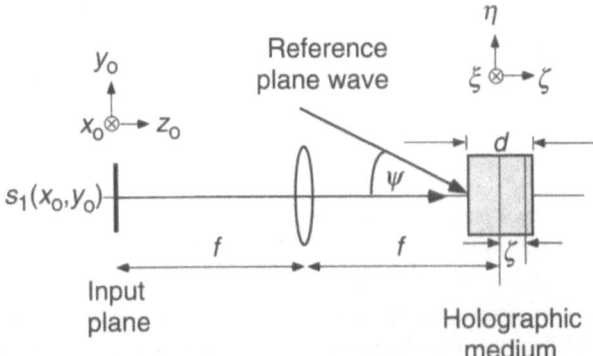

Fig. 5.13. Construction of the matched filter using a volume holographic medium situated at the Fourier plane. The scene to be stored is Fourier transformed and interferes with a plane wave incident at an angle ψ. The plane $\zeta = 0$ corresponds to the back focal plane of the lens

where C is a complex constant. The amplitude $S(\eta, \zeta)$ in (5.24) is obtained with the analogous expression to (5.25) and R is the amplitude of the plane wave reference wave used for storing the matched filter. Expression (5.24) is valid under the assumption that the phase hologram amplitude is linearly proportional to the light intensity. Note that for photorefractive materials this condition is only valid in the small modulation regime [5.78].

After Fourier transforming by the second lens in the VanderLugt scheme (Fig. 5.14) the complex amplitude $K(y_1)$ at the output plane can be calculated by integration of the contributions of the different hologram layers,

$$K(y_1) = \int_{-d/2}^{d/2} \left\{ \frac{1}{\mathrm{i}\lambda f} \exp(\mathrm{i}kf) \exp\left(\mathrm{i}k\frac{\eta^2}{2f}\right) \mathcal{F}(\delta E) \right\} \mathrm{d}\zeta \qquad (5.26)$$

where $\mathcal{F}$ denotes the Fourier transformation with respect to the coordinate η. Considering that the Fourier transformation of a plane wave is a delta

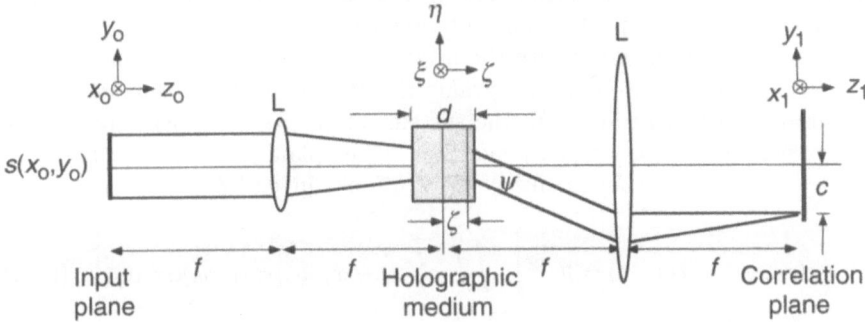

Fig. 5.14. Correlation of an input pattern $s(x_0, y_0)$ with a matched filter in a VanderLugt-type correlator. The reconstructed reference beam is focused in the correlation plane

function $\mathcal{F}(R) = |R|\delta(y_1 - c)$ and inserting the expressions for $S_1^*(\eta, \zeta)$ and $S(\eta, \zeta)$, the above integration leads to

$$K(y_1) = C_3 \int\limits_{\text{Aperture}} s_1^*(y_0 - y_1 + c)s(y_0) \text{sinc}\left(\frac{\pi dy_0(y_1 - c)}{f^2 n \lambda}\right) dy_0 \quad (5.27)$$

where C_3 is again a complex constant. The distribution $K(y_1)$ is centered around $y_1 = c$ due to the convolution with the delta function. In (5.27) we accounted for refraction into the crystal, which results in the term n appearing in the denominator of the sinc function. The constant $c = f \tan \psi$ depends on the angle between the object beam and the reference beam during recording of the matched filter. Equation (5.27) indicates that the Bragg effect, which reduces the shift invariance, introduces an additional sinc function in the correlation plane. In the case of a large crystal size, i.e. the width of the sinc function is much smaller than the input scene dimension, the sinc function vanishes for $y_1 \neq c$. The intensity distribution, which is the square of the amplitude distribution $K(y_1)$, is then given by

$$\lim_{d \to \infty} = \left| C_3 \int\limits_{\text{Aperture}} s_1^*(y_0)s(y_0) dy_0 \right|^2 \quad (5.28)$$

and in this limit the cross-correlation is obtained only for an exact positioning of the input image $s_1(y_0)$.

The influence of the crystal thickness d on the output intensity distribution can be seen in Fig. 5.15. The output intensity which is given by $|K(y_1)|^2$ is calculated using (5.27). The parameters are as follows: wavelength $\lambda = 500 \, \text{nm}$, refractive index $n = 2.23$, focal length $f = 30 \, \text{cm}$, input aperture $2y_{0,\text{max}} = 2 \, \text{cm}$. In Fig. 5.15 the loss of shift invariance for increasing hologram thickness can be easily recognized. A further increase in crystal thickness narrows the final intensity peak at $y_1 = c$. The lack of shift invariance allows us to distinguish between objects that are shifted in the input plane. An intensity peak is only observed if the input image shows some overlap of its bright areas with the bright areas of the stored scene.

While for thick volume holograms the Bragg-condition eliminates shift invariance in the y_0-direction, the system retains its shift invariance in the transversal (x_0) direction perpendicular to the plane of incidence. Thus the two-dimensional output intensity distribution is given by

$$I(x_1, y_1) = I(x_1, c) = |C|^2 \left| \iint\limits_{\text{Aperture}} s_1^*(x_0 - x_1, y_0)s(x_0, y_0) dx_0 dy_0 \right|^2. \quad (5.29)$$

Therefore, the effect of the thick hologram is to mask off the two-dimensional correlation pattern except for one vertical stripe perpendicular to the plane of

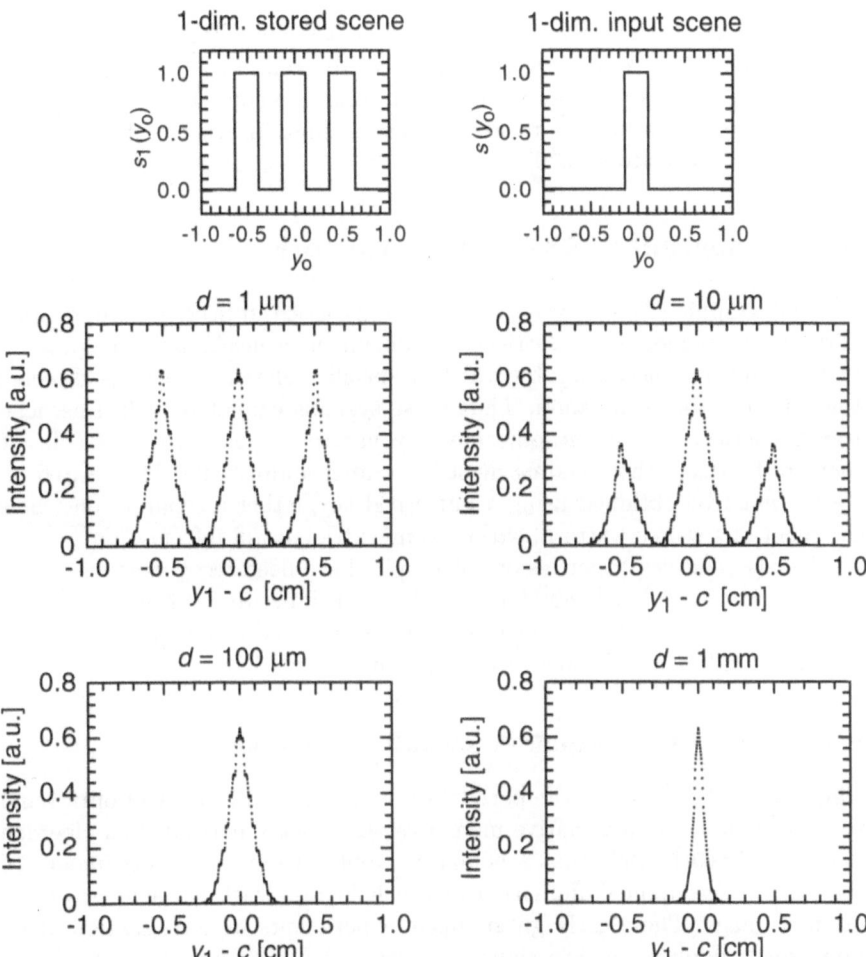

Fig. 5.15. Output intensity distribution as a function of the output dimension. Increasing the crystal thickness d leads to a single intensity peak at $y_1 = c$

the beams (x_1) whose position (c) depends on the angle of the reference beam. If we change the angle of the reference beam and record a different hologram with it, the one-dimensional stripe will be produced at a different location. One can store N different templates and get a system of N correlators with one-dimensional shift invariance. Although the shift invariance is lost in the one dimension, the utilization of volume holograms in the VanderLugt type correlator is favorable due to the large storage density that can be reached. Instead of storing only one matched filter we can store many templates in one crystal by angular encoding. The shift invariance is lost, but a much greater amount of data can be processed in parallel.

If the loss of shift invariance represents a serious problem for a given system, another type of optical correlator, the reflection-type wavelength-multiplexed correlator, may be used. In volume media the reflection type correlator has been shown to allow a larger shift invariance than JTC or VanderLugt correlators [5.79].

5.3   Holographic Associative Memories

The holographic storage systems that are discussed in Sect. 5.1 are location addressable memories. To retrieve one document a mechanical or optoelectronic control is necessary by which a specific reference beam is directed towards the storage medium. Thus these systems cannot be called associative memories. If a holographic memory has to implement the associative memory function, then reading must be content-addressable. The desired response must be obtained using a patterned key, either a separate one, or a fragment of a stored pattern, without any extra control information.

The associative properties of holography have been recognized ever since its invention by Gabor [5.80]. One can distinguish between linear and nonlinear holographic associative memories, the second kind possessing a nonlinear thresholding function for better discrimination.

5.3.1   Linear Holographic Associative Memories

The good match between the parallelism and interconnectivity of optics and the requirements of associative memories was quickly noticed. Van Heerden [5.81] predicted in 1963 that a hologram would produce a "ghost image" of a stored image upon illumination of the hologram with a fragment of the original image. These early ghost image experiments were characterized by poor image quality and low signal-to-noise ratios. The invention of off-axis holography greatly improved the SNR by angularly separating the desired signal term from the undesired noise due to self-interference among scattered waves from the original image. Some of the early experiments in holographic associations were performed by Collier and Pennington [5.82]. In these experiments a hologram was formed from two object waves. By illumination of the hologram with part of wavefront A, a complete version of wavefront B was reconstructed. If exposures from several pairs of different patterns (A, B) are recorded (superimposed), an information pattern associated with a particular key pattern can be reconstructed from the stored holograms with reasonable selectivity. The reconstructed wavefront is a linear mixture of the images of the B patterns, with relative intensities that depend on the degree of matching of the stored field patterns with the field pattern that is used during reading. Thus one pattern will dominate, whereas the other terms in the mixture will represent superimposed noise. These holographic memories suffered from distortion, poor SNR and low storage capacity. More recently

Paek et al. [5.83] proposed a ghost image-type associative memory using the second-order diffraction from a thin hologram to simplify the optical system and improve the output intensity, which usually suffers from the necessity to perform two optical correlation operations in series.

As mentioned above, when a content-addressable memory is used to store arbitrary patterns, large storage capacity cannot be achieved by a superposition method because of the cross-talk noise between the patterns. Therefore page-oriented holographic memories were developed which used mechanical or acousto-optic deflection of reference beams to read out one of many spatially separated holograms [5.84]. This was one of the earliest applications of the optical VanderLugt-type correlator [5.69] for optical associative memories. In this application, memory data were stored in a large number of spatially multiplexed holograms. The small holograms, corresponding to the various stored entries, had to be written using narrow coherent beams confined to them only. During recording, different data planes or "pages" were recorded in a hologram by sequentially shifting a plane wave reference beam over the entire recording medium. Content-addressable reading, on the other hand, must be made in an all-parallel fashion, by interrogating all entries simultaneously and in parallel. In the readout phase the light from the input data page illuminated the entire set of holograms. A parallel associative search of all of the stored data could be performed simultaneously. A detector matrix determined the location of the resultant correlation peak, which determined the location of the hologram containing the matching data. This information was used to shift a readout reference to the proper hologram for readout of the associated data. The system could also be used for heteroassociation by shifting the readout beam to a hologram different from the matching one. Associations could be made by processing the correlation plane with lookup tables. Such page-oriented associative holographic memories are capable of large storage capacities, but they handle multiple associations serially because of the mechanical scanning of the readout beam.

In response to the need for highly parallel architectures for neural network models, a new class of associative memories has been developed which is also based on the VanderLugt-type correlator, unlike page-oriented associative memories these nonlinear holographic associative memories (NHAM) use nonlinear gain and feedback to implement competition between stored data [5.85].

5.3.2   Nonlinear Holographic Associative Memories

The NHAM is an optical associative memory which combines the fully parallel image-to-image heteroassociative capabilities of ghost image holography with the high SNR, processing gain and storage capacity of thresholded Vander-Lugt correlators. They are potentially superior to linear associative memory approaches, because nonlinearities allow the selection of a particular stored image over all others on the basis of incomplete input data. As a consequence the storage capacity is increased. Phase conjugation is often used to implement the features of gain, nonlinear feedback and competition. The concept of NHAM can be also expanded to the optical implementation of neural network models. A schematic diagram of a representative system is shown in Fig. 5.16.

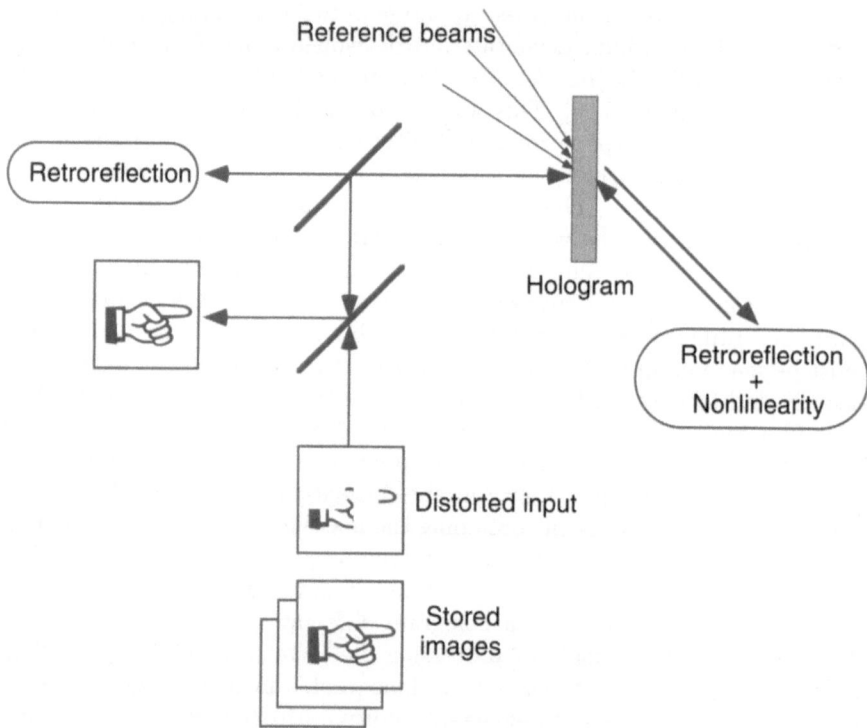

Fig. 5.16. Recording and readout of reference-based nonlinear holographic associative memories (redrawn after [5.85])

The heart of the system is a hologram in which Fourier transforms of objects are recorded sequentially using angularly multiplexed reference beams. For readout of the NHAM, phase-conjugating mirrors or another means of forming retroreflected time-reversed beams are positioned on both sides of the hologram. When a partial or distorted version of an object addresses the hologram via the beamsplitter, a set of partially reconstructed reference beams is generated. Each reconstructed reference beam corresponds to the correlation of the input with the stored object associated with that particular reference beam. This part of the system is identical to a matched VanderLugt correlator. The distorted reconstructed reference beams are phase conjugated by the reference arm phase-conjugating mirror (PCM) and retrace their paths to the hologram. These beams reconstruct the complete stored objects. The reconstructed objects are phase conjugated by the object arm PCM and the process is iterated until the system settles into a self-consistent solution or eigenmode, assuming that the gain of the PCMs is sufficient for oscillation.

The most important common feature of NHAMs is nonlinearity. Without it these associative memories could not "choose" a particular image over all others and the output would be a linear superposition of multiple recalled images. The nonlinear response and multistable operation allow selections between patterns to be made on the basis of incomplete data, since gain will exceed loss only for the stored pattern with the largest correlation with the input pattern. Nonlinearities also improve the signal-to-noise ratio and storage capacity over ghost image holography or linear matched filter correlators.

For thin Fourier holograms the effects of nonlinearities in the reference arm on the SNR and storage capacity of NHAMs has been extensively discussed by Owechko [5.85]. In this case, since the thresholded reference beam reconstructs not only the object, which has been written with that specific reference beam, centered on the input object, but also partially all other stored objects, an aperture has to be placed over the input plane to block these displaced objects. This aperture prevents ambiguities in the output plane at the cost of a reduced amount of shift invariance. An estimation of the number of stored objects N as a function of the degree of shift invariance leads to $N \approx (\Delta\phi f/z)^2$ [5.85], where $\Delta\phi$ is the range of reference-object beam angles for which the hologram have good diffraction efficiency, f is the focal length of the Fourier transform field and z is the amount of shift invariance. For parameter values of $f = 200\,\mathrm{mm}$, $z = 10\,\mathrm{mm}$ and $\Delta\phi = 30°$, the maximum number of stored images is $N = 110$.

For volume holograms Bragg selectivity prevents reconstruction of an object written at a different angle from the original. In this case the storage capacity or storage density of an NHAM depends on the storage density of the storage crystal. If we assume that cross-talk during readout with the thresholded phase-conjugated reference beam is the only noise source, i.e. perfect thresholding, we can directly apply the results of Sect. 5.1.4 for the case of angular multiplexing. With (5.18) the number of images that can be stored and compared in parallel at the same time is $N \approx (2d\,f/\lambda Y (\mathrm{SNR})_{\min})$, where Y is the linear dimension of the stored objects and $(\mathrm{SNR})_{\min}$ is the minimum tolerable signal-to-noise ratio. For a readout wavelength of $\lambda = 500\,\mathrm{nm}$, an output dimension of $Y = 2\,\mathrm{cm}$, a focal length of $f = 200\,\mathrm{mm}$ and a minimum signal-to-noise ratio of 100 we can store and associate in parallel 400 images for a 1 mm thick crystal and 4000 images for a 10 mm thick crystal. However, in practical NHAM it is often difficoult to obtain perfect thresholding, so that the size of the stored image library is limited further.

Implementations of nonlinear holographic associative memories can be categorized on the basis of the method used for generating the reference beams used in recording the holograms and on the resonator geometry. All-optical implementations will be discussed in Sect. 5.6. The basic principles of hybrid optical–electronic NHAMs are the same as for all-optical ones, but the implementation of the nonlinearity and the input are quite different. Owechko implemented a hybrid NHAM that used a pseudoconjugation system consisting of video detectors and liquid crystal light valves (LCLV) [5.86]. This approach allowed the programming of general nonlinear feedback functions. Between the reference loop video detector and the LCLV, the correlation plane is nonlinearly processed in electronic form using digital lookup tables in a PC board-level image processor. With this image processor heteroassociations or multilayer optical neural networks can also be programmed [5.87].

5.3.3 Ring Resonator Associative Memories

Another type of associative memory is the ring resonator NHAM, which was first described by Anderson [5.88]. He demonstrated a ring phase conjugate resonator containing the holograms and an externally pumped two-wave mixing configuration to introduce gain to offset the loop loss (Fig. 5.17). This system was designed to have two eigenmodes representing orthogonal patterns defined by a hologram prestored in a bleached silver halide emulsion. For a plane wave imput the system would usually resonate with equal strength in each of the two modes. However, if the input object is similar to one of the resonator modes, competition for the pump energy in the gain element will decrease amplification of the second mode below threshold and suppress its oscillation. Therefore, by injection of a portion of one of the original patterns

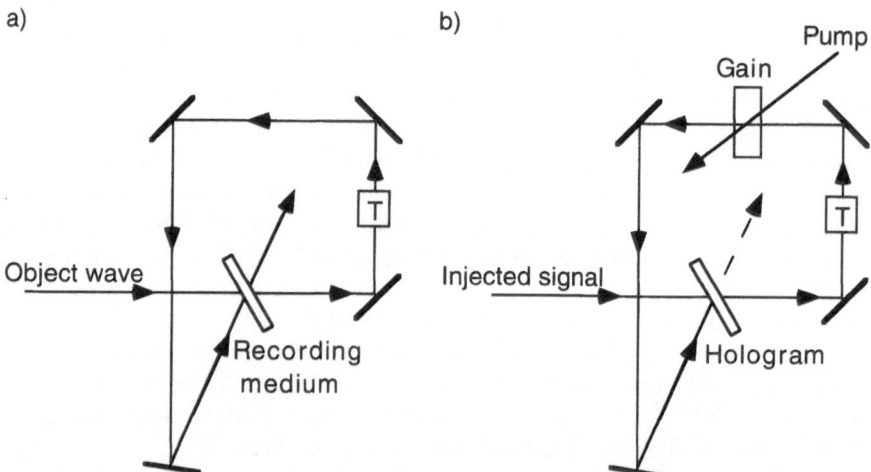

Fig. 5.17a,b. Holographic ring resonator memory. (**a**) Recording of hologram. (**b**) Recall by injected signal. Gain is supplied by a pumped photorefractive medium. The general transformation of the original object in the ring is represented by the black-box operator T (redrawn after [5.88])

the function of association is performed. The system locks so that the resonance of the other eigenmode is suppressed [5.89]. Note that this approach is different from the previously described NHAM architectures. Here, during recording of the hologram, the reference beam is derived from the object beam itself (Fig. 5.17). During readout there is no individual thresholding on the reconstructed reference beam. Instead, a nonlinear gain competition mechanism favours one reconstruction over other possible ones. This results in a simpler design and automatic generation of reference beams during recording, but at the cost of losing storage capacity. These systems also lack some of the discrimination obtainable using separate reference beams, but they do incorporate competition between stored modes using nonlinear gain saturation. In related systems, the sharp thresholding characteristics of resonators and the competition between modes could be used to implement self-learning and different neural networks operations on an all-optical basis [5.90–91].

5.4   Photorefractive Materials as Volume Storage Media

An ideal material for volumetric data storage would have a fast response during writing (submicroseconds), be as sensitive as photographic film (microjoules/cm^2), retain information for long time periods (> 10 years), have a broad sensitivity range extending into the near-infrared region of the spectrum, and be available in large quantities at good optical quality and large sizes. High optical quality and low scatter are required to insure that the signal bearing wavefront is not adversely distorted and that the noise level from scattered light is manageable. Also a large refractive index modulation is needed so that there is sufficient dynamic range to multiplex the many holograms. Unfortunately such a material does not exist today. Photopolymers offer a good sensitivity (mJ/cm^2) and a long storage time (years), but the fixing of the holograms leads to film shrinkage that degrades the quality of the reconstructed pages and shifts the Bragg angle. The main problem with these materials is the limited film thickness of a few tens of micrometers which limits the number of multiplexed holograms [5.92–93]. In a different approach photochemical hole-burning is used for a large-capacity memory system. But high capacities have only been demonstrated for very low temperatures (a few K) [5.94].

Most promising candidates for volume storage media are photorefractive crystals [5.60], which have been widely investigated for real-time holography, phase conjugation and data storage [5.9]. These crystals offer a large effective optical nonlinearity available at low continuous optical power densities of a few $mWcm^{-2}$ and long storage times for fixed holograms. It is generally accepted that the photorefractive phase gratings in electro-optic materials arise from optically generated charge carriers that migrate when the crystal is exposed to a spatially varying illumination pattern. A thorough introduction to the photorefractive effect (PE) has been given in Chap. 4. The effect has been observed in many photoconductive inorganic electro-optic crystals including $LiNbO_3$, $KNbO_3$, $BaTiO_3$, BSO, GaAs, and CdTe, as well as in the organic crystals COANP:TCNQ, MNBA and DAST [5.49, 5.95–101]. Beside crystalline materials, photorefractive polymers (see Chap. 4) have also been suggested for holographic data storage. The major limitation in achieving useful device performance in these materials is the thinness ($< 350\,\mu m$) and the large applied fields that are necessary for the observation of the PE.

Some of the important system characteristics and experimental techniques which are relevant for applications of photorefractive materials within associative memory systems or as data storage elements are shortly discussed in this section.

5.4.1 Recording Schemes

One potential problem with photorefractive materials when used as storage media is the partial erasure of previously stored holograms while storing a new hologram page. Therefore holograms stored with equal energies may reconstruct images of unequal brightness, those recorded later being brighter than holograms recorded earlier. To maximize the storage capacity all the stored holograms should be of almost equal diffraction efficiency. Also, for many applications, such as neural network implementations or associative memories, it is crucial that each hologram is stored with a distinct diffraction efficiency. To ensure constant hologram strength two storage schemes have been proposed: sequential recording and incremental recording. Both can be applied to each of the three multiplexing techniques for volume holograms of Sect. 5.1.2.

A   Sequential Storage Procedure

An obvious method used to avoid excessive erasure of the previously stored holograms is to record subsequent holograms for shorter and shorter times. In this way, the diffraction efficiency decrease of "old" holograms is reduced and at the same time the initial diffraction efficiency of "new" holograms diminishes with hologram number. This kind of sequential approach was used in early times by Burke et al. [5.102]. However, their exposure time sequence was such that an absolutely constant diffraction efficiency over 18 hologram pages could not yet be achieved.

Accurate formulas for the sequence of recording times needed to reach equal refractive index changes and thus equal diffraction efficiencies over multiple pages may be obtained by taking into account the write and erase characteristics of the material. To a first approximation (see Chap. 4), the erasure of the amplitude of the refractive index change Δn of a hologram

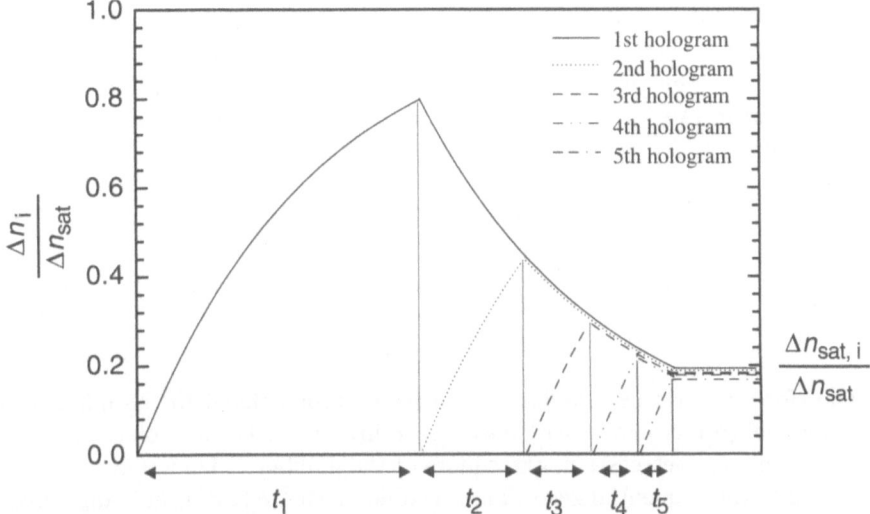

Fig. 5.18. Relative refractive index change of five holograms as a function of time. The recording of the next hologram in the sequence is stopped when its refractive index change is equal to the previous recorded holograms

recorded for a time t during subsequent recording of another hologram for a time t' can be expressed as

$$\Delta n = \Delta n_{\text{sat}} \left[1 - \exp \left(-\frac{t}{\tau_{\text{w}}} \right) \right] \exp \left(-\frac{t'}{\tau_{\text{e}}} \right), \tag{5.30}$$

where Δn_{sat} is the saturation amplitude of the index modulation when only one hologram is recorded, τ_{w} is the exponential time for hologram recording, and τ_{e} is the exponential erasure time. For equal illumination intensities the time constants τ_{w} and τ_{e} are usually equal; however, in some materials, such as LiNbO$_3$, the erasure time τ_{e} can be significantly longer than τ_{w}.

As illustrated in Fig. 5.18, in order to obtain a constant diffraction efficiency for all pages, the recording of the ith hologram must be stopped whenever its diffraction efficiency equals that of the ith -1 hologram which is being erased. For a material with equal recording and erasure time constants ($\tau_{\text{w}} = \tau_{\text{e}} = \tau$) the required recording time for the ith hologram can be calculated making use of (5.30), so that one finds the time t_i as [5.103]

$$t_i = \tau \ln \left[\frac{1 + (i-1)\beta}{1 + (i-2)\beta} \right] \qquad (i > 1), \tag{5.31}$$

where $\beta = \Delta n_1 / \Delta n_{\text{sat}}$ is the ratio of the index of refraction modulation recorded during the first exposure to the saturation index modulation. In Fig. 5.18 the saturation parameter β was set to $\beta = 0.8$. At the end of the recording schedule all holograms show the same refractive index change amplitude. For full recording of the first hologram ($\beta = 1$), (5.31) reduces to the expression $t_i = \tau \ln[i/(i-1)]$ given earlier [5.42, 5.104].

It is important to consider how the saturation amplitude of the refractive index change $\Delta n_{\mathrm{sat},i}$ depends on the total number of stored holograms N. One finds

$$\Delta n_{\mathrm{sat},i} = \Delta n_{\mathrm{sat},N} = \Delta n_{\mathrm{sat}} \frac{\beta}{1 + (N-1)\beta}, \tag{5.32}$$

which for $N \gg 1/\beta$ reduces to

$$\Delta n_{\mathrm{sat},i} = \Delta n_{\mathrm{sat},N} \cong \Delta n_{\mathrm{sat}} \frac{1}{N}. \tag{5.33}$$

Therefore the grating amplitude is inversely proportional to the number of stored holograms and, with (5.8), the diffraction efficiency of each single hologram page decreases as the square of the number of hologram pages N.

Using the sequential recording schedule given by (5.31), 500 angle-multiplexed holograms with 0.01% diffraction efficiency have been stored in an LiNbO$_3$ crystal [5.105].

B   Incremental Storage Procedure

Since the above sequential recording schedules are all calculated from the material's response times and the maximum attainable index modulation, any small error in material characterization can result in highly nonuniform diffraction efficiencies. Making the long initial exposures also introduces problems with beam fanning and coupling between the recording beams, both of which tend to restrict the maximum attainable index modulation below the theoretical value. These problems can be avoided by recording with a series of short exposures, so-called incremental recording.

For incremental storage [5.106–108], each hologram i out of N holograms is recorded with a series of exposures δt_i, each relatively short compared with the material's response time. During recording, the object and reference pair are sequentially displayed, repetitively cycling through all N pairs. As this process is repeated, the diffraction efficiencies of all the holograms gradually increase. After many cycles, the recording process approaches saturation, where the increase of the recording increment is exactly matched by the decrease during the $(N-1)$ erasure steps. This approach puts more severe constraints on the stability of the holographic setup than sequential recording. The interference pattern must be stable over all the time needed to record the hologram stack, not only during recording of a single page.

The principal aspects of incremental recording are visualized by Fig. 5.19. After many cycles the system asymptotically converges towards a state for which the amplitude of each hologram shows only small variations around a constant average level and recording can be stopped. For a short incremental recording time δt_i compared with the photorefractive time constants the

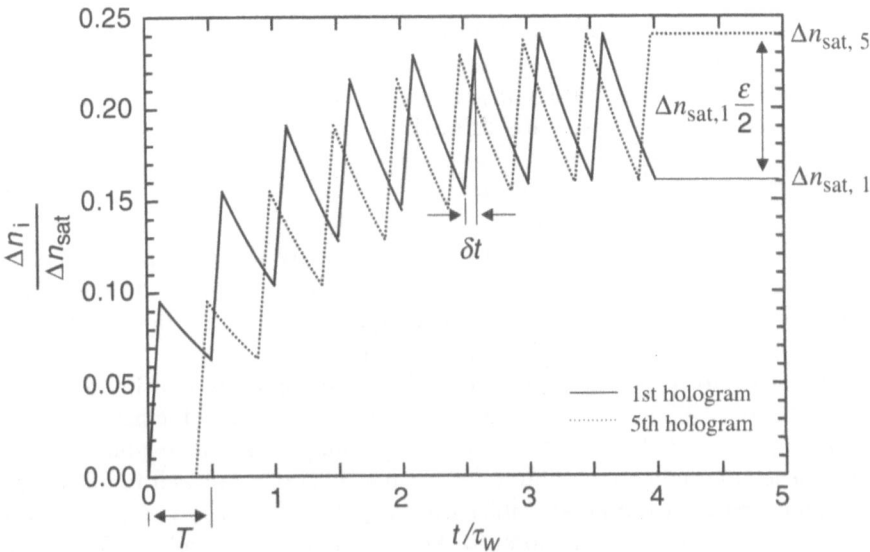

Fig. 5.19. Relative amplitude of the refractive index change as a function of the writing time for the first and the last image (out of five) in the recording cycle. The refractive index change increases until the increase of the recording increment is exactly matched by the decrease during the erasure steps. The maximum difference in the saturation refractive index change for each hologram is given by the parameter ε. Parameters: $\tau_e \equiv \tau_w$; $\delta t_i = \delta t = 0.1\tau_w$; $N = 5$; $T/\tau_w = 0.5$

asymptotic refractive index change amplitude $\Delta n_{sat,i}$ for hologram i is found to be

$$\Delta n_{sat,i} = \Delta n_{sat} \frac{\delta t_i}{\delta t_i + (T - \delta t_i)\,\tau_w/\tau_e}, \tag{5.34}$$

where τ_w and τ_e are again the writing and erasure time constants of the material and $T = \sum \delta t_i$ is the total cycle time through all holograms. For $\tau_w = \tau_e$ (5.34) simplifies to

$$\Delta n_{sat,i} = \Delta n_{sat} \left(\frac{\delta t_i}{T}\right) = \frac{\Delta n_{sat}}{N}, \tag{5.35}$$

by analogy with (5.33) valid for sequential recording. Here the second equality holds only if equal incremental times δt are chosen for every hologram page. It should be remarked that sequential and incremental storage achieve the same final grating modulation level in exactly the same time period [5.107, 5.109]. The same also holds for a third technique, parallel storage using N mutually incoherent but pairwise coherent beam pairs [5.109–111], a technique that has also been proposed for hologram copying from an archival storage medium to a secondary storage medium.

As seen in Fig. 5.19, at equilibrium, for incremental recording the hologram diffraction efficiency η_i shows a small variation ε because holograms incremented more recently are stronger. Of course, this variation can be minimized by shortening the recording increment time δt [5.106,108]. If we require that all the holograms satisfy $\eta_{sat} \leq \eta_i \leq \eta_{sat}(1 + \varepsilon)$, then the maximum δt is determined from the erasure during time $(N - 1)\delta t$ to be

$$\delta t = \Delta n_{sat} \frac{\tau_e}{2(N - 1)} \ln(1 + \varepsilon). \tag{5.36}$$

Thus the signal uniformity can be made to satisfy the system requirement by using the single parameter δt. For the sake of visualization, in Fig. 5.19 a relatively long incremetal time with consequently large variation $\varepsilon/2$ around the average value $\Delta n_{sat,i} = \Delta n_{sat}/5$ was chosen.

In order to prove the validity of (5.35), five angularly multiplexed holograms were recorded with a distinct ratio of their diffraction efficiencies in a nominally pure BaTiO$_3$ crystal, where the assumptions of an exponential rise and decay and equal record and decay time constants are well fulfilled. The normalized target diffraction efficiencies $\eta_{sat,i}/\eta_{sat} \cong (\Delta n_{sat,i}/\Delta n_{sat})^2$ of the five holograms have been chosen to be $\eta_1 = 1$, $\eta_2 = 0.5$, $\eta_3 = 0.8$, $\eta_4 = 1$, $\eta_5 = 0.2$. The ratio of the writing times δt_i is equal to the ratio of the square root of the normalized diffraction efficiencies. Each incremetal writing time δt_i is selected to be short compared with the material's response time. Setting $\delta t_1 = 4\,s$ the other must be selected as $\delta t_2 = 2.8\,s$, $\delta t_3 = 3.6\,s$, $\delta t_4 = 4\,s$ and $\delta t_5 = 1.8\,s$. The recording procedure was stopped after 50 cycles, when saturation was reached. In Fig. 5.20 the normalized diffraction efficiency is shown as a function of the angle of the readout beam. The experimental values of the diffraction efficiencies at the Bragg angle of each of the five holograms are quite close to the predicted ones.

It should be noted that these recording schemes assume that the space charge field of each hologram depends only on the record and erase time and that τ_w and τ_e are the same for all holograms, i.e. the dependence of the space charge field and of the photorefractive response time on the grating spacing or on the wavelength of the incident light are not considered. Cross-talk effects and beam-coupling are also not considered. These effects become important if many holograms are stored using angle or wavelength multiplexing and deviations of the predicted diffraction efficiencies occur which are due to these effects.

5.4.2   Storage Capacity of Photorefractive Holographic Media

For the case of photorefractive materials, besides cross-talk and the optical system (see Sect. 5.1.4), further mechanisms limit the achievable storage capacity. These limitations are due to the finite dynamic range of the photorefractive material [5.112] and to noise from crystal imperfections and have to be taken into account when determining the storage density of photorefractive crystals.

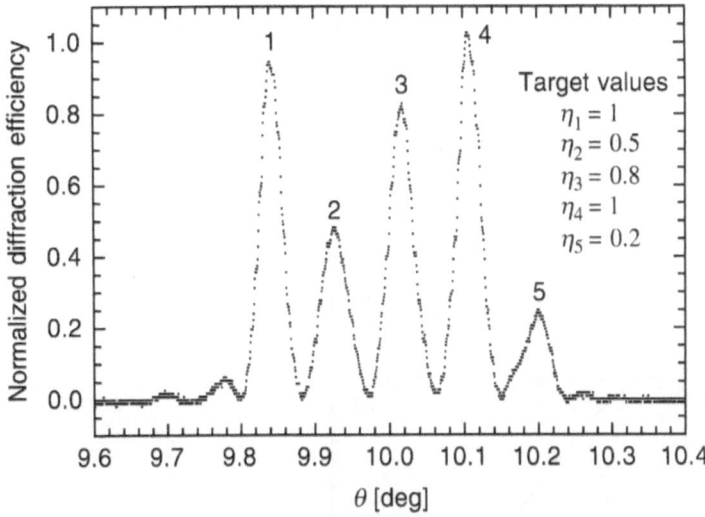

Fig. 5.20. Normalized diffraction efficiency η_{sat} of five angle-multiplexed holograms recorded with different writing times δt_i in an incremental recording scheme as a function of the readout angle of the reference beam

In Sect. 5.4.1 we have seen that the saturation value of the refractive index change decreases with $1/N$, where N is the number of stored holograms. In the case of unequal writing and erasure time constants $(\tau_w \neq \tau_e)$ the saturation refractive index change amplitude falls off for large N as [5.103]

$$\Delta n_{\text{sat},i} = \frac{\Delta n_{\text{sat}}}{N} \frac{\tau_e}{\tau_w} , \tag{5.37}$$

which is valid for both sequential and incremental recording. Unequal recording and erase time constants can be observed for instance in photorefractive LiNbO$_3$ where the bulk photogalvanic effect is important in the formation of the space charge field. The refractive index change and diffraction efficiency for a strong volume hologram as a function of the number of stored holograms is shown in Fig. 5.21.

While the signal-to-noise ratio related to cross-talk noise decreases in inverse proportion to the number of stored pages (Sect. 5.1.4), the signal-to-noise ratio related to scattering or detector noise decrease more rapidly and essentially quadratically with the number of pages N. This is because the noise amplitude can be considered to be essentially independent on N, while the intensity of the useful signal (proportional to the hologram diffraction efficiency) decreases as $1/N^2$ because of (5.37) and (5.8).

One may define the minimum allowable signal diffraction efficiency $\eta_{\min}$ which is sufficient to overcome a constant noise level in the apparatus, such

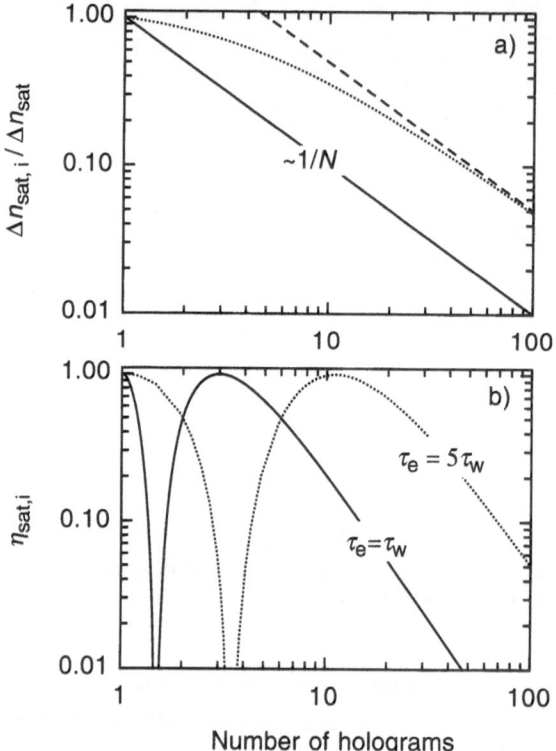

Fig. 5.21. (a) Normalized refractive index change as a function of the number of stored images. Solid line: symmetric recording/erasure ($\tau_w = \tau_e$); dotted line: asymmetric recording/erasure ($\tau_e = 5\tau_w$) from (5.34); dashed line: asymptotic fall-off as $1/N$ for large N and $\tau_e = 5\tau_w$ (5.37). (b) Saturation diffraction efficiency as a function of the number of stored images for $\phi_{max} = 3\pi/2$

as scattering or detector noise. The maximum number of pages N_{max} that can be stored can then be easily calculated as

$$N_{max} = 1 + \frac{\tau_e}{\tau_w}\left(\frac{\phi_{max}}{\phi_{min}} - 1\right),\tag{5.38}$$

where $\phi_{min} = (\eta_{min})^{1/2}$ and the maximum strength parameter $\phi_{max} = (\pi\Delta n_{sat}d)/(\lambda\cos\theta)$ is defined by the argument of the sin function in (5.8) and contains the maximum achievable refractive index change Δn_{sat} and the material thickness d.

In Table 5.1 we show an example of the maximum number of storable holograms N_{max} and corresponding storage density D. An (arbitrary) minimum diffraction efficiency η_{min} of 0.001% ($10\,\text{mWcm}^{-2}$ read beam resulting in a $100\,\text{nWcm}^{-2}$ diffracted beam) and a maximum strength parameter $\phi_{max} = 3\pi/2$ are assumed. The results are summarized for the case of $\tau_e = \tau_w$ and for $\tau_e = 5\tau_w$. In this example the storage density reaches a few tens of

Table 5.1. Number of stored holograms N_{max} and the storage density D assuming a maximum strength parameter of $\phi_{max} = 3\pi/2$ and a minimum allowable diffraction efficiency of 0.001% for a $d = 1$ cm thick crystal. The storage density is calculated using (5.15) and the parameter values $F = 1$, $f = 50$ mm, $n_r = 1.22$, $\lambda = 0.5\,\mu$m and $\alpha_L = 10^{-3}$

$\phi_{max} = 3\pi/2$	N_{max}	D (Gbit/cm^3)
$\tau_e = \tau_w$	1490	18
$\tau_e = 5\tau_w$	7450	91

Gbit/cm^3, which is less than predicted in Fig. 5.6 from limitations due to the optical system, diffraction and cross-talk. This is because the additional limitation due to scattering and detector noise implicitly contained in the value of η_{min} further reduces the maximum number of holograms N_{max}. Note that the effect of dynamically forming noise gratings [5.113], arising by interference of a new reference wave with the Bragg-mismatched diffraction from older gratings, have been neglected both here and in the treatment of Sect. 5.1.4.

It should be kept in mind that the minimum diffraction efficiency that one can afford in a given system is limited not only by noise, but also by speed and data rate issues [5.114]. At the detector a sufficient number of photons has to be integrated for each bit, thus a too-weak diffracted beam goes at the expense of processing speed. As an example, for the 5000 320 × 220 pixel holograms of [5.21], which were read out with a 50 mW/cm^2 and diffraction efficiency of $\eta = 4 \times 10^{-6}$, one can compute a necessary readout time of the order of 1 ms [5.9]. Despite the low diffraction efficiency, this readout time still corresponds to the fastest data rates of available CCD cameras, indicating that a fast data rate might represent one of the most important advantages of holographic memories over conventional optical point storage devices.

To further increase the storage capacity of a photorefractive memory conventional error encoding techniques can be used to reduce the minimum signal-to-noise ratio required to achieve an acceptable bit-error rate. Neifield et al. [5.115] suggested that a factor of 2 improvement in capacity can be achieved even when error-correction bits diminish the number of storage bits per page. Thus photorefractive memory systems with a storage capacity of few tens of Gbit/cm^3 seem to be feasible.

Recent research in academia and industry (primarily in the USA) has resulted in a number of impressive laboratory demonstrations, most of which are based on photorefractive materials. Holographic memories now seem to possess the storage density, readout speed and fidelity required for competitive commercial products. The first fully digital holographic storage system using a photorefractive crystal was demonstrated by Heanue et al. [5.116]. The total useful capacity of that system based on an LiNbO$_3$ crystal was 163 kbyte. In a later prototype demonstrator constructed by McMichael et al. [5.117] 20 000 holograms each containing ≈ 70 kbits of data were stored in

20 different locations of an $LiNbO_3$ crystal, corresponding to a raw capacity of $\approx 1.4\,\text{Gbit}$. Fully digital holographic storage systems capable of real-time retrieval and electronic decoding of MPEG compressed video has been demonstrated recently [5.118]. The often-cited goal of storage of megabit data pages with gigabit-per-second data rates has recently been demonstrated by an IBM group [5.119]. Pixel-matched reconstruction of 1024×1024 arrays of binary pixels with $1\,\text{ms}$ optical exposure of the holograms gave a raw bit-error rate as low as 2.4×10^{-6} using global threshold detection. A digital holographic storage system incorporating thermal fixing or optical fixing by means of a two-color gating technique has also been demonstrated [5.120–121].

5.4.3 Hologram Fixing and Nondestructive Readout

The use of photorefractive crystals in holographic storage data systems, neural networks and associative memories requires that the stored holograms are not erased during readout. The storage time in the dark is also finite and depends mainly on the dark conductivity of the materials. At least three possibilities are exploited to overcome these problems: the use of long-lived complementary gratings (often achieved through heating of the crystal (thermal fixing)), nondestructive readout using a wavelength at which the photorefractive material is not photoconductive, and sustained readout using refresh and low-intensity probe techniques.

The thermal fixing procedure takes advantage of mobile ions resident in the crystal as a consequence of the growth process, which replicate a hologram or a series of holograms formed by trapped electrons or holes. This process can be described schematically as follows [5.122–124]: During or after hologram recording, the sample is heated to temperatures in the 100–200°C range. At these temperatures the mobile ions move to form another grating and neutralize the electronic grating (fixing stage). On cooling to room temperature, homogeneous illumination is used to partially erase the electronic grating and bring out the replica (developing stage). The complementary grating remains and cannot be optically erased at lower temperatures. This technique was first demonstrated in $LiNbO_3$ [5.125–126] and later in $Bi_{12}SiO_{20}$ [5.127], $KNbO_3$ [5.128–130] and in $BaTiO_3$ [5.131]. With this process an erase time constant at room temperature for the fixed gratings of several years ($LiNbO_3$) or a few days ($KNbO_3$) instead of minutes or seconds, respectively, can be achieved even under illumination of the crystal.

Another mechanism relies on domain patterns in ferroelectrics. Replication of a space charge field due to trapped charge is achieved by domain formation. The exact configuration taken by domain walls to compensate the space charge distribution is not yet known. Three models have been proposed: local switching of the spontaneous polarization wherever the total field (the sum of the applied and space charge fields) exceeds the coercive field (Fig. 5.22a) [5.132], needle-shaped domains nucleated at one electrode propagate towards the other electrode until they are stopped by the space

a) 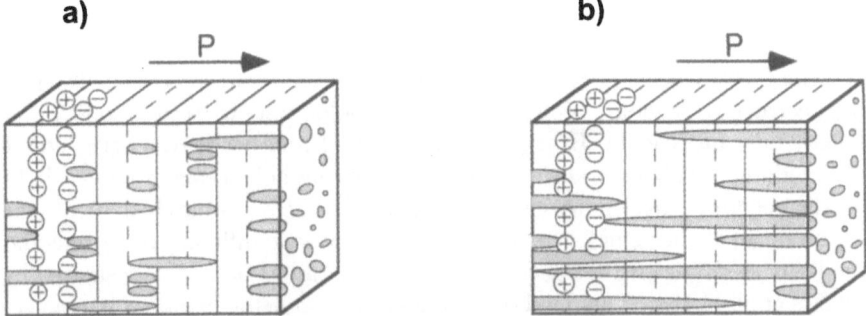 b)

Fig. 5.22. (a) Local switching of domains wherever the applied and the space charge field are larger than the coercive field. (b) Needle shaped domains nucleate at one electrode and propagate through the crystal until they are stopped by the space charge field. The charges shown are those redistributed by means of the photorefractive effect. These charges are compensated by the bound surface charges at the boundaries of the domains (charge density $\sigma = 2\boldsymbol{P}\cdot\boldsymbol{n}$, where $\boldsymbol{n}$ is the domain wall normal)

charge distribution (see Fig. 5.22b) [5.133], and deformation of the domain walls by the space charge field. These domain gratings are not erased by light and they can be reconverted into photorefractive gratings by applying an electric field. It has been demonstrated that the diffraction efficiency of the revealed photorefractive grating is higher than that of the photorefractive grating that originally generated the domain grating.

Deformation of the domain walls by the space-charge field such that this field is screened (Fig. 5.23) has been proposed by Cudney et al. [5.134]. In this case it is possible to record and simultaneously compensate an optically induced space charge pattern by depoling the crystal before the grating is created. By repoling the crystal a diffracted beam reappears.

Evidence of these effects have been identified in $Sr_xBa_{(1-x)}Nb_2O_6$ [5.135–136], $BaTiO_3$ [5.133] and $KNbO_3$ [5.137]. Recently, storage and fixing of 1000 angle-multiplexed holograms by ferroelectric domain switching has been

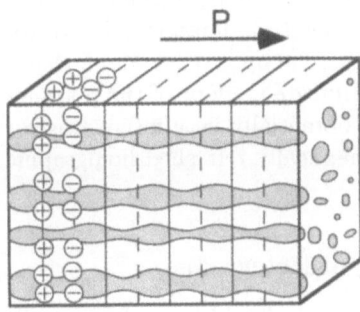

Fig. 5.23. The photorefractive space charge field induces a deformation of domain walls in an unpoled crystal so that this field is screened

demonstrated in $Sr_{0.75}Ba_{0.25}Nb_2O_6$ [5.24]. The lifetime of the fixed gratings was of the order of 14 hours.

The spectral sensitivity of the photorefractive material can also be used for nondestructive readout. Simple plane wave holograms written at one wavelength may be read by Bragg-matched light of another wavelength, where the medium is not photoactive. For image-bearing holograms, however, light at the readout wavelength cannot in general be Bragg-matched to all the components of the signal. A nonvolatile memory with different recording and read-out wavelengths that uses preformatted data has been demonstrated by Psaltis et al. [5.138]. This kind of techniques is also being explored in connection with shift multiplexing using spherical reference waves [5.36]. Recently, two-color illumination has also attracted considerable interest because of the possibility of obtaining two-photon photorefractive recording at cw power levels. As in previous observations with pulsed light [5.139–140], the gating light at wavelength λ_2 populates certain defect levels and increases the absorption at another (usually longer) wavelength λ_1. A hologram can therefore be recorded with wavelength λ_1, to which the material would be otherwise insensitive. Readout at wavelength λ_1 only does not affect the hologram and is therefore nondestructive [5.141–143]. While this effect was first attributed to the presence of Pr^{3+} or other rare earth ions, the real origin of the involved defect levels is still unknown. The sensitivity of the two-color recording effect is enhanced [5.143] by the use of $LiNbO_3$ crystals with a composition close to stoichiometric [5.144] instead of the more common congruently grown crystals.

In addition to these techniques, which rely on the properties of the materials, there are feedback techniques that refresh and sustain holograms during their use [5.145–148]. In one of these techniques, holograms are periodically copied when their diffraction efficiency falls below a certain level [5.145]. Holograms may also be circulated between two photorefractive crystals by transferring data from one to the other before the holograms in the first are erased during readout, and then reversing the roles of the crystals [5.146]. However, the effectiveness of many of these techniques is limited by noise effects; noisy gratings are copied and amplified as well. As an alternative to all-optical approaches, hybrid prototype systems where refreshing is performed by using an optoelectronic device acting as both spatial light modulator and detector array have been demonstrated recently [5.148]. The optoelectronic device is integrated in an optical system with phase conjugate readout of the photorefractive memory, which makes this approach potentially very interesting for the construction of very compact lensless dynamically refreshed holographic storage systems.

5.4.4   Coherent Erasure and Updating of Holograms

High storage density optical holographic memories should have the capability of updating pages of information. In photorefractive crystals erasure of recorded holograms is usually performed either by uniform illumination of the storage crystal at the recording wavelength or by heating the crystal if thermal fixing has been used. These bulk erasure processes apply indiscriminately to all multiplexed holograms in a stack. Selective erasure of a single stored hologram has been demonstrated by using a coherent erasure technique [5.149]. This technique is based on superimposed recording of a π-phase-shifted image. The selective erasure is shown to be much faster than the incoherent erasure owing to the coincidence of the previous charge distribution and the π-phase-shifted new light illumination pattern. In another demonstration, incremental recording was used for recording of the initial binary holograms as well as for the replacing (selective erasure and updating) of a single hologram out of four and selective erasure and updating of a single hologram out of four [5.150]. This method can not only be used to erase a single hologram but also to erase a single bit or part of a hologram by applying the phase-shift not to the reference beam but to the part of the image that has to be erased. Note also that when, after storing a binary image a second image is stored with a π-phase-shifted reference beam, some parallel logical operations such as XOR or OR can be implemented [5.151].

5.5 Optical Correlators Using Photorefractive Crystals

In Sect. 5.2 we discussed the general features of Joint Fourier Transform (JTC) or VanderLugt-type optical correlators and analyzed the consequences of using thick photorefractive materials for the necessary optical processing. Despite the fact that the Bragg condition in volume media destroys shift invariance and partly limits the applicability of such systems, recent implementations of optical correlators (see Sect. 5.2) based on photorefractive nonlinearities are numerous [5.68, 5.71, 5.74–76, 5.79, 5.152–165]. While many of these works are laboratory optical-table size demonstrations, some compact devices have been constructed as well.

Rajbenbach et al. [5.68, 5.158] have constructed a compact photorefractive JTC system based on a $Bi_{12}SiO_{20}$ (BSO) designed for an industrial application, namely to sort mechanical tools of different basic two-dimensional geometrical shapes. The reference set of objects is presented to a CCD camera, while the unknown object is viewed by a second camera. The two video signals are mixed so as to generate a single video image that is displayed by a liquid crystal display in order to build a single-axis JTC. To write the grating in the photorefractive BSO crystal a visible green beam from a frequency-doubled Nd:YAG minilaser is used. The real-time phase hologram is read-out by a Bragg-matched plane wave from an He–Ne laser. A CCD sensor in the correlation plane analyzes the diffracted signal. The response time ranges between 35 and 100 ms for illumination intensities of the order of $10 \, mW/cm^2$. Recognition of tools out of a set of four has been demonstrated, but the system can easily be expanded to more objects. In another demonstration that uses a BSO crystal as well, Yu et al. [5.75] employed a Galilean telescope to decrease the angles between the object beams in the JTC, which partly relaxes the limitations due to Bragg diffraction.

The use of photorefractive crystals as a buffer memory for JTC-type correlators has been suggested by Alves et al. [5.166]. In order to get the full advantage of using thin holograms in the Fourier plane they proposed using a photorefractive crystal as a buffer memory for a fast data access time and to store many filters. To achieve correlation at video rate each image has to be compared to a set of filters in less than 40 ms. The Fourier transform of this image is recorded in a nonlinear material by interfering with a plane wave. A third beam carrying the Fourier transform of the filters reads out the hologram. If the whole set of filters can be read out from the photorefractive crystal in less than 40 ms, dead time exists between each video image which can be used to refresh data. Alves et al. analyze the refreshing scheme and show also erasure and updating of the stored filters in the memory. Nevertheless, fast correlation with such a system has not yet been implemented. If a thin real-time holographic medium is used to record the images which have to be compared, fast image correlation with a large set of filters seems to be possible.

A photorefractive crystal for permanent storage of matched filters in a VanderLugt-type hybrid digital/optical correlator has been demonstrated by Duelli et al. [5.164]. In contrast to most photorefractive crystal based correlators, where the thick photorefractive material is placed in the filter plane, in this system the optical memory is used as an input device for the reference library and a phase modulating liquid crystal television is placed in the filter plane. Thus the correlator is fully shift invariant. During the recognition process the input scene is Fourier transfomed electronically and displayed on the LCTV and a sequential search of the input scene within the library of reference objects is performed by rotating the storage crystal and reading out the angularly multiplexed matched filters. This correlator shows high discrimination, sharp correlation peaks and high light efficiency, and is fully shift-invariant in spite of the photorefractive crystal thickness.

Besides inorganic crystals and organic polymers, another class of thin photorefractive materials, semiconductor multiple quantum well (MQW) devices, has been employed for optical correlation in a JTC configuration [5.67]. In this class of materials the photorefractive effect is mediated by resonant effects such as the Franz–Keldysh effect or the quantum-confined Stark effect [5.167–168]. Owing to their large carrier mobility these materials exhibit a fast response time and a correlation operation can be performed in a time of the order of 1 μs. On the other hand, the high mobility is detrimental to the achievable resolution, which is limited to a feature size of about 10 μm. All-optical correlation has also been demonstrated using infrared wavelength and bulk photorefractive semiconductor materials as a real-time holographic medium in both types of optical correlator [5.71, 5.160].

5.6   All-optical Nonlinear Associative Memories

In this section we mention some reported associative memory systems based on the general architecture described in Sect. 5.3.2. We distinguish between implementations using thin and thick holograms. An associative memory containing a photochromic saturable absorber as a nonlinear threshold element realized recently in our laboratory is treated in more depth at the end of this section. For optical associative memories based on optical resonators we refer to Sect. 5.3.3.

5.6.1   Thin Storage Media Implementations

All-optical implementations of NHAM can be differentiated by the form and implementations of the nonlinearities and by the storage medium used. The use of thin Fourier transform holograms for storing the pages of information has the advantage of shift invariance and the capability of heteroassociation by manipulating the correlation plane. A great disadvantage is the lack of Bragg selectivity, which results in low storage capacity compared with volume holograms. Most implementations use a single iteration nonresonant configuration, where the object arm phase-conjugating mirror (PCM) is missing (see Fig. 5.16).

The first all-optical associative memory was demonstrated by Soffer et al. [5.169–170]. They used thin thermoplastic Fourier transform holograms as the storage medium and a PCM based on degenerate four-wave mixing (DFWM) in BaTiO$_3$. The output of the hologram acted as a probe for the DFWM system, generating an amplified phase conjugate of the correlation plane. A partial version of one of the stored objects was used to address the hologram. The conjugated backward propagating beam illuminated the hologram, recreating a complete version of the stored objects. Thresholding was not demonstrated in this system, which was basically a linear associative memory. The discrimination of the two stored images was due to the coding of the objects using high spatial frequency diffusers.

Paek and Psaltis also demonstrated a single nonresonant associative memory that used a simple mirror to retroreflect the reconstructed reference beams. In their system four spatially multiplexed objects are recorded simultaneously using a single reference beam. This approach is equivalent to sequentially recording objects centered in the same aperture but with angularly shifted plane wave reference beams. During readout the presence of a correlation peak in a particular subregion is a unique label for which the stored object is being recognized. A pinhole array acts as the thresholding element; only the peaks of the correlations pass and the sidelobe noise is suppressed [5.171]. Using a PCM instead of a simple mirror in connection with a pinhole array association of two stored objects has also been demonstrated [5.172]. However, the discrimination in these implementations was not complete, as faint images of other stored holograms could still be observed in the output plane. Moreover, the quality of the reconstructed images was quite poor. This was attributed to the relatively large size of the pinholes. Although a thin hologram was used which allows shift invariance, this advantage was lost due to the fixed pinhole array. Paek and Psaltis discussed aproaches for restoring the shift invariance by eliminating the pinhole array and using quadratic nonlinearities in the correlation plane.

A novel method of thresholding was demonstrated by Wang et al. [5.173]. They used a bistable Fabry–Perot etalon. This nonlinear element was situated in the correlation plane and holding beams were used to bias the etalon. If the peak of the autocorrelation function was sufficient to switch the etalon, the holding beam at that point would be transmitted. Since the holding beams were aligned to be counterpropagating to the reference beams, after switching the transmitted holding beam read out the hologram and reconstructed the associative image. The need for a PCM was therefore avoided, but the system complexity was increased. Although auto- and heteroassociation could be implemented by directing the holding beams to the same or different holograms, the use of a separate holding beam for each stored image is not practical for large data storage.

5.6.2   Volume Storage in Associative Memories

The use of volume holograms in associative memories to extend the storage capacity was first pointed out by Yariv et al. [5.174]. They demonstrated all-optical association among two images stored in a photorefractive $BaTiO_3$ crystal. The reconstructed reference waves from the holographic memory were coupled to a photorefractive bistable ring resonator by means of multimode optical fibers. The ring resonator performed the function of thresholding; only one of the two waves survives and illuminates a transparency similar to that used to record the holographic memory. Xu et al. [5.175–176] stored four objects in a photorefractive KNSBN:Co crystal and liquid crystal electro-optical switches served as the (optoelectronic) thresholding element. Similar to the bistable Fabry–Perot etalon, counterpropagating holding beams have to be used in order to bias the switches and to read out the associated image. Lu et al. [5.177] presented a multilayer associative memory which is basically similar to a Hopfield network [5.4] with nonlinear processing on the inner product between the input and the stored patterns. The hybrid optical–electronic implementation involved storage of volume holograms in two photorefractive crystals while thresholding of the reconstructed reference waves was performed electronically. Reconstruction of one out of four images was demonstrated. Heteroassociation was also demonstrated. Kang et al. [5.178], Tanaka et al. [5.179] and Chen et al. [5.180] used a $LiNbO_3$ crystal as a thick holographic recording medium and a PCM that provided optical feedback, thresholding and gain. Unfortunately the thresholding behavior was not analysed and associative readout was only demonstrated for a single image or for orthonormal images.

Association has also been demonstrated by Ingold et al. [5.181] using a nonresonant cavity with image-bearing beams. The nonresonant cavity consists of an optical feedback configuration containing a photorefractive crystal and a nematic liquid crystal. The photorefractive $KNbO_3$ crystal is used as an optical amplifier for several beams and for image comparison and the nematic liquid crystal as a thresholding element. Associative readout, error correction and image completion are demonstrated for two and three stored black-and-white images.

Associative recall of stored images has been demonstrated for one, mostly two and at maximum five stored images in almost all published all-optical nonlinear holographic associative memories. The maximum number of angularly multiplexed holograms was limited by the high loss through the hologram and other optical components. A further limit was imposed by the low gain of the nonlinear elements. The thresholding capabilities as well as the storage capacity and practical limitations of the proposed or demonstrated systems were mostly not described.

In the following we describe in more detail an all-optical nonlinear holographic associative memory that was developed recently in our laboratory [5.182]. It is similar to the one of Soffer et al. [5.169] in that it is a single-pass NHAM that uses a phase conjugating mirror for retroreflection of the reconstructed reference beams. A schematic setup is shown in Fig. 5.24. The images are stored in a volume holographic medium consisting of a photorefractive $LiNbO_3$ crystal by angle multiplexing. Since the phase conjugation is realized via degenerate four-wave mixing in a photorefractive $KNbO_3$ crystal, its reflectivity is essentially linear. Therefore an additional nonlinear element has to be added in order to provide thresholding. This is achieved by inserting a saturable absorber with a tunable threshold into the correlation plane. The absorber consists of a photochromic colorant-doped polymethylmetacrylate (PMMA) and the threshold intensity for absorption saturation can be adjusted by simultaneous illumination with UV incoherent light [5.183].

As mentioned earlier (Sect. 5.2.2.B), the thickness of the photorefractive crystal makes the holograms Bragg-selective, which destroys the shift invariance in the plane of incidence and with it the main advantage of using Fourier holograms. Furthermore, shift invariance is also destroyed by the presence of a stationary diffuser which is attached to the liquid crystal display present in the object arm. A two-dimensional analysis (see (5.29)) shows that for thick media the intensity distribution $I_{\mathrm{ref},i}(u)$ of the reconstructed reference beam i can be approximated by the square of an overlap integral of the transmission functions of the partial input image $t_{\mathrm{input}}(x,y)$ and the stored image $t_i(x,y)$ corresponding to that reference beam, i.e. the cross-correlation sampled at the point zero:

$$I_{\mathrm{ref},i}(u) = I_{\mathrm{ref},i}(u = u_i) \propto \left| \iint_{\mathrm{Aperture}} t^*_{\mathrm{input}}(x,y)t_i(x,y)\mathrm{d}x\mathrm{d}y \right|^2 \qquad (5.39)$$

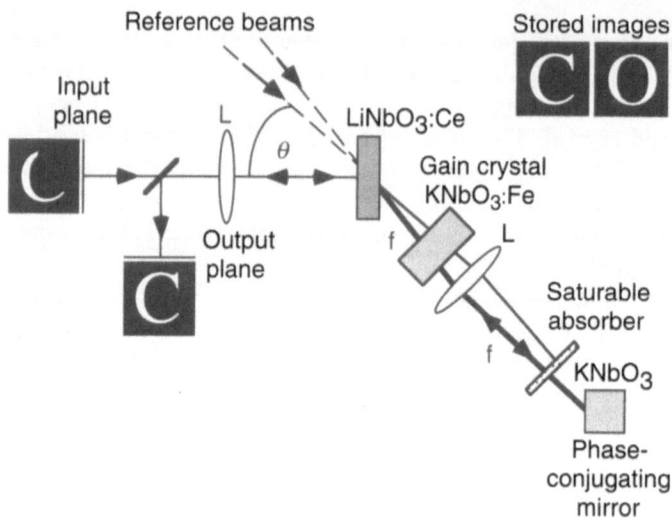

Fig. 5.24. Schematic of our all-optical associative memory. L: lenses, f: focal length of the lenses

where u is the spatial coordinate in the plane of incidence at the saturable absorber, u_i is the place of the focus of the reference beam i at the absorber, and x and y are the spatial coordinates at the input plane. The integration runs over the aperture of the input scene.

During associative readout the reconstructed reference beams pass through the saturable absorber, are phase conjugated and retrace their paths to the hologram passing through the saturable absorber again. This double-pass through the absorber greatly enhances the intensity difference between the beams, such that only one of the reconstructed reference beams has a non-negligible intensity by the time it addresses the $LiNbO_3$ crystal from the right side. Due to the strong angular selectivity of the thick hologram only the stored image which is most similar to the input is reconstructed and observed in the output plane. The number of stored images that can be associated in parallel is increased by amplifying the reference beams. A gain crystal transfers energy from a pump beam to the reconstructed reference beams via photorefractive two-beam coupling. This is necessary, since the diffraction efficiency decreases with the number of images that are stored in the $LiNbO_3$ crystal and thus the signal-to-noise ratio increases at the output plane.

The thresholding effect of the saturable absorber can be seen in Fig. 5.25 where the intensity ratio of two reference beams after the thresholding process is shown as a function of the intensity ratio before thresholding. The saturable absorber strongly enhances the intensity ratio and suppresses the weaker reference beam almost completely. An intensity ratio of one to two of the reconstructed reference beams before thresholding is necessary to reach a

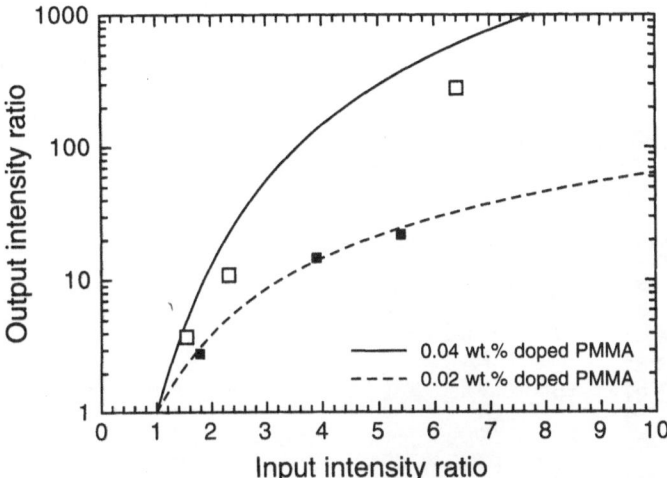

Fig. 5.25. Thresholding characteristic of the saturable absorber. The full squares are measured data for 0.02 wt.% and the open squares are data for 0.04 wt.% colorant-doped PMMA. The curves give the theoretically expected transmission ratios according to a simple model presented in [5.183]

signal to noise ratio of one to ten in the output plane. The readout time of our implementation of an associative memory is limited by the response time of the amplifier and the saturable absorber and is about 1 second for the intensities we used.

Ten black-and-white images (letters) with an area overlap of 20–80% have been stored in the LiNbO$_3$ crystal and successfully readout using a partial input. The exposure time per image per cycle during recording was 10 s and 60 cycles were necessary to reach saturation using an intensity for the object and reference beam of approximately 25 mW/cm^2. Due to the long writing time the erasure of the recorded holograms during associative readout is negligible. Photographs of the observed output using a partial input are shown in Fig. 5.26.

The image completion properties can be seen in Fig. 5.27. In this case five objects have been stored and the output has been viewed with a CCD camera. This was necessary due to the very small amount of light that was used for the associative readout. Only 80 pixels (out of 85 000 of the entire screen and out of 40 000 of the entire associated image) are necessary to get full reconstruction of the associated image. The small dot that is used as the input has a large overlap with the letter A and a smaller overlap with the letter M. Notice that cross-talk is virtually absent.

One serious problem encountered by associative memories is to distinguish between strongly correlated images and, in particular, images that are partially or totally enclosed by others. Recently we have presented a storage scheme for amplitude modulated images that eliminates this problem

stored images

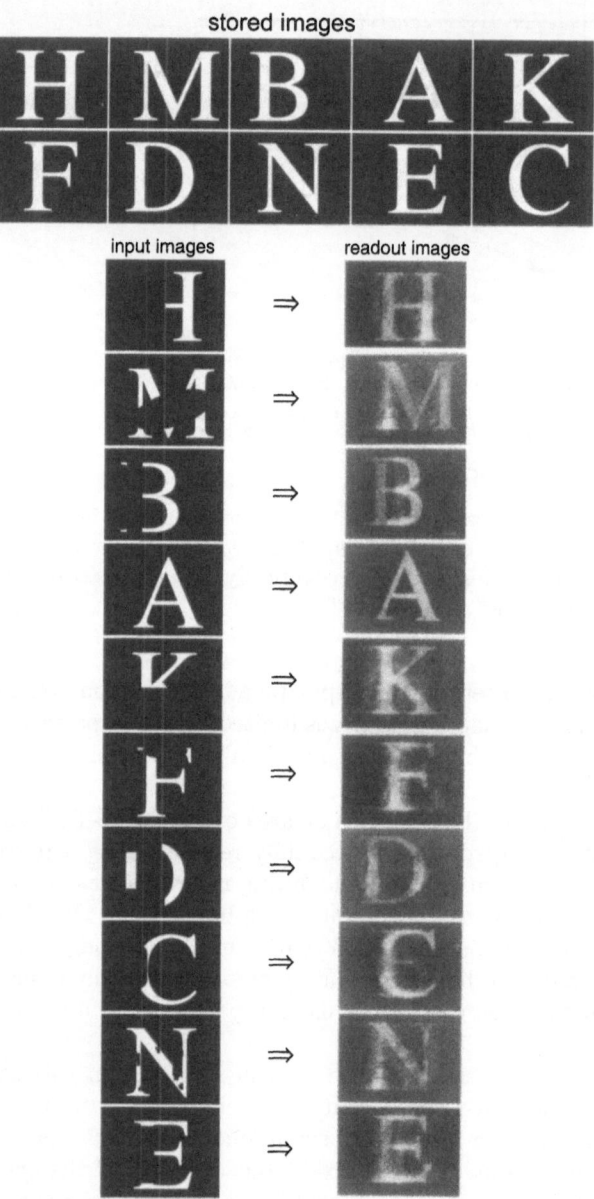

Fig. 5.26. Ten letters have been stored; associative readout using only a partial input image

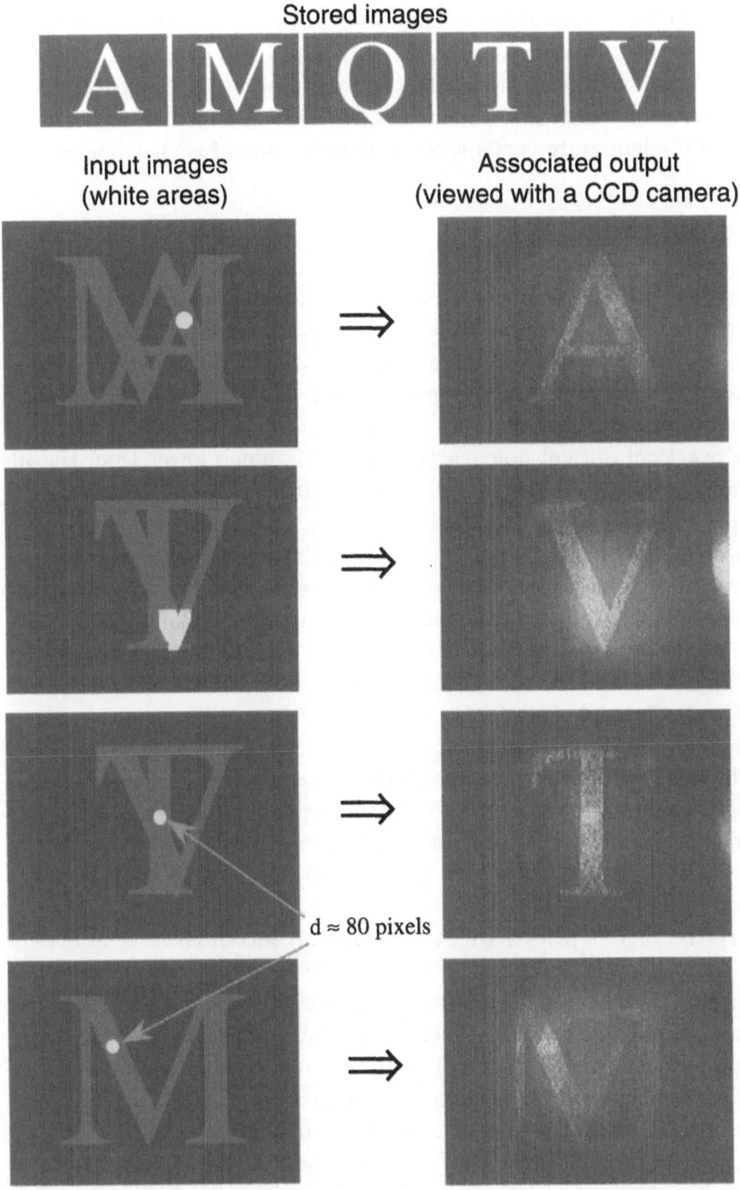

Stored images

Input images
(white areas)

Associated output
(viewed with a CCD camera)

$d \approx 80$ pixels

Fig. 5.27. Reconstruction of five stored images. Only 80 pixels are necessary to get full reconstruction of the stored images. The white area was used as input, while the gray letters are shown only in order to indicate the overlap of the small input area with the stored images

in nonlinear holographic associative memories [5.184]. The method is based on controlling the diffraction efficiency of each hologram corresponding to a stored image. It was demonstrated that the weighted storage method allows the discrimination of highly correlated images. The weighted storage can be accomplished by adequately varying the exposure time of each hologram.

5.7   Summary

Holographic data storage has received renewed interest in the last few years. In this chapter we have discussed some recent advances in holographic memory systems, and have placed the main emphasis on their applications in optical processing using photorefractive media, such as pattern recognition and associative memories. Several theoretical studies have shown that the storage density of a practical optical holographic system is much lower than the ultimate diffraction limit, but it can still reach $\approx 20\,\mathrm{Gbit/cm}^{-3}$ for a bit-error rate of of the order of 10^{-9}. The limitations are introduced by the optical system and cross-talk during readout. Among the most promising materials to implement a holographic storage device are photorefractive inorganic crystals. They offer a large effective optical nonlinearity available at low continuous optical power densities of a few mWcm^{-2} and long storage times for fixed holograms. Due to the erasure of previous recorded holograms special recording schemes must be used to achieve a constant diffraction efficiency of all stored holograms. In these materials nowadays the storage capacity is limited by the intrinsic scattering noise. Still, photorefractive storage systems with a storage density of few tens of $\mathrm{Gbit/cm}^{-3}$ seem to be feasible. Although this number has not been achieved yet, it has been shown that recording of 1 Mbit data pages with a 1 Gbps data transfer rate upon readout is possible in practice.

These large storage densities are of great importance if these systems are applied to perform image processing. We have discussed the influence of the volume nature of the stored data pages on the performance characteristics of the two common types of optical correlator (Joint Fourier Transform correlator and VanderLugt-type correlator). The latter is especially attractive in using angularly multiplexed holograms for storage of matched templates and thus for performing several pattern recognitions in parallel. Also, nonlinear holographic associative memories gain from the recent advances in photorefractive storage systems. The latest results for this kind of optical processing task have been summarized and an associative memory system constructed in our laboratory has been described in deeper detail.

Nevertheless, while a few holographic storage devices with limited capacity are present in niche-markets, practical commercial systems for mass storage or associative memories do not yet exist. Material problems are still the main obstacle. In the case of holographic associative memories, association at TV frame rates with a large set of stored images is still a distant goal.

Moreover, in order to implement higher-order tasks (like rotation and scale invariance) these systems must be incorporated in general neural network architectures.

References

5.1 P. J. vanHeerden, Appl. Opt. **2**, 393 (1963).

5.2 K. Iizuka, in *Engineering Optics* (ed. T. Tamir) (Springer-Verlag, Berlin, 1986).

5.3 T. Kohonen, in *Self-Organization and Associative Memory* (ed. T. S. Huang, M. R. Schroeder) (Springer, Berlin, 1987).

5.4 J. J. Hopfield, Proc. Natl. Acad. Sci. USA **79**, 2554 (1982).

5.5 G. Pauliat, G. Roosen, Int. J. Opt. Comp. **2**, 271 (1991).

5.6 L. Hesselink, M. C. Bashaw, Opt. Quantum Electr. **25**, S611 (1993).

5.7 C. Alves, P. Aing, G. Pauliat, G. Roosen, Optical Memory & Neural Networks **3**, 167 (1994).

5.8 J. H. Hong, I. McMichael, T. Y. Chang, W. Christian, P. Eung Gi, Opt. Engin. **34**, 2193 (1995).

5.9 C. Alves, G. Pauliat, G. Roosen, in *Insulating Materials for Optoelectronics* (ed. F. Agulló-López) (World Scientific, Singapore, 1995), p. 277.

5.10 N. H. Farhat, D. Psaltis, A. Prata, E. Paek, Appl. Opt. **24**, 1469 (1985).

5.11 E. Kim, S. Lee, W. Lee, Jap. J. Appl. Phys. **29**, 1304 (1990).

5.12 M. Takino, N. Ohtsu, T. Yatagi, Jap. J. Appl. Phys. **29**, 1317 (1990).

5.13 K. J. Weible, G. Pedrini, W. Xue, R. Thalmann, Jap. J. Appl. Phys. **29**, L1301 (1990).

5.14 S. Jutamulia, F. T. S. Yu, Optics and Laser Technology **28**, 59 (1996).

5.15 T. K. Gaylord, M. G. Moharam, Appl. Opt. **20**, 3271 (1981).

5.16 H. Kogelnik, Bell Syst. Tech. J. **48**, 2909 (1969).

5.17 G. Montemezzani, M. Zgonik, Phys. Rev. E **35**, 1035 (1997).

5.18 E. Guibelalde, Opt. Quantum Electr. **16**, 173 (1984).

5.19 K. Curtis, A. Pu, D. Psaltis, Opt. Lett. **19**, 993 (1994).

5.20 F. H. Mok, M. C. Tackitt, H. M. Stoll, Opt. Lett. **16**, 605 (1991).

5.21 F. H. Mok, Opt. Lett. **18**, 915 (1993).

5.22 S. Tao, R. Selviah, J. E. Midwinter, Opt. Lett. **18**, 912 (1993).

5.23 G. W. Burr, F. H. Mok, D. Psaltis, Conference on Lasers and Electro-optics, Anaheim, CA (1994), p. 9.

5.24 J. Ma, T. Chang, J. Hong, R. Neurgaonkar, G. Barbastathis, D. Psaltis, Opt. Lett. **22**, 1116 (1997).

5.25 S. Naruse, A. Shiratori, M. Obara, Appl. Phys. Lett. **71**, 4 (1997).

5.26 J. T. L. Macchia, D. L. White, Appl. Opt. **7**, 91 (1968).

5.27 C. C. Sun, R. H. Tsou, W. Chang, J. Y. Chang, M. W. Chang, Opt. Quantum Electr. **28**, 1551 (1996).

5.28 C. Denz, G. Pauliat, G. Roosen, T. Tschudi, Opt. Commun. **85**, 171 (1991).

5.29 C. Denz, G. Pauliat, G. Roosen, T. Tschudi, Appl. Opt. **31**, 5700 (1992).

5.30 J. Lembcke, C. Denz, T. Tschudi, Opt. Mat. **4**, 428 (1995).

5.31 C. Alves, G. Pauliat, G. Roosen, Opt. Lett. **19**, 1894 (1994).

5.32 J. Lembcke, C. Denz, T. H. Barnes, T. Tschudi, *Photorefractive Materials, Effects and Devices*, Kiev, Ukraine (Technical Digest, 1993), p. 574.

5.33 C. Denz, T. Dellwig, J. Lembcke, T. Tschudi, Opt. Lett. **21**, 278 (1996).

5.34 D. Z. Anderson, D. M. Lininger, Appl. Opt. **26**, 5031 (1987).

5.35 D. Psaltis, M. Levene, A. Pu, G. Barbastathis, K. Curtis, Opt. Lett. **20**, 782 (1995).

5.36 G. Barbastathis, D. Psaltis, Opt. Lett. **21**, 432 (1996).

5.37 G. A. Rakuljic, V. Leyva, A. Yariv, Opt. Lett. **17**, 1471 (1992).

5.38 F. T. S. Yu, F. Zhao, H. Zhou, S. Yin, Opt. Lett. **18**, 1849 (1993).

5.39 S. Yin, H. Zhou, F. Zhao, M. Wen, Z. Yang, J. Zhang, F. T. S. Yu, Opt. Commun. **101**, 317 (1993).

5.40 D. Lande, J. F. Heanue, M. C. Bashaw, L. Hesselink, Opt. Lett. **21**, 1780 (1996).

5.41 W. T. Cathey, in *Optical Information Processing and Holography* (ed. S. S. Ballard) (John Wiley & Sons, New York, 1974).

5.42 K. Bløtekjaer, Appl. Opt. **18**, 57 (1979).

5.43 T. Tanaka, S. Kawata, J. Opt. Soc. Am. **B 13**, 935 (1996).

5.44 J. R. Wullert, Y. Lu, Appl. Opt. **33**, 2192 (1994).

5.45 T. Jannson, Optica Acta **27**, 1335 (1980).

5.46 D. Brady, D. Psaltis, Opt. Quantum Electr. **25**, 596 (1993).

5.47 P. Graf, M. Lang, Appl. Opt. **11**, 1382 (1972).

5.48 A. VanderLugt, Appl. Opt. **12**, 1675 (1973).

5.49 P. Günter, U. Flückiger, J. P. Huignard, F. Micheron, Ferroelectrics **13**, 297 (1976).

5.50 R. G. Ramberg, RCA Rev. **33**, 5 (1972).

5.51 G. Akos, G. Kiss, P. Varga, Opt. Commun. **20**, 63 (1977).

5.52 C. Gu, J. Hong, I. McMichael, R. Saxena, J. Opt. Soc. Am. **A 9**, 1978 (1992).

5.53 K. Curtis, C. Gu, D. Psaltis, Opt. Lett. **18**, 1001 (1993).

5.54 M. C. Bashaw, J. F. Heanue, A. Aharoni, J. F. Walkup, L. Hesselink, J. Opt. Soc. Am. **B 11**, 1820 (1994).

5.55 K. Curtis, D. Psaltis, J. Opt. Soc. Am. **A 10**, 2547 (1993).

5.56 X. Yi, P. Yeh, C. Gu, Opt. Lett. **19**, 1580 (1994).

5.57 X. Yi, S. Campbell, P. Yeh, C. Gu, Opt. Lett. **20**, 779 (1995).

5.58 K. Curtis, D. Psaltis, Opt. Lett. **19**, 1774 (1994).

5.59 M. P. Bernal, H. Coufal, R. K. Grygier, J. A. Hoffnagle, C. M. Jefferson, R. M. MacFarlane, R. M. Shelby, G. T. Sincerbox, P. Wimmer, G. Wittmann, Appl. Opt. **35**, 2360 (1996).

5.60 G. T. Sincerbox, Opt. Mat. **4**, 370 (1995).

5.61 N. Collings, *Optical Pattern Recognition Using Holographic Techniques,* (Addison-Wesley, Wokingham, UK, 1988).

5.62 C. S. Weaver, J. W. Goodman, Appl. Opt. **5**, 1248 (1966).

5.63 T. C. Lee, J. Rebholz, P. Tamura, Opt. Lett. **4**, 121 -123 (1979).

5.64 Q. Zhan, T. Minemoto, Jpn. J. Appl. Phys. **32**, 3471 (1993).

5.65 R. Thoma, N. Hampp, Opt. Lett. **17**, 1158 (1992).

5.66 I. Biaggio, J. P. Partanen, B. Ai, R. J. Knize, R. W. Hellwarth, Nature **371**, 318 (1994).

5.67 A. Partovi, A. M. Glass, T. H. Chiu, D. T. H. Liu, Opt. Lett. **18**, 906 (1993).

5.68 H. Rajbenbach, S. Bann, P. Réfrégier, P. Joffre, J.-P. Huignard, H. Buchkremer, A. S. Jensen, E. Rasmussen, K. Brenner, G. Lohman, Appl. Opt. **31**, 5666 (1992).

5.69 A. VanderLugt, IEEE Trans. Inf. Theory IT **10**, 139 (1964).

5.70 M. G. Nicholson, I. R. Cooper, M. W. McCall, C. R. Petts, Appl. Opt. **26**, 278 (1987).

5.71 G. Gheen, L. J. Cheng, Appl. Opt. **27**, 2756 (1988).

5.72 I. Biaggio, B. Ai, R. J. Knize, J. P. Partanen, R. W. Hellwarth, Opt. Lett. **19**, 1765 (1994).

5.73 R. A. Athale, K. Raj, Opt. Lett. **17**, 880 (1992).

5.74 J. O. White, A. Yariv, Appl. Phys. Lett. **37**, 5 (1980).

5.75 F. T. S. Yu, S. Wu, S. Rajan, D. A. Gregory, Appl. Opt. **31**, 2416 (1992).

5.76 D. Pepper, J. Yeung, D. Fekete, A. Yariv, Opt. Lett. **3**, 7 (1978).

5.77 F. T. S. Yu, *Optical Information Processing* (John Wiley & Sons, New York, 1983).

5.78 N. V. Kuhktarev, V. B. Markov, S. G. Odulov, M. S. Soskin, V. L. Vinetskii, Ferroelectrics **22**, 949 (1979).

5.79 F. T. S. Yu, S. Yin, Opt. Engin. **34**, 2224 (1995).

5.80 D. Gabor, Nature **161**, 777 (1948).

5.81 P. J. vanHeerden, Appl. Opt. **2**, 387 (1963).

5.82 R. J. Collier, K. S. Pennington, Appl. Phys. Lett. **8**, 44 (1966).

5.83 E. G. Paek, E. C. Jung, Opt. Lett. **16**, 1034 (1991).

5.84 G. R. Knight, Appl. Opt. **13**, 904 (1974).

5.85 Y. Owechko, IEEE J. Quantum Electr. **25**, 619 (1989).

5.86 Y. Owechko, Appl. Opt. **26**, 5104 (1987).

5.87 R. A. Athale, H. H. Szu, C. B. Friedlander, Opt. Lett. **11**, 482 (1986).

5.88 D. Z. Anderson, Opt. Lett. **11**, 56 (1986).

5.89 D. Z. Anderson, M. C. Erie, Opt. Engin. **26**, 434 (1987).

5.90 M. Saffman, C. Benkert, D. Z. Anderson, Opt. Lett. **16**, 1993 (1991).

5.91 G. Montemezzani, G. Zhou, D. Z. Anderson, Opt. Lett. **19**, 2012 (1994).

5.92 K. Curtis, D. Psaltis, Appl. Opt. **33**, 5396 (1994).

5.93 H. S. Rhee, H. J. Caulfield, C. S. Vikram, J. Shamir, Appl. Opt. **34**, 846 (1995).

5.94 B. Kohler, S. Bernet, A. Renn, U. P. Wild, Opt. Lett. **18**, 2144 (1993).

5.95 A. Ashkin, G. D. Boyd, J. M. Dziedzic, R. G. Smith, A. A. Ballman, J. J. Levinstein, K. Nassau, Appl. Phys. Lett. **25**, 233 (1966).

5.96 R. L. Townsend, J. T. LaMacchia, J. Appl. Phys. **41**, 5188 (1970).

5.97 M. Peltier, F. Micheron, J. Appl. Phys. **48**, 3683 (1977).

5.98 A. M. Glass, A. M. Johnson, D. H. Olson, W. Simpson, A. A. Ballmann, Appl. Phys. Lett. **44**, 948 (1984).

5.99 K. Sutter, J. Hulliger, P. Günter, Solid State. Commun. **74**, 867 (1990).

5.100 K. Sutter, J. Hulliger, R. Schlesser, P. Günter, Opt. Lett. **18**, 778 (1993).

5.101 S. Follonier, C. Bosshard, F. Pan, P. Günter, Opt. Lett. **21**, 1655 (1996).

5.102 W. J. Burke, P. Sheng, J. Appl. Phys. **48**, 681 (1977).

5.103 E. S. Maniloff, K. M. Johnson, J. Appl. Phys. **70**, 4702 (1991).

5.104 D. Psaltis, D. Brady, K. Wagner, Appl. Opt. **27**, 1752 (1988).

5.105 F. Mok, M. Tackitt, H. M. Stoll, OSA Technical Digest Series **18**, 76 (1989).

5.106 Y. Taketomi, J. E. Ford, H. Sasaki, J. Ma, Y. Fainman, S. H. Lee, Opt. Lett. **16**, 1774 (1991).

5.107 E. S. Maniloff, K. M. Johnson, Opt. Lett. **17**, 961 (1992).

5.108 Y. Taketomi, J. E. Ford, H. Sasaki, J. Ma, Y. Fainman, S. H. Lee, Opt. Lett. **17**, 962 (1992).

5.109 S. Campbell, Y. Zhang, P. Yeh, Opt. Commun. **123**, 27 (1996).

5.110 S. Piazzolla, B. K. Jenkins, J. A. R. Tanguay, Opt. Lett. **17**, 676 (1992).

5.111 V. A. Vanin, Sov. J. Quantum Electron. **8**, 809 (1978).

5.112 J. H. Hong, P. Yeh, D. Psaltis, D. Brady, Opt. Lett. **15**, 344 (1990).

5.113 C. Gu, J. Hong, Opt. Commun. **93**, 213 (1992).

5.114 P. Aing, C. Alves, G. Pauliat, G. Roosen, J. Phys. III France **4**, 2427 (1994).

5.115 M. A. Neifield, M. McDonald, Opt. Lett. **19**, 1483 (1994).

5.116 J. F. Heanue, M. C. Bashaw, L. Hesselink, Science **265**, 749 (1994).

5.117 I. McMichael, W. Christian, D. Pletcher, T. Y. Chang, J. H. Hong, Appl. Opt. **35**, 2375 (1996).

5.118 L. Hesselink, S. G. Orlov, A. Akella, D. Lande, A. Liu, Proc. Waseda International Symposium on Phase Conjugation & Wave Mixing, Tokyo, Japan (1997), p. 36.

5.119 R. M. Shelby, J. A. Hoffnagle, G. W. Burr, C. M. Jefferson, M. Bernal, H. Coufal, R. K. Grygier, H. Günther, R. M. Macfarlane, G. T. Sincerbox, Opt. Lett. **22**, 1509 (1997).

5.120 J. F. Heanue, M. C. Bashaw, A. J. Daiber, R. Snyder, L. Hesselink, Opt. Lett. **21**, 1615 (1996).

5.121 D. Lande, S. S. Orlov, A. Akella, L. Hesselink, R. R. Neurgaonkar, Opt. Lett. **22**, 1722 (1997).

5.122 M. Carrascosa, F. Agulló-López, J. Opt. Soc. Am. B **7**, 2317 (1990).

5.123 L. Arizmendi, P. D. Townsend, M. Carrascosa, J. Baquedano, J. M. Cabrera, J. Phys., Condensed Matter **3**, 5399 (1991).

5.124 V. V. Kulikov, S. I. Stepanov, Sov. Phys. Solid State **21**, 1849 (1979).

5.125 D. L. Staebler, W. J. Burke, W. Phillips, J. J. Amodei, Appl. Phys. Lett. **26**, 182 (1975).

5.126 P. Hertel, K. H. Ringhofer, R. Sommerfeldt, Phys. stat. sol. (a) **104**, 855 (1987).

5.127 L. Arizmendi, J. Appl. Phys. **65**, 423 (1989).

5.128 G. Montemezzani, P. Günter, J. Opt. Soc. Am. B **7**, 2323 (1990).

5.129 G. Montemezzani, M. Zgonik, P. Günter, J. Opt. Soc. Am. B **10**, 171 (1993).

5.130 X. L. Tong, M. Zhang, A. Yariv, A. Agranat, Appl. Phys. Lett. **69**, 3966 (1996).

5.131 D. Kirillov, J. Feinberg, Opt. Lett. **16**, 1520 (1991).

5.132 F. Micheron, G. Bismuth, Appl. Phys. Lett. **20**, 79 (1972).

5.133 R. S. Cudney, J. Fousek, M. Zgonik, P. Günter, M. H. Garrett, D. Rytz, Appl. Phys. Lett. **63**, 3399 (1993).

5.134 R. S. Cudney, J. Fousek, M. Zgonik, P. Günter, M. H. Garrett, D. Rytz, Phys. Rev. Lett. **72**, 3883 (1994).

5.135 F. Micheron, G. Bismuth, Appl. Phys. Lett. **23**, 71 (1973).

5.136 Y. Qiao, S. Orlov, D. Psaltis, R. R. Neurgaonkar, Opt. Lett. **18**, 1004 (1993).

5.137 R. S. Cudney, P. Bernasconi, M. Zgonik, J. Fousek, P. Günter, Appl. Phys. Lett. **70**, 1339 (1997).

5.138 D. Psaltis, F. Mok, H. S. Li, Opt. Lett. **19**, 210 (1994).

5.139 K. Buse, F. Jermann, E. Kratzig, Appl. Phys. A **58**, 191 (1994).

5.140 K. Buse, F. Jermann, E. Kratzig, Opt. Mat. **4**, 237 (1995).

5.141 Y. S. Bai, R. R. Neurgaonkar, R. Kachru, Opt. Lett. **21**, 567 (1996).

5.142 Y. S. Bai, R. Kachru, Phys. Rev. Lett. **78**, 2944 (1997).

5.143 S. G. Orlov, A. Akella, L. Hesselink, R. R. Neurgoankar, Conference on Lasers and Electro-Optics (CLEO 97), Baltimore (1997), Postdeadline paper CPD29 1–3.

5.144 Y. Furukawa, K. Kitamura, Y. Ji, G. Montemezzani, M. Zgonik, C. Medrano, P. Günter, Opt. Lett. **22**, 501 (1997).

5.145 D. Brady, K. Hsu, D. Psaltis, Opt. Lett. **15**, 817 (1990).

5.146 H. Sasaki, Y. Fainman, J. E. Ford, Y. Taketomi, S. H. Lee, Opt. Lett. **16**, 1874 (1991).

5.147 S. Boj, G. Pauliat, G. Roosen, Opt. Lett. **17**, 438 (1992).

5.148 J. J. P. Drolet, E. Chuang, G. Barbastathis, D. Psaltis, Opt. Lett. **22**, 552 (1997).

5.149 J. P. Huignard, J. P. Herriau, F. Micheron, Appl. Phys. Lett. **26**, 256 (1975).

5.150 H. Sasaki, J. Ma, Y. Fainman, S. H. Lee, Y. Taketomi, Opt. Lett. **17**, 1468 (1992).

5.151 J. V. Alvarez Bravo, L. Arizmendi, Opt. Mat. **4**, 419 (1995).

5.152 N. Peyghambarian, K. Meerholz, B. L. Volodin, Sandalphon, B. Kippelen, in *Proc. Photoactive Organic Materials, Science and Applications*, Avignon, France, 25–30 June 1995, p. 281.

5.153 R. C. D. Young, C. R. Chatwin, Opt. Engin. **36**, 2754 (1997).

5.154 Z. Q. Wen, X. Y. Yang, Opt. Commun. **135**, 212 (1997).

5.155 R. Tripathi, J. Joseph, K. Singh, Opt. Commun. **143**, 5 (1997).

5.156 J. Rodolfo, H. Rajbenbach, J. P. Huignard, Opt. Engin. **34**, 1166 (1995).

5.157 H. Rajbenbach, C. Touret, J. P. Huignard, M. Curon, C. Bricot, Proc. SPIE **2752**, 214 (1996).

5.158 H. Rajbenbach, Proc. SPIE **2237**, 329 (1994).

5.159 G. S. Pati, A. Roy, K. Singh, Opt. Commun. **129**, 81 (1996).

5.160 D. T. H. Liu, L. Cheng, Appl. Opt. **31**, 5675 (1992).

5.161 G. Asimellis, M. Cronin Golomb, J. Khoury, J. Kane, C. Woods, Appl. Opt. **34**, 8154 (1995).

5.162 G. Asimellis, J. Khoury, C. L. Woods, Opt. Engin. **36**, 2392 (1997).

5.163 M. D. Lasprilla, S. Granieri, N. Bolognini, Optik **105**, 61 (1997).

5.164 M. Duelli, A. R. Pourzand, N. Collings, R. Dandliker, Opt. Lett. **22**, 87 (1997).

5.165 B. L. Volodin, B. Kippelen, K. Meerholz, B. Javidi, N. Peyghambarian, Nature **383**, 58 (1996).

5.166 C. Alves, G. Pauliat, G. Roosen, Opt. Mat. **4**, 423 (1995).

5.167 D. D. Nolte, D. H. Olson, G. E. Doran, W. H. Knox, A. M. Glass, J. Opt. Soc. Am. **B 7**, 2217 (1990).

5.168 A. Partovi, A. M. Glass, D. H. Olson, G. J. Zydzik, K. T. Short, R. D. Feldmann, R. F. Austin, Opt. Lett. **17**, 655 (1992).

5.169 B. H. Soffer, G. J. Dunning, Y. Owechko, E. Marom, Opt. Lett. **11**, 118 (1986).

5.170 G. J. Dunning, E. Marom, Y. Owechko, B. H. Soffer, Opt. Lett. **12**, 346 (1987).

5.171 E. G. Paek, D. Psaltis, Opt. Engin. **26**, 428 (1987).

5.172 H. J. White, N. B. Aldrige, I. Lindsay, Opt. Engin. **27**, 30 (1988).

5.173 L. Wang, V. Esch, R. Feinleib, L. Zhang, R. Jin, H. M. Chou, R. W. Sprague, H. A. Macleod, G. Khitrova, H. M. Gibbs, K. Wagner, D. Psaltis, Appl. Opt. **27**, 1715 (1988).

5.174 A. Yariv, S. Kwong, K. Kyuma, Appl. Phys. Lett. **48**, 1114 (1986).

5.175 H. Xu, Y. Yuan, Y. Yu, K. Xu, Y. Xu, Appl. Opt. **29**, 3375 (1990).

5.176 H. Xu, Y. Yuan, J. Zhang, K. Xu, Opt. Commun. **92**, 326 (1992).

5.177 G. Lu, M. Lu, F. T. S. Yu, Appl. Opt. **34**, 5109 (1995).

5.178 H. Kang, C. Yang, G. Mu, Z. Wu, Opt. Lett. **15**, 637 (1990).

5.179 Y. Tanaka, Z. Chen, T. Kasamatsu, T. Shiosaki, Jap. J. Appl. Phys. **30**, 2359 (1991).

5.180 Z. Chen, T. Kasamatsu, T. Shiosaki, Jap. J. Appl. Phys. **31**, 3178 (1992).

5.181 M. Ingold, M. Duelli, P. Günter, J. Opt. Soc. Am. **B 9**, 1327 (1992).

5.182 M. Duelli, R. S. Cudney, P. Günter, Opt. Engin. **34**, 2044 (1995).

5.183 M. Duelli, G. Montemezzani, C. Keller, F. Lehr, P. Günter, JEOS Pure Appl. Opt. **3**, 215 (1994).

5.184 M. Duelli, R. S. Cudney, P. Günter, Opt. Commun. **123**, 49 (1996).

6 Second-Harmonic Generation in Ferroelectric Waveguides

T. Pliska, D. Fluck, and P. Günter

This chapter reviews optical frequency doubling in waveguides, focusing on blue and green light generation. The main research efforts in this area have been aimed at the realization of compact all-solid state blue lasers based on near-infrared diode lasers, e.g. AlGaAs and InGaAs diode lasers, and of diode-pumped solid state lasers, e.g. Nd:YAG lasers, frequency doubled in a nonlinear waveguide. Waveguides provide ideal structures for efficient frequency conversion because of the high optical intensity that can be maintained over a long interaction distance [6.1]. Despite the rapid development in blue–green semiconductor lasers and first reports on ZnSe- and GaN-based diode lasers [6.2–3], frequency-doubled lasers based on near-infrared diode lasers still offer advantages regarding output power scalability, reliability and lifetime.

In general, the suitability of a material for guided-wave nonlinear optical interactions depends critically on its optical, mechanical and thermal properties. In addition, the possibility of forming waveguides by a reliable, reproducible and simple technique is a key issue in nonlinear integrated optics. Despite remarkable progress in the past two decades in the development of new nonlinear optical materials, practical devices for second-harmonic generation have remained limited to a small number of materials: the four inorganic ferroelectrics lithium niobate ($LiNbO_3$), lithium tantalate ($LiTaO_3$), potassium niobate ($KNbO_3$) and potassium titanyl phosphate ($KTiOPO_4$, KTP). Therefore the major part of this chapter will be devoted to waveguide devices in these ferroelectrics. We shall not focus only on the optical properties of the waveguides themselves, but will also discuss material technology and structural analysis of nonlinear optical waveguide materials, as far as these issues are related to the optical properties. In addition, where we feel that reference to other than ferroelectric materials, e.g. poled polymers, might provide further insight into the concepts, techniques, and phenomena of nonlinear integrated optics, we will extend our discussion to these other material classes.

6.1 Second-Harmonic Generation in Waveguides: Basic Concepts

In this section, we give an introduction to second-harmonic generation in waveguides. It will provide a basis for the discussion of experimental results in the subsequent sections. Although we will focus on second-harmonic generation, it should be noted that the concepts for any other guided-wave second-order nonlinear optical process, such as difference frequency generation, sum frequency generation and optical parametric oscillation, are identical. For a more comprehensive overview of second-order nonlinear optical effects we refer to Sects. 3.1 and 3.2.

6.1.1   Planar and Channel Waveguides

The principles of nonlinear integrated optics have been outlined in a number of review papers, e.g. [6.1, 6.4–6]. The key to efficient frequency conversion is to maintain an intensity as large as possible over a distance as long as possible because the second-harmonic field grows with intensity of the fundamental wave and interaction length. Waveguides provide the optimum geometry for the following reasons:

1. An optical waveguide confines light to cross-sectional transverse dimensions of the order of the wavelength of the radiation, i.e. close to the diffraction-limited minimum.
2. For phase matched interactions, the efficiency grows with the square of the interaction length. This is in contrast to Gaussian beams, where the efficiency grows only linearly with length because of diffraction.

The three most common forms of waveguides are planar (slab) waveguides, channel (strip) waveguides and fibers. Here, we shall concentrate on planar and channel waveguides which are the most important types for second-order processes, whereas fibers have been studied mainly in relation to third-order interactions.

Planar waveguides are specified by the following properties:

1. The optical field is confined in the x-direction (perpendicular to the surface) and has at least one maximum in the guiding layer, i.e. the region of high refractive index. In the y-direction (parallel to the surface) the field may be described as a plane wave or a diffracting Gaussian beam.
2. The fields in the bounding media decay exponentially with distance from the guiding layer.
3. The modes are identified by a single integer m which indicates the number of nodes of the field in the x-direction.
4. There are two orthogonally polarized families of modes, designated transverse electric TE$_m$ ($E_y \neq 0$, $E_x = E_z = 0$) and transverse magnetic TM$_m$ ($E_x \neq 0$, $E_z \neq 0$, $E_y = 0$) modes.
5. The effective index N_m of mode m lies between the index of the guiding layer and the highest index of the bounding media.

In *channel waveguides* the situation is more complicated, requiring a few simplifying assumptions:

1. The field is guided in both x- and y-directions with at least one maximum in the guiding core and exponentially decaying fields in the surrounding media.
2. The modes are specified by two integers m and n as a consequence of the fact that two resonance conditions need to be satisfied for two-dimensional confinement. Each component of the electrical field is written as

$$E_{mn}(x,y,z,t) = \frac{1}{2}\left(E_{mn}(z)\varepsilon_{mn}(x,y)\mathrm{e}^{\mathrm{i}(\omega t - \beta_{mn}z)} + \mathrm{c.c.}\right).\qquad (6.1)$$

$E_{mn}(z)$ denotes the amplitude of the field component, and $\varepsilon_{mn}(x,y)$ is the transverse field distribution associated mode mn, normalized such that

$$\iint \varepsilon_{mn}\varepsilon_{mn}^* \mathrm{d}x\mathrm{d}y = 1 .\qquad (6.2)$$

To allow for a simpler mathematical treatment the transverse components are often decoupled so that the field distribution may be approximated by

$$\varepsilon_{mn}(x,y) = \varepsilon_m(x)\varepsilon_n(y) .\qquad (6.3)$$

3. Again, there are two orthogonally polarized families of modes. For channel waveguides the modes labeled TE_{mn} have the dominant field component parallel to the y-axis, and the modes denoted by TM_{mn} parallel to the x-axis. Both types of mode also have smaller field components along the other two axes.
4. The effective index lies between the core refractive index and the highest index of the bounding media.

The detailed characteristics of a waveguide depend on the materials and the waveguide preparation technology used. For example, diffused or implanted waveguides are characterized by graded-index profiles rather than by sharp boundaries between the different regions. Barrier waveguides, typically formed by ion implantation, may have leaky modes, as the confinement is provided by a bounding region of finite width only. Figure 6.1 sketches five different types of channel waveguide geometry. For simplicity, the figure shows abrupt transitions of the refractive indices. In all five examples the light is essentially confined in the material with refractive index n_{g}. The refractive indices of the cover n_{c}, substrate n_{s} and strip-load n_{l} are smaller than the guide index n_{g}. Figure 6.1a shows a buried channel waveguide with a uniform cladding. Figure 6.1b displays the cross-section of an embedded channel waveguide which is, for example, typical of waveguides fabricated with masked diffusion. The raised channel waveguide, shown in Fig. 6.1c, can be fabricated by depositing a planar slab waveguide on a substrate, masking the strip and subsequently removing of the surrounding film by etching. The ridge waveguide displayed in Fig. 6.1d is formed similarly to the raised channel guide, with the only difference that the surround of the channel is not removed completely. Finally, the strip-loaded waveguide sketched in Fig. 6.1e

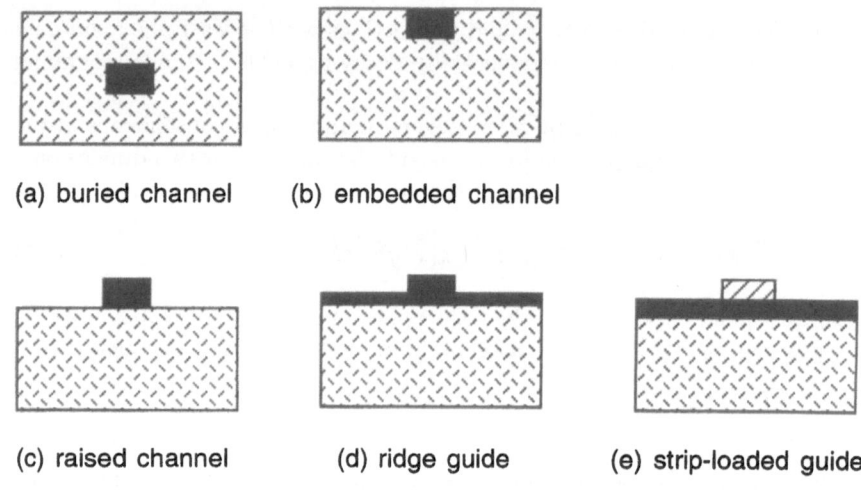

(a) buried channel   (b) embedded channel

(c) raised channel   (d) ridge guide   (e) strip-loaded guide

Fig. 6.1a–e. Cross-sections of five different channel waveguide types. ■ n_g (waveguide), ⬡ n_s (substrate), ⬡ n_l (load) (adapted from [6.291])

is fabricated by depositing a strip of lower index $n_g < n_g$ on a planar waveguide. The ridge waveguide and strip-loaded waveguide use a "propagating surround", which means that the planar waveguides on both sides of the channel support at least one guided mode.

6.1.2 Figures of Merit for Second-Harmonic Generation in Waveguides

When two incident optical fields interact within a medium the induced second-order nonlinear optical polarization P^{NL} can be written, as discussed in Sect. 3.1, as a sum of the products of the interacting fields:

$$P_i^{NL}(\omega_3) = \varepsilon_0 d_{ijk} E_j(\omega_1) E_k(\omega_2) \ , \tag{6.4}$$

where we use Einstein's sum notation. ε_0 denotes the permittivity constant and d_{ijk} the nonlinear optical susceptibility tensor element. The nonlinear polarization provides coupling between the waves E_i, E_j and E_k that oscillate at different frequencies in general. For second-harmonic generation E_j and E_k both oscillate at frequency $\omega_1 = \omega_2 \equiv \omega$, inducing a field E_i at frequency $\omega_3 = 2\omega$. Substituting the nonlinear polarization into Maxwell's equations yields the coupled-mode formalism giving the exchange rate between the amplitudes of the different waves involved in the process [6.7]. For the sake of clarity we will make a few simplifying assumptions in the following, which, however, do not affect the generality of our discussion:

1. There is only one single input beam ($j = k$), implying that we consider only type I second-harmonic generation. We therefore drop the vectorial subscripts.

2. We consider only the interaction between one fundamental and one second-harmonic mode. However, both modes involved in the process may be of different order.
3. We assume interactions between TE modes, expressed by (6.1). The extension to TM modes merely requires proper normalization of the relevant field components of the magnetic field.
4. We make use of the non-depleted pump approximation, i.e. we neglect the depletion of the fundamental beam due to the energy transfer into the second-harmonic beam. This approximation holds for most continuous wave interactions on a sub-Watt power level, and therefore applies to most of the experiments described in the forthcoming sections.

Under these assumptions the coupled mode theory yields the following growth rate of the second-harmonic amplitude:

$$\frac{\mathrm{d}E_{2\omega}(z)}{\mathrm{d}z} = -\mathrm{i}K E_{\omega}^2(z)\mathrm{e}^{-\mathrm{i}\Delta\beta z} , \tag{6.5}$$

where the expression

$$\Delta\beta = \beta_{2\omega} - 2\beta_{\omega} \tag{6.6}$$

with

$$\beta_i = \frac{2\pi}{\lambda_i} N_i \tag{6.7}$$

describes the phase velocity mismatch between the fundamental and second-harmonic waves. λ_i is the vacuum wavelength of the corresponding wave. The key parameter in (6.5) is the coupling constant K given by

$$K = \frac{\omega}{c} \frac{d_{\mathrm{mat}}}{N_{2\omega}} \iint \overline{d}(x,y)\varepsilon_{2\omega}(x,y)\varepsilon_{\omega}^2(x,y)\mathrm{d}x\mathrm{d}y , \tag{6.8}$$

where c is the speed of light in vacuum. The nonlinear optical susceptibility is written in the form

$$\overline{d}(x,y) = \frac{d_{\mathrm{wg}}(x,y)}{d_{\mathrm{mat}}} , \tag{6.9}$$

where d_{mat} is the effective nonlinear coefficient of the material, while $d_{\mathrm{wg}}(x,y)$ is the nonlinearity of the waveguiding layer, which may differ from the bulk material and may depend on the spatial coordinates. $\overline{d}(x,y)$ is thus the normalized nonlinear susceptibility. Integrating (6.5) yields the second-harmonic power $P_{2\omega}(L)$ at the end of the waveguide

$$P_{2\omega}(L) = \frac{2\omega^2}{\varepsilon_0 c^3} \frac{d_{\mathrm{mat}}^2}{N_{2\omega} (N_{\omega})^2} L^2 [P_{\omega}(0)]^2 h(\Delta\beta, \alpha_{\omega}, \alpha_{2\omega}, L) \Gamma , \tag{6.10}$$

where L denotes the guide length and $P_\omega(0)$ the fundamental power coupled into the waveguide. The function

$$h\left(\Delta\beta, \alpha_\omega, \alpha_{2\omega}, L\right) = \exp\left(-\left(\frac{\alpha_{2\omega}}{2} + \alpha_\omega\right)L\right)$$
$$\times \frac{(\sinh(\Delta\alpha\, L/2)\cos(\Delta\beta\, L/2))^2 + (\cosh(\Delta\alpha\, L/2)\sin(\Delta\beta\, L/2))^2}{(\Delta\alpha^2 + \Delta\beta^2)\, L^2/4} \quad (6.11)$$

with

$$\Delta\alpha = \frac{\alpha_{2\omega}}{2} - \alpha_\omega \quad (6.12)$$

contains the dependence of the second-harmonic power upon the phase mismatch $\Delta\beta$ and the linear losses α_i at each of the two wavelengths. For a loss-free waveguide, $\alpha_\omega = \alpha_{2\omega} = 0$ the function h is given by the characteristic interference term

$$h\left(\Delta\beta, L\right) = \frac{\sin^2\left(\Delta\beta L/2\right)}{(\Delta\beta L/2)^2}\, . \quad (6.13)$$

On the other hand, in the phase matched case, $\Delta\beta = 0$, h may be expressed as

$$h\left(\alpha_\omega, \alpha_{2\omega}, L\right) = \exp\left(-\left(\frac{\alpha_{2\omega}}{2} + \alpha_\omega\right)L\right)\frac{(\sinh(\Delta\alpha\, L/2))^2}{((\Delta\alpha\, L/2))^2}\, . \quad (6.14)$$

Finally, the overlap integral

$$\Gamma = \left(\iint \overline{d}(x,y)\varepsilon_{2\omega}(x,y)\varepsilon_\omega^2(x,y)\mathrm{d}x\mathrm{d}y\right)^2 \quad (6.15)$$

represents the inverse effective cross-section of the interaction, provided the normalization condition (6.2) is satisfied. Such an effective area is common to all nonlinear guided wave phenomena.

The direct comparison between the theoretical prediction and experimental results of a second-harmonic measurement is complicated by the uncertainty of several parameters, including the relevant material properties such as the nonlinearity d_{mat} and the waveguide nonlinearity profile $d_{\mathrm{wg}}(x,y)$, propagation loss, overlap integral and the spectral properties of the pump laser. It is common to introduce figures of merit that allow comparison of different waveguide devices. However, these figures of merit, or normalized conversion efficiencies, are not used consistently throughout the literature. For the sake of clarity it is useful to distinguish between internal and external conversion efficiencies. Equation (6.10) relates the internal powers $P_\omega(0)$ and $P_{2\omega}(L)$ which denote the fundamental power at the beginning of the waveguide and the second-harmonic power at the end of the waveguide, respectively. They are related to the external powers P_ω^{in} and $P_{2\omega}^{\mathrm{out}}$ by

$$P_\omega(0) = T_c T_f^{\mathrm{in}} P_\omega^{\mathrm{in}} \quad (6.16)$$

$$P_{2\omega}(L) = \left(T_f^{\text{out}}\right)^{-1} P_{2\omega}^{\text{out}} , \tag{6.17}$$

where T_c denotes the coupling efficiency into the waveguide without Fresnel reflection loss, T_f^{in} the Fresnel transmission coefficient of the fundamental at the front face of the guide, and T_f^{out} the Fresnel transmission coefficient of the second-harmonic at the end face of the guide. From a strictly application-oriented viewpoint the relevant parameter to characterize a device is the overall efficiency η_{device} defined by the ratio of the second-harmonic output to the input power P_ω^{in}, i.e.

$$\eta_{\text{device}} \equiv \frac{P_{2\omega}^{\text{out}}}{P_\omega^{\text{in}}} . \tag{6.18}$$

Many laboratory experiments, however, are not optimized with respect to this figure. A quantity reflecting the internal properties of the waveguide is more appropriate for its characterization. Because of the quadratic dependence of the second-harmonic power on the fundamental power and guide length, an internal normalized conversion efficiency can be introduced by

$$\eta \equiv \frac{1}{L^2} \frac{P_{2\omega}(L)}{[P_\omega(0)]^2} \tag{6.19}$$

in units of $\% \text{W}^{-1}\text{cm}^{-2}$. Substituting (6.19) into (6.10) yields

$$\eta = \frac{2\omega^2}{\varepsilon_0 c^3} \frac{d_{\text{mat}}^2}{N_{2\omega} (N_\omega)^2} h\, \Gamma. \tag{6.20}$$

For an ideal loss-free waveguide and under phase matched conditions, we find from (6.14) $h = 1$, so that the maximum normalized conversion efficiency is given by

$$\eta_0 = \frac{2\omega^2}{\varepsilon_0 c^3} \frac{d_{\text{mat}}^2}{N_{2\omega} (N_\omega)^2} \Gamma. \tag{6.21}$$

The difficulty associated with the normalized conversion efficiency defined by (6.19) is due to the fact that the incoupled fundamental power $P_\omega(0)$ is often unknown, because it is difficult to determine the coupling efficiency T_c experimentally. Therefore, it has become customary to reference the normalized conversion efficiency to the internal fundamental power $P_\omega(L)$ at the guide end by defining

$$\tilde{\eta} \equiv \frac{1}{L^2} \frac{P_{2\omega}(L)}{[P_\omega(L)]^2} = \frac{2\omega^2}{\varepsilon_0 c^3} \frac{d_{\text{mat}}^2}{N_{2\omega} (N_\omega)^2} h\, \Gamma \exp\left(2\alpha_\omega L\right) , \tag{6.22}$$

where we have used $P_\omega(L) = P_\omega(0) \exp\left(-\alpha_\omega L\right)$. $P_\omega(L)$ is defined by analogy with $P_{2\omega}(L)$, given by (6.17). The normalized conversion efficiency introduced by (6.22) is unsatisfactory from a physical viewpoint because it neglects the effect of propagation loss and tends to overestimate the conversion efficiency if the guide is lossy. Nevertheless, it can provide useful information if $\alpha_{2\omega} = 2\alpha_\omega$. In this case, $h = \exp(-2\alpha_\omega L)$ according to (6.14), so that with (6.21) and (6.22) we find

$$\tilde{\eta} = \eta_0. \tag{6.23}$$

Equation (6.23) states, rather surprisingly, that for a waveguide with $\alpha_{2\omega} = 2\alpha_\omega$, the measured internal conversion efficiency normalized to the fundamental power at the guide end is equal to the normalized conversion efficiency of the same waveguide without propagation loss. Thus $\tilde{\eta}$ provides an experimental measure of the maximum attainable conversion efficiency. The insensitivity of the ratio $\tilde{\eta}/\eta_0$ to propagation loss may be explained by the fact that the effects of propagation loss of the second-harmonic and fundamental waves balance each other with respect to this ratio. Indeed, the condition $\alpha_{2\omega} \approx 2\alpha_\omega$ holds as a good approximation for many waveguides, which allows one to experimentally estimate the loss-free conversion efficiency η_0 with good accuracy.

However, for full characterization of the waveguide properties the effect of propagation loss may not be neglected, and the normalized conversion efficiency should be referenced to $P_\omega(0)$. Moreover, scaling the figure of merit with respect to the square of the guide length can be unjustified if material and waveguide nonuniformity, attenuation, thermal effects or optical damage limit the useful guide length. Therefore, we prefer to use the normalized conversion efficiency defined by

$$\overline{\eta} \equiv \frac{P_{2\omega}(L)}{[P_\omega(0)]^2} = \eta\, L^2 \tag{6.24}$$

in units of $\%\text{W}^{-1}$.

Finally, when analyzing second-harmonic generation, the spectral properties of the pump laser must be taken into account. Equation (6.10) has been derived for a laser oscillating on exactly one axial mode at the fundamental wavelength, i.e. a single-frequency laser. According to Ducuing and Bloembergen [6.8] an axial multimode pump leads to an enhancement of the measured second-harmonic power by a factor g given by

$$g = 2 - \frac{1}{m}, \tag{6.25}$$

where m is the number of independently oscillating laser modes. For $m \gg 1$, we find $g \approx 2$, implying that with a multimode pump the second-harmonic efficiency is twofold enhanced as long as the laser linewidth is within the material acceptance bandwidth. This enhancement can be explained by an additional contribution from sum frequency generation between different axial modes.

On the basis of (6.10) we can discuss some important issues for efficient frequency conversion in waveguides:

Table 6.1. Nonlinear optical coefficients and figures of merit (f.o.m.) of ferroelectric oxides commonly used for blue light second-harmonic generation. All values were measured at a wavelength of 1.064 μm and are referenced to d_{11}(quartz) = $0.3\,\mathrm{pmV^{-1}}$

	d_{ij} $(\mathrm{pmV^{-1}})$	d_{ii} $(\mathrm{pmV^{-1}})$	d_{eff} $(\mathrm{pmV^{-1}})^a$	n	f.o.m. d^2/n^3 $(\mathrm{pm^2V^{-2}})$
LiNbO₃	4.6 (d_{31})			2.2	2
[6.51]		25.2 (d_{33})	16.0	2.2	24
KNbO₃	11.3 (d_{31})			2.2	12
[6.50]	12.8 (d_{32})			2.2	15
		19.5 (d_{33})	12.4	2.2	14
LiTaO₃	0.85 (d_{31})			2.2	0.007
[6.51]		13.8 (d_{33})	8.9	2.2	7
KTiOPO₄	1.4 (d_{31})			1.8	0.3
[6.217]	2.65 (d_{32})			1.8	1
		10.7 (d_{33})	6.8	1.8	8

^{a} First-order quasi-phase-matching, $d_{\mathrm{eff}} = (2/\pi)d_{33}$ according to (6.29)

1. The expression

$$\frac{d^2_{\mathrm{mat}}}{N_{2\omega}N^2_{\omega}}$$

is the material figure of merit. Obviously, the quadratic dependence upon the nonlinear susceptibility implies the use of materials with high optical nonlinearities. Table 6.1 compares the relevant nonlinear optical coefficients and the figures of merit of the four ferroelectric oxides considered in this chapter. Furthermore, for guided wave applications, it is essential that the nonlinearity of the complete guiding region is preserved during the guide fabrication process so that $d_{\mathrm{wg}} \approx d_{\mathrm{mat}}$. As we will discuss in Sects. 6.3.2 and 6.6.1 in more detail, standard procedures like proton exchange, ion diffusion or ion implantation were observed to affect the material structure resulting in partial or total loss of the nonlinearity in the waveguide.

2. The term

$$L^2 h\left(\Delta\beta, \alpha_{\omega}, \alpha_{2\omega}, L\right)$$

includes the dependence of the second-harmonic power on the linear properties of the waveguide, i.e. propagation constants, attenuation and guide length. In practice, efficient frequency conversion is possible only under phase matching, $\Delta\beta = 0$. Methods to meet this condition such as birefringent phase matching, quasi-phase-matching, modal-dispersion-phase-matching and Cerenkov-type phase matching will be discussed in Sect. 6.1.3. Satisfying the phase matching condition requires detailed knowledge of the propagation constants as a function of the guide parameters such as thickness and width.

Moreover, phase matching must be maintained over a distance as long as possible. Nonuniformities of $\Delta\beta$, resulting e.g. from thickness and width variations in the guide geometry or from material index inhomogeneities, reduce the phase matched interaction length. This often imposes stringent requirements on fabrication technology. Another important parameter for efficient interactions is the linear loss at both the fundamental and second-harmonic wavelength, usually given in units of cm^{-1} or $dB\,cm^{-1}$ ($1\,cm^{-1} = 4.34\,dB\,cm^{-1}$). In dielectric waveguides, attenuation originates from absorption, volume and surface scattering, and leakage, depending on the waveguide type. An attenuation of $2\,dB\,cm^{-1}$ at both wavelengths reduces the conversion efficiency already by a factor of two in a 1 cm long waveguide. Minimizing the loss requires the identification of its origin, optimizing the guide fabrication process and adjusting the guide geometry for minimum loss.

3. The overlap integral, given by (6.15), represents the inverse effective cross-section of the waveguide. Its value is given by the specific interaction, the details of the index profile and the interacting modes. The larger the effective area, the less efficient the conversion process. The overlap integral usually limits the number of modes which lead to efficient interactions. An example of two possible scenarios for second-harmonic generation in a planar waveguide is shown in Fig. 6.2. The conversion process $TE_0^\omega + TE_0^\omega \rightarrow TE_0^{2\omega}$ gives rise to a large overlap. In contrast, for the conversion process $TE_0^\omega + TE_0^\omega \rightarrow TE_1^{2\omega}$ there are both positive and negative contributions to the overlap integral which cancel each other almost completely, leading to a large effective cross-section and a small efficiency.

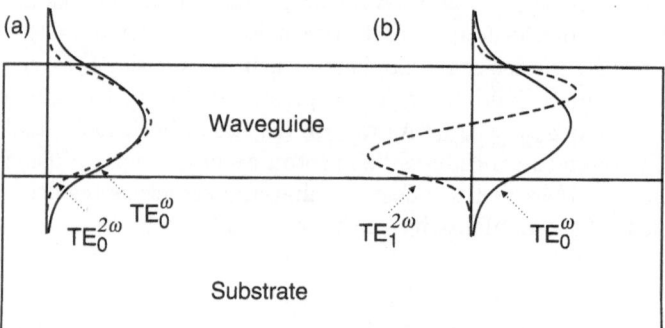

Fig. 6.2. Schematic illustrating the field distributions of two mode conversion processes for second-harmonic generation: (**a**) $TE_0^\omega + TE_0^\omega \rightarrow TE_0^{2\omega}$ interaction; (**b**) $TE_0^\omega + TE_0^\omega \rightarrow TE_1^{2\omega}$ interaction (adapted from [6.1])

6.1.3   Phase Matching Schemes for Second-Harmonic Generation in Waveguides

A technique for maintaining the relative phase between the interacting waves is a prerequisite for efficient nonlinear frequency conversion, as outlined in Sect. 3.5.1. If the interacting waves are not phase matched, the second-harmonic wave oscillates periodically along the waveguide axis due to destructive interference. This spatial interference pattern is given by the factor $\sin^2(\Delta\beta L/2)$ as can be deduced from (6.10) and (6.13). Two adjacent peaks of this spatial oscillation are separated by twice the so-called coherence length l_c, which we define by

$$l_c \equiv \frac{\pi}{\Delta\beta} = \frac{\lambda_\omega}{4\left(N_{2\omega} - N_\omega\right)} \ . \tag{6.26}$$

The coherence length gives a measure of the waveguide length that is useful for second-harmonic generation. Under non-phase matched circumstances l_c is of the order of 10^{-2} cm. For example, for $\Delta N = 10^{-2}$ and $\lambda_\omega = 1\,\mu\text{m}$, we find $l_c = 50\,\mu\text{m}$. This is far too short for efficient frequency conversion. In the phase matched case, $\Delta\beta = 0$, and thus $l_c \to \infty$.

A   Birefringent Phase Matching

According to (6.26) the coherence length is infinite if both waves travel at the same phase velocity, i.e.

$$N_\omega = N_{2\omega}. \tag{6.27}$$

The most direct approach to phase matching takes advantage of the natural birefringence of the material. In normally dispersive materials the index of refraction increases with the frequency of light, making it impossible to satisfy (6.27) if both the fundamental and second-harmonic waves are of identical polarization. However, if the waves have different polarization phase matching can be achieved under specific circumstances. Mixed polarizations of the fundamental and second-harmonic imply the use of off-diagonal nonlinear tensor elements.

Figure 6.3a illustrates the scheme for birefringent phase matching. We consider second-harmonic generation in a KNbO$_3$ channel waveguide as an example. Because KNbO$_3$ is biaxial there are two principal configurations, denoted A and B. A fundamental TE mode at a wavelength of 0.876 μm polarized along the b-axis can be converted to a second-harmonic mode at 0.438 μm polarized along c (configuration A). This interaction is provided by the nonlinear optical coefficient d_{32}. Similarly, phase matching is possible between a fundamental mode at 1.014 μm polarized along a and a second-harmonic mode at 0.507 μm polarized along c (configuration B, coefficient d_{31}). Figure 6.3b displays the wave vector diagram for birefringent phase matching.

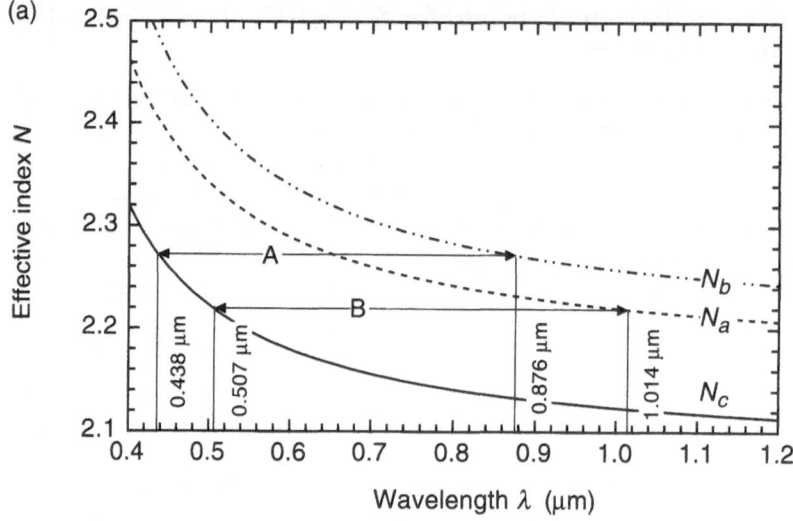

(a)

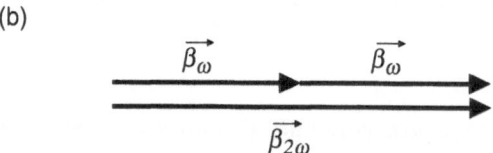

(b)

Fig. 6.3a,b. Schematic illustrating birefringence phase matched second-harmonic generation in a KNbO$_3$ waveguide. (**a**) A fundamental wave at a wavelength of 0.876 µm polarized parallel to the b-axis is phase matched to a second-harmonic wave polarized parallel to the c-axis (A). Similarly, phase matching is obtained at 1.014 µm for a fundamental wave polarized parallel to a (B). (**b**) Corresponding wave vector diagram (calculated from the data given in [6.256–257, 6.276])

The phase matching condition (6.27) can be satisfied by various means. Temperature tuning makes use of the temperature dependence of the refractive indices (thermo-optic coefficients) which allows us to obtain phase matching at a fixed fundamental wavelength. Alternatively, at a fixed waveguide temperature, tuning of the fundamental wavelength may provide phase matching. For the design of devices with phase matching at a specific temperature and wavelength, phase matching can be obtained by adjustment of the direction of propagation relative to the crystal axes. This will be illustrated in Sect. 6.6.4, when we discuss angle-tuned phase matching in KNbO$_3$ channel waveguides.

The great benefit of birefringent phase matching lies in that it makes use of the intrinsic material properties and does not require any additional processing step, such as periodic domain reversal for quasi-phase-matching.

Despite its simplicity birefringent phase matching often suffers from two limitations imposed by the material properties:

1. In many materials, e.g. $LiNbO_3$, $LiTaO_3$ and $KTiOPO_4$, the off-diagonal nonlinear tensor elements are modest, typically below $5\,pm\,V^{-1}$ (Table 6.1), resulting in small conversion efficiencies. This does not apply to $KNbO_3$, where the off-diagonal tensor elements are larger, allowing one to make efficient use of birefringent phase matching.
2. Among the four ferroelectrics discussed here, only $KNbO_3$ has sufficiently large birefringence that permits birefringent phase matching for blue light generation. In $LiNbO_3$, $LiTaO_3$ and $KTiOPO_4$ the birefringence is too small. Thus, in these materials, quasi-phase-matching for frequency up-conversion into the blue spectral range is mandatory.

B   Modal-Dispersion-Phase-Matching

An approach related to birefringence phase matching is modal-dispersion-phase-matching. This technique takes advantage of the fact that modes of different order obey different dispersion relations. An example is displayed in Fig. 6.4 where the effective index of a planar polymer waveguide is plotted as a function of guide thickness [6.9]. The process considered here is second-harmonic generation at a fundamental wavelength of $1.55\,\mu m$. Equal effective

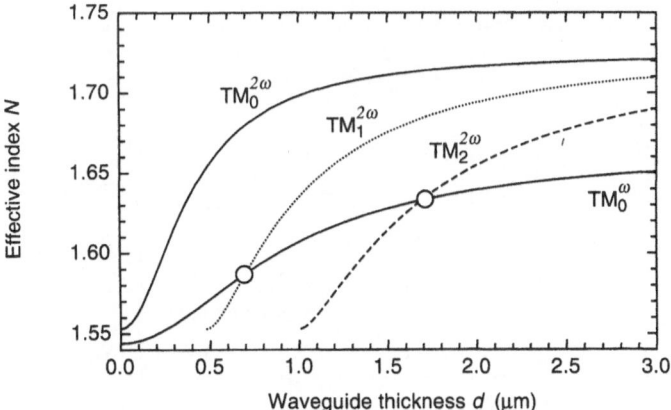

Fig. 6.4. Schematic showing modal-dispersion-phase-matching. The curves indicate the effective indices as a function of guide thickness of a poly(styrene-maleic anhydride) polymer waveguide used for second-harmonic generation at $1.55\,\mu m$ wavelength. Because of the dependence of the effective indices on the guide thickness, interactions between modes of different order and identical polarization can be phase matched at specific guide thicknesses, as indicated by open circles. The wave vector diagram for modal-dispersion-phase-matching is the same as for birefringence phase matching (Fig. 6.3b) (adapted from [6.9])

indices are obtained for a guiding layer thickness of $0.7\,\mu m$ and $1.7\,\mu m$ for the $TM_0^\omega + TM_0^\omega \to TM_1^{2\omega}$ and $TM_0^\omega + TM_0^\omega \to TM_2^{2\omega}$ interactions, respectively.

Modal-dispersion-phase-matching is attractive for materials with large diagonal nonlinear susceptibilities, e.g. polymers, because it does not require the use of mixed polarizations, in contrast to birefringent phase matching. An intrinsic drawback, however, is that the second-harmonic is carried in a higher-order mode which is often undesirable, because the overlap integral is considerably reduced for interactions between modes of different order due to interference effects along the guide cross-section, as illustrated in Fig. 6.2. Modal-dispersion-phase-matching therefore needs optimization of the waveguide structure to improve the overlap. This has been demonstrated in polymers where multilayered waveguides consisting of nonlinear optically active polymers with different glass transition temperatures have been selectively poled, as schematically shown in Fig. 6.5 [6.9]. By this means the sign of

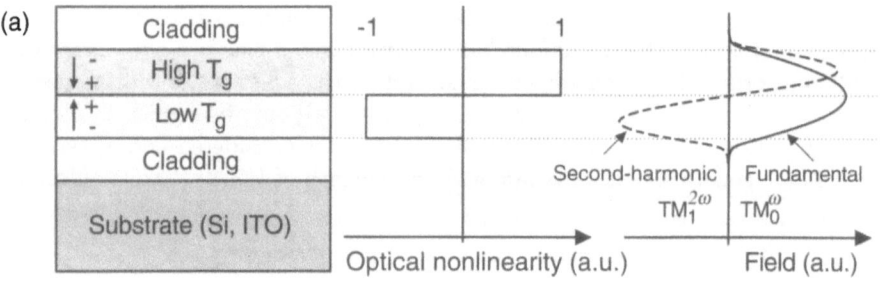

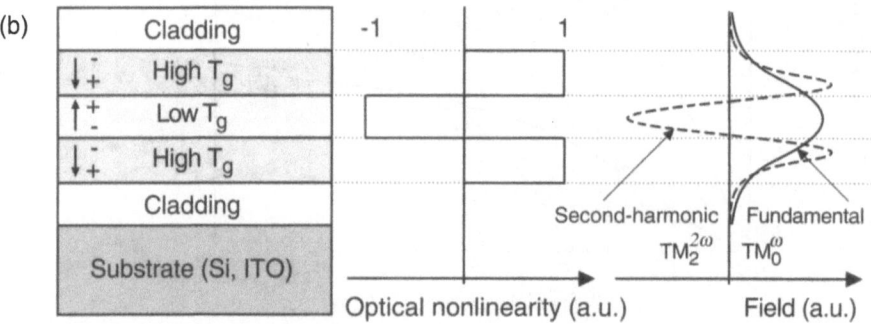

Fig. 6.5a,b. Multilayer polymer waveguide structures for optimized field overlap. The core of the waveguide consists of either (**a**) two or (**b**) three layers of nonlinear optically active polymers with different glass transition temperatures. Selective electric field poling allows fabrication of a structure where the orientation of the nonlinear chromophores of the low T_g layer is opposite to that of the high T_g layer. Hence the sign of the nonlinearity is reversed yielding optimized overlap integrals for (**a**) the $TM_0^\omega + TM_0^\omega \to TM_1^{2\omega}$ interaction and (**b**) the $TM_0^\omega + TM_0^\omega \to TM_2^{2\omega}$ interaction, respectively. (Adapted from [6.9])

the nonlinearity changes each time the second-harmonic field changes sign. As a result, the contribution from different layers to the overlap integral do no longer cancel, leading to a significant improvement of the conversion efficiency. Conversion efficiencies of up to $14\,\%\mathrm{W}^{-1}\mathrm{cm}^{-2}$ at $1.55\,\mu\mathrm{m}$ have been demonstrated in such waveguides [6.10].

C   Anomalous-Dispersion-Phase-Matching

Although anomalous-dispersion-phase-matching is applicable to polymer waveguides mainly, we shall give a brief outline here because it provides an interesting concept for phase matching. In this scheme a nonlinear chromophore with an absorption maximum between the fundamental and the second harmonic wavelengths is incorporated into a host polymer matrix that exhibits normal dispersion. At an appropriate chromophore concentration, the normal dispersion of the host material is canceled by the anomalous dispersion of the chromophore resulting in equal refractive indices for both waves.

An experimental realization of this concept was reported in [6.11]. Figure 6.6a shows the dispersion of the refractive indices and absorption, respectively, of the system with poly(methyl methacrylate) (PMMA) as the host and a thiobarbituric acid derivative (TBA) as the nonlinear chromophore

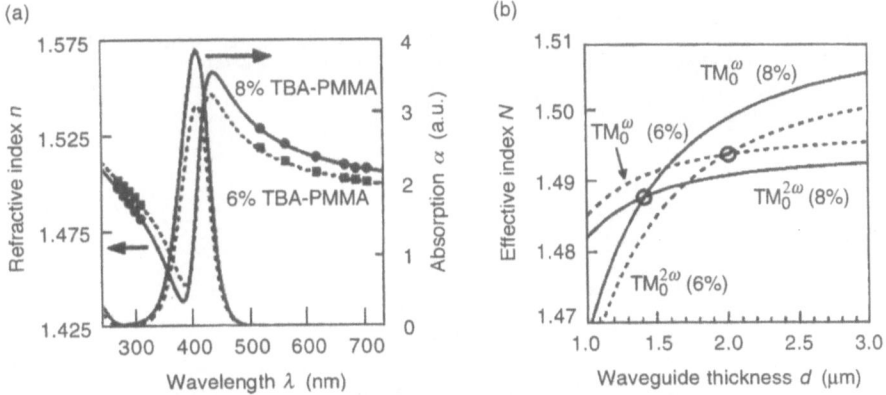

Fig. 6.6a,b. Schematic of anomalous-dispersion-phase-matching. (a) Refractive index as a function wavelength of the TBA-PMMA polymer film (poly(methyl methacrylate) (PMMA) doped with thiobarbituric acid (TBA) chromophores). The data were taken for two different chromophore concentrations (squares: 6%; points: 8%). The curves indicate fits according to a two-term Sellmeier equation. The two curves without data points indicate the measured absorption spectra. (b) Effective mode indices as a function of film thickness. Phase matching for $\mathrm{TM}_0^\omega + \mathrm{TM}_0^\omega \to \mathrm{TM}_0^{2\omega}$ is obtained for a film thickness of $1.4\,\mu\mathrm{m}$ (6% TBA concentration) and $2.0\,\mu\mathrm{m}$ (8% TBA concentration). Again, the wave vector diagram for anomalous-dispersion-phase-matching is the same as for birefringence phase matching (Fig. 6.3b). (Adapted from [6.11])

(TBA-PMMA), used in [6.11]. Figure 6.6b shows the dependence of the effective index on the guide thickness of a waveguide formed from this material by a film floating technique. The phase matched length in this experiment was only 32 µm resulting in a rather modest conversion efficiency. 1.2 µW at 407 nm were generated with an average fundamental input power of 133 mW from a pulsed optical parametric oscillator.

In contrast to modal-dispersion-phase-matching, anomalous-dispersion-phase-matching allows phase matched interactions between the lowest-order modes at both wavelengths. The chromophores used for this technique can have large nonlinearities because their first electronic transition is in the visible region of the spectrum. However, great care has to be taken to avoid residual absorption in the transparency windows.

D   Quasi-Phase-Matching

Quasi-phase-matching (QPM) was devised in the early days of nonlinear optics [6.12–16]. This technique corrects the relative phase mismatch between the interacting waves by means of a built-in structural periodicity in the nonlinear material, as shown in Fig. 6.7a, resulting in a repeated inversion of the relative phase of the interacting waves. The phase is thus reset periodically so that the proper relation for growth of the generated wave is maintained

(a)
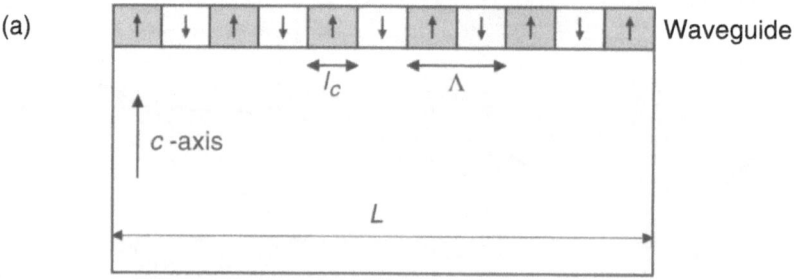

(b)
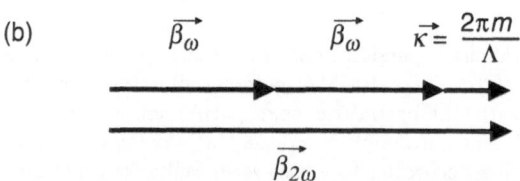

Fig. 6.7a,b. Schematic illustrating quasi-phase-matching in polar materials. (a) The material polarity in the waveguide is modulated with a period Λ that corresponds to the coherence length l_c of the nonlinear interaction (first-order quasi-phase-matching). (b) Wavevector diagram for quasi-phase-matching (adapted from [6.1])

on average. Viewed in the wave vector picture, the periodicity introduces an additional vector that compensates the phase mismatch (Fig. 6.7b). Thus the wave vector mismatch for second-harmonic generation may be written as

$$\Delta\beta = \beta_{2\omega} - 2\beta_{\omega} - \frac{2\pi m}{\Lambda} \, , \qquad (6.28)$$

where Λ is the grating period and m is an integer indicating the order of the quasi-phase-matched process.

The most effective type of periodic structure is one in which the sign or magnitude of the nonlinear optical coefficient is changed after each coherence length throughout the material. In ferroelectric crystals this technique usually involves forming regions of periodically reversed spontaneous polarization, as shown in Fig. 6.7a. In polymers where the second-order nonlinearity is induced by an applied electric field, the periodic structure can be obtained by periodically reversing the poling field.

The most rapid growth of the second-harmonic signal is obtained if the periodicity is equal to the coherence length. This case is referred to as first-order quasi-phase-matching. Alternatively, quasi-phase-matching may be accomplished by choosing a modulation period that is an odd integer times the coherence length. For this so-called higher-order quasi-phase-matching the grating fabrication tolerances are more relaxed than for first-order quasi-phase-matching as the grating period increases. On the other hand, the second-harmonic signal grows less rapidly and, therefore, higher-order quasi-phase-matching is less efficient.

The use of a periodic structure, giving rise to an additional vector component in the wave vector mismatch relation (6.28), introduces an additional degree of freedom for achieving phase matching. By properly setting the grating period phase matching can be achieved at any wavelength. This is particularly interesting for materials where a small or large birefringence limits the range for birefringent phase matching. Quasi-phase-matching also allows access to the diagonal nonlinear optical susceptibility tensor elements which in many materials are the only ones with sufficient magnitude for efficient frequency conversion, e.g. $LiNbO_3$, $LiTaO_3$, $KTiOPO_4$ and polymers.

For quasi-phase-matching the effective nonlinear optical coefficient is given by

$$d_{\mathrm{eff}} = \frac{2}{\pi m} d_{\mathrm{mat}} \sin(\pi m D) \, , \qquad (6.29)$$

where D is the duty cycle defined by the ratio of the length of the positively poled section and the total grating period Λ. The maximum nonlinear coefficient is thus obtained for first-order quasi-phase-matching and a duty cycle of 0.5. In this case the conversion efficiency is reduced by a factor of $(2/\pi)^2$ compared with birefringence phase matching with a nonlinear coefficient of identical magnitude. Note that with a duty cycle $\neq 0.5$, even-order quasi-phase-matching is also possible.

Although the concept of quasi-phase-matching had been developed early on, experimental implementation of the technique proved to be difficult for a long time because of the lack of appropriate techniques to prepare crystals. From a technological viewpoint quasi-phase-matching is a most challenging technique because it requires the fabrication of a regular grating with a period of a few microns over a length of several millimeters or centimeters. This applies in particular to blue light generation, where periods smaller than $4\,\mu\text{m}$ are required. Precise control of the domain growth, completeness of poling over the whole device length, and temporal and thermal stability of the domain structure are issues of concern related to this technique. Fabrication tolerances and tuning possibilities have been discussed in [6.17–18].

Quasi-phase-matching was initially demonstrated in a number of bulk experiments, for example with a stack of differently oriented GaAs plates [6.19–20], in LiNbO$_3$ and LiTaO$_3$ crystals grown by special techniques [6.21–24], or LiNbO$_3$ fibers [6.25]. It has been demonstrated in GaAs waveguides with a surface grating [6.26]. However, only after the successful fabrication of a Ti-diffused domain grating with an annealed proton-exchanged LiNbO$_3$ waveguide in 1989 [6.27] did it become a widely used technique in ferroelectric materials. We will present an outline of domain grating fabrication techniques in LiNbO$_3$ LiTaO$_3$ and KTiOPO$_4$ in Sects. 6.3.3, 6.4.2 and 6.5.2, respectively.

E   Cerenkov-Type Phase Matching

Cerenkov-type phase matching makes use of the coupling between a guided fundamental mode and a radiative second-harmonic mode. While guided modes are discrete, radiation modes have a continuous spectrum. The process is schematically illustrated in Fig. 6.8a. Points A and B indicate positions of identical phase modulo 2π of the fundamental wave. The second-harmonic waves, generated in A and B by the nonlinear polarization of the material and radiated at an angle θ into the substrate, interfere constructively if the condition

$$v_\omega^{\text{waveguide}} \cos \theta = v_{2\omega}^{\text{substrate}} \qquad (6.30)$$

is satisfied, where $v_\omega^{\text{waveguide}}$ and $v_{2\omega}^{\text{substrate}}$ denote the phase velocities of the fundamental wave in the waveguide and the second-harmonic wave in the substrate, respectively. Equation (6.30) immediately leads to the Cerenkov phase-matching condition

$$N_\omega = n_{2\omega}^{\text{substrate}} \cos \theta \,, \qquad (6.31)$$

where N_ω denotes the effective index of the guided fundamental wave and $n_{2\omega}^{\text{substrate}}$ the refractive index seen by the second-harmonic in the substrate. The corresponding wave vector diagram is shown in Fig. 6.8b. Equation (6.31) implies that for a guided fundamental mode ($N_\omega > n_\omega^{\text{substrate}}$) the allowed effective indices of the fundamental wave are given by the condition

$$n_{2\omega}^{\text{substrate}} > N_\omega > n_\omega^{\text{substrate}} \,. \qquad (6.32)$$

(a)

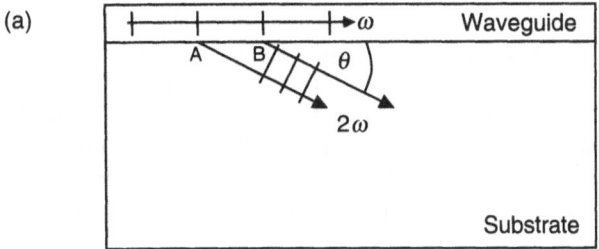

(b)

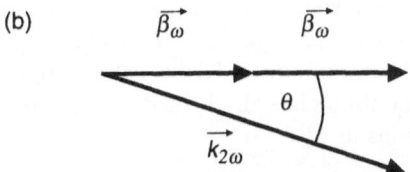

Fig. 6.8. (a) Schematic showing Cerenkov-type phase matching. The fundamental wave is confined in the waveguide, while the second-harmonic is radiated into the substrate. (b) Wave-vector diagram for Cerenkov-type phase matching (adapted from [6.1])

Thus phase matching is satisfied and maintained as long as the phase velocity of the guided fundamental mode is faster than that of the second-harmonic wave in the substrate.

Analysis of Cerenkov-type second-harmonic generation is not simple because both guided and radiative modes are involved. Most often either approximate analytical or numerical approaches are used. Cerenkov second-harmonic generation in planar waveguides was analyzed using a radiating antenna model [6.28], direct solution of propagation equations [6.29], and numerical analysis using a finite-element approach [6.30]. A Green's function method was used to describe Cerenkov second-harmonic generation in fibers [6.31]. Rectangular channel waveguides with step-index profiles used for Cerenkov second-harmonic generation have been analyzed by solving wave equations [6.32] and the effective index method [6.33]. Coupled-mode theory was used to investigate Cerenkov second-harmonic generation in embedded LiNbO$_3$ waveguides [6.34]. A more general coupled-mode approach yielding analytical expressions for the conversion efficiency and the near- and far-field patterns was used by Suhara et al. [6.35]. This analysis was further extended by Fluck et al. [6.36]. Krijnen et al. theoretically investigated Cerenkov second-harmonic generation in the strong conversion limit using a beam-propagation formalism and found interesting effects as a consequence of generation of coupled second-harmonic and fundamental solitary waves because of cascaded nonlinearities [6.37].

Cerenkov-type interactions are inherently less efficient than guided mode interactions because the process includes the overlap between a guided and radiated mode resulting in a smaller overlap integral. Moreover, because a radiation field is involved, the radiated power scales only linearly with length. On the other hand, the tolerances regarding guide quality are more relaxed. The far field of the radiated field has the shape of a semicircle that has to be transformed into a circular beam. This requires the design of special optical elements [6.38].

Experimentally, the first observation of Cerenkov second-harmonic generation was reported by Tien et al. in a ZnS film waveguide on a ZnO substrate [6.28]. Results on Cerenkov second-harmonic generation have been reported for proton-exchanged $LiNbO_3$ waveguides [6.29, 6.32, 6.39], strip-loaded $KTiOPO_4$ waveguides [6.40], and ion-implanted $KNbO_3$ waveguides [6.36, 6.41–42]. Cerenkov-type frequency doubling has also been demonstrated in a number of organic thin film waveguides [6.43].

6.2 Ferroelectric Waveguides: Overview

The suitability of ferroelectric materials for nonlinear optics was recognized by several authors in the 1960s [6.44–46]. The origin of the large nonlinearities in these materials is the high hyperpolarizability of the oxygen–transition metal (Nb, Ta, Ti) bonds. DiDomenico and Wemple described the nonlinearity in oxygen-octahedra ferroelectrics by polarization-induced Stark-like energy band shifts using the polarization potential tensor concept and proposed electro-optical and nonlinear optical device applications [6.47–48]. Jerphagnon introduced a description of the nonlinear susceptibilities on the basis of the tensor decomposition into irreducible parts and recognized the proportionality between the spontaneous polarization and the vector part of the nonlinearities [6.49]. Comprehensive overviews of standard nonlinear materials were given by Roberts [6.50] and recently by Shoji et al. [6.51].

The suitability of waveguides for electro-optical and nonlinear optical applications has stimulated research activities in integrated optics in ferroelectric materials. Three different approaches have been followed to form ferroelectric waveguides for nonlinear frequency conversion: chemical methods such as ion diffusion and ion exchange, ion implantation and thin film growth. While diffusion and implantation techniques use single crystals as the starting substrate, which is then subject to waveguide formation close to the surface layer, thin film technology requires the growth of a uniform and oriented ferroelectric film on a host substrate. Ferroelectric thin films fabricated by various deposition techniques such as rf sputtering, liquid-phase epitaxy, molecular beam epitaxy and pulsed laser deposition are reviewed in [6.52]. Growth of single-domain ferroelectric thin films of optical quality is most challenging because of the problems associated with lattice matching between film and substrate, film orientation, uniformity and surface roughness. Despite considerable effort and progress in the past years, thin film deposition techniques for ferroelectric waveguides are not as mature as diffusion and implantation methods at the moment.

Three mechanisms cause a refractive index change when a single crystal is subject to waveguide formation by ion diffusion or ion implantation: volume expansion, change of the atomic bond polarizability, and change of the structural factor. To a first approximation the refractive index n is linked to the polarizability α and particle density N via the Clausius–Mossotti equation given by

$$\frac{n^2 - 1}{n^2 + 2} = (1 - \kappa) \frac{N\alpha}{3\varepsilon_0} , \tag{6.33}$$

where the parameter κ is introduced to account for structural effects in crystalline materials [6.53]. The composite correction factor κ vanishes in an amorphous solid or a cubic crystal, and is positive in a crystal such as $LiNbO_3$ or $KNbO_3$. Differentiation of (6.33) leads to

$$\frac{\Delta n}{n} = \frac{(n^2 - 1)(n^2 + 2)}{6n^2} \left(-\frac{\Delta V}{V} + \frac{\Delta \alpha}{\alpha} + \frac{\Delta \kappa}{1 + \kappa} \right) , \tag{6.34}$$

where we made use of the relation $\Delta N/N = -\Delta V/V$. Which of the mechanisms prevails in the index modification process depends on the material and waveguide fabrication technique.

A large variety of chemical in- and outdiffusion and ion exchange techniques have been reported to produce waveguides in ferroelectric materials. The waveguide formation involves a reaction at the solid–liquid or solid–gaseous interface between the substrate and the surrounding medium. Kaminow and Carruthers reported waveguiding in Li_2O out-diffused layers in $LiNbO_3$ and $LiTaO_3$ [6.54]. Schmidt and Kaminow were the first to produce waveguides in $LiNbO_3$ by in-diffusion of transition metals such as Ti, V and Ni [6.55]. Noda et al. fabricated waveguides in $LiTaO_3$ by Cu diffusion [6.56]. Proton exchange in $LiNbO_3$ [6.57] and $LiTaO_3$ [6.58] was reported to produce waveguiding layers in these materials. The fabrication of waveguides in $KTiOPO_4$ by ion exchange processes was first reported by Bierlein et al. [6.59].

Diffusion and exchange techniques do not usually require sophisticated equipment and are thus relatively cost-effective. Nevertheless, tight control of the processing conditions is mandatory to achieve high reproducibility. In particular, because of the exponential Arrhenius-like temperature dependence of the diffusion constants, the diffusion temperature is a critical parameter. Further issues of concern are the purity and composition of the exchange melt or gas, substrate quality, and control of the diffusion time.

Ion implantation is a technique essentially based on physical effects rather than chemical processes. It makes use of structural changes associated with a refractive index modification induced by irradiation of a single crystal with usually light ions (H^+, He^+). It was shown to produce waveguides in a large variety of nonlinear optical materials including the ferroelectrics [6.60–61].

The main process parameters, ion energy and ion dose, that determine the index profile of the waveguide, are relatively easy to adjust and control, which guarantees good reproducibility and a high degree of freedom for tailoring waveguides. On the other hand, because the necessary ion energies are in the MeV range, ion implantation requires the use of accelerators, leading to higher process cost.

6.3   Lithium Niobate Waveguides

Among the ferroelectric waveguide materials, lithium niobate (LiNbO$_3$) is the most widely investigated. A vast amount of literature has been published over the past 25 years reflecting the fact that LiNbO$_3$ technology has developed into a separate field within integrated optics. The physics and chemistry of LiNbO$_3$ were outlined in the classical review by Raeuber [6.62]. Today, LiNbO$_3$ waveguides are used in electro-optic and acousto-optic switches and modulators, directional couplers, waveguide lasers and frequency converters [6.63]. The development of LiNbO$_3$ technology has been driven by the relative ease of growth of this material, its mechanical stability and the relatively large nonlinear optical figures of merit. While the off-diagonal nonlinear tensor elements of LiNbO$_3$ are relatively small, the d_{33} figure of merit is the highest among the ferroelectric oxides considered in this chapter (Table 6.1). However, LiNbO$_3$ is known to suffer from several problems. It has a relatively low damage threshold for pulsed laser radiation compared with other nonlinear materials. Particularly detrimental is its susceptibility to photorefractive effects in the visible and near-infrared spectral range, which can cause a detoriation of the beam profile, a shift of the phase matching wavelength, and short- and long-term throughput power fluctuations, often leading to a complete failure of the device. This might be a reason why, despite the high conversion efficiencies reported in the literature, frequency converters based on LiNbO$_3$ have been supplanted by devices based on other materials.

Kaminov and Carruthers were the first to report on low-loss optical waveguides in x-cut LiNbO$_3$ in 1973 [6.54, 6.64]. Their procedure to form a waveguiding layer was based on the observation that in Li$_2$O-deficient layers the extraordinary index n_e increases compared with stoichiometric LiNbO$_3$. Consequently, Li$_2$O outdiffusion from the surface layer was induced by heating LiNbO$_3$ crystals to 1100°C and keeping them at this temperature for up to 64 hours. However, the most successful devices up to now have been based on Ti-indiffused and proton-exchanged waveguides. Both methods take advantage of doping a layer close to the surface by a diffusion process, which results in either place exchange between dopant and lattice ions or the occupation of interstitial sites by the dopant, associated with a refractive index increase in the doped layer. Titanium indiffusion was the standard process in the 1970s and early 1980s, and many devices based on this technique were demonstrated. With the development of the proton exchange method after

1982, Ti-indiffused waveguides have been more and more replaced by proton-exchanged waveguides.

Refractive index changes by implantation of Ar or Ne ions into $LiNbO_3$ were reported for the first time by Wei et al. in 1974 [6.53]. Destefanis et al. studied the effect of implantation of various light and heavy ions at energies ranging from 7 keV to 2 MeV [6.65]. They found that large negative index changes for both n_0 and n_e were produced by the energy deposited in nuclear collisions between the implanted ions and the lattice. This index change was observed to follow a universal curve as a function of deposited energy, independent of the ion species. These authors pointed out that, in particular, the light ions H^+ and He^+ were attractive for the fabrication of integrated optical devices, as a negligible index change was produced with these ions in the surface layer where the energy loss was mainly due to electronic interactions. This allowed the writing of an index barrier inside the material, defining a waveguide in the surface layer. Further reports concentrated on the damage and index profiles induced by ion implantation into $LiNbO_3$ [6.66–68]. Naden and Weiss investigated the optical properties of He^+ ion-implanted planar $LiNbO_3$ waveguides [6.69], while Reed and Weiss reported on the fabrication of channel waveguides by ion implantation [6.70]. Reviews of ion-implanted waveguides were given by Buchal et al. [6.60] and Townsend et al. [6.61, 6.71].

6.3.1   Titanium-Indiffused Lithium Niobate Waveguides

A typical fabrication process for a Ti-indiffused $LiNbO_3$ waveguide comprises the following steps. An oxidized Ti layer of 20–100 nm thickness is deposited on top of an $LiNbO_3$ substrate, usually by a lift-off procedure if channel waveguides are to be defined. The indiffusion is carried out in a furnace at a temperature between 950 and 1100°C, mostly in a dry or wet oxygen or argon atmosphere. The diffusion time is of the order of several hours (typically 8 hours) depending on the temperature and the desired diffusion depth. The diffusion of Ti induces an increase of both the ordinary and extraordinary indices of refraction which is of the order of 3×10^{-2} for n_e and 1×10^{-2} for n_0 [6.72], depending on the specific diffusion conditions.

The first report on waveguide formation by metal ion indiffusion goes back to 1974 [6.55]. Indiffusion of Ti, V and Ni metal ions was observed to form a layer where both the ordinary n_0 and extraordinary n_e increased, leading to the formation of waveguides that supported both TE and TM modes. Three mechanisms were proposed to explain the incorporation of the metal ions into the material: occupation of vacant Li^+ sites, occupation of interstitial sites present at a relatively high density in $LiNbO_3$, and place exchange with Nb^{5+} ions. The optical loss of the waveguides was estimated to be $1\,dB\,cm^{-1}$. A substantial amount of work was directed towards determination of the diffusion constants and establishing a relation between Ti concentration and refractive index [6.72–83]. The analytical methods used include electron microprobe analysis, X-ray microanalysis, secondary ion mass spectroscopy

(SIMS), transmission electron microscopy (TEM), X-ray photoelectron spectroscopy (XPS), and optical near-field methods. Wide investigations on the optical properties have been performed, including measurement of the modal field profiles, propagation constants, and optical loss [6.84–85]. Losses as low as $0.5\,\mathrm{dB\,cm^{-1}}$ have been reported [6.86].

6.3.2   Proton-Exchanged Lithium Niobate Waveguides

A   Fabrication

Annealed proton exchange is today the most frequently used method of fabricating waveguides in LiNbO$_3$. The standard process for the fabrication of annealed proton-exchanged waveguides starts with the formation of a metal mask (from Al or Ta, alternatively SiO$_2$) on top of an LiNbO$_3$ wafer, where the strip-like mask openings define the position of the channel guides. These mask openings are usually prepared by a dry etching process. The sample is then immersed into a bath of benzoic acid (C$_6$H$_5$COOH), typically at a temperature between 180 and 249°C (boiling point of benzoic acid) for 15 to 90 minutes, depending on the desired exchange degree and depth. In this way a high proton concentration layer of a few tenths of a micrometer thickness is formed below the surface. The subsequent annealing in air, typically carried out at 350°C for several hours, transforms the initial narrow step-like index profile with a large index change into a broad diffused profile with a decreased index change at the surface. Figure 6.9 schematically displays the index profiles after exchange and after annealing.

As we will discuss in more detail in the next paragraph, the Li$_{1-x}$H$_x$NbO$_3$ compound undergoes a series of structural phase transitions as a function of growing hydrogen content and exchange conditions. Immediately after the exchange the waveguiding layer is in a high proton concentration β-phase, which usually displays degraded optical properties. It is desirable to transform the waveguiding layer into the low proton concentration α-phase, which is achieved by subsequent annealing at an elevated temperature.

Attempts to form waveguides by immersing LiNbO$_3$ substrates into molten AgNO$_3$ and TlNO$_3$ suggested a large increase of n_e ($\Delta n_e \approx 0.12$) [6.87]. It was assumed that the addition of Ag$^+$ and Tl$^+$ associated with loss of Li$^+$ was responsible for this large index change. However, it was not until 1982 that Jackel and Rice recognized that the index modification did not result from the introduction of the heavy Ag$^+$ and Tl$^+$ ions but from lithium–hydrogen exchange due to the presence of hydrogen as an impurity in the melt [6.88]. Subsequently, the same authors described a process to form waveguides by proton exchange [6.57]. It was observed that in strong acids such as nitric acid (HNO$_3$) and sulfuric acid (H$_2$SO$_4$) the lithium–hydrogen substitution was complete, resulting in the undesired formation of the cubic compound HNbO$_3$. However, in less acidic environments partial exchange was achieved.

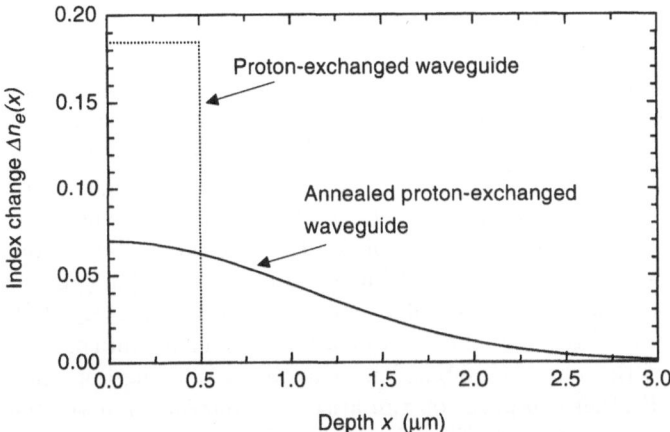

Fig. 6.9. Index profile of a proton-exchanged ($\cdots\cdots$) and an annealed proton-exchanged (——) LiNbO$_3$ waveguide at a wavelength of 0.458 μm. The initial exchange depth was 0.5 μm ($\Delta n_e = 0.18$ at 0.458 μm) and annealing was carried out at 333°C for 5.3 hours. (Adapted from [6.92])

Benzoic acid turned out to be particularly suitable because of its high boiling point of 249°C, allowing the process to be carried out at temperatures where diffusion is rapid. A change of n_e as large as $\Delta n_e \approx 0.12$ at a wavelength of 0.633 μm was observed in the surface layer of x-cut substrates, while n_0 decreased by $\Delta n_0 \approx -0.05$. Losses were measured to be about 0.5 dB cm^{-1}. Because benzoic acid does not attack most metals, Jackel and Rice suggested using metal masks to define channel waveguides by proton exchange.

The effect of annealing of proton-exchanged waveguides was initially studied by De Micheli et al. [6.89]. Conversion of the initial step-like index profile after the exchange process ($\Delta n_e \approx 0.12$ at 0.633 μm) into a gradient index profile with a reduced index change at the surface (down to $\Delta n_e \approx 0.01$) and significantly larger depth was observed after annealing at 400°C for 4 hours in an oxygen atmosphere. In addition, the same authors also investigated the effect of diluting the benzoic acid melt with lithium benzoate (C$_6$H$_5$COOLi) and found that the surface index change decreased with increasing lithium content in the melt. The possibility of controlling the surface index change and index profile shape almost independently by varying the lithium concentration in the melt and the annealing conditions, respectively, was pointed out.

Several publications focused on establishing a relation between hydrogen concentration, annealing conditions and the refractive index profile. Clark et al. investigated the proton exchange process in z-cut lithium niobate and confirmed the validity of the step-index assumption for proton-exchanged waveguides [6.90]. Using secondary ion mass spectroscopy (SIMS) of the concentration profiles of protons and lithium, Vohra et al. studied the diffusion processes occurring during the formation of proton-exchanged and annealed proton-exchanged waveguides [6.91]. The concentration profiles of annealed proton-exchanged waveguides were fitted to the solutions of diffusion equations obtained from standard diffusion theory. Index profiles were then calculated from these equations. Bortz and Fejer established a model that predicted the modal properties of annealed proton-exchanged waveguides on z-cut substrates from processing parameters [6.92]. The index profiles were derived by use of the inverse WKB technique. These authors pointed out several complications associated with the modeling of the diffusion process. A complete diffusion model in a polyphase system such as $Li_{1-x}H_xNbO_3$ involves the solid solubilities and the concentration dependent diffusion coefficients in each of the phases combined in a nonlinear moving-boundary diffusion problem. Further complication can arise from interface kinetic limitations influencing the transport. Therefore the authors introduced an empirical concentration dependent continuous-diffusion coefficient to predict the index profiles and modal field distributions. Cao et al. performed a detailed study of annealed proton-exchanged planar and channel waveguides fabricated on z-cut substrates from pure benzoic acid at 200°C and annealed at 350°C for 2 to 14 hours [6.93]. Their model provided closed form expressions for the mode indices and mode field profiles and yielded close agreement with the measured data.

On the basis of the idea of De Micheli et al. to use a lithium rich benzoic acid solution to achieve better control of the proton exchange process [6.89], recent work on proton-exchanged waveguides has focused on the development of fabrication methods that avoid the formation of a high proton concentration layer in the initial stage prior to annealing. The use of less aggressive procedures can lead to the direct formation of an α-phase waveguide, in contrast to the traditional methods where the α-phase is reached only after annealing. These methods make use of proton sources other than benzoic acid, e.g. pyrophosphoric acid ($H_4P_2O_7$) [6.94–95], octanoic acid ($CH_3(CH_2)_6COOH$) [6.96], stearic acid ($CH_3(CH_2)_{16}COOH$) [6.97], m-toluic acid (m-$CH_3C_6H_4COOH$) [6.98] and glycerol and $KHSO_4$ [6.99–100], diluted with lithium benzoate. Two groups independently reported on the fabrication of proton-exchanged waveguides in z-cut crystals from benzoic acid vapor [6.101–102]. The vapor exchange was carried out at temperatures of 290–350°C and 250–375°C, respectively, for up to 24 hours.

The proton exchange method with subsequent annealing is a low-temperature process with a maximum temperature of about 350°C compared with more than 1000°C necessary for Ti indiffusion. Therefore, proton exchange is compatible with the formation of domain gratings for quasi-phase-matching by Ti indiffusion, which we will discuss in Sect. 6.3.3. In addition to its relative ease of fabrication, proton exchange offers a number of benefits concerning the optical properties, namely a larger refractive index change and convenience of index tailoring. Proton-exchanged waveguides also have the reputation of being less susceptible to photorefractive effects than their Ti-indiffused counterparts [6.103].

B Structural Properties

It was recognized soon after the first report on proton-exchanged waveguides in LiNbO$_3$ [6.57] that Li$_{1-x}$H$_x$NbO$_3$ exhibits a complicated structural behavior [6.104–105]. When replacing Li$^+$ with H$^+$ the compound undergoes a transformation from the rhombohedral LiNbO$_3$ structure to the cubic perovskite HNbO$_3$ structure [6.105]. Yi-Yan found significant index instabilities in proton-exchanged waveguides as a function of time measured over a period of 20 days [6.106]. He suggested that these profile changes were caused by the continuous migration of protons within the guiding layer and that some modification might be introduced due to structural changes as a function of exchange time. It was also found in early work on proton-exchanged waveguides that their optical properties, such as refractive index change, mode propagation loss, electro-optic properties and temporal stability, depended on the specific fabrication conditions, in particular if dilute or undilute acids were used [6.89, 6.107]. Adding lithium to the melt was found to reduce the number of hydrogen atoms that exchanged places with lithium, resulting in a less aggressive exchange process. The close correlation between structural and optical properties demanded detailed structural studies of the Li$_{1-x}$H$_x$NbO$_3$ system. Many analytical methods have been used to reveal the structural behavior of this compound, including the following:

1. Diffractional methods, namely X-ray diffraction [6.100, 6.104–105, 6.108–109] and transmission electron microscopy (TEM) [6.110], have been used to determine structural properties such as lattice constants, strain, defects and dislocations.
2. Rutherford backscattering spectroscopy (RBS) can be used to investigate lattice distortions associated with niobium dislocations [6.109, 6.111]. Owing to the quadratic dependence of the scattering cross-section on the atomic number, RBS shows poor sensitivity to hydrogen and lithium.
3. Induction of nuclear reactions is an ion beam technique complementary to RBS [6.109], based on the analysis of nuclear reaction products generated by bombardment with charged particles. In particular, it allows locally resolved detection of hydrogen and lithium, e.g. by the reaction ^{15}N+^{1}H→^{12}C+α + γ.
4. Secondary ion mass spectroscopy (SIMS) allows local probing of the concentration of all the elements in the periodic table, from hydrogen to uranium, and is thus a powerful tool to evaluate the concentration profiles of specific atomic species. In particular, it is the only surface analytical technique capable of detecting hydrogen directly and it also has an excellent sensitivity for lithium [6.91, 6.109]. If a standard is used, even absolute quantitative information can be obtained from SIMS measurements.

5. Infrared (IR) spectroscopy has been utilized to detect the OH stretching vibrational modes in the band between 3000 and 4000 cm^{-1} [6.105, 6.111–112]. IR spectra allow determination of the hydrogen content and the presence of interstitial hydrogen. IR spectroscopy was combined with hydrogen–deuterium isotopic exchange reactions to study the reactivity of annealed proton-exchanged waveguides with atmospheric water vapor [6.112].

6. Micro-Raman spectroscopy has been used to locally monitor the Nb–O bend (Raman shift $\sim 250\,\mathrm{cm}^{-1}$), Nb–O stretch ($\sim 700\,\mathrm{cm}^{-1}$) and O–H stretch ($\sim 3500\,\mathrm{cm}^{-1}$) bands as a function of annealing time [6.113–114]. Again, this allows determination of the local hydrogen concentration and, moreover, because the nonlinear optical properties are related to the Nb–O octahedra, the nonlinearity of the exchanged layer. A theoretical and experimental study of the correlation between optical nonlinearity and Raman spectra was earlier given by Johnston [6.115].

7. Scanning electron microscopy (SEM) (SEM) allows investigation of the topology of the exchanged and annealed layer [6.109].

Broadly speaking, one can distinguish between two different situations when analyzing the $Li_{1-x}H_xNbO_3$ system: low proton concentration layers and high proton concentration layers. It was found already by Rice and Jackel that waveguide layers in a low proton concentration phase exhibit the best optical properties [6.104]. This low proton concentration phase with $x < 0.12$, designated as the α-phase, is characterized by a modest change of the cell constants compared to stoichiometric $LiNbO_3$. Layers in this phase can be obtained either by dilute melt exchange in benzoic acid (more than 3.5 mol.% lithium benzoate in the melt) [6.89, 6.104], annealing after exchange in pure benzoic acid [6.89, 6.107], or use of acids weaker than benzoic acid. α-phase layers are structurally matched to the substrate so that scattering and strain are minimized. As a consequence the waveguides exhibit low losses, of the order of 0.5 dB cm^{-1}, almost preserved electro-optic coefficients, and good temporal stability [6.107]. On the other hand, the maximum index change in α-phase layers is only about 0.03 at 0.633 µm, typically four to five times less than in highly exchanged layers, which results in a decrease of the mode confinement compared with high proton concentration waveguides. With increasing proton concentration the strain in the layer grows as a result of the change in lattice constants associated with the formation of high proton concentration phases, designated as β-phase layers. It has been suggested that this strain causes a deformation of the diamond building block (parallelogram) appearing in the (012) plane [6.108]. The diamond becomes more square-like and thus more centrosymmetric, leading to a decrease of the electro-optic and nonlinear-optical activity. The high proton concentration layers, in general, show strongly degraded optical properties in terms of loss and nonlinearity, although for some β-phase layers losses below 1 dB cm^{-1} have been observed [6.116]. The complexity of the structure of $Li_{1-x}H_xNbO_3$ has been investigated by Korkishko et al. who found as many as seven different phases whose occurrence depends on the specific fabrication conditions [6.100, 6.116].

C   Nonlinear Susceptibility

A key issue in nonlinear integrated optics is the nonlinear susceptibility of the waveguiding layer. The waveguide formation process can considerably distort or even completely destroy the nonlinearity of the material. However, the task of probing the nonlinear optical properties of a layer only a few micrometers thick is difficult. Guided-wave interactions sample the waveguiding layer but probe the average value of the nonlinear coefficient weighted by the modal fields, i.e. the overlap integral. Thus a quantitative analysis is difficult. Therefore second-harmonic generation in a reflection geometry sensitive to the surface layer is most often used to monitor the nonlinearity of the waveguiding region relative to the bulk crystal. The theoretical basis for second-harmonic generation in reflection was worked out by Bloembergen and Pershan [6.117].

A substantial decrease of the electro-optic coefficient by a factor of 2.7 was observed by Becker in an integrated optical Mach–Zehnder interferometer, shortly after the first report on fabrication of proton-exchanged waveguides [6.118]. Although the electronic and ionic polarizations contribute differently to the electro-optic effect and second-harmonic generation, the ionic contribution being smaller in second-harmonic generation than in the electro-optic effect, a substantial reduction could also be expected for the d coefficient.

The results reported on measurements of nonlinear optical coefficients of annealed proton-exchanged LiNbO$_3$ waveguides are not always consistent. However, two general trends may be noted: the proton exchange process strongly reduces the nonlinearity of the exchanged layer, while annealing tends to restore the nonlinearity to a degree which depends on annealing time and temperature. The recovery of the nonlinearity is a result of the formation of an α-phase layer during annealing. Table 6.2 provides an overview of results obtained from measurements of nonlinear optical coefficients of proton-exchanged and annealed proton-exchanged LiNbO$_3$ waveguides. A technique using a proton-exchanged grating allowing the observation of the second-harmonic power in different diffraction orders was developed by Suhara et al. [6.119] and adopted by Keys et al. [6.120]. Both groups reported a d_{33} coefficient in the proton-exchanged layer of roughly one half of unexchanged LiNbO$_3$. However, Bortz and Fejer pointed out that this technique requires a complicated analysis that depends sensitively on both the linear and nonlinear properties of the grating. Hence interpretation of the results was thought to be difficult [6.121]. The measurements done by Cao et al., yielding a high nonlinearity already in the proton-exchanged waveguide and almost full recovery after annealing [6.122], were later questioned by Laurell et al. and Bortz et al., who pointed out that the substrate contribution to the surface reflectivity measurement had been overlooked and the analysis of the mea-

Table 6.2. Measured relative nonlinear optical coefficients $d_{33}^{(wg)}/d_{33}^{(bulk)}$ of proton-exchanged and annealed proton-exchanged LiNbO$_3$ waveguides

Reference	Proton exchange	$d_{33}^{(wg)}/d_{33}^{(bulk)}$ after proton exchange	Annealing	$d_{33}^{(wg)}/d_{33}^{(bulk)}$ after annealing	Measurement method
[6.119]	Pure benzoic acid, 230 °C, 2 h	0.5	–	–	Grating diffraction
[6.120]	Pure benzoic acid, 235 °C, 1 h	0.45	275 °C, 10 min 350 °C, 20 min	0.65	Grating diffraction
[6.122]	Pure benzoic acid, 180 °C, 1 h	0.62	350 °C, 10 h	~ 0.9	Reflected SHG, $\lambda_\omega = 1.064$ μm
[6.123]	Pure benzoic acid, 180 °C, 1 h	< 0.03	350 °C, 28 h 500 °C, 10 h	< 0.03, within a layer of 50 nm thickness	Reflected SHG, $\lambda_\omega = 0.532$ μm
[6.121]	Pure benzoic acid, 220 °C, 2 h	< 0.01	–	–	Reflected SHG, $\lambda_\omega = 1.064$ μm
[6.124]	Pure benzoic acid, pyrophosphoric acid 180 °C, 0.5 h 180 °C, 1, 1.5 h	0	310–350 °C, up to 17 h	~ 0.55 0	Reflected SHG, $\lambda_\omega = 0.532$ μm
[6.125]	Pure benzoic acid, 173 °C, 66 min	0	333 °C, up to 63 h	0.8–1	Reflected SHG, $\lambda_\omega = 0.532$ μm, wedged sample
[6.126]	Dilute melt exchange, glycerol/KHSO$_4$/lithium benzoate 230 °C	> 0.9	–	–	Reflected SHG, $\lambda_\omega = 0.532$ μm
[6.102]	Benzoic acid vapor 250–375 °C 0.25–24 h	> 0.9 (α- and β-phase waveguides)	–	–	Non-phase matched transmitted SHG, $\lambda_\omega = 0.532$ μm

surements had been incomplete [6.121, 6.123]. The following publications by
Laurell et al., Bortz et al. and Hsu et al. consistently reported an almost
complete extinction of the second-harmonic activity in the initial proton-
exchanged layer [6.121, 6.123–125]. Laurell et al. did not observe any recovery of the nonlinearity after annealing in a 50 nm thick layer below the
surface that was probed by their measurement [6.123]. However, these authors pointed out that this observation did not exclude a partial or complete
recovery of the d coefficient in deeper layers. This was confirmed by Bortz and
Fejer, who used a polished wedge-shaped sample to achieve a high depth resolution for their surface second-harmonic generation measurements [6.125].
These authors never observed any second-harmonic generation activity from
the initially proton-exchanged surface. In annealed samples, an abrupt increase from 0 to 80% of the bulk $LiNbO_3$ d coefficient was observed at a position corresponding to the interface between the original proton-exchanged
film and the $LiNbO_3$ substrate. The position of this interface moved towards
the sample surface with increasing annealing time, indicating continuous recovery of the exchanged material. At depths larger than 1 μm the measured
second-harmonic signal was found to be indistinguishable from that of bulk
$LiNbO_3$. These reports illustrate once more the importance of producing
an α-phase waveguiding layer. Only such waveguides display considerable
second-harmonic activity while $\alpha + \beta$ mixed-phase or β-phase layers show
a strongly degraded nonlinearity. This has been confirmed recently by Veng
et al., who investigated waveguides fabricated by a dilute proton-exchange
method using glycerol which contained $KHSO_4$ and lithium benzoate [6.126].
For undilute or weakly dilute exchange, a strongly reduced d_{33} coefficient was
measured, while for strongly dilute exchange, d_{33} of the waveguide reached
more than 90% of that of bulk $LiNbO_3$. Recently, Rams et al. reported a high
($> 90\%$) d_{33} coefficient in vapor phase proton-exchanged waveguides [6.102].
In contradiction to previous reports, these authors reported similarly high
d_{33} values also for β-phase waveguides. The structural properties of these
waveguides were investigated in [6.127].

6.3.3 Domain Inversion and Quasi-Phase-Matching in Lithium Niobate

The investigation of domain reversal in $LiNbO_3$ has attracted interest for two
reasons. First, early devices based on Ti-indiffused waveguides often showed
unexpectedly degraded electro-optic properties which made a careful study of
the domain structure in the waveguide necessary. Second, it was recognized
that for the development of efficient frequency converters based on $LiNbO_3$,
access to the nonlinear optical coefficient d_{33} was mandatory. The figure
of merit for interactions driven by d_{33} is roughly ten times larger than for
d_{31} (Table 6.1). However, d_{33} can be involved only in quasi-phase-matched
interactions. This latter fact triggered the development of poling techniques
and fabrication of periodic domain gratings.

In its ferroelectric state $LiNbO_3$ has a spontaneous polarization P_s that has been found to be reversible by a variety of techniques: heat treatment [6.128–129], diffusion of Ti into the c^+ face [6.130], outdiffusion of Li_2O [6.131], heat treatment with an SiO_2 cladding on the c^+ face [6.132], bombardment of the c^- face with high-energy electrons [6.133–134], electron beam writing on the c^--face [6.135–137], proton exchange [6.138] and applying strong electric fields [6.139]. Some of these techniques are schematically displayed in Fig. 6.10. Most of the studies on domain inversion are based on the observation that the polarity can be easily revealed by etching, typically in a mixture of HNO_3 and HF. The c^- face (negative P_s end) shows a deeply etched pattern, while the c^+ face (positive P_s end) is barely attacked [6.140–143].

Houé and Townsend recently reviewed the methods used for periodic poling of nonlinear crystals [6.144]. While the physical mechanisms associated with domain inversion by application of an electric field are straightforward and the effect of domain inversion by electron beam irradiation may be explained in terms of the electric field generated between the injected electrons and a ground plate, the mechanisms related to in- and outdiffusion processes leading to domain reversal have been widely discussed in the literature.

An important step in the development of domain inversion techniques was the observation reported by Miyazawa that Ti-indiffused waveguides on a c^+ face are domain inverted with respect to the substrate, while on the c^- face no domain inversion occurs for diffusion between 950 and 1100°C [6.130], a result later confirmed by Thaniyavarn et al. [6.145]. While Miyazawa explained the domain inversion occurring during Ti indiffusion in terms of a polarization gradient [6.130], Peuzin considered the concentration gradient of the impurities resulting in an electric field that drives the reversal [6.146]. This electric field is assumed to be antiparallel to the impurity gradient, which explains that domain inversion occurs at the c^+ face but not at the c^- face. Today it is generally assumed that a concentration gradient will cause a corresponding electric field so that when the substrate is cooled down after quick heat treatment in a diffusion furnace, domain inversion will appear in regions where a dopant has been introduced. Such a gradient may be present due to Ti indiffusion, proton exchange or Li_2O outdiffusion [6.147]. The formation of a periodic grating by Ti indiffusion is displayed in Fig. 6.10a.

The first successful fabrication of a quasi-phase-matching waveguide device in a split step procedure was demonstrated by Lim et al. [6.27]. First, a periodic domain inverted grating was fabricated by local Ti indiffusion at 1100°C. Thereafter, an annealed proton-exchanged waveguide was produced by a 30 minute exchange in pure benzoic acid at 200°C followed by a 4 hour anneal in oxygen at 350°C. Second-harmonic generation at 1.064 µm in a planar waveguide with a 1 mm long third-order quasi-phase-matching grating was demonstrated with an efficiency of 0.05% W^{-1}. This procedure

(a) Titanium indiffusion

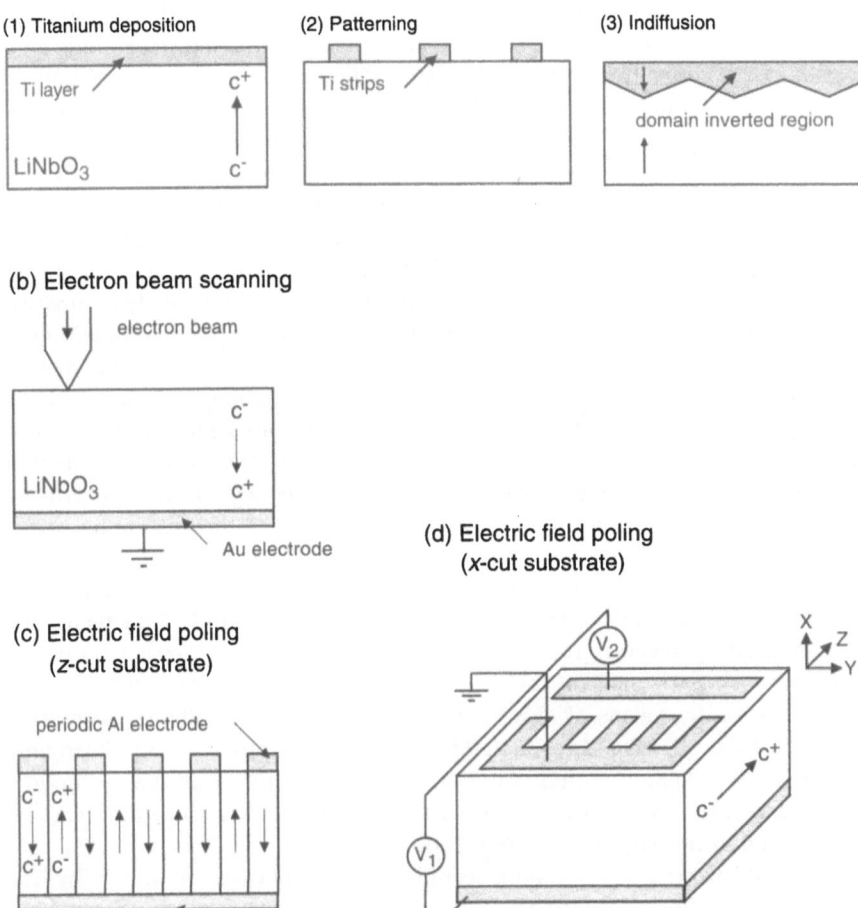

Fig. 6.10a–d. Examples of techniques for periodic poling of LiNbO$_3$. (**a**) Titanium indiffusion. The process comprises the following steps: (1) deposition of a Ti layer, (2) patterning of the Ti layer by photolithography and etching, and (3) indiffusion of Ti at $\sim 1100°$C causing domain reversal in the surface close region (adapted from [6.149]). (**b**) Electron beam scanning (adapted from [6.135]). (**c**) Electric field poling on z-cut substrates (adapted from [6.139]). (**d**) Electric field poling on x-cut substrates (adapted from [6.154])

became the standard process for fabrication of quasi-phase-matched devices in LiNbO$_3$ until the advent of electric field poling (cf. Table 6.3).

Although annealed proton-exchanged waveguides with a domain grating formed by Ti indiffusion have been frequently used for second-harmonic generation, they suffer from major drawbacks. The domain inversion is limited to a shallow region below the surface and the shape of the domains is triangular (schematically illustrated in Fig. 6.10a), significantly differing from the ideal laminar domain structure required for maximum conversion efficiency. Optimizing the second-harmonic output of these waveguides requires precise control of the guide thickness and domain inversion depth over the whole waveguide length, which is often difficult to achieve. The influence of deviations from the ideal structure on the second harmonic has been investigated by several authors [6.17–18, 6.148]. Helmfrid et al., for example, found a difference of 26 dB between their best waveguide and the theoretical prediction [6.148], of which 6 dB was attributed to the triangular shape of the domains. Although the fabrication of annealed proton-exchanged waveguides with Ti-indiffused domain gratings has been further developed in the meantime [6.149], the conversion efficiency in waveguides with a domain grating produced by electric field poling is almost one order of magnitude higher at present (c.f. Table 6.3).

Domain reversal by use of electron beams, schematically displayed in Fig. 6.10b, was introduced by Haycock and Townsend [6.133]. This technique was adopted by several groups to produce domain reversal, either by use of a metal mask patterned on the c^- face of a LiNbO$_3$ substrate [6.134] or by direct writing by means of an electron beam [6.135–137]. The domain gratings were found to extend through 0.5–1 mm thick samples. Second-harmonic generation was demonstrated in devices periodically poled by this technique, but with a rather low conversion efficiency, mainly because of imperfections of the domain inverted structure [6.150].

A breakthrough in the poling technique of LiNbO$_3$ was reported by Yamada et al., who succeeded in forming a first-order quasi-phase-matching grating by direct application of an electric field at room temperature [6.139]. Domain reversal occurs when the applied field exceeds the coercive field, which is about 20 kV mm^{-1} in LiNbO$_3$ at room temperature. In the initial work by Yamada et al. a periodic electrode consisting of a metal grating was patterned on the c^+ face, while a plain electrode was applied on the c^- face, as shown in Fig. 6.10c. A short voltage pulse of 100 μs length was then applied to the LiNbO$_3$ wafer. Annealed proton-exchanged waveguides were prepared after the poling process. This report stimulated new research activity in the field of periodically poled nonlinear materials. The method has been extended to thicker substrates using liquid electrolyte contacts [6.151]. Today, periodically poled bulk LiNbO$_3$ is emerging as a new nonlinear material. Recently, Li et al. investigated the compatibility of the electric field poling process with annealed proton exchange [6.152]. The domain inversion was found to sustain the annealed proton exchange process, proving the versatility of electric field poling compared with other poling methods.

The main advantage of electric field poling of z-cut (c-cut) LiNbO$_3$ over other poling techniques lies in the fact that the domains are deep and show a laminar structure with walls perpendicular to the waveguide surface, as required for optimized quasi-phase-matched interactions. However, fabrication of annealed proton-exchanged waveguides on z-cut substrates implies that the waveguide modes are TM polarized, which is often undesirable for coupling the waveguide to diode lasers, which usually operate in a TE mode with polarization parallel to the plane of the diode laser. Therefore a technique for periodically poling x-cut substrates and subsequent formation of annealed proton-exchanged waveguides supporting TE modes has been developed recently [6.153–154], schematically displayed in Fig. 6.10d. Initially, the depth of domain inversion was only about 1 μm. By using slightly oblique cut substrates, the domain inversion depth was increased to 2.5 μm [6.155].

Despite its benefits, however, electric field poling is not completely free of problems. High fields can cause catastrophic breakdown of the LiNbO$_3$ substrate during poling when the applied pulse is too long or the sample too thick. Therefore, precise control of the pulse length by monitoring the poling current flow is necessary. In addition, the domains tend to grow beyond the width of the metal electrode as a result of fringing fields along the edges. This effect requires a careful design of the grating electrodes in order to obtain the ideal 1:1 line-to-space ratio (duty cycle of 0.5). Investigation of the lateral domain growth's dependence on the poling parameters is a prerequisite for obtaining large-scale reproducibility of the poling process.

6.3.4   Optical Damage in Lithium Niobate Waveguides

One of the most severe problems encountered when using ferroelectric oxides, e.g. LiNbO$_3$ and LiTaO$_3$, as frequency-doubling materials is optical damage, in general attributed to light-induced index changes. Optical damage can be particularly detrimental at the high intensities in waveguides. The effect manifests itself through distortions of the optical field distribution and throughput power fluctuation and degradation, often leading to a complete failure of the device after a few seconds or minutes. The effect was discovered when focusing an intense blue or green laser beam into LiNbO$_3$ or LiTaO$_3$ crystals [6.156].

Broadly speaking, a material should exhibit the following two properties in order to be insensitive to optical damage: a small number of defect or impurity centers and a large dark conductivity. Evidently a small number of impurities will prevent the excitation of a large number of free charge carriers, and a high dark conductivity inhibits the accumulation of charges in non-illuminated regions of the crystal, preventing the build-up of space charge fields that induce index changes through the electro-optic effect. In addition, it should be noted that photorefractive effects are only present if an electro-optic coefficient is active in the chosen configuration. In $LiNbO_3$ this is the case in any interaction where the d_{33} nonlinear optical coefficient is involved, since in this case electro-optic interactions can be provided via the r_{33} coefficient.

An early investigation by Jackel et al. suggested that the photorefractive effect in proton-exchanged $LiNbO_3$ waveguides was at least four orders of magnitude lower than that of Ti-indiffused guides [6.103]. However, because the lithium–hydrogen exchange degree was up to 70%, the cause of the low photorefractive sensitivity was mainly the strongly reduced electro-optic coefficient. This was later confirmed by Fujiwara et al., who found a three times higher photorefractive sensitivity in annealed proton-exchanged $LiNbO_3$ waveguides compared with proton-exchanged guides [6.157]. This behavior could be explained by the recovered r_{33} coefficient after annealing.

In order to decrease the sensitivity to optical damage, several studies on doping $LiNbO_3$ with various ions were performed. It was recognized that doping with Mg by adding MgO to the melt improved the resistance of $LiNbO_3$ to optical damage [6.158]. Volk et al. investigated Mg, Zn and In-doped $LiNbO_3$ crystals [6.159–160]. They concluded that the increased resistance to optical damage originated from an increase in photoconductivity. These authors proposed two scenarios to explain this observation. First, in the absence of any Fe impurities, the addition of Mg, Zn and In leads to a cancellation of Nb–Li defects (Nb occupying vacant Li sites) that govern the photoconductivity in undoped $LiNbO_3$. Second, in the presence of Fe impurities, doping with Mg, Zn and In was thought to sharply decrease the capture cross-section of Fe^{3+} ions [6.160]. Young et al. reported on the fabrication of Zn-diffused planar $LiNbO_3$ waveguides with increased damage resistance [6.161]. A comprehensive overview of defects in $LiNbO_3$ with particular emphasis on centers containing hydrogen was given by Schirmer et al. [6.162]. Electron spin resonance (ESR) techniques have been shown to be a powerful tool in the investigation of defect centers, e.g. for studying Fe defects in $LiNbO_3$ [6.163]. Steinberg et al. compared photorefractive index changes in annealed proton-exchanged waveguides in MgO-doped and congruent $LiNbO_3$ [6.164]. They found that the refractive index change at $0.633\,\mu m$ in 7% MgO-doped $LiNbO_3$ was reduced by two orders of magnitude. This effect was found to be caused by a reduced photovoltaic current density in MgO-doped $LiNbO_3$.

The photorefractive effect in periodically poled ferroelectrics was theoretically modeled by Taya et al. [6.165]. The calculation yielded that perturbations that are due to the photogalvanic effect are reduced compared with homogeneously poled crystals by approximately the square of the product of the poling grating vector and a characteristic transverse dimension of the irradiance. Eger et al. experimentally investigated photorefractive effects in Ti-poled and electric field-poled $LiNbO_3$ annealed proton-exchanged waveguides [6.166]. They showed that annealed proton-exchanged $LiNbO_3$ waveguides were distinctly more susceptible to optical damage than Rb-exchanged KTiOPO$_4$ waveguides.

6.3.5 Second-Harmonic Generation in Lithium Niobate Waveguides

Table 6.3 provides an overview of results for second-harmonic generation in LiNbO$_3$ waveguides. We have listed the waveguide fabrication method, the waveguide type (planar or channel) and length, the phase matching technique, the grating fabrication method if quasi-phase-matching was used, the laser source used for the measurement, the second-harmonic wavelength, the relevant fundamental and second-harmonic power, and the conversion efficiency. If not specially indicated, the measurements were carried out with continuous wave lasers. In some reports neither the waveguide loss nor the in-coupled fundamental power were indicated, so we calculated the in-coupled fundamental power $P_\omega(0)$ using the guide length and an estimated waveguide loss of $1\,\mathrm{dB\,cm^{-1}}$, a value typically observed in periodically poled Ti-indiffused and annealed proton-exchanged waveguides. This enabled us to calculate the normalized conversion efficiency according to the definition given by (6.24) with normalization referenced to the in-coupled internal power.

Some interesting developments regarding waveguide fabrication, grating formation, and reported conversion efficiencies can be deduced from Table 6.3. In the 1970s and 1980s all the successful devices for frequency conversion were based on Ti-indiffused waveguides. In all experiments the interaction was provided by the nonlinear coefficient $d_{31} = 4.4\,\mathrm{pm\,V^{-1}}$ at a wavelength of $1.064\,\mathrm{\mu m}$ [6.51] using the birefringence phase matching scheme. Owing to the relatively small birefringence of LiNbO$_3$, this type of phase matching is limited to green wavelengths, while the generation of blue light is impossible. At a fundamental wavelength of $1.064\,\mathrm{\mu m}$, phase matching between the fundamental TE^ω_{00} and the second-harmonic $\mathrm{TM}^{2\omega}_{00}$ was found to occur at around $-2°\mathrm{C}$ in z-cut undoped LiNbO$_3$, while in x-cut MgO-doped LiNbO$_3$ the $\mathrm{TM}^\omega_{00} + \mathrm{TM}^\omega_{00} \to \mathrm{TE}^{2\omega}_{00}$ interaction was observed at $19°\mathrm{C}$ [6.167]. However, in the same report, the less favorable $\mathrm{TM}^\omega_{00} + \mathrm{TM}^\omega_{00} \to \mathrm{TE}^{2\omega}_{10}$ interaction, phase matched at $54°\mathrm{C}$, was found to be the most efficient. In a planar waveguide on an x-cut MgO-doped substrate, Fejer et al. found the $\mathrm{TM}^\omega_0 + \mathrm{TM}^\omega_0 \to \mathrm{TE}^\omega_2$ at $102°\mathrm{C}$ to yield the highest second-harmonic output [6.168].

Second-harmonic generation of a continuous wave Nd:YAG laser at $1.064\,\mathrm{\mu m}$ in a 1 cm long waveguide was reported as early as in 1976 by Uesugi and Kimura [6.169]. $0.3\,\mathrm{\mu W}$ at 532 nm wavelength was generated with an input power of $2\,\mathrm{mW}$, yielding a normalized conversion efficiency of $7.5\,\%\mathrm{W^{-1}}$. The most remarkable result reported for birefringence phase matched devices was the demonstration of a 4 cm long waveguide resonator with a conversion efficiency of $1000\,\%\mathrm{W^{-1}}$ and an output power of $0.1\,\mathrm{mW}$ at 546 nm by Regener and Sohler [6.170]. However, despite the high conversion efficiency, the operation of the device was restricted by stability problems, which would

Table 6.3. Second-harmonic generation in LiNbO$_3$ waveguides

Year Ref.	Waveguide fabrication	Waveguide type	L (cm)	Phase matching technique	Domain inversion technique	Laser[a]	$\lambda_{2\omega}$ (nm)	$P_\omega(0)$[b] (mW)	$P_\omega(L)$ (mW)	$P_{2\omega}(L)$ (mW)	η^{c} (%W^{-1})
1976 [6.169]	Ti indiff.	channel	1.0	birefr.	–	Nd:YAG	532	2	–	0.0003	7.5
1978 [6.315]	Ti indiff.	channel	1.7	birefr.	–	OPO (pulsed)	540	4.5e4 (peak)	–	1.1e4 (peak)	0.54
1986 [6.316]	Ti indiff.	channel	2.7	birefr.	–	Nd:YAG (pulsed)	532	120 (peak)	–	5.0 (peak)	35
1986 [6.168]	Ti indiff.	planar	1.9	birefr.	–	Nd:YAG (cw) (pulsed)	532 532	400 –	– –	0.87 22 (peak)	0.54 –
1988 [6.170]	Ti indiff.	channel	4.7 4.0	birefr.	–	Ar$^+$ ion	546.2 546.2	0.5 0.1	– –	7 × 10^{-5} 1 × 10^{-4}	28.7 1000 (resonant)
1988 [6.167]	Ti indiff.	channel	1.6	birefr.	–	Nd:YAG (pulsed)	532	1350	–	43	2.36
1989 [6.27]	a.p.e.	planar	0.1	q.p.m. (3rd order)	Ti indiff.	Nd:YAG	532	(1.1)	1	5 × 10^{-7}	4 × 10^{-2}
1989 [6.317]	a.p.e.	channel	0.1	q.p.m. (3rd order)	Ti indiff.	Styryl-9 dye	410	(17.5)	17.1	1.1 × 10^{-3}	0.36
1989 [6.318]	a.p.e.	channel	1.4	q.p.m. (1st order)	Li$_2$O outdiff.	DL	416	(4.1)	3	6 × 10^{-5}	0.35
1989 [6.131]	Ti indiff.	channel	0.7	q.p.m. (1st order)	–	Nd:YAG (pulsed)	532	800 (peak)	160 (peak)	0.36 (peak)	0.056 (peak)

1991 [6.173] a.p.e.	ridged channel	0.3	q.p.m. (1st order)	Ti indiff.	Nd:YAG	532	0.3	–	1.5×10^{-5}	16.6
1991 [6.237] a.p.e.	channel	0.2	q.p.m. (1st order)	Ti indiff.	DL (0.06 mW)	662.4	–	–	6.52×10^{-1}	0.16
1991 [6.132] a.p.e.	channel	0.4	q.p.m. (1st order)	masked heat treatment	Nd:YAG	532	0.4	–	2.6×10^{-5}	16.3
1991 [6.319] p.e.	channel	0.9	q.p.m. (3rd order)	masked Li$_2$O outdiff.	Ti:Al$_2$O$_3$	417	9.1	5	1.5×10^{-3}	1.81
1992 [6.320] a.p.e.	channel	1.0	q.p.m. (1st order)	Ti indiff.	Ti:Al$_2$O$_3$	435	(98)	77.9	0.65	6.8
1992 [6.148] a.p.e.	channel	1.05	q.p.m.	Li$_2$O outdiff.	DL	416	–	–	–	2.3
1992 [6.321] a.p.e.	channel	0.33	q.p.m. (1st order)	e.b.s.	Nd:YAG	532	(1.8)	1.7	1.3×10^{-3}	40
1992 [6.322] a.p.e.	channel	0.5	q.p.m. (2nd order)	Ti indiff.	Ti:Al$_2$O$_3$ (200 mW)	432	55	52	0.257	8.5
1992 [6.174] a.p.e.	channel	0.6	q.p.m. (2nd order)	Ti indiff.	DL (130 mW)	425.1	40.2	36	0.3	18.5

Table 6.3. (Continued)

Year Ref.	Waveguide fabrication	Waveguide type	L (cm)	Phase-matching technique	Domain inversion technique	Laser[a]	$\lambda_{2\omega}$ (nm)	$P_\omega(0)$[b] (mW)	$P_\omega(L)$ (mW)	$P_{2\omega}(L)$ (mW)	η^c (%W^{-1})
1992 [6.323]	a.p.e.	channel	0.6	q.p.m. (2nd order)	Ti indiff.	Ti:Al$_2$O$_3$ Nd:YAG	433 532	50.5 33.4	45.2 30	0.45 0.25	17.6 22.4
1992 [6.324]	a.p.e.	channel	0.33	q.p.m. (1st order)	e.b.s.	Ti:Al$_2$O$_3$	440	(2.7)	2.5	4×10^{-3}	55
1993 [6.139]	a.p.e.	channel	0.3	q.p.m. (1st order)	e.f.p.	Ti:Al$_2$O$_3$	425.9	(210)	195.9	20.7	47
1993 [6.150]	a.p.e.	channel	0.33	q.p.m. (1st order)	e.b.s.	Ti:Al$_2$O$_3$	440	(2.7)	2.5	0.004 5	62
1994 [6.175]	a.p.e.	channel	1.05	q.p.m. (1st order)	Ti indiff.	Ti:Al$_2$O$_3$	488	< 1	–	–	(126)
1996 [6.178]	a.p.e	channel	0.3 0.3	q.p.m. (1st order)	e.f.p.	Ti:Al$_2$O$_3$ diode laser	432 432.5	(36.4) (0.86)	34 0.8	1.7 4.8×10^{-4}	128 66
1996 [6.325]	a.p.e.	channel	0.3 0.3	q.p.m. (1st order)	e.f.p.	Ti:Al$_2$O$_3$ (300 mW) DL (80 mW)	383 383	56 14.5	– –	1.9 0.072	61 34
1997 [6.326]	Ti indiff.	channel	0.5	q.p.m. (3rd order)	e.f.p.	Ti:Al$_2$O$_3$	422.5	45	–	0.0079	0.39
1997 [6.154]	a.p.e.	channel	1.0	q.p.m. (1st order)	e.f.p.	Ti:Al$_2$O$_3$	434	(70.5)	56	19	382

1997 [6.153] a.p.e.	channel	0.7	q.p.m. (1st order)	e.f.p.	Ti:Al$_2$O$_3$ 475	(105)	90	2.2	20
1997 [6.176] a.p.e.	channel	0.5	q.p.m. (1st order)	e.f.p.	Ti:Al$_2$O$_3$ 434	(47)	42	5.5	249
1997 [6.177] a.p.e.	channel	1.0	q.p.m. (1st order)	e.f.p.	DL (250 mW, peak) 426	–	–	17 (peak)	–
1997 [6.149] a.p.e.	channel	1.0	q.p.m. (1st order)	Ti indiff.	Ti:Al$_2$O$_3$ 421.5	(151)	120	25	110
a.p.e.	channel	1.0	q.p.m. (1st order)	e.f.p.	Ti:Al$_2$O$_3$ 551.5	–	–	–	(177)
1997 [6.155] a.p.e.	channel	1.0	q.p.m. (1st order)	e.f.p.	Ti:Al$_2$O$_3$ 475	(135)	107	37	203

[a] Values in brackets indicate the laser output power if given by the author.
[b] Values in brackets indicate an estimated in-coupled value for $P_\omega(0)$ if the author gave only $P_\omega(L)$. $P_\omega(0)$ was calculated using $P_\omega(0) = P_\omega(L) \exp(\alpha_\omega L)$ assuming a loss of 1 dB cm^{-1} (0.23 cm^{-1}) which is typically observed in LiNbO$_3$ waveguides.
[c] Defined by (6.24): $\bar\eta = \frac{P_{2\omega}(L)}{[P_\omega(0)]^2}$

Laser types
OPO   optical parametric oscillator
DL   diode laser (mostly AlGaAs)
Ti:Al$_2$O$_3$ titanium:sapphire laser

Legend
a.p.e.   annealed proton exchange
p.e.   proton exchange
q.p.m.   quasi-phase-matching
e.b.s.   electron beam scanning
e.f.p.   electric field poling

have required active stabilization of the resonator, and the onset of optically induced changes of the refractive indices even at the modest power level at which the experiment was performed. Reviews of birefringent phase matched integrated-optical LiNbO$_3$ parametric devices were published by Sohler et al. [6.4, 6.171].

Quasi-phase-matching allows access to the nonlinear coefficient d_{33}, which has the highest nonlinear figure of merit (Table 6.1). The split-step procedure, introduced by Lim et al., comprising periodic domain reversal by Ti indiffusion followed by the fabrication of an annealed proton-exchanged waveguide [6.27, 6.172], was adopted by a number of groups in the following years. Although remarkably high normalized conversion efficiencies were reported for some devices [6.173–175], the absolute second-harmonic powers remained surprisingly low, typically on a sub-milliwatt power level. An inherent drawback of Ti-indiffused periodic gratings is their triangular shape, the shallow depth of domain inversion, and its position about 1 μm below the surface (Fig. 6.9a). As pointed out by Bortz et al., the maximum overlap for the grating structure shown in Fig. 6.9c is obtained for the $TM_{00}^\omega + TM_{00}^\omega \rightarrow TM_{01}^{2\omega}$ interaction, while the overlap integral for the $TM_{00}^\omega + TM_{00}^\omega \rightarrow TM_{00}^{2\omega}$ process is significantly smaller [6.175]. Thus the conversion efficiency to the $TM_{01}^{2\omega}$ mode greatly exceeds that to the $TM_{00}^{2\omega}$ mode, which would be more desirable for device applications.

A new impetus for further development of periodically poled LiNbO$_3$ waveguides arose from the report by Yamada et al. on electrical field poled LiNbO$_3$ waveguides [6.139]. The combination of electric field poled grating fabrication with the annealed proton exchange technique yields several advantages over annealed proton-exchanged waveguides with Ti-indiffused gratings, namely processing at relatively low temperatures (below 350°C), the possibility of freely adjusting the grating period, and fabrication of deep laminar gratings resulting in an improved overlap for the $TM_{00}^\omega + TM_{00}^\omega \rightarrow TM_{00}^{2\omega}$ interaction. The technique of electric field poling was adopted and further developed by a number of groups (c.f. Table 6.3). Annealed proton-exchanged waveguides with electric field poled domain gratings are so far the most successful LiNbO$_3$ waveguide, and blue output powers and normalized conversion efficiencies exceeding 15 mW and 200 %W^{-1}, respectively, have been reported [6.139, 6.154–155, 6.176–177]. A remarkable output power of 37 mW and conversion efficiency of 200 %W^{-1} was recently obtained in a waveguide fabricated on an oblique x-cut MgO-doped LiNbO$_3$ substrate [6.155].

Despite the progress in waveguide fabrication and poling techniques aiming at the realization of compact visible laser sources, recently reviewed by Webjörn et al. [6.149], a number of problems associated with LiNbO$_3$ waveguides remain to be solved. As already discussed in Sect. 6.3.4, the most prominent among these is the susceptibility of LiNbO$_3$ to optical damage. Although electric field poled annealed proton-exchanged waveguides were found to be less sensitive to optical damage than Ti-poled waveguides, the effect is still considerable manifesting itself in long-term decay of the output power and power fluctuations. A further problem arises from lateral domain wall growth during electric field poling which complicates control of the poling process [6.178].

6.4 Lithium Tantalate Waveguides

Lithium tantalate ($LiTaO_3$) is isomorphic to $LiNbO_3$ and has similar electro-optic and nonlinear optical properties. Thus many developments in material technology and processing progressed in parallel with $LiNbO_3$. However, there have been substantially fewer efforts to utilize $LiTaO_3$, mainly because of its significantly lower Curie temperature of about 600°C. As a waveguide material for blue light second-harmonic generation, $LiTaO_3$ was recognized only in the 1990s with the development of a suitable domain inversion technique for quasi-phase-matching, because the small birefringence and small off-diagonal nonlinear coefficients (Table 6.1) prevent birefringent phase matching.

6.4.1 Fabrication and Properties
of Proton-Exchanged Lithium Tantalate Waveguides

The fabrication techniques for $LiTaO_3$ waveguides are very similar to those used in $LiNbO_3$. Optical waveguiding layers were prepared by Li_2O outdiffusion at 1150 and 1400°C [6.54, 6.64]. Inverted single-domain layers were observed after heat treatment at temperatures of 800–1200°C. Attempts to form waveguides by metal ion indiffusion were performed above and below the Curie temperature of about 600°C, including indiffusion of Nb at 1100°C [6.179], Cu indiffusion at 500 and 800°C [6.56], Ti indiffusion at temperatures ranging from 1075–1450°C [6.180] and at 1200°C [6.181], vapor phase Zn indiffusion at 800°C [6.182] and Ag indiffusion at 240–370°C [6.183]. The most successful technique for waveguide fabrication to date, as in $LiNbO_3$, is proton exchange [6.58, 6.184–185]. Most commonly, the exchange is carried out in a hot bath of either benzoic acid [6.58, 6.185] or pyrophosphoric acid [6.186–188]. Diffusion characteristics, the effects of annealing, refractive index profiles, and waveguide losses have been studied by several groups [6.187, 6.189–193]. Propagation losses as low as $0.25\,\mathrm{dB\,cm^{-1}}$ at a wavelength of 820 nm were measured [6.185].

Yuhara et al. observed in $LiTaO_3$ an anomalous increase of the extraordinary refractive index after annealing, in contrast to $LiNbO_3$ where annealing is accompanied by a continuous decrease of the index of the proton-exchanged layer [6.191]. This result was later confirmed by Åhlfeldt et al. [6.193]. This behavior may be understood by considering the phase diagram of the $Li_{1-x}H_xTaO_3$ system. The structural properties of this compound have been investigated by Ganshin et al. [6.194], and a comprehensive study of proton-exchanged $LiTaO_3$ solid solutions was presented by Fedorov and Korkishko [6.195]. Five different phases were found depending on exchange and annealing conditions and crystal cut. Guidelines to control the exchange process by considering the phase diagram of the $Li_{1-x}H_xTaO_3$ layers were given by El Hadi et al. [6.196]. The α-phase, displaying the best optical properties, was not observed on samples prepared by direct exchange. It was obtained only by annealing samples initially displaying different phases. The fabrication of α-phase layers is also important in view of the long-term stability of the waveguide. Significant changes of the index profile and effective indices were observed for certain exchange and annealing conditions [6.193, 6.197], while α-phase layers were found to be stable [6.196].

One of the major advantages of annealed proton-exchanged $LiTaO_3$ waveguides is thought to be their smaller susceptibility to optical damage compared with their $LiNbO_3$ counterparts. The damage threshold of Ti-indiffused $LiTaO_3$ waveguides was measured to be around $11\,kW\,cm^{-2}$ at a wavelength of $0.515\,\mu m$ and an elevated temperature of 150°C [6.180]. Howerton and Burns demonstrated channel waveguides and Mach–Zehnder interferometers capable of handling 35–75 mW of power (power density $100–400\,kW\,cm^{-2}$) at a wavelength of $0.83\,\mu m$ [6.198]. They attributed this better power-handling capability to a higher dark conductivity and reduced photoconductivity compared to $LiNbO_3$. Matthews et al. observed the onset of optical damage at a guided power of about 55 mW at $0.86\,\mu m$ ($100\,kW\,cm^{-2}$) [6.192] which was claimed to be about four orders of magnitude higher than in Ti-indiffused $LiNbO_3$ waveguides and one order of magnitude higher than in annealed proton-exchanged $LiNbO_3$ waveguides [6.103]. At higher power levels the modal field distribution was observed to degrade rapidly after about one minute, indicating the onset of optical damage. Although these results might suggest that $LiTaO_3$ is advantageous over $LiNbO_3$ regarding optical damage, $LiTaO_3$ is also not free of this problem, and the onset of photorefractive effects starts at relatively modest guided powers.

6.4.2   Domain Inversion and Quasi-Phase-Matching in Lithium Tantalate

Domain inversion techniques to form periodic gratings for quasi-phase-matching in $LiTaO_3$ have to take the relatively low structural phase transition temperature of about 600°C into account, in contrast to $LiNbO_3$ ($T_c \approx 1100$°C). Titanium indiffusion and Li_2O outdiffusion, both successful techniques in $LiNbO_3$, were found to fail to produce domain inverted layers at 590°C [6.199], because of the small diffusion constants at this temperature. Therefore, the initial effort concentrated on the proton exchange technique to form an Li-deficient layer below the surface which was thought to be essential for triggering domain reversal [6.199]. $LiTaO_3$ samples were proton-exchanged in benzoic acid at 220°C and then submitted to a heat treatment at 590°C, slightly below the Curie temperature. Domain inverted layers with a thickness of up to $100\,\mu m$ were observed to form after this procedure on the c^- face, in remarkable contrast to $LiNbO_3$ where domain reversal occurred on the c^+ face. A key feature of the domain inversion process based on proton exchange and heat treatment was recognized by Mizuuchi et al. who found a threshold-like behavior of the inversion process as a function of the heat treatment temperature [6.186]. Domain reversal was observed only for a heat treatment temperature above 450°C, while no change in the domain structure occurred below this temperature. This makes domain reversal by proton exchange compatible with processes carried out below the domain reversal threshold temperature, in particular with the process of forming a waveguide by proton exchange and subsequent annealing. The first fabrication of a first-order quasi-phase-matching grating was reported in [6.200]. The domain shape obtained by this procedure is semicircular. The process requires precise control of the quick heat treatment time, of the order of seconds, to prevent lateral spreading of the domains. Although the semicircular shape of the domains is not ideal, it allows a better overlap with the waveguide modes than do the triangular domains produced by Ti indiffusion in LiNbO₃. Further details on the domain inversion process were discussed in [6.189, 6.201]. Experiments performed to clarify the domain inversion process by proton exchange and heat treatment were described in [6.202–203].

The fabrication of a LiTaO₃ waveguide device for second-harmonic generation using proton exchange for domain inversion and waveguide formation typically comprises the following sequence of steps [6.201]: (1) formation of a mask defining the periodic grating; (2) proton exchange in pyrophosphoric acid at 260°C for 20 minutes; (3) quick heat treatment at 540°C for 0.5 minute (heating rate 80°C s⁻¹); (4) formation of a mask defining channel waveguides; (5) proton exchange at 260°C for 14 minutes; and (6) annealing at 420°C for 1 minute (heating rate 40°C s⁻¹). Steps (1)–(3) lead to the formation of a periodic domain grating, while steps (4)–(6) produce a channel waveguide. An alternative method has recently been used by Yi et al. with both proton exchange steps carried out at a lower temperature of 220°C for 40 minutes [6.204]. In this way, the initial degree of proton exchange is reduced, which is thought to result in an enhanced waveguide quality. Proton outdiffusion during the heat treatment and annealing steps, respectively, is inhibited by an SiO₂ mask deposited on the waveguide surface. This promotes proton indiffusion.

Electric field poling at about 600°C, close to the Curie temperature, with a voltage of 1.4 V was used by Matsumoto et al. to form a third-order quasi-phase-matching grating with a period of 14 μm [6.205]. Domain inversion by electron beam writing was demonstrated by Hsu and Gupta [6.206]. Bombardment with 25 keV electrons of the c^- face was observed to reverse the polarity to a depth of several hundred micrometers. Domain inversion was explained in terms of an electric field induced by the injected electrons. Electric field poling at room temperature of up to 300 μm thick LiTaO₃ plates was demonstrated by Mizuuchi and Yamamoto [6.207]. Nucleation of inverted domains by application of an electric field starts at the c^+ face and grows towards the c^- face. To inhibit lateral spreading during domain growth, the samples were proton exchanged after fabrication of a periodically patterned Ta electrode on the c^+ face, as domain nucleation is suppressed in the proton-exchanged regions. A flat electrode was deposited on the c^- face. Application of 3 ms voltage pulses generating a field of 29 kV mm^{-1} over the sample resulted in the formation of regular gratings with a period of 3.8 μm and a length of 10 mm. The same authors reported a period as short as 3.3. μm over a 10 mm length fabricated by the same technique [6.208].

6.4.3   Second-Harmonic Generation in Lithium Tantalate Waveguides

A number of earlier measurements using electro-optic waveguide modulators suggested that the r_{33} electro-optic coefficient showed no change in its value from that of bulk LiTaO$_3$ [6.185, 6.191, 6.209]. Åhlfeldt et al. measured the 266 nm second-harmonic signal from proton-exchanged and annealed proton-exchanged LiTaO$_3$ layers in a reflection geometry [6.210]. They concluded that the nonlinearity of the proton-exchanged layer was strongly reduced, while a partial recovery due to annealing was observed. A more detailed study by Åhlfeldt revealed that the nonlinear coefficient of the exchanged layer practically vanishes [6.211]. Almost full recovery was observed after annealing at 400°C for 16 hours. This depth profiling measurement of the d_{33} coefficient yielded a very similar result as that performed by Bortz et al. in annealed proton-exchanged LiNbO$_3$ waveguides [6.125].

Results on second-harmonic generation in proton-exchanged LiTaO$_3$ waveguides are summarized in Table 6.4. Early devices were based on third-order quasi-phase-matching because of the lack of a technique that allowed the controlled fabrication of gratings with a period smaller than 4 μm necessary for first-order quasi-phase-matching. In first-order quasi-phase-matching a high second-harmonic power of 31 mW at 435 nm was reported by Yamamoto et al. [6.212], and normalized conversion efficiencies exceeding 100 %W^{-1} were reported several times by the same group. The highest normalized conversion efficiency of 650 %W^{-1} was reported by Yi et al., however, with a relatively modest blue output power of 1 mW only [6.204].

6.5   Potassium Titanyl Phosphate Waveguides

Potassium titanyl phosphate (KTiOPO$_4$, KTP) is a relatively new nonlinear optical material, reported for the first time in 1976 by Zumsteg et al. [6.213]. In its bulk form it is widely used for frequency doubling of Nd lasers, in particular the 1.064 μm Nd:YAG laser, and for optical parametric oscillators [6.214], because of its relatively high nonlinear coefficients, high damage threshold, and wide angular and temperature acceptance range. The essential material properties are reviewed in [6.215]. There has been some controversy regarding

Table 6.4. Second-harmonic generation in LiTaO$_3$ waveguides

Year Ref.	Waveguide fabrication	Waveguide type	L (cm)	Phase matching technique	Domain inversion technique	Laser[a]	$\lambda_{2\omega}$ (nm)	$P_\omega(0)^{\text{b}}$ (mW)	$P_\omega(L)$ (mW)	$P_{2\omega}(L)$ (mW)	$\bar{\eta}^{\text{c}}$ (%W^{-1})
1991 [6.186]	p.e.	channel	0.9	q.p.m. (3rd-order)	p.e./h.t.	Ti:Al$_2$O$_3$	421	32.7	26.6	0.13	12.2
1991 [6.327]	p.e.	channel	0.4	q.p.m. (3rd-order)	p.e./h.t.	Styryl-9 dye	435	(5.7)	4.2	1.4×10^{-4}	0.43
1991 [6.319]	p.e.	channel	0.9	q.p.m. (3rd-order)	p.e./h.t.	Ti:Al$_2$O$_3$	421	50.7	42.0	0.32	12.4
1991 [6.205]	p.e.	planar	0.1	q.p.m. (3rd-order)	e.f.p. (600 °C)	Ti:Al$_2$O$_3$	453	–	20.8	7.5×10^{-6}	1.7×10^{-3}
		channel	0.1				458	–	41	1.3×10^{-3}	0.077
1992 [6.328]	p.e.	channel	0.9	q.p.m. (1st-order)	p.e./h.t.	Ti:Al$_2$O$_3$	433	145	–	15	71
1992 [6.329]	p.e.	channel	1.0	q.p.m. (1st-order)	p.e./h.t.	Ti:Al$_2$O$_3$	435	145	121	23	110
						DL (77 mW)	435	37	–	1.1	80
1992 [6.330]	p.e.	channel	0.9	q.p.m. (3rd-order)	p.e./h.t.	Ti:Al$_2$O$_3$	424	210	–	12	27
1993 [6.212]	p.e.	channel	1.0	q.p.m. (1st-order)	p.e./h.t.	Ti:Al$_2$O$_3$	435	174	145	31	103
						DL (140 mW)	435	72	–	8	154
1994 [6.201]	p.e.	channel	1.0	q.p.m. (1st-order)	p.e./h.t.	Ti:Al$_2$O$_3$	435	115	–	23	174

Table 6.4. (Continued)

Year Ref.	Waveguide fabrication	Waveguide type	L (cm)	Phase matching technique	Domain inversion technique	Laser[a]	$\lambda_{2\omega}$ (nm)	$P_\omega(0)$[b] (mW)	$P_\omega(L)$ (mW)	$P_{2\omega}(L)$ (mW)	$\bar{\eta}^c$ (%W^{-1})
1994 [6.190]	p.e.	channel	1.0	q.p.m. (1st-order)	p.e./h.t.	Ti:Al$_2$O$_3$	434	120	–	22	153
						DL (80 mW)	434	35	–	1.3	106
1994 [6.331]	p.e.	channel	0.1	q.p.m. (1st-order)	e.b.s.	Ti:Al$_2$O$_3$	420	(31)	30	0.026	2.8
1995 [6.332]	p.e.	channel	1.0	q.p.m. (1st-order)	p.e./h.t.	DL (pulsed, 54 mW average)	428	34	–	4.5 (av.)	398
						DL (cw, 54 mW)	428	34	–	1.0	87
1995 [6.333]	p.e.	channel	1.0	q.p.m. (1st-order)	p.e./h.t.	Ti:Al$_2$O$_3$	429	–	–	–	196
						DL	429	40	–	2.0	125
1996 [6.204]	p.e.	channel	0.79	q.p.m. (1st-order)	p.e./h.t.	Ti:Al$_2$O$_3$	429	(12.4)	10.3	1.0	1042

[a] Values in brackets indicate the laser output power if given by the author.
[b] Values in brackets indicate an estimated in-coupled value for $P_\omega(0)$ if the author gave only $P_\omega(L)$. $P_\omega(0)$ was calculated using $P_\omega(0) = P_\omega(L)\exp(\alpha_\omega L)$ assuming a loss of $1\,\mathrm{dB\,cm^{-1}}$ ($0.23\,\mathrm{cm^{-1}}$) which is typically observed in LiTaO$_3$ waveguides.
[c] Defined by (6.24): $\bar{\eta} = \dfrac{P_{2\omega}(L)}{[P_\omega(0)]^2}$

Legend
q.p.m.   quasi-phase-matching
p.e.   proton exchange
p.e./h.t. proton exchange/heat treatment
e.f.p.   electric field poling
e.b.s.   electron beam scanning

Laser types
DL   AlGaAs diode laser
Ti:Al$_2$O$_3$ titanium:sapphire laser

the nonlinear coefficients of KTiOPO$_4$. While Bierlein and Vanherzeele reported a relatively high value of 16.9 pm V^{-1} for the d_{33} nonlinear coefficient at 1.064 μm wavelength [6.216], Boulanger et al. measured a value of only 10.7 pm V^{-1} [6.217]. The most recent measurement by Shoji et al. yielded values of $d_{33} = 14.6$ pm V^{-1} at 1.064 μm and 16.6 pm V^{-1} at 0.852 μm. Taking the latter value and accounting for the relatively small refractive index of KTiOPO$_4$ ($n = 1.8$), the nonlinear figure-of-merit of KTiOPO$_4$ is roughly 60% of that of LiNbO$_3$ (Table 6.1).

Single crystals of KTiOPO$_4$ can be grown by means of either flux methods or hydrothermal techniques. The main difference between these two growth methods is reflected in the ion conductivity of the obtained crystals. Generally, a flux-grown crystal has an ionic conductivity in the range $10^{-5} - 10^{-8}$ Ω^{-1} cm^{-1}, while a hydrothermally grown KTiOPO$_4$ crystal has a lower ionic conductivity in the range of $10^{-10} - 10^{-11}$ Ω^{-1} cm^{-1}. Morris et al. developed a method to reduce the ionic conductivity of flux-grown crystals utilizing the doping with trivalent ions (Ga, Al) on the Ti site and tetravalent ions (Si) on the P site in KTiOPO$_4$ [6.218].

6.5.1 Fabrication and Properties of Rubidium-Exchanged Potassium Titanyl Phosphate Waveguides

Waveguide fabrication in KTiOPO$_4$, reported for the first time by Bierlein et al., is based on the high ionic conductivity along the polar axis [6.59]. The relatively large mobility of the potassium ions along this axis allows efficient ion exchange with other monovalent cations. Solid solutions exist in the MTiOPO$_4$ structure, where M can be K, Rb, Tl or Cs. The increase in the surface refractive index scales, in general, with the electronic polarizability of the exchanged ion relative to potassium. Most often the K–Rb exchange process is used for waveguide fabrication. The high anisotropy of the diffusion constant, with virtually no diffusion in the xy-plane, allows the fabrication of waveguides with sharp edges, and restricts practical devices to z-cut substrates [6.59].

Typically, the exchange process is carried out at a temperature between 300 and 350°C for a few tens of minutes in an RbNO$_3$/Ba(NO$_3$)$_2$ melt, with a few mol.% of Ba(NO$_3$)$_2$ [6.219]. The diffusion is strongly temperature dependent, being about 2.5 times faster at 350°C than at 310°C. Because the ion exchange rate depends on the ionic conductivity of the substrate, variations in conductivity will result in variations of the diffusion constants, making control of the guide fabrication difficult. It was found that the addition of divalent cations with appropriate radii to the melt, in particular Ba^{2+}, enhances the ionic conductivity near the surface and provides a means of controlling the diffusion rate and the resulting index profile [6.215]. The increase in the ionic conductivity is due to the substitution of a small amount of K$^+$ by Ba^{2+} in the KTiOPO$_4$ lattice, resulting in the formation of potassium vacancies. By adding 20 mol.% of Ba^{2+} to the exchange melt, the exchange depth was increased by one to two orders of magnitude, allowing waveguide formation with a depth of up to 100 µm [6.215]. The ion exchange process in KTiOPO$_4$ was studied in detail in [6.220–221]. The effects of K–Rb exchange on the crystal structure of KTiOPO$_4$ were investigated, for example, by micro-Raman spectroscopy [6.222]. It was suggested that disruption and tilting of the TiO$_6$ octahedra, evaluated by micro-Raman spectroscopy, was correlated with changes of specific nonlinear optical susceptibilities.

One of the major problems of KTiOPO$_4$ is considered to be optical damage, often referred to as 'gray tracking'. As pointed out by Loiacono et al. [6.223], there is some confusion about the definition of this term. In general, the effect becomes apparent as dark regions are formed in the crystal when a light beam passes through, resulting in beam profile degradation and throughput power dissipation. In particular, gray tracking is present in green and blue second-harmonic generation often leading to breakdown of the device. The effect was observed to be both reversible and irreversible. Most of the reports studying gray tracking relate this phenomenon to the presence of Ti^{3+} defect centers in the KTiOPO$_4$ crystal. Bierlein and Vanherzeele found evidence that hydrothermally grown KTiOPO$_4$ appears to have a higher optical damage threshold than flux-grown material [6.215]. This view was supported by Roloefs who performed a spectroscopic study and found four related Ti^{3+} centers in KTiOPO$_4$, which were much more difficult to produce in hydrothermally grown KTiOPO$_4$ than in flux-grown KTiOPO$_4$ [6.224]. Loiacono et al. did not observe any correlation of gray track susceptibility with crystal growth temperature, ionic conductivity or hydroxyl concentration, in contradiction to earlier reports [6.223]. In addition, they found that gray tracking occurred using only 0.53 µm radiation, while the fundamental wave at 1.064 µm was absent. They concluded that gray tracking in KTiOPO$_4$ might be induced by a two-photon absorption process involving two second-harmonic photons. Evidence for similarity between gray tracking in KTiOPO$_4$ and optical damage due to photorefractive effects in LiNbO$_3$ was found in this study. Jongerius et al. studied optical damage effects with continuous wave lasers in undoped and Sc^{3+}-doped KTiOPO$_4$ waveguides [6.225]. They found that the sensitivity to optical damage was approximately the same. Optical damage appeared in the form of strong degradation of the milliwatt-level blue output on a time-scale of minutes and a pronounced linewidth broadening of the second-harmonic output as a function of wavelength. However, the effect was reversible in Sc^{3+}-doped KTiOPO$_4$ waveguides after decreasing the fundamental power, while it appeared to be permanent in undoped KTiOPO$_4$.

6.5.2 Second-Harmonic Generation in Potassium Titanyl Phosphate Waveguides

Birefringent phase matching in bulk KTiOPO$_4$ crystals is possible to a minimum fundamental wavelength of about 0.99 µm by a type-II process

[6.226–227]. Risk investigated birefringent phase matching in Rb-exchanged KTiOPO$_4$ waveguides via a TE$_m^\omega$ + TM$_n^\omega$ → TE$_p^{2\omega}$ interaction [6.219]. He found that the most favorable interaction with all three waves in the lowest-order mode ($m = n = p = 0$) was phase matched at 1.13 µm, hence with a second-harmonic wavelength outside the blue–green spectral range. Generation of shorter wavelengths involved a higher-order second-harmonic mode, and below 0.905 µm, the second-harmonic appeared in the form of Cerenkov radiation. Bierlein et al. demonstrated balanced phase matching for the second harmonic at 1.064 µm in a segmented KTiOPO$_4$ system [6.228]. Each segment consisted of two different sections of different length, one of which was Rb-exchanged and the other was bulk KTiOPO$_4$. The concept of balanced phase matching is based on the condition that the phase mismatch acquired in the first section is compensated in the second, resulting in a zero net phase mismatch over the whole segment. Type-II phase matching with all three waves in the lowest-order mode was observed in this structure. The idea behind balanced phase matching is essentially identical to quasi-phase-matching with a modulation of the refractive index rather than of the sign of the nonlinear coefficient. Quasi-phase-matching provided by a type-I interaction was demonstrated by van der Poel et al. in the same mixed guided–unguided structure earlier used for balanced phase matching [6.229]. Second-harmonic outputs at wavelengths between 380 and 480 nm were observed. In this report the authors discovered that the Rb/Ba ion exchange process also produced domain inversion in KTiOPO$_4$. This observation was confirmed by Laurell et al., who detected domain reversal by surface second-harmonic generation, toning, etching and piezoelectric measurements [6.230]. Domain reversal was observed in Rb-exchanged KTiOPO$_4$ if Ba(NO$_3$)$_2$ was added to the melt, while no reversal occurred for exchange in pure RbNO$_3$. As a result, in the exchanged sections the refractive index increases by replacement of K by Rb, and the spontaneous polarization is reversed by the replacement of K by Ba. Segmented waveguides for second-harmonic generation were characterized by Eger et al. [6.231]. These waveguides are considered to suffer from several limitations. The domain-inverted region may not be as deep as the waveguide [6.232], the losses are relatively high (about 4 dB cm^{-1} [6.228]) because of the guided–unguided segmentation, and the reliability and reproducibility of domain inversion produced by the ion exchange process is questionable [6.233]. Therefore other methods were investigated to induce domain inversion in KTiOPO$_4$. Similar to LiNbO$_3$ and LiTaO$_3$, electron beam scanning was shown to produce thick domain gratings in KTiOPO$_4$ [6.234]. A beam with 30 keV electrons was used to write a 20 µm period grating for fifth-order quasi-phase-matching. Domain reversal occurred through the whole sample thickness of 1 mm. Electric field poling, similar to that used for LiNbO$_3$ [6.139], was developed for KTiOPO$_4$ by Chen and Risk [6.235]. Voltage pulses of 0.5 ms to 3 s duration producing a field of 2 kV mm^{-1} periodically poled the KTiOPO$_4$ crystal with a period of 4 µm for first-order quasi-phase-matching. The required field is about one order of magnitude smaller than for electric field poling of LiNbO$_3$. The same authors demonstrated second-harmonic generation in an Rb-exchanged waveguide in electric field poled KTiOPO$_4$ [6.233]. Electric field poling of flux-grown bulk KTiOPO$_4$ has recently been investigated by Karlsson and Laurell [6.236].

Segmented waveguides may simultaneously act as quasi-phase-matching devices and distributed Bragg reflectors that provide controlled feedback to stabilize the diode laser used for second-harmonic generation, as proposed and demonstrated by Shinozaki et al. in a proton-exchanged LiNbO$_3$ waveguide [6.237]. For Rb/Ba-exchanged KTiOPO$_4$ waveguides this concept was originally studied by Risk and Lau [6.238]. It was implemented in a number of devices [6.239–240].

Results for second-harmonic generation in KTiOPO$_4$ waveguides are summarized in Table 6.5. It is noteworthy that most waveguides reported in the literature were relatively short, typically 0.5 cm. For segmented Rb-exchanged waveguides, the highest second-harmonic powers were reported by Eger et al. In particular, the generation of 3.6 mW at 429 nm is among the highest powers reported for diode laser frequency doubling [6.240–241]. Jongerius et al. reported blue power at the milliwatt-level; however, they observed considerable power decrease as a function of time [6.225]. The highest blue output power of 6.3 mW at 425 nm could be generated for a short period of time only. Recently, a remarkable blue output power of 12 mW at 431 nm from a continuous Rb/Ba-exchanged waveguide in an electric field poled KTiOPO$_4$ crystal was demonstrated by Chen and Risk [6.233]. These authors observed stable blue output of 5 mW for 50 hours, which might suggest an improved optical damage threshold in this type of waveguide. The fabrication, thermal stability and nonlinear properties of electric field-poled Rb-exchanged waveguides in flux-grown KTiOPO$_4$ were investigated by Eger et al. [6.242].

An interesting approach for frequency doubling using KTiOPO$_4$ was followed by Doumuki et al., who fabricated an approximately 200 nm thick Ta$_2$O$_5$ high-index core layer on a KTiOPO$_4$ substrate and a low-index SiO$_2$ strip load for lateral guiding. Highly efficient Cerenkov and modal-dispersion-phase-matching with a normalized efficiency of almost 100 %W^{-1} and an output power of up to 28 mW was reported [6.40, 6.243]. To obtain optimum conversion efficiency, however, the low-pressure chemical vapor deposition (LPCVD) growth of the Ta$_2$O$_5$ core layer required tight thickness control on a sub-nanometer scale.

6.6   Potassium Niobate Waveguides

This section is devoted to our own work on second-harmonic generation in ion-implanted potassium niobate (KNbO$_3$) waveguides. Although the problems associated with characterization and optimization of the optical properties with respect to second-harmonic generation are identical to those encountered

Table 6.5. Second-harmonic generation in KTiOPO$_4$ waveguides

Year	Ref.	Waveguide fabrication	Waveguide type	L (cm)	Phase matching technique	Domain inversion technique	Laser[a]	$\lambda_{2\omega}$ (nm)	$P_\omega(0)$[b] (mW)	$P_\omega(L)$ (mW)	$P_{2\omega}(L)$ (mW)	$\bar{\eta}$[c] (%W^{-1})
1990	[6.228]	Rb exchange	segmented channel	0.5	balanced p.m.	-	Nd:YAG	532	48	25	0.023	1.4
1990	[6.229]	Rb exchange	segmented channel	0.5	q.p.m.	Rb exchange	Ti:Al$_2$O$_3$	425	(140)	85	1.0	5.1
1992	[6.334]	Rb exchange	planar	0.5	Cerenkov	-	-	435	-	50	1×10^{-6}	4×10^{-7}
1993	[6.335]	Rb exchange	segmented channel	0.5	q.p.m.	Rb exchange	DL (140 mW)	424	(92)	56	1.35 (resonant)	15.9
1993	[6.336]	Rb exchange	channel	0.82	birefr. (type-II)	Rb exchange	Ti:Al$_2$O$_3$	509	(94) (TM$_{00}$) (118) (TE$_{00}$)	41.4 / 52.1	0.25	2.25
1993	[6.337]	Rb exchange	segmented channel	0.45	q.p.m.	Rb exchange	DL	425	(107)	68.3	1.2	10.5
1993	[6.231]	Rb exchange	segmented channel	0.36	q.p.m.	Rb exchange	Ti:Al$_2$O$_3$	420	(136)	95	3.0	16.2
1994	[6.225]	Rb exchange	segmented channel	0.45	q.p.m.	Rb exchange	Ti:Al$_2$O$_3$	425	125	80	3.7	23.7
1994	[6.241]	Rb exchange	segmented channel	0.38	q.p.m.	Rb exchange	Ti:Al$_2$O$_3$	425	(75)	51	3.0	53
1994	[6.40]	KTiOPO$_4$ substrate/ Ta$_2$O$_5$ core/ SiO$_2$ strip load	strip loaded	0.45	Cerenkov	-	Ti:Al$_2$O$_3$	414	370	332	28	20

Table 6.5. (Continued)

Year	Ref.	Waveguide fabrication	Waveguide type	L (cm)	Phase matching technique	Domain inversion technique	Laser[a]	$\lambda_{2\omega}$ (nm)	$P_\omega(0)$[b] (mW)	$P_\omega(L)$ (mW)	$P_{2\omega}(L)$ (mW)	$\bar{\eta}$[c] (%W^{-1})
1994	[6.338]	Rb exchange	segmented channel	0.7	q.p.m.	Rb exchange	Ti:Al$_2$O$_3$	502	(201)	100	2.1	5.2
1994	[6.243]	KTiOPO$_4$ substrate/ Ta$_2$O$_5$ core/ SiO$_2$ strip load	strip loaded	0.41	m.d.p.m.	–	Ti:Al$_2$O$_3$	413	117	106	13	96
1995	[6.240]	Rb exchange	segmented channel	1.4	q.p.m.	Rb exchange	DL (128 mW)	429	83	51	3.6	52
1996	[6.233]	Rb exchange	channel	0.36	q.p.m. (1st-order)	e.f.p.	Ti:Al$_2$O$_3$	431	(159)[d]	146	12	48
1996	[6.339]	Rb exchange	segmented channel	0.68	q.p.m.	e.f.p.	–	427	(179)	91	0.9	2.8
1997	[6.149]	Rb exchange	segmented channel	0.8	q.p.m.	Rb exchange	Ti:Al$_2$O$_3$	425	(35)	16	0.8	65

[a] Values in brackets indicate the laser output power if given by the author.
[b] Values in brackets indicate an estimated incoupled value for $P_\omega(0)$ if the author gave only $P_\omega(L)$. $P_\omega(0)$ was calculated using $P_\omega(0) = P_\omega(L) \exp(\alpha_\omega L)$ assuming a loss of $4\,\mathrm{dB\,cm^{-1}}$ ($1\,\mathrm{cm^{-1}}$) which is typically observed in segmented KTiOPO$_4$ waveguides.
[c] Defined by (6.24): $\bar{\eta} = \frac{P_{2\omega}(L)}{[P_\omega(0)]^2}$
[d] Estimated loss $1\,\mathrm{dB\,cm^{-1}}$ ($0.23\,\mathrm{cm^{-1}}$)

Legend

q.p.m.   quasi-phase-matching
m.d.p.m.   modal-dispersion-phase-matching
e.f.p.   electric field poling

Laser types

DL   AlGaAs diode laser
Ti:Al$_2$O$_3$ titanium:sapphire laser

in indiffused and proton-exchanged waveguides, ion implantation as a method significantly differs from chemical fabrication techniques, implying a number of differences regarding structural properties, index profiles, loss mechanisms and nonlinear properties.

KNbO$_3$ is one of the most attractive materials for nonlinear optical applications due to its outstanding nonlinear optical and electro-optical properties. The high optical nonlinearity was recognized by Kurtz and Perry [6.46], and early studies of the optical properties of KNbO$_3$ crystals were carried out by Wiesendanger [6.244–246]. Uematsu reported the first measurements of the optical nonlinearities [6.247–248]. The first measurements of the electro-optic and polarization-optic coefficients were performed by Günter [6.249]. More recently, several authors reported values for the nonlinear optical [6.50–51, 6.250] and electro-optical coefficients [6.251].

While the other ferroelectric nonlinear materials discussed in this chapter require quasi-phase-matching by periodic poling for blue light generation, KNbO$_3$ allows birefringent phase matched blue light second-harmonic generation in combination with the relatively high off-diagonal nonlinear coefficients. These favorable phase matching properties of KNbO$_3$ for near-infrared wavelengths were perceived by Wiesendanger who discovered the possibility of phase matched second-harmonic generation for wavelengths above 855 ± 10 nm [6.244]. First experiments on phase matched second-harmonic generation of a 1.064 µm Nd:YAG laser in KNbO$_3$ were reported by Uematsu and Fukuda [6.252, 253]. Noncritically phase matched second-harmonic generation at 0.86 µm in KNbO$_3$ at room temperature was reported for the first time by Günter [6.254]. Günter et al. were also the first to report on noncritically phase matched second-harmonic generation in KNbO$_3$ using near-infrared AlGaAs laser diodes [6.255]. The phase matching possibilities in the whole transparency range of KNbO$_3$ between 0.4 µm and 4.5 µm for second-harmonic generation, sum and difference frequency generation, and optical parametric oscillation were worked out by Zysset et al. and Biaggio et al. [6.256–257].

The first report on waveguiding in KNbO$_3$ was by Baumert et al. [6.258]. An electro-optically induced waveguide was realized by fabricating two electrodes on the top of a KNbO$_3$ crystal. Applying a voltage of larger than 10 V shifted the index n_c by the electro-optic effect in the region between the electrodes. By this means light could be guided in the index-raised region. While chemical methods, e.g. indiffusion, to produce waveguides in KNbO$_3$ fail because of the small diffusion constants at temperatures below the structural phase transition at 220°C, permanent waveguides in KNbO$_3$ can be formed by ion implantation, as shown for the first time by Bremer et al. [6.259]. Further reports on ion-implanted waveguides concentrated on the index depth profiles [6.260–261]. The influence of the ion dose on the loss in planar waveguides was investigated by Strohkendl et al. [6.262]. Fluck et al. [6.263–264] were the first to report on the fabrication of channel waveguides. Continuous progress of the guide fabrication technology allowed the demonstration of efficient blue light second-harmonic generation in Cerenkov-type [6.36, 6.41] and phase matched [6.265–266] configurations.

Waveguiding epitaxial $KNbO_3$ thin films were fabricated by rf sputtering on MgO [6.267]. Rather low losses were measured in some of these films; however, the films exhibited a tetragonal crystal structure, and the nonlinear optical susceptibility was about three to four times lower than in orthorhombic $KNbO_3$. $KNbO_3$ thin films were also obtained by ion beam sputtering on $KTaO_3$, $MgAl_2O_4$ and MgO [6.268], and by pulsed laser deposition on MgO [6.269–270]. The growth of $KNbO_3$ films with the favorable nonlinear optical properties of single domain crystals is difficult, while the growth of oriented barium titanate $(BaTiO_3)$ films was shown to be easier [6.271]. So far, no results on frequency conversion in $KNbO_3$ films comparable to those achieved in waveguides fabricated in single-domain crystals have been reported.

6.6.1   Fabrication of Ion-Implanted Waveguides in Potassium Niobate

A   Effects of Ion Implantation

The effect of MeV He^+ ion implantation has been discussed for many optical materials [6.61]. In general, the refractive index profile is characterized by a buried optical barrier confining a guiding layer. However, a closer investigation shows that the detailed features of the refractive index in the guide region as well as the index modification in the barrier depend strongly on the material.

Implantation of MeV He^+ ions into a $KNbO_3$ single crystal has an impact on the lattice itself as well as on the refractive index [6.272]. The change of the material properties is determined by the energy deposited from the incident He^+ ions into the target material. The energy is transferred from the ion beam to the crystal lattice by two mechanisms: electronic excitation, dominating at high ion velocity (energy $> 10\,keV$), and nuclear collision, becoming effective as the ions slow down (energy $< 10\,keV$). For $KNbO_3$ the process was simulated by the TRIM '89 Monte Carlo code (TRIM: *TR*ansport of *I*ons in *M*atter [6.273]). According to [6.272] there is a distinct difference between the refractive index and lattice damage profiles. The refractive index is about 10 times more sensitive to the irradiation than the lattice itself, i.e. the onset of significant index change starts at a much lower dose than the onset of lattice damage. Figure 6.11 displays the refractive index change $\Delta n(x)$ and the percentage lattice damage $s(x)$ of an He^+ ion-implanted $KNbO_3$ waveguide. By comparing the refractive index and the lattice damage profiles, shown in the inset of Fig. 6.11, one notes that the relative change of the refractive index in the electronic stopping region ($x < 4\,\mu m$) is much larger than the lattice damage. This behavior reflects the fact that the refractive index is significantly more sensitive to irradiation than the lattice itself.

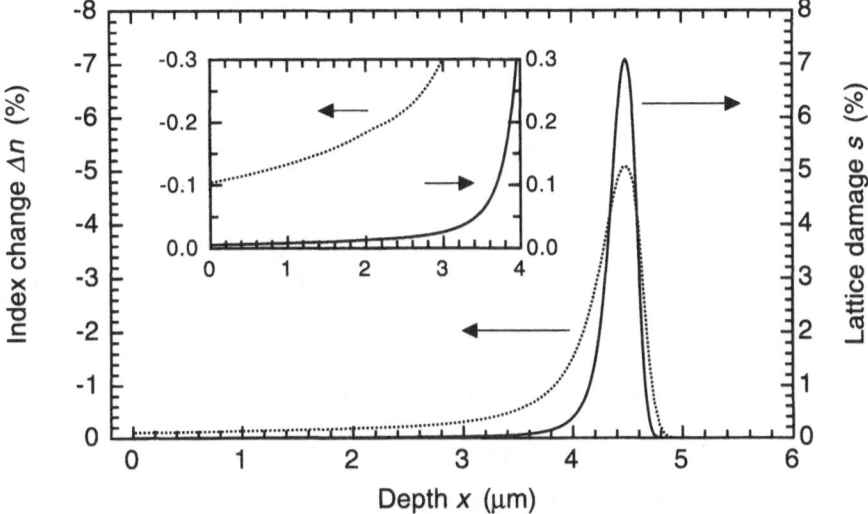

Fig. 6.11. Refractive index profile of the index n_b ($\cdots\cdots$) and lattice damage (——) as a function of depth of an ion-implanted KNbO$_3$ waveguide formed with an ion dose of 1.5×10^{15} cm^{-2} and an ion energy of 2 MeV. Notice that the maximum relative index change and lattice damage are 9% and 100%, respectively. The index decrease is thus already halfway to its saturation value, whereas the lattice damage is still relatively small. The inset shows the refractive index change and lattice damage in the region where energy loss of the ions due to electronic excitation is dominant (electronic stopping region, $x < 4$ μm). (Adapted from [6.274])

B Refractive Index Profile

The refractive index modification in KNbO$_3$ due to ion irradiation is thoroughly discussed in [6.274–276]. A quantitative model based on TRIM calculations was proposed to reconstruct the index profile from mode index measurements using the prism coupling technique [6.274]. Figure 6.12 shows examples of profiles of the three principle refractive indices of KNbO$_3$ as a function of depth.

The ionization mechanism cannot be neglected when considering the refractive index modification. In many materials where electron beam irradiation alone does not produce any modification on the index profile, the refractive index is considerably altered by light ion implantation in a region close to the surface where nuclear energy deposition is negligible. Thus, an occurrence of a synergistic effect between ionization and defects formed by ionic collision was proposed to explain the anomalous index increase in the surface-close tail of the ion track [6.61, 6.71]. In KNbO$_3$, in the region close to the crystal surface where electronic excitation prevails, a change of the bulk value was reported for all three indices: n_a and n_b decrease slightly, while n_c shows an anomalous increase [6.277]. To illustrate the impact of electronic excitation on the refractive index change, we plotted the index profiles on a

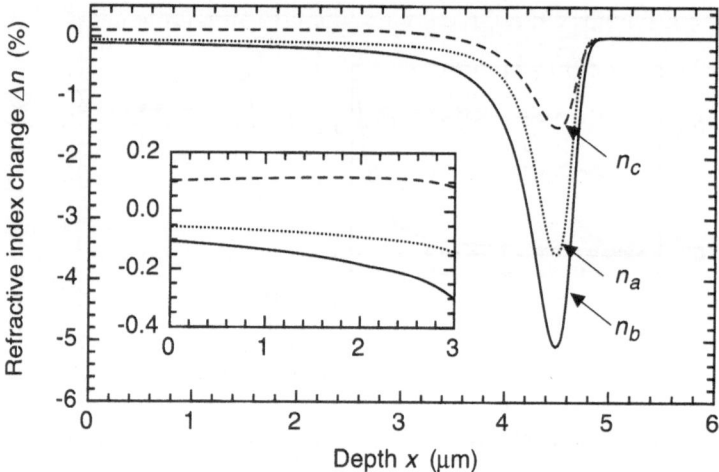

Fig. 6.12. Calculated refractive index profiles (in percentage index change with respect to the bulk index) of n_a ($\cdots\cdots$), n_b (———), and n_c (– – – – –) as a function of depth. The inset shows the profiles in the electronic stopping region. The profiles were calculated for an ion energy of 2 MeV and an ion dose of $1.5 \times 10^{15}\,\mathrm{cm}^{-2}$. (Adapted from [6.276])

magnified scale to a depth of 3 μm that approximately covers the electronic stopping region (inset of Fig. 6.12). In this region the index change of n_c is positive, whereas the indices n_a and n_b decrease slightly. This behavior of the n_a/n_b index pair and the index n_c in KNbO$_3$ is similar to the impact of ion implantation on the ordinary n_0 and extraordinary n_e index of LiNbO$_3$. In this material, n_0 decreases and n_e increases in the electronic excitation layer [6.278].

The position of the buried barrier can be controlled by means of the ion energy. At higher energies, the ions penetrate deeper into the material before their speed is reduced such that nuclear collisions become effective. Figure 6.13a shows the refractive index profiles of n_a of waveguides formed with different ion energies. The empirical relation between the position of the barrier d and the implantation energy E can be described by a parabola:

$$d = a_0 + a_1 E + a_2 E^2 , \tag{6.35}$$

where $a_0 = 0.87\,\mu\mathrm{m}$, $a_1 = 1.08\,\mu\mathrm{m/MeV}$, $a_2 = 0.36\,\mu\mathrm{m/MeV}^2$, d is in μm and E in MeV. The waveguide thickness is defined by d. Since the accuracy of the energy in a tandetron accelerator is better than $\pm 10\,\mathrm{keV}$, the resulting uncertainty in the waveguide thickness is smaller than $\pm 1\%$.

Figure 6.13b shows the n_a index profiles of waveguides formed with an ion energy of 2 MeV and different doses. The index decrease in the barrier grows with increasing ion dose. It saturates at a dose of about $10^{16}\,\mathrm{cm}^{-2}$, where all

(a)   (b)

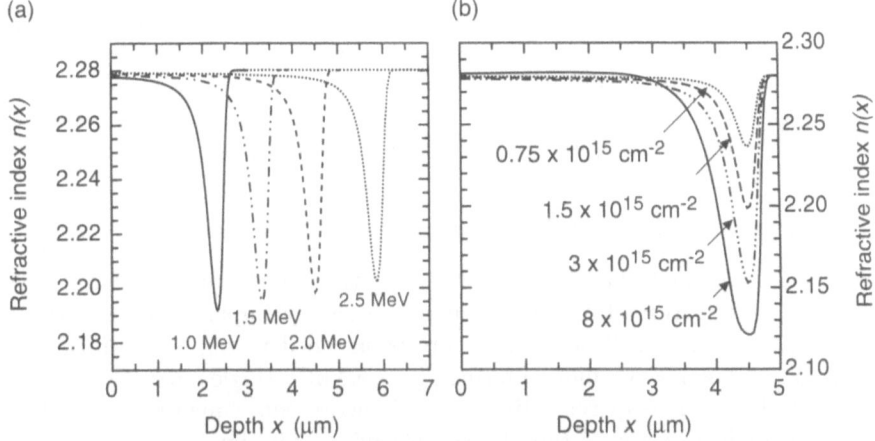

Fig. 6.13. (a) Refractive index profiles (n_a, wavelength $0.633\,\mu m$) of ion-implanted $KNbO_3$ waveguides formed with an ion dose of $1.5 \times 10^{15}\,cm^{-2}$ and different ion energies. (b) Refractive index profiles (n_a, wavelength $0.633\,\mu m$) of waveguides formed with an ion energy of $2\,MeV$ and different ion doses. Saturation of the index change in the barrier occurs at a dose of about $1 \times 10^{16}\,cm^{-2}$, with a minimum index of 2.12 at $0.633\,\mu m$. (Calculated from the data given in [6.276])

three indices merge to a common value of 2.12 at $0.633\,\mu m$, indicating the formation of an isotropic damage layer [6.277].

C Lattice Damage

The generation of damage by light ion implantation in ternary oxides is thought to be substantially determined by the energy density deposited into nuclear processes, whereas the deposition of electronic energy has a much smaller effect (e.g. Götz in [6.279]). Rutherford backscattering spectroscopy (RBS) was used in $KNbO_3$ to measure the percentage lattice damage s [6.272], defined as the ratio of aligned peak yield at its maximum to random yield. When normalized to $100\% = $ RBS-amorphous ($s = 1$), s indicates the percentage of target ions displaced from their original lattice sites. The lattice damage's dependence on the depth x was found to follow a function given by

$$s(x) = 1 - \left(1 + \frac{G_n(x)}{G_{0,\text{RBS}}}\right) \exp\left(-\frac{G_n(x)}{G_{0,\text{RBS}}}\right),\tag{6.36}$$

where $G_{0,\text{RBS}}$ was measured to be $6.5 \times 10^{23}\,eV cm^{-3}$. An example of a lattice damage profile $G_n(x)$ is shown in Fig. 6.11.

The density of defects is closely related to the optical loss of the waveguide. For a typical He^+ implantation dose of $1.5 \times 10^{15}\,cm^{-2}$, the barrier layer is about 6% RBS-amorphous [6.272] (density of lattice defects $\sim 10^{21}\,cm^{-3}$). In the waveguiding region, according to TRIM calculations, the density of vacancies is expected to be of the order of $2 \times 10^{20}\,\mathrm{cm}^{-3}$ (2600 ppm). Thus, the creation of absorption centers (point defects, e.g. vacancies) in the waveguide is inherently related to the formation of the barrier waveguide itself. The optical loss induced by these defect centers can be reduced by annealing the waveguides after ion irradiation, as we will discuss below.

D   Formation of Channel Waveguides

Embedded Channel Waveguides. The formation of channel waveguides in $KNbO_3$ by ion irradiation using a multiple energy implantation technique was reported in [6.264]. The procedure is sketched in Fig. 6.14a. In a first step (1) a planar waveguide is formed in a $KNbO_3$ single crystal by irradiation with He^+ ions of typically 2–3 MeV energy. Subsequently, a shielding mask is applied on top of the planar waveguide. Initially, tungsten wires or fibers were used as masks. Alternatively, shielding can be provided by a photolithographically formed photoresist strip. In the next step (2), the sample is irradiated with He^+ ions of lower energy than in step (1) to decrease the refractive index in the unshielded region. This process is repeated two to four times, each time with ions of lower energy, until the index in the whole surround region is modified. As a result an embedded channel waveguide with a rectangular cross-section is obtained (3).

Embedded Barrier Waveguides. An alternative method to fabricate channel waveguides by ion implantation using structured photoresist with oblique edges as an implantation mask was reported in [6.266]. The process is sketched in Fig. 6.14b. Exposing and developing of the photoresist is adapted such that the edges of the mask are not normal to the crystal surface but tilted by an angle of 30° to 60° (1). Subsequently, the crystal is irradiated by He^+ ions that are stopped within the photoresist in the shielded region and penetrate into the $KNbO_3$ in the open part of the mask (2). The oblique edges of the mask lead to the formation of a channel waveguide with a trapezoidal cross-section (3). This channel waveguide may be considered as a modification of the embedded channel guide.

Ridged Channel Waveguides. A technique to form ridged channel waveguides (ridge waveguides) in $KNbO_3$ by combining He^+ ion implantation and Ar^+ ion sputtering was outlined in [6.280]. The fabrication process is schematically shown in Fig. 6.14c. The first step (1), irradiation of a $KNbO_3$ single crystal with MeV He^+ ions, leads to the formation of a planar waveguide. Then, a positive photoresist mask defining strips of a few micrometers width is formed on top of the planar waveguide. Subsequently, Ar^+ ion sputtering in a plasma ion source is used to etch the planar waveguide from the top in the unshielded regions (2). The resulting ridge waveguide is shown in (3).

The sputtering process provides a convenient method for thin-film removal applicable to a wide variety of materials. In a plasma sputtering process

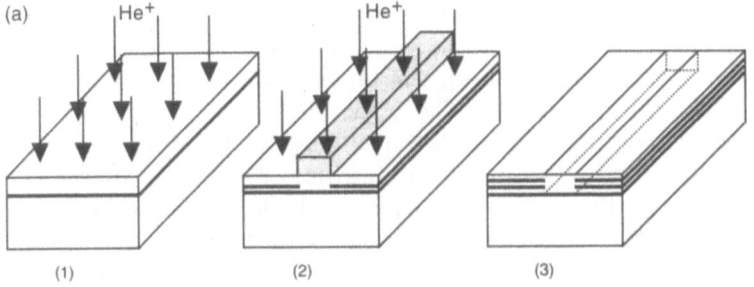

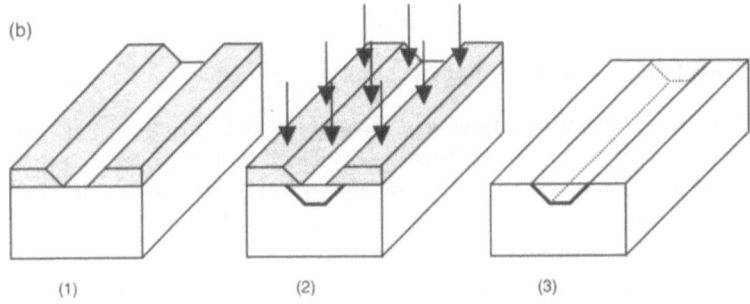

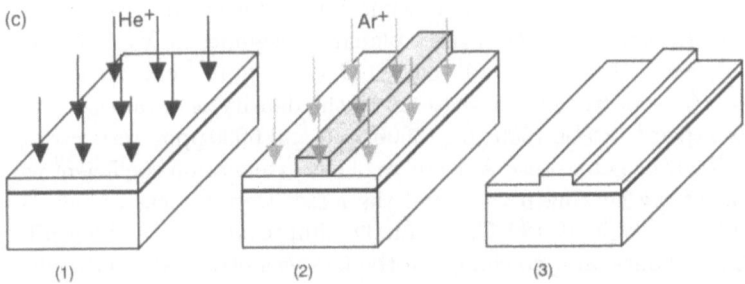

Fig. 6.14a–c. Fabrication of $KNbO_3$ channel waveguides by ion implantation. (a) Fabrication of an *embedded channel waveguide* by a multiple energy implantation process. (1) Formation of a planar waveguide by He^+ ion implantation, (2) shadow masking and subsequent implantation steps with ions of lower energy, (3) resulting embedded channel waveguide. (**b**) Fabrication of an *embedded barrier waveguide* by a single-energy implantation process. (1) Formation of a photolithographic mask with wedged edges, (2) implantation of He^+ ions, (3) resulting embedded barrier waveguide. (**c**) Fabrication of a *ridge waveguide*. (1) Formation of a planar waveguide by He^+ ion implantation, (2) shadow masking and subsequent Ar^+ ion plasma sputtering, (3) resulting ridge waveguide

an ionized inert gas is produced for fabrication purposes in a high-vacuum plasma chamber by a dc or rf discharge, accelerated, and directed either with or without final collimation on the target surface. The ion energies are usually in the range 0.2–20 keV. The impact of the ions causes sputtering ejection of atoms from the surface of the target material. By means of contact or shadow masking, patterns can be machined into the target surface. The directional impact of the ions permits etching of sharply defined patterns with perpendicular smooth side walls, in contrast to wet etching where the problem of mask undercut often limits the precision of structuring. The choice of the gas is determined by the fact that noble gases are the most efficient, the sputtering yield increasing with mass. Inert gas sputtering is most often performed with Ar since the heavier gases Xe and Kr are very expensive.

E Post-Implantation Annealing and Repoling

Annealing. Ion-irradiated waveguides usually suffer from optical loss if they are not annealed after implantation because of defect centers generated during irradiation. The effect of annealing of He^+ implanted waveguides was investigated in several oxides such as $LiNbO_3$ and $LiTaO_3$ [6.65–66, 6.281], $KTaO_3$ [6.282] and $Ba_2NaNb_5O_{12}$ (BNN) [6.283].

The structural phase transition of $KNbO_3$ at 220°C [6.244] prevents the use of a high annealing temperature. In our annealing experiments the waveguides were heated from room temperature to 180°C at a rate of 30°C/h and kept at 180°C for several hours in normal atmosphere [6.284].

Figure 6.15 shows the waveguide attenuation coefficient as a function of annealing time measured *in situ* in three planar waveguides at $\lambda = 0.457\,\mu m$. The ion doses were $D = 0.75 \times 10^{15}\,cm^{-2}$, $D = 1.5 \times 10^{15}\,cm^{-2}$ and $D = 3.0 \times 10^{15}\,cm^{-2}$, respectively. For these doses the density of vacancies in the waveguide is expected to be 1500 ppm, 3000 ppm and 6000 ppm, respectively, according to TRIM calculations. A decrease of the attenuation coefficient was observed during the heating period, and the losses were further reduced as the waveguide was kept at 180°C. No further improvement was seen after approximately 8 hours, and no change of the loss was observed after cooling the waveguide down to room temperature.

Figure 6.16a shows the attenuation coefficient as a function of the ion dose measured at 0.457 µm wavelength before and after annealing, respectively. The attenuation was significantly decreased by annealing. At an ion dose of $D = 3.0 \times 10^{15}\,cm^{-2}$ a decrease of more than $2\,cm^{-1}$ was observed. The decrease $\Delta\alpha$ of the loss coefficient due to annealing as a function of the wavelength is plotted in Fig. 6.16b. Obviously, the loss reduction was higher for the higher implantation dose and was stronger at blue and green wavelengths. For $D = 1.5 \times 10^{15}\,cm^{-2}$, the loss reduction at $\lambda = 0.457\,\mu m$ was $\Delta\alpha = 1.5\,cm^{-1}$ and at $\lambda = 0.860\,\mu m$ $\Delta\alpha = 0.3\,cm^{-1}$.

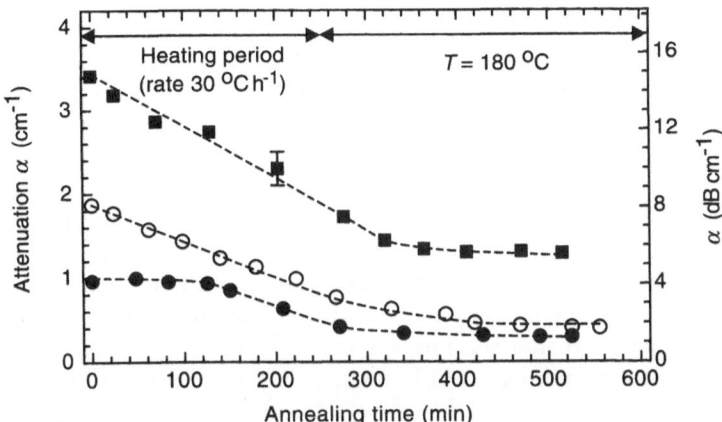

Fig. 6.15. Waveguide attenuation at $0.457\,\mu m$ measured *in situ* as a function of annealing time of three planar $KNbO_3$ waveguides. The dashed lines are guides to the eye. Waveguide fabrication parameters: thickness $d = 4.5\,\mu m$, • ion dose $D = 0.75 \times 10^{15}\,cm^{-2}$, o $D = 1.5 \times 10^{15}\,cm^{-2}$, ■ $D = 3.0 \times 10^{15}\,cm^{-2}$ (adapted from [6.284])

Repoling. In ferroelectric ABO_3-type perovskites the displacement of the B-ion from its center position in the oxygen octahedron gives rise to the spontaneous polarization and, because of the proportionality between the

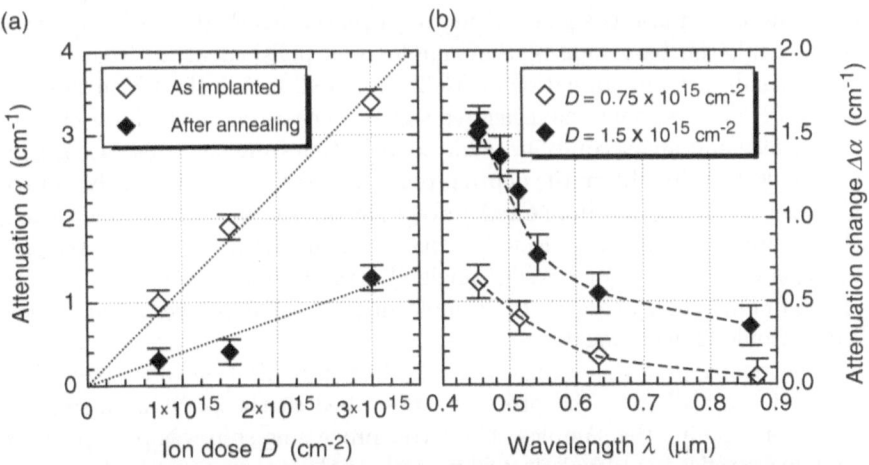

Fig. 6.16. (a) Waveguide attenuation at $0.457\,\mu m$ of planar waveguides (thickness $d = 4.5\,\mu m$) as a function of ion dose after implantation (◊) and after annealing (◆), respectively. The lines indicate the best fit with a linear function. (b) Reduction of the attenuation by annealing as a function of wavelength. ◊ $D = 0.75 \times 10^{15}\,cm^{-2}$, ◆ $D = 1.5 \times 10^{15}\,cm^{-2}$. The dashed lines are guides to the eye. (Adapted from [6.284])

spontaneous polarization and the vector part of the second-order susceptibility, to the optical nonlinearity [6.47, 6.49]. For KNbO$_3$ there exist a total
of 12 possible directions for the niobium ion to be displace when the crystal
is transformed from its cubic (high temperature) to the orthorhombic (room
temperature) structure [6.244]. Although the electronic energy loss of the
implant ions in the region close to the surface causes negligible permanent
damage to the crystal lattice [6.272], it is more than sufficient to excite the
oxygen octahedrons. Subsequently, the niobium atom in excited octahedrons
can relax into any of the 12 possible directions causing a microscopic depolarization and, as a consequence, a decrease of the optical nonlinearity. In the
nuclear collision damage region the target material is partially amorphized
and the material density reduced which suggests a strong decrease of the
optical nonlinearity.

A depth profiling measurement of the nonlinear optical susceptibility, similar to the one reported in [6.125] for annealed proton-exchanged LiNbO$_3$
waveguides, was performed on an He$^+$ ion-implanted KNbO$_3$ waveguide
[6.285]. Potassium niobate samples were prepared in the following way. After formation of a planar waveguide by He$^+$ ion implantation, the samples
were polished at a wedge angle of 1°, as shown in Fig. 6.17a. This geometry improves the resolution of the waveguiding structure when measuring
the reflected second-harmonic signal from the sample surface. For this purpose, the 0.684 μm wavelength beam of a Q-switched frequency-doubled and
Raman-shifted Nd:YAG laser was focused on the wedge surface. The reflected
second-harmonic signal at 0.342 μm was measured with a photomultiplier. To
eliminate any spurious signal arising from reflections at the back surface of
the crystal, the back side of the samples was polished at an angle of 8°.
The second-harmonic radiation at 0.342 μm is well within the UV absorption
band of KNbO$_3$ so that the measured signal is generated in the layer close to
the surface and any second-harmonic contribution from the substrate is suppressed [6.123]. To obtain the depth profile along the x-direction of the optical
nonlinearity of the ion-implanted waveguide the sample was scanned along
the x'-direction, and the reflected second-harmonic power was measured as a
function of position. For a wedge angle of 1° and a focus diameter of 10 μm
of the fundamental beam on the sample surface, the depth resolution of the
index profile along x was 0.2 μm.

The depth profile was investigated after implantation, after annealing
at 180°C according to the procedure described in the previous paragraph
and after repoling the samples. The latter processing step was performed by
applying a pulsed external electric field along the polar c-axis of the crystal
of 2–3 kVcm^{-1} at a temperature of 180°C.

Figure 6.17b shows the reflected second-harmonic power at 0.342 μm versus depth measured on the as-implanted, annealed and repoled samples, respectively. The fundamental and second-harmonic beams were polarized normal to the plane of incidence so that the coefficient d_{33} was monitored. In the

(a)

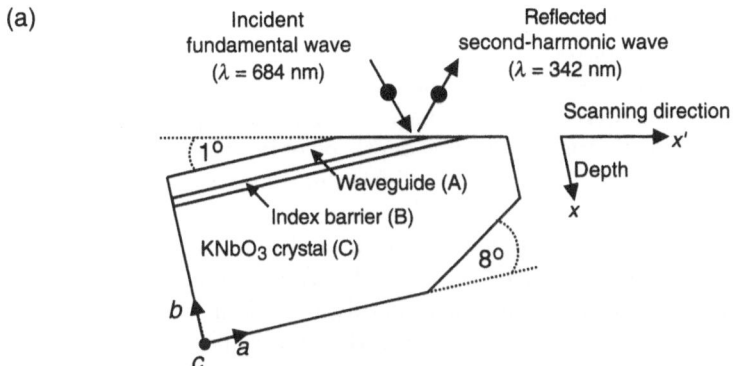

(b)

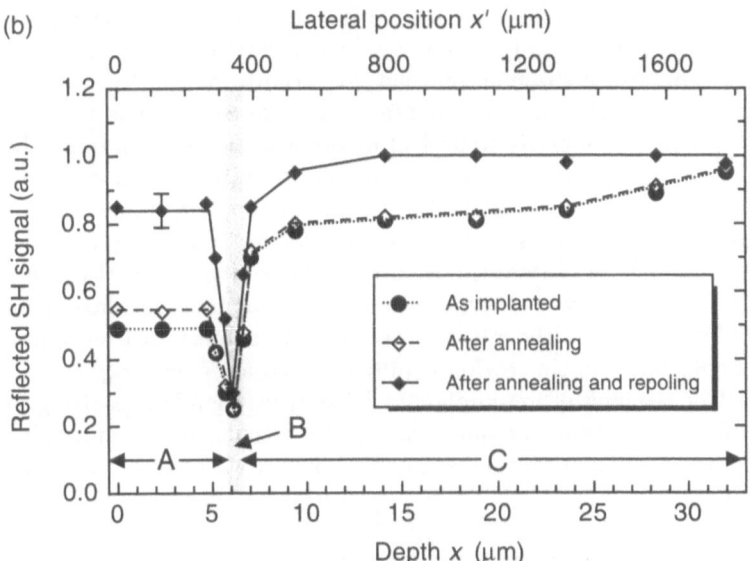

Fig. 6.17. (a) Schematic of the wedge-shaped polished waveguide used to measured the depth profile of the optical nonlinearity. The wedge of 1° on the top surface resolves the depth profile of the waveguide along x while scanning the sample along the x'-direction. The depth profile of the optical nonlinearity is measured in reflection with a fundamental beam at $0.684\,\mu m$ with polarization normal to the plane of incidence (parallel to the crystal c-axis). The angle of 8° eliminates spurious reflections from the bottom face. (b) Second-harmonic power as a function of depth measured by the surface reflectivity technique with a wedged polished crystal. The reflected second-harmonic power is normalized to the signal from the undamaged bulk crystal. The three curves represent the measurement after implantation $(\cdots\bullet\cdots)$, after annealing $(-\Diamond-)$ and after repoling $(-\blacklozenge-)$. Waveguide fabrication parameters: He$^+$ ion energy $E = 2.6\,MeV$ ($d = 6.2\,\mu m$), dose $D = 0.75 \times 10^{15}\,cm^{-2}$. A: waveguide, B: refractive index barrier, C: bulk KNbO$_3$ crystal (adapted from [6.285])

as-implanted sample the second-harmonic power in the waveguide region (A) dropped to about 50% of the second-harmonic power reflected from the non-irradiated $KNbO_3$ surface, and in the index barrier region (B) to about 25%. In the region immediately below the barrier the signal increased strongly; however, it reached its maximum at a depth of only about $30\,\mu m$ (region C). A similar depth profile was measured after annealing, the only difference being a slight increase in the second-harmonic power to about 55% in the waveguide region. The depth profile significantly changed after repoling the sample. In the waveguiding layer the second-harmonic power increased to about 85% of the bulk value, dropped sharply to 25% in the barrier, and increased immediately behind the barrier to the bulk value.

These results suggest the following interpretation [6.285]. In the waveguiding layer the irradiation reduces the nonlinear optical coefficient to 70% of its bulk value due to the microscopic depolarization caused by the electronic excitation. Because there is only a little permanent lattice damage after annealing, almost all of the niobium ions can be forced to realign by applying an external electric field along the spontaneous polarization. The repoling recovers the optical nonlinearity to 95% of its original value. In contrast, the strong lattice damage in the barrier region caused by nuclear interactions cannot be recovered by either thermal annealing or repoling. However, this is of little practical interest, since the optical fields are essentially confined in the waveguiding layer. Interestingly, even though of little practical significance, the region below the barrier seems to be partially depolarized to a depth of about $30\,\mu m$, although it is not directly subject to irradiation. Again, the nonlinearity in this region is fully restored after repoling.

The effect of repoling of ion-implanted $KNbO_3$ waveguides, resulting in almost complete recovery of the nonlinear optical coefficient, is comparable to the effect of annealing of proton-exchanged $LiNbO_3$ waveguides, as shown earlier in Table 6.2.

Effects of Annealing and Repoling on Second-Harmonic Generation. Figure 6.18 illustrates the effect of annealing and repoling on second-harmonic generation as measured in an ion-implanted $KNbO_3$ planar waveguide. The curves indicate the second-harmonic power generated with a fundamental input power of $300\,mW$ as a function of guide length for as-implanted, annealed and repoled waveguides, respectively. The curves were calculated on the basis of the measured decrease of the waveguide loss because of annealing and the measured improvement of the nonlinear coefficient because of repoling. The data points represent measurements in a $0.75\,cm$ long waveguide of $6.2\,\mu m$ thickness fabricated with an ion dose of $1.5 \times 10^{15}\,cm^{-2}$. A more than three-fold improvement of the second-harmonic conversion efficiency was observed after annealing and repoling compared with the as-implanted waveguide, illustrating the importance of post-implantation annealing and repoling for second-harmonic generation in these ion-implanted waveguides.

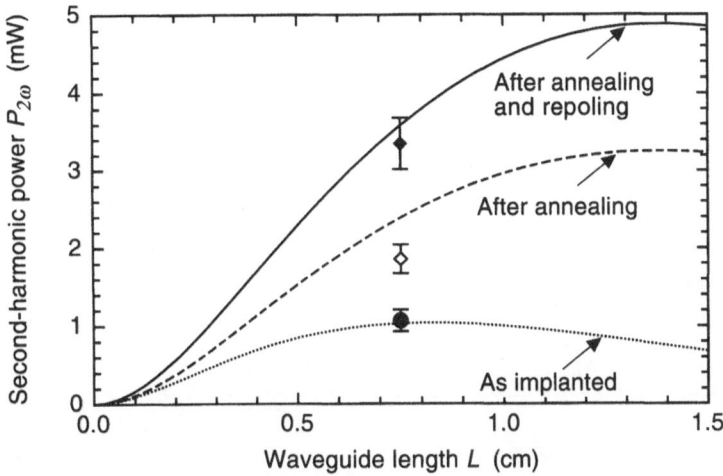

Fig. 6.18. Second-harmonic power as a function of guide length generated in a planar waveguide after implantation, annealing and repoling, respectively. Waveguide fabrication parameters: thickness $d = 6.2 \, \mu m$, dose $D = 1.5 \times 10^{15} \, cm^{-2}$, length $L = 0.75 \, cm$ (adapted from [6.284])

6.6.2 Linear Properties of Potassium Niobate Waveguides

A Propagation Constants

Precise knowledge of the mode propagation constants is a prerequisite for calculation of phase matching configurations in nonlinear optical waveguides. We have discussed the effects of ion implantation into KNbO$_3$ and the resulting refractive index profiles in Sect. 6.6.1. The effective mode indices of a planar waveguide can be calculated on the basis of the reconstructed index profile. Mathematical approaches to analyze the properties of modes of rectangular channel waveguides have generally approximated the two-dimensional problem by replacing it with two one-dimensional problems [6.286], or have attempted to be more precise with extensive computer calculations [6.287]. Marcatili's method [6.286] has been extended to strip-loaded waveguides [6.288] and to ridge waveguides [6.289] where the lateral surround of the channel guide supports guided modes and thus precludes the applications of Marcatili's original analysis. This extension has been achieved by replacing the refractive indices of the waveguide structure by effective mode indices. Therefore, this technique has been referred to as the effective index method by Kogelnik [6.290]. A detailed description of the effective index method and its application to Ti:LiNbO$_3$ ridge waveguides is given in [6.291]. We have used the effective index method to calculate the effective indices of ion-implanted KNbO$_3$ channel waveguides fabricated by the procedures shown in Fig. 6.14.

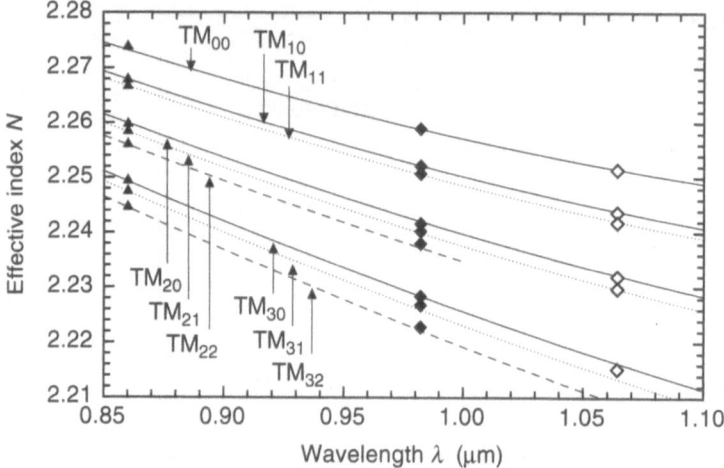

Fig. 6.19. Wavelength dispersion of the effective mode indices of a KNbO$_3$ ridge waveguide (TM modes with polarization parallel to the crystallographic b-axis). The lines indicate the calculated dispersion, and the symbols represent the values measured at 0.860 μm (▲), 0.982 μm (♦), and 1.064 μm (◇). Notice that the cut-off wavelength for the TM$_{22}$ mode is about 1 μm. Waveguide fabrication parameters: thickness $d = 5.3$ μm, width $w = 5.9$ μm, ridge height $h = 0.64$ μm, dose $D = 1.5 \times 10^{15}$ cm^{-2}. (Adapted from [6.297])

Figure 6.19 shows the comparison between calculated and measured wavelength dispersion of the TM mode indices (polarized along the b-axis) of a KNbO$_3$ ridge waveguide. All measured effective indices correspond within an accuracy of 8×10^{-4} to the calculated values, which confirms that the effective index method is suited to calculate the effective indices of these ridge waveguides.

B Propagation Losses

Loss Mechanisms. Losses in ion-implanted waveguides originate from three contributions: scattering, absorption and tunneling [6.61]. Scattering may occur either in the bulk of the waveguide or at the waveguide surface. Surface scattering is the loss due to interaction of the optical wave with the interface between the waveguide and its surround region. Its amount is determined by the roughness of the interface. Volume scattering is caused by material inhomogeneities. Such imperfections can originate from local index fluctuations, contaminant atoms and lattice defects within the bulk of the waveguide. In ion-implanted waveguides, scattering centers in the waveguide can be generated by collision or electronic interactions between the implant ions and the atoms of the crystal lattice. A small number of such point defects formed during the irradiation process are found in the bulk of the waveguiding layer, while the density of defects is highest in the barrier region formed at the end of the ion track.

Absorption in crystalline ferroelectric materials is due to contaminant atoms or vacancies present in the crystal lattice. This intrinsic material absorption is in most cases very small compared to the other contributions. However, in ion-implanted waveguides additional absorption loss can arise from defects generated by the implantation process.

Leakage or tunneling of light out of the waveguide is an inherent property of barrier waveguides where the optical confinement is provided by a layer of decreased refractive index of finite width. Such a structure is typical of ion-implanted waveguides where the implant ions create a damage layer of a few tenths of a micron thickness that separates the waveguiding layer form the substrate. The width of the barrier is determined by the straggling of the ions in the target material. For He^+ ion implantation in $KNbO_3$ the barrier width slightly increases with ion energy in the range between 1 and 3 MeV. Typically, the barrier width is 0.25 μm for a waveguide formed with a single-energy implantation.

The losses of ion-implanted $KNbO_3$ waveguides have been investigated in detail in [6.292]. A complex refractive index profile has been introduced where the real part is given by the reconstructed index profile, while the imaginary part is represented by a loss profile which is related to the lattice damage profile (Fig. 6.11). This procedure allows us to calculate numerically the mode propagation and attenuation constants simultaneously. The main contributions to the total propagation loss arise from tunneling and absorption by defect centers induced during irradiation. The calculation shows that tunneling and absorption can be approximated by power-law functions with respect to wavelength, ion dose and guide thickness. Table 6.6 gives a survey of the different sources of loss and summarizes their trends with increasing wavelength, ion dose and guide thickness. It is noteworthy that the dominating loss mechanisms, tunneling and absorption, follow opposite trends with respect to wavelength and ion dose. While absorption dominates at blue and green wavelengths, tunneling causes most of the attenuation in the near-infrared. Therefore, optimization of the loss characteristics with respect to second-harmonic generation with a near-infrared fundamental and a blue second-harmonic wavelength is a trade-off between tunneling and absorption which requires careful adjustment of the ion dose.

Loss Measurements. Several methods have been reported for measuring the propagation loss of waveguides in crystalline materials, such as calorimetric methods [6.293], pyroelectric measuring techniques [6.294] and Fabry–Perot resonance methods [6.295]. The latter provides a simple and accurate method to determine the propagation loss. However, it is applicable to single-mode waveguides only, and thus cannot be used with our multi-mode waveguides.

We have developed a method based on optimum end-fire coupling by phase conjugation of the out-coupled waveguide mode. This technique is de-

Table 6.6. Loss mechanisms in ion-implanted $KNbO_3$ waveguides and their dependence on wavelength and the guide fabrication parameters

Loss mechanism	Cause	Trend with increasing wavelength λ	Trend with increasing ion dose D	Trend with increasing waveguide thickness d
Tunneling through finite index barrier layer	Leakage of light through index barrier	increase $\sim \lambda^7$	decrease $\sim D^{-2}$	decrease $\sim d^{-4}$
Absorption at implantation induced point defects	Point defects in guiding layer and index barrier	decrease $\sim \lambda^{-3}$	increase $\sim D^3$	decrease $\sim d^{-2}$
Material absorption	Impurities, contaminant atoms	decrease $\sim \lambda^{-4}$	constant	constant
Surface scattering	Roughness of air–waveguide interface	constant	constant	decrease $\sim d^{-3}$

scribed in detail in [6.296] and was used to measure the propagation loss at blue and green wavelengths, i.e. at the laser lines of an Ar^+ ion laser between $0.454\,\mu m$ and $0.515\,\mu m$. In the red and near-infrared part of the spectrum the loss was derived from waveguide transmission measurements.

Waveguides were formed by ion implantation in $KNbO_3$ with losses below $0.2\,cm^{-1}$ ($1\,dB\ cm^{-1}$) at a specific wavelength [6.292]. Waveguides used for second-harmonic generation exhibit a higher overall loss because of the above-mentioned competing trends of absorption and tunneling with respect to wavelength and ion dose (Table 6.6). Figure 6.20 displays the wavelength dependence of the attenuation constant of a $KNbO_3$ ridge waveguide optimized for second-harmonic generation. The loss is a minimum in the red part of the spectrum ($\alpha = 0.5\,cm^{-1}$ at $0.633\,\mu m$), and increases towards the blue because of absorption ($\alpha = 1.3\,cm^{-1}$ at $0.454\,\mu m$) and towards the infrared because of tunneling ($\alpha = 0.8\,cm^{-1}$ at $1.064\,\mu m$).

6.6.3 Power-Handling Capabilities of Potassium Niobate Waveguides

A rough method to detect signs of optical damage is to observe the output near field and the output power as a function of increasing input power, usually referred to as power-handling capability measurement. At the onset of optical damage, waveguides tend to guide less power and the mode tends

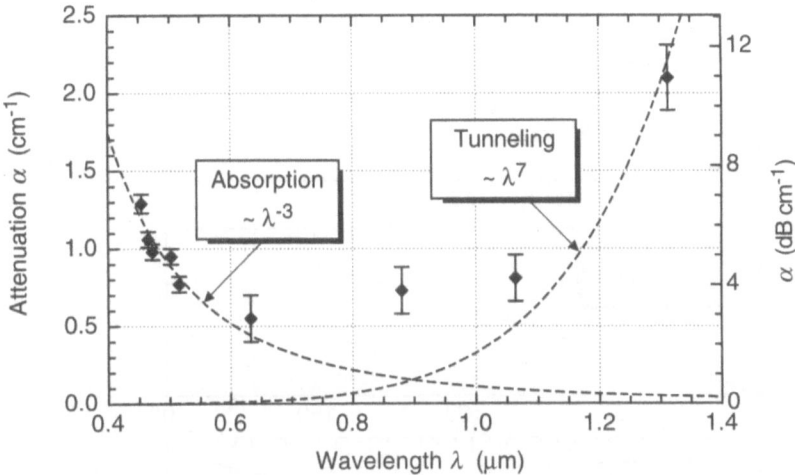

Fig. 6.20. Attenuation as a function of wavelength of a KNbO$_3$ ridge waveguide. The dashed lines indicate the calculated power-law dependence of absorption and tunneling, respectively. Waveguide fabrication parameters: thickness $d = 5.3\,\mu m$, width $w = 2.6\,\mu m$, ridge height $h = 0.91\,\mu m$, dose $D = 1.5 \times 10^{15}\,cm^{-2}$. (Adapted from [6.297])

to show distortions. Figure 6.21 shows the transmitted power as a function of in-coupled power in the TM$_{00}$ mode (polarized parallel to the b-axis) of a KNbO$_3$ ridge waveguide measured at the wavelengths of 457, 488 and 880 nm. The maximum intensities in the guide were 100 kW cm^{-2} at 457 nm, 540 kW cm^{-2} at 488 nm, and 1.1 MW cm^{-2} at 880 nm. The maximum in-coupled powers were limited by the available laser power at all three wavelengths. No sign of deviation from the expected linear behavior, indicated by fitted curves, was observed. Furthermore, no significant change of the mode profile occurred when changing the in-coupled power from 1 to 162 mW at a wavelength of 488 nm [6.297].

The very good power handling capability of KNbO$_3$ waveguides is mainly attributed to the relatively high dark conductivity of KNbO$_3$. Measurements of the conductivity of oxidic materials depend critically on various parameters such as growth conditions, impurity and defect density, and measurement conditions. Reported dark conductivity values may extend over several orders of magnitude for one material. For KNbO$_3$, a typical value is $10^{-12}\,\Omega^{-1}cm^{-1}$, while for LiTaO$_3$ and LiNbO$_3$, well known as optical storage materials, typically values of 10^{-18}–$10^{-19}\,\Omega^{-1}cm^{-1}$ are reported [6.298].

Similar power throughput measurements were carried out in annealed proton-exchanged LiNbO$_3$ waveguides where onset of optical damage was observed at about 5 mW at a wavelength of 633 nm [6.299], and in proton-exchanged LiTaO$_3$ where throughput degradation occurred at about 50 mW power at 860 nm [6.192]. The authors of [6.299] observed an improved input–

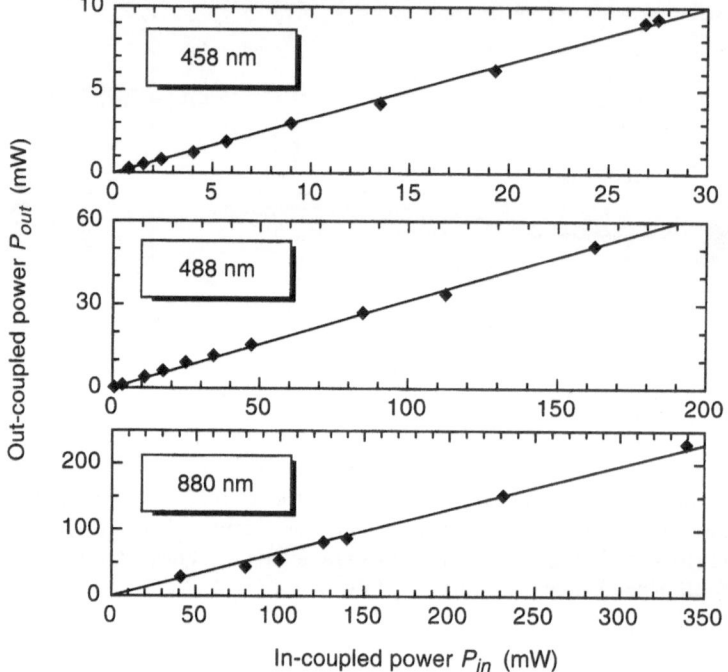

Fig. 6.21. Measured out-coupled power as a function of in-coupled power at the wavelengths of 457, 488 and 880 nm in a KNbO$_3$ ridge waveguide. The curves indicate fitted linear functions. (Adapted from [6.297])

output behavior of their annealed proton-exchanged LiNbO$_3$ waveguides after irradiation with 1 MeV protons. The power where optical damage started increased to 16 mW at 633 nm after ion beam irradiation, which was attributed to a decreased photovoltaic current.

6.6.4   Second-Harmonic Generation in Potassium Niobate Waveguides

A   Phase Matching Configurations

The refractive indices of KNbO$_3$ are given in [6.256], while phase matching configurations for second-harmonic generation in bulk KNbO$_3$ are discussed in detail in [6.257]. For the case with the beam propagating in a plane perpendicular to a principal axis, there are three possible situations:

  I Polarization of the second-harmonic wave along c, with the beam propagating in the ab-plane (d_{31} and d_{32} used).

IIA Polarization of the fundamental wave along a, with the beam propagating in the bc-plane (d_{31} used).

IIB Polarization of the fundamental wave along b, with the beam propagating in the ac-plane (d_{32} used).

The most suitable configuration for blue light second-harmonic generation is case I. For this particular geometry the nonlinear optical coefficient is always large. It varies from $d_{32} = 14.6\,\mathrm{pm}\ \mathrm{V}^{-1}$ at $0.86\,\mathrm{\mu m}$ to $d_{31} = 11.8\,\mathrm{pm}$ V^{-1} at $0.98\,\mathrm{\mu m}$ when taking the values listed in [6.50] and Miller's rule [6.44] into account. In addition, it is relatively easy to prepare a crystal for angle tuning in the ab-plane, since the spontaneous polarization is perpendicular to the top and bottom surface of the crystal, and thus the crystal can be poled rather easily.

To a first approximation, the configurations for birefringence phase matched second-harmonic generation in ion-implanted $KNbO_3$ waveguides are similar to those in $KNbO_3$ bulk crystals. In more detail, the phase matching wavelengths, temperatures and angles for waveguides are shifted with respect to the bulk crystals due to the modification of the material during guide fabrication. The ion implantation, i.e. the formation of the underlying planar waveguide, has a relatively strong impact on the phase matching parameters, whereas the formation of the lateral confinement, e.g. the etching of a ridge guide, modifies the phase matching configuration only slightly.

The dispersion of the effective indices in the visible and near-infrared part of the spectrum of an ion-implanted $KNbO_3$ waveguide is plotted in Fig. 6.3a as an example for birefringence phase matched second-harmonic generation. From Fig. 6.3a it is obvious that any fundamental wavelength between $0.876\,\mathrm{\mu m}$ and $1.014\,\mathrm{\mu m}$ can be phase matched if the fundamental wave has the polarization and propagation direction in the ab-plane so that a second-harmonic wave polarized along c is generated. The effective nonlinear coefficient is a combination of d_{31} and d_{32}. Figure 6.22 shows the phase matching angle φ_{pm}, defined as the angle between the propagation direction (waveguide axis), designated by β, and the a-axis (inset of Fig. 6.22), as a function of wavelength for the $\mathrm{TE}_{00}^{\omega} + \mathrm{TE}_{00}^{\omega} \to \mathrm{TM}_{00}^{2\omega}$ interaction at room temperature for case I. The phase matching angle φ_{pm} can be derived from the relation

$$\frac{1}{(N_c^{2\omega})^2} = \frac{\sin^2 \varphi_{\mathrm{pm}}}{(N_a^{\omega})^2} + \frac{\cos^2 \varphi_{\mathrm{pm}}}{(N_b^{\omega})^2} = \frac{1}{(N_{ab}^{\omega}(\varphi_{\mathrm{pm}}))^2}. \tag{6.37}$$

For an ion dose of $D = 1.5 \times 10^{15}\,\mathrm{cm}^{-2}$, typical for the formation of waveguides for second-harmonic generation, the curve for the $\mathrm{TE}_{00}^{\omega} + \mathrm{TE}_{00}^{\omega} \to \mathrm{TM}_{00}^{2\omega}$ interaction is shifted by about $20\,\mathrm{nm}$ with respect to the bulk crystal.

Second-harmonic generation is only efficient if the phase matching condition (6.27) is satisfied over a long interaction length. Variations of the effective indices along the guide axis resulting in birefringence nonuniformities will reduce the effective interaction length, and hence the second-harmonic power. We investigated the birefringence uniformity of ion-implanted $KNbO_3$ waveguides by means of collinear second-harmonic generation in combination with

Second-harmonic wavelength $\lambda_{2\omega}$ (µm)

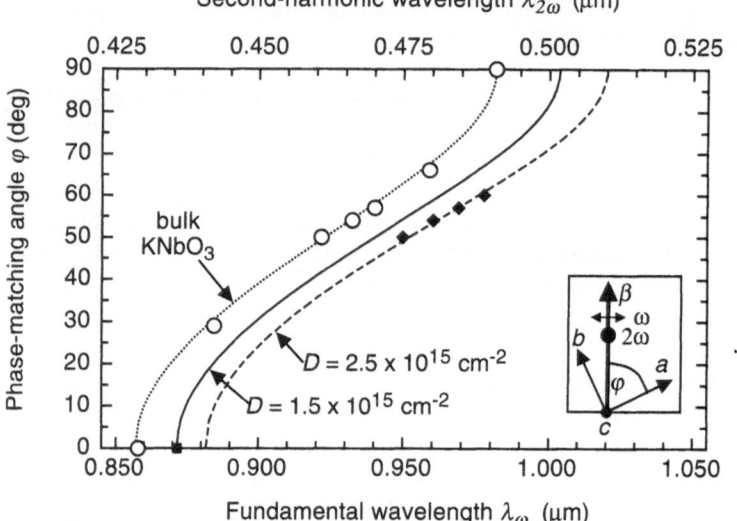

Fig. 6.22. Phase matching angle as a function of wavelength for case I second-harmonic generation in an ion-implanted KNbO$_3$ waveguide and bulk crystal at room temperature ($T = 23°$C). The dotted line ($\cdots\cdots\cdots$) represents the case for a bulk KNbO$_3$ crystal, while the full (———) and dashed (— — — —) lines show the dependence for ridge waveguides implanted with ion doses of 1.5×10^{15} cm^{-2} and 2.5×10^{15} cm^{-2}, respectively. The phase matching angle is calculated for the TE$_{00}^{\omega} \rightarrow$ TM$_{00}^{\omega}$ interaction. The inset displays the geometry with β denoting the direction of propagation, a, b and c, the crystal axes, and ω and 2ω the direction of polarization of the fundamental and second harmonic, respectively. The data points represent measurements in differently cut crystals. (Adapted from [6.297])

analysis of wavelength and temperature tuning curves of the second-harmonic signal and noncollinear second-harmonic generation in planar waveguides [6.300]. The latter technique, which is based on second-harmonic generation with two beams intersecting at an angle such that the second-harmonic signal is generated in a small region, allows local probing of the phase matching temperature or wavelength and hence detection of birefringence variations along the guide. From the comparison of the optical homogeneity measurements in waveguides and their substrate bulk crystals it was found that the birefringence uniformity is not or is only slightly affected by the implantation process.

B   Second-Harmonic Generation Measurements

Table 6.7 summarizes the results of second-harmonic generation in KNbO$_3$ waveguides. We listed the waveguide type according to our classification of Sect. 6.6.1, the guide length, the phase matching scheme, the laser source used, the second-harmonic wavelength, the fundamental in- and outcoupled

Table 6.7. Second-harmonic generation in KNbO₃ waveguides

Year Ref.	Waveguide type	L (cm)	Phase matching technique	Laser[a]	$\lambda_{2\omega}$ (nm)	$P_\omega(0)$ (mW)	$P_\omega(L)$ (mW)	$P_{2\omega}(L)$ (mW)	$\bar\eta$[b] (%W^{-1})
1992 [6.265]	planar	0.84	birefringent	Ti:Al₂O₃ (pulsed)	434	1.3e6 (peak)	0.8e6 (peak)	0.39e6 (peak)	0.02 (peak)
1992 [6.41]	embedded channel	0.56	Cerenkov	Ti:Al₂O₃ DL (50 mW)	430 430	266 35	97 13	1.1 0.02	1.56 1.63
1993 [6.340]	planar	0.56	birefringent	Ti:Al₂O₃	433	452	165	2.4	1.2
1993 [6.340]	embedded channel	0.56	Cerenkov	Ti:Al₂O₃	428	482	176	7.1	3.1
1994 [6.301]	embedded channel	1.12	Cerenkov	Ti:Al₂O₃	432	471	196	12	5.4
1994 [6.42]	embedded channel	1.12	Cerenkov	DL (150 mW)	430	110	50	0.5	4.2
1995 [6.284]	planar	0.75	birefringent	Ti:Al₂O₃	436	400	320	1.73	1.7
1996 [6.36]	embedded channel	1.12	Cerenkov	Ti:Al₂O₃	430	270	110	4.1	5.6
1996 [6.266]	embedded barrier	0.58	birefringent	Ti:Al₂O₃	441	280	190	2.6	3.3
1998 [6.280]	ridge	0.73	birefringent	Ti:Al₂O₃	438	340	204	14	12.1

[a] Values in brackets indicate the laser output power.
[b] Defined by (6.24): $\bar\eta = \frac{P_{2\omega}(L)}{[P_\omega(0)]^2}$

Laser types
DL   AlGaAs diode laser
Ti:Al₂O₃   titanium:sapphire laser

power, and the normalized conversion efficiency according to the definition given by (6.24).

The first observation of birefringence phase matched second-harmonic generation in a $KNbO_3$ waveguide was reported for a planar waveguide with a pulsed Ti:sapphire laser [6.265]. In embedded channel waveguides (Fig. 6.14a) second-harmonic generation was found to occur in the Cerenkov-type scheme [6.36, 6.41]. Birefringence phase matched second-harmonic generation with both modes guided was demonstrated for the first time in embedded barrier waveguides (Fig. 6.14b) [6.266]. The highest conversion efficiency in a $KNbO_3$ waveguide reported up to now was for a ridge waveguide (Fig. 6.14c) [6.280].

Embedded Channel Waveguides. In embedded channel waveguides, fabricated by the multiple-energy ion implantation procedure shown in Fig. 6.14a, second-harmonic generation occurred in the Cerenkov configuration where the fundamental wave is guided while the second harmonic is radiated into the substrate (Fig. 6.8). Results were reported in [6.41], and frequency doubling of diode lasers in these waveguides was discussed in [6.42]. A theoretical discussion and experimental results are given in [6.36].

A maximum second-harmonic power of 12 mW of blue power at 432 nm with 471 mW of fundamental power from a Ti:sapphire laser coupled into a 1.12 cm long waveguide was generated. This corresponds to a normalized conversion efficiency of 5.4 %W^{-1} [6.301]. The loss of the fundamental in the waveguide was 0.8 cm^{-1}, and of the second-harmonic in the substrate about 0.15 cm^{-1}. Using a single-mode AlGaAs diode laser with 150 mW output power, 0.5 mW of blue power at 430 nm was obtained [6.42]. Stabilization of the diode laser emission was provided by an external cavity configuration, as shown in Fig. 6.23. The Cerenkov radiation angle was about 5°. Although the Cerenkov scheme is less efficient, it is attractive because it offers the advantage of a larger wavelength and temperature acceptance bandwidth.

Embedded Barrier Waveguides. Embedded barrier waveguides were prepared by the single-energy ion implantation process shown in Fig. 6.14b [6.266]. Birefringence phase matched second-harmonic generation with both waves confined in the waveguide was observed.

The highest conversion efficiency was measured in a waveguide that was fabricated with an ion dose of 3×10^{15} cm^{-2}. The guide thickness and width were 4.5 and 16 μm, respectively. A blue output power of 2.6 mW at 441 nm was measured with a fundamental input power of 280 mW in this 0.58 cm long waveguide, yielding a normalized conversion efficiency of 3.3 %W^{-1}.

Ridge Waveguides. Among the $KNbO_3$ waveguides discussed in this section, ridge waveguides (Fig. 6.14c) were found to provide the highest second-harmonic figure of merit [6.280]. Details of the linear and nonlinear optical properties of these waveguides were recently discussed in [6.297]. Figure 6.24 displays the second-harmonic power as a function of the fundamental input

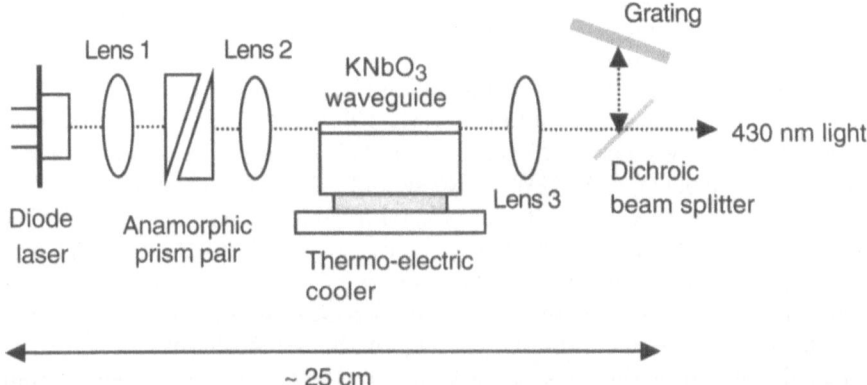

Fig. 6.23. Setup for frequency doubling of an AlGaAS diode laser in a KNbO$_3$ channel waveguide. The output beam of the diode laser is collimated with lens 1 and circularized with an anamorphic prism pair. Lens 2 focuses the light into the waveguide which is mounted on a thermo-electric cooler. Lens 3 is used for collimating the outcoupled light. The fundamental is split off with a dichroic mirror onto a grating and reflected back into the diode laser to provide feedback for stabilization of the emission wavelength. (Adapted from [6.42])

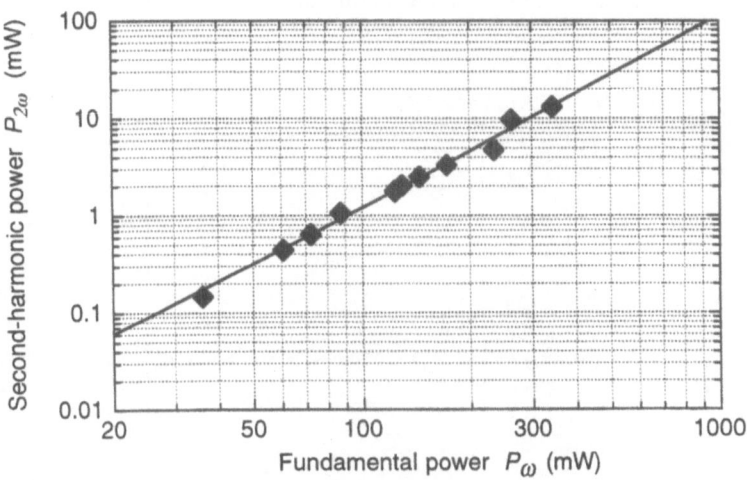

Fig. 6.24. Second-harmonic power at $0.438\,\mu$m as a function of the fundamental power coupled into a ridge waveguide. The fundamental input power was derived by measuring the laser power after the coupling lens and multiplying this power by the Fresnel transmission coefficient of 0.85 and the launch efficiency of 0.8. (Adapted from [6.280])

power measured in a ridge waveguide fabricated with the following parameters: thickness $d = 5.3\,\mu m$, width $w = 2.6\,\mu m$, ridge height $h = 0.9\,\mu m$ and ion dose $D = 1.5 \times 10^{15}\,cm^{-2}$. A maximum blue power of $14\,mW$ at $0.438\,\mu m$ at room temperature was obtained with $340\,mW$ fundamental power coupled into the waveguide of $0.73\,cm$ length. This corresponds to a normalized conversion efficiency of $12\ \%W^{-1}$. No sign of roll-off from the quadratic dependence was observed. The losses at the fundamental and second-harmonic wavelength were $0.7\,cm^{-1}$ ($3\,dB\ cm^{-1}$) and $1.3\,cm^{-1}$ ($6\,dB\ cm^{-1}$), respectively (Fig. 6.20). To investigate the stability of the blue output the second-harmonic power was monitored as a function of time for 6 hours, shown in Fig. 6.25. It was found that the average of the second-harmonic power remained stable over the whole period and no degradation of the blue output was observed.

The good ability to control the guide fabrication process, the excellent power handling capability, the high second-harmonic power of up to $14\,mW$ and the stability of the second-harmonic output demonstrate the suitability of $KNbO_3$ waveguides for efficient frequency conversion. On the basis of our estimates, further improvement in waveguide fabrication technology, focusing on waveguides with lower loss and better modal confinement, should allow the generation of second-harmonic power of 15 to $25\,mW$ with an in-coupled power of $200\,mW$ and a normalized conversion efficiency of about $50\ \%W^{-1}$ [6.297].

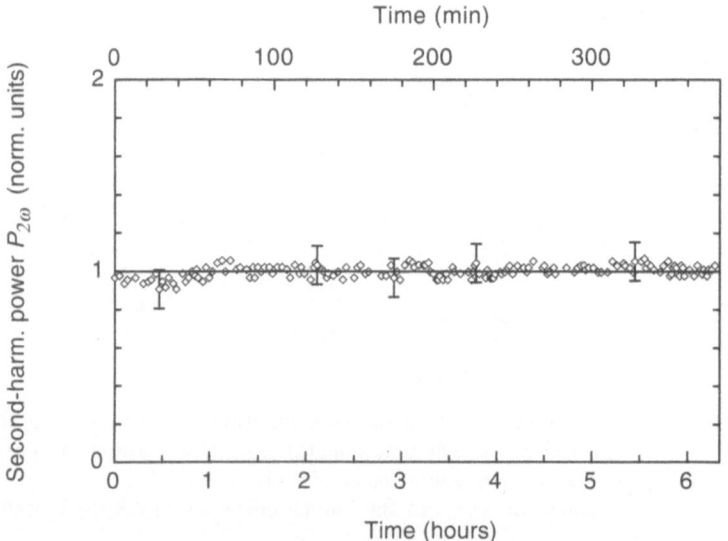

Fig. 6.25. Normalized second-harmonic output power from a $KNbO_3$ ridge waveguide as a function of time, measured for more than 6 hours. The line indicates the average second-harmonic power. The blue power was $7\,mW$ with $240\,mW$ of fundamental power coupled into the waveguide. (Adapted from [6.280])

6.7   Discussion and Concluding Remarks

Tables 6.3, 6.4, 6.5 and 6.7 provide the basis for a comparison of second-harmonic generation in $LiNbO_3$, $LiTaO_3$, $KTiOPO_4$ and $KNbO_3$ waveguides. Generally it is evident that there has been a trend towards higher figures of merit, reflecting the fact that frequency doubling in ferroelectric waveguides has become an increasingly mature technique over the past two decades. Blue lasers based on frequency doubling in waveguides are ideal to span the 1 to 100 mW range for applications requiring compact, efficient, low-noise and long-lived lasers. However, despite the fact that many of the reported figures of merit would be sufficient for practical devices, very few attempts to implement ferroelectric waveguides into commercial products have been made so far. There are several issues that have still to be addressed for large-scale implementation of ferroelectric waveguides into devices: noise and long-term power fluctuations of the blue light, temporal stability of the waveguide structure, reproducibility of waveguide fabrication, availability of suitable diode lasers with sufficient power and good beam quality, efficient coupling of the diode laser output to the waveguide, packaging of the diode laser and waveguide with sufficient robustness, and cost of production.

Future research activities in the field of waveguide frequency doubling can be expected to concentrate on the generation of shorter ultraviolet wavelengths below 0.4 µm. Research in this area can be expected to focus on the fabrication of waveguides in ultraviolet-transparent materials, e.g. borates, and on investigation of nonlinear optical effects in these waveguides.

Finally, we shall briefly mention cascaded second-order effects in ferroelectric waveguides which have recently attracted some attention. Cascaded second-order nonlinearities, e.g. the combined process of second-harmonic generation ($\omega \rightarrow 2\omega$) and subsequent difference frequency generation between the second-harmonic and fundamental waves ($2\omega \rightarrow \omega$), are closely related to second-harmonic generation as the material figure of merit for cascading is identical to that for second-harmonic generation, given by the ratio d^2/n^3. As discussed in detail in Chap. 2, cascaded second-order effects mimic third-order (Kerr) nonlinearities, and hence can be used in all-optical switching devices. Reviews of cascading phenomena were given by Bosshard [6.302] and Stegeman et al. [6.303]. Schiek et al. reported on interferometric measurements of nonlinear phase shifts due to cascaded nonlinearities in Ti-indiffused $LiNbO_3$ waveguides [6.304]. Recently, Schiek et al. also investigated birefringent phase matched second-harmonic generation in Ti-indiffused $LiNbO_3$ waveguides at a fundamental wavelength of 1.32 µm with respect to cascaded second-order nonlinear effects [6.305]. All-optical switching by cascading using directional waveguide couplers and Mach–Zehnder interferometers in Ti-indiffused $LiNbO_3$ waveguides was reported in [6.306–308]. One-dimensional spatial solitary waves due to cascaded second-order nonlinearities were observed in Ti-indiffused planar $LiNbO_3$ waveguides [6.309]. Large nonlinear phase modulation as a result of cascading was reported for quasi-phase-matched $KTiOPO_4$ waveguides by Sundheimer et al. [6.310]. Cascaded effects and spatial solitary waves were also observed in a number of experiments performed in bulk ferroelectrics, such as periodically poled $LiNbO_3$ [6.311–312], $KTiOPO_4$ [6.313] and $KNbO_3$ [6.314].

References

6.1 G.I. Stegeman, in *Proceedings of NATO Advanced Study Institute on Guided Wave Nonlinear Optics*, Cargese, France, 12–24 August, (eds.) Ostrowsky, D.B., Reinisch, R. (Kluwer Academic, Dordrecht 1992) p. 11.

6.2 Haase, M.A., Qiu, J., Puydt, J.M., Cheng, H., Appl. Phys. Lett. **59**, 1272 (1991).

6.3 Nakamura, S., Senoh, M., Nagahama, S., Iwasa, N., Yamada, T., Matsushita, T., Kiyoku, H., Sugimoto, Y., Jpn. J. Appl. Phys. **35**, L74 (1996).

6.4 Sohler, W., in *New directions in guided wave and coherent optics* (eds.) Ostrowsky, D.B., Spitz, E., (Martinus Nijhoff, The Hague 1984).

6.5 Stegeman, G.I., Seaton, C.T., J. Appl. Phys. **58**, R57 (1985).

6.6 Stegeman, G.I., Stolen, R.H., J. Opt. Soc. Am. B **6**, 652 (1989).

6.7 Yariv, A., IEEE J. Quantum Electron. **9**, 919 (1973).

6.8 Ducuing, J., Bloembergen, N., Phys. Rev. A **133**, A1493 (1964).

6.9 Wirges, W., Yilmaz, S., Brinker, W., Bauer-Gogonea, S., Bauer, S., Jäger, M., Stegeman, G.I., Ahlheim, M., Stählin, M., Zysset, B., Lehr, F., Diemeer, M., Flipse, M.C., Appl. Phys. Lett. **70**, 3347 (1997).

6.10 Jäger, M., Stegeman, G.I., Flipse, M.C., Diemeer, M., Möhlmann, G., Appl. Phys. Lett. **69**, 4139 (1996).

6.11 Kowalczyk, T.C., Singer, K.D., Cahill, P.A., Opt. Lett. **20**, 2273 (1995).

6.12 Armstrong, J.A., Bloembergen, N., Ducuing, J., Pershan, P.S., Phys. Rev. **127**, 1918 (1962).

6.13 Franken, P.A., Ward, J.F., Rev. Mod. Phys. **35**, 23 (1963).

6.14 Miller, R.C., Phys. Rev. **134**, A1313 (1964).

6.15 Bloembergen, N., Sievers, A.J., Appl. Phys. Lett. **17**, 483 (1970).

6.16 Somekh, S., Yariv, A., Opt. Commun. **6**, 301 (1972).

6.17 Helmfrid, S., Arvidsson, G., J. Opt. Soc. Am. B **8**, 797 (1991).

6.18 Fejer, M.M., Magel, G.A., Jundt, D.H., Byer, R.L., IEEE J. Quantum Electron. **28**, 2631 (1992).

6.19 Thompson, D.E., McMullen, J.D., Anderson, D.B., Appl. Phys. Lett. **29**, 113 (1976).

6.20 Szilagyi, A., Hordvik, A., Schlossberg, H., J. Appl. Phys. **47**, 2025 (1976).

6.21 Feng, D., Ming, N.-B., Hong, J.-F., Yang, Y.-S., Zhu, J.S., Yang, Z., Wang, Y.-N., Appl. Phys. Lett. **37**, 607 (1980).

6.22 Feisst, A., Koidl, P., Appl. Phys. Lett. **47**, 1125 (1985).

6.23 Jundt, D.H., Magel, G.A., Fejer, M.M., Byer, R.L., Appl. Phys. Lett. **59**, 2657 (1991).

6.24 Magel, G.A., Fejer, M.M., Byer, R.L., Appl. Phys. Lett. **56**, 108 (1990).

6.25 Luh, Y.S., Fejer, M.M., Byer, R.L., Feigelson, R.S., J. Crystal Growth **85**, 264 (1987).

6.26 Van der Ziel, J.P., Ilegems, M., Foy, P.W., Mikulyak, R.M., Appl. Phys. Lett. **29**, 775 (1976).

6.27 Lim, E.J., Fejer, M.M., Byer, R.L., Electron. Lett. **25**, 174 (1989).
6.28 Tien, P.K., Ulrich, R., Martin, R.J., Appl. Phys. Lett. **17**, 447 (1970).
6.29 Li, M.J., De Micheli, M., He, Q., Ostrowsky, D.B., IEEE J. Quantum Electron. **26**, 1384 (1990).
6.30 Hayata, K., Koshiba, M., Electron. Lett. **25**, 376 (1989).
6.31 Chikuma, K., Umegaki, S., J. Opt. Soc. Am. B **7**, 768 (1990).
6.32 Sanford, N.A., Connors, J.M., J. Appl. Phys. **65**, 1429 (1989).
6.33 Hayata, K., Sugawara, T., Koshiba, M., IEEE J. Quantum Electron. **26**, 123 (1990).
6.34 Tamada, H., IEEE J. Quantum Electron. **26**, 1821 (1990).
6.35 Suhara, T., Morimoto, T., Nishihara, H., IEEE J. Quantum Electron. **29**, 525 (1993).
6.36 Fluck, D., Pliska, T., Günter, P., Beckers, L., Buchal, C., IEEE J. Quantum Electron. **32**, 905 (1996).
6.37 Krijnen, G.J.M., Hoekstra H.J.W.M., Stegeman, G.I., Torruellas, W., Opt. Lett. **21**, 851 (1996).
6.38 Tatsuno, K., Yanagisawa, H., Andou, T., McLoughlin, M., Appl. Opt. **31**, 305 (1992).
6.39 Yamamoto, K., Yamamoto, H., Taniuchi, T., Appl. Phys. Lett. **58**, 1227 (1991).
6.40 Doumuki, T., Tamada, H., Saitoh, M., Appl. Phys. Lett. **64**, 3533 (1994).
6.41 Fluck, D., Moll, J., Günter, P., Fleuster, M., Buchal, Ch., Electron. Lett. **28**, 1092 (1992).
6.42 Fluck, D., Pliska, T., Günter, P., Fleuster, M., Buchal, Ch., Rytz, D., Electron. Lett. **30**, 1937 (1994).
6.43 Bosshard, C., Sutter, K., Prêtre, Ph., Hulliger, Flörsheimer, M., Kaatz, P., Günter, P., Organic Nonlinear Optical Materials (Gordon & Breach, London 1995).
6.44 Miller, R.C., Appl. Phys. Lett. **5**, 17 (1964).
6.45 DiDomenico Jr, M., Wemple, S.H., Appl. Phys. Lett. **12**, 352 (1968).
6.46 Kurtz, S.K., Perry, T.T., J. Appl. Phys. **39**, 3798 (1968).
6.47 DiDomenico Jr, M., Wemple, S.H., J. Appl. Phys. **40**, 720 (1969).
6.48 DiDomenico Jr, M., Wemple, S.H., J. Appl. Phys. **40**, 735 (1969).
6.49 Jerphagnon, J., Phys. Rev. B **2**, 1091 (1970).
6.50 Roberts, D.A., IEEE J. Quantum Electron. **28**, 2057 (1992).
6.51 Shoji, I., Kondo, T., Kitamoto, A., Shirane, M., Ito, R., J. Opt. Soc. Am. B **14**, 2268 (1997).
6.52 Prokhorov, A.M., Kuz'minov, Y.S., Khachaturyan, O.A., *Ferroelectric Thin-film Waveguides in Integrated Optics and Optoelectronics* (Cambridge International Science Publishing, Cambridge, 1996).
6.53 Wei, D.T.Y., Lee, W.W., Bloom, L. R., Appl. Phys. Lett. **25**, 329 (1974).
6.54 Kaminow, I.P., Carruthers, J.R., Appl. Phys. Lett. **22**, 326 (1973).
6.55 Schmidt, R.V., Kaminov, I.P., Appl. Phys. Lett. **25**, 458 (1974).
6.56 Noda, J., Saku, T., Uchida, N., Appl. Phys. Lett. **25**, 308 (1974).
6.57 Jackel, J.L., Rice, C.E., Veselka, J.J., Appl. Phys. Lett. **41**, 607 (1982).
6.58 Spillman, J., W.B., Sanford, N.A., Soref, R.A., Opt. Lett. **8**, 497 (1983).
6.59 Bierlein, J.D., Feretti, A., Brixner, L.H., Hsu, W.Y., Appl. Phys. Lett. **50**, 1216 (1987).
6.60 Buchal, C., Withrow, S.P., White, C.W., Poker, D.B., Annual Review of Materials Science **24**, 125 (1994).

6.61 Townsend, P.D., Chandler, P.J., Zhang, L., *Optical Effects of Ion Implantation* (Cambridge University Press, Cambridge 1994).

6.62 Raeuber, A., in *Current Topics in Materials Science* (ed.) Kaldis, E. (North–Holland, Amsterdam 1978), vol. 1

6.63 Sanford, N.A., (ed.) *Feature issue on recent advances in lithium niobate optical technology*, IEEE J. Quantum Electron. **33**, 1625 (1997).

6.64 Carruthers, J.R., Kaminov, I.P., Stulz, L.W., Appl. Opt. **13**, 2333 (1974).

6.65 Destefanis, G.L., Gailliard, J.P., Ligeon, E.L., Valette, S., Farmery, B.W., Townsend, P.D., Perez, A., J. Appl. Phys. **50**, 7898 (1979).

6.66 Wenzlik, K., Heibei, J., Voges, E., Phys. Stat. Sol. (a) **61**, K207 (1980).

6.67 Karge, H., Goetz, G., Jahn, U., Schmidt, S., Nucl. Instrum. Meth. **182/183**, 777 (1981).

6.68 Götz, G., Karge, H., Nucl. Instrum. Meth. **209/210**, 1079 (1983).

6.69 Naden, J.M., Weiss, B.L., J. Lightwave Technol. **LT-3**, 855 (1985).

6.70 Reed, G.T., Weiss, B.L., Electron. Lett. **23**, 792 (1987).

6.71 Townsend, P.D., Nucl. Instrum. Meth. **B46**, 18 (1990).

6.72 Vollmer, J., Nisius, J.P., Hertel, P., Krätzig, E., Appl. Phys. A **32**, 125 (1983).

6.73 Fukuma, M., Noda, J., Iwasaki, H., J. Appl. Phys. **49**, 3693 (1978).

6.74 Minakata, M., Saito, S., Shibata, M., Miyazawa, S., J. Appl. Phys. **49**, 4677 (1978).

6.75 Minakata, M., Saito, S., Shibata, M., J. Appl. Phys. **50**, 3063 (1979).

6.76 Burns, W.K., Klein, P.H., West, E.J., Plew, L.E., J. Appl. Phys. **50**, 6175 (1979).

6.77 Griffiths, G.J., Esdaile, R.J., IEEE J. Quantum Electron. **20**, 149 (1984).

6.78 Ctyroky, J., Hofman, M., Janta, J., Schröfel, J., IEEE J. Quantum Electron. **20**, 400 (1984).

6.79 Fouchet, S., Carenco, A., Daguet, C., Guglielmi, R., Riviere, L., J. Lightwave Technol. **5**, 700 (1987).

6.80 Sjöberg, A., Arvidsson, G., Lipovskii, A.A., J. Opt. Soc. Am. B **5**, 285 (1988).

6.81 Kip, D., Gather, B., Bendig, H., Krätzig, E., Phys. Stat. Sol. (a) **139**, 241 (1993).

6.82 Caccavale, F., Gonella, F., Quaranta, A., Mansour, I., Electron. Lett. **31**, 1054 (1995).

6.83 Twigg, M.E., Maher, D.M., Nakahra, S., Sheng, T.T., Homes, R.J., Appl. Phys. Lett. **50**, 501 (1987).

6.84 Strake, E., Bava, G.P., Montrosset, I., J. Lightwave Technol. **6**, 1126 (1988).

6.85 Irrera, F., Valli, M., J. Appl. Phys. **64**, 1704 (1988).

6.86 Fukuma, M., Noda, J., Appl. Opt. **19**, 591 (1980).

6.87 Shah, M.L., Appl. Phys. Lett. **26**, 652 (1975).

6.88 Jackel, J.L., Rice, C.E., Appl. Phys. Lett. **41**, 508 (1982).

6.89 De Micheli, M., Botineau, J., Neveu, S., Sibillot, P., Ostrowsky, D.B., Papuchon, M., Opt. Lett. **8**, 114 (1983).

6.90 Clark, D.F., Nutt, A.C.G., Wong, K.K., Laybourn, P.J.R., De La Rue, R.M., J. Appl. Phys. **54**, 6218 (1983).

6.91 Vohra, S.T., Mickelson, A.R., Asher, S.E., J. Appl. Phys. **66**, 5161 (1989).

6.92 Bortz, M.L., Fejer, M.M., Opt. Lett. **16**, 1844 (1991).

6.93 Cao, X.F., Ramaswamy, R.V., Srivastava, R., J. Lightwave Technol. **10**, 1302 (1992).

6.94 Goto, N., Yip, G.L., Appl. Opt. **28**, 60 (1989).

6.95 Yamamoto, K., Taniuchi, T., J. Appl. Phys. **70**, 6663 (1991).
6.96 Pun, E.Y., Tse, Y.O., Chung, P.S., IEEE Photon. Technol. Lett. **3**, 522 (1991).
6.97 Pun, E.Y., Zhao, S.A., Loi, K.K., Chung, P.S., IEEE Photon. Technol. Lett. **3**, 1006 (1991).
6.98 Shen, C.-Y., Wang, S.T., Jpn. J. Appl. Phys. **35**, L1333 (1996).
6.99 Veng, T., Skettrup, T., Appl. Opt. **36**, 5941 (1997).
6.100 Korkishko, Y.N., Fedorov, V.A., De Micheli, M., Baldi, P., El Hadi, K., Leycuras, A., Appl. Opt. **35**, 7056 (1996).
6.101 Masalkar, P.J., Fujimura, M., Suhara, T., Nishihara, H., Electron. Lett. **33**, 519 (1997).
6.102 Rams, J., Olivares, J., Cabrera, J.M., Appl. Phys. Lett. **70**, 2076 (1997).
6.103 Jackel, J., Glass, A.M., Peterson, G.E., Rice C.E., Olson, D.H., Veselka, J.J., J. Appl. Phys. **55**, 269 (1984).
6.104 Rice, C.E., Jackel, J.L., Mat. Res. Bull. **19**, 591 (1984).
6.105 Rice, C.E., J. Solid State Chem. **64**, 188 (1986).
6.106 Yi-Yan, A., Appl. Phys. Lett. **42**, 633 (1983).
6.107 Suchoski, P.G., Findakly, T.K., Leonberger, F.J., Opt. Lett. **13**, 1050 (1988).
6.108 Minakata, M., Kumagai, K., Kawakami, S., Appl. Phys. Lett. **49**, 992 (1986).
6.109 Canali, C., Carnera, A., Della Mea, G., Mazzoldi, P., Al Shukri, S.M., Nutt, A.C.G., De La Rue, R.M., J. Appl. Phys. **59**, 2643 (1986).
6.110 Lee, W.E., Sanford, N.A., Heuer, A.H., J. Appl. Phys. **59**, 2629 (1986).
6.111 De Micheli, M., Ostrowsky, D.B., Barety, J.P., Canali, C., Carnera, A., Mazzi, G., Papuchon, P., J. Lightwave Technol. **LT-4**, 743 (1986).
6.112 Loni, A., Hay, G., De La Rue, R.M., Winfield, J.M., J. Lightwave Technol. **7**, 911 (1989).
6.113 Paz-Pujalt, G.R., Tuschel, D.D., Appl. Phys. Lett. **62**, 3411 (1993).
6.114 Paz-Pujalt, G.R., Tuschel, D.D., Braunstein, G., Blanton, T., Lee, S.T., Salter, L.M., J. Appl. Phys. **76**, 3981 (1994).
6.115 Johnston Jr, W.D., Phys. Rev. B **1**, 3494 (1970).
6.116 Korkishko, Y.N., Fedorov, V.A., IEEE J. Selected Topics in Quantum Electronics **2**, 187 (1996).
6.117 Bloembergen, N., Pershan, P.S., Phys. Rev. **128**, 606 (1962).
6.118 Becker, R.A., Appl. Phys. Lett. **43**, 131 (1983).
6.119 Suhara, T., Tazaki, H., Nishihara, H., Electron. Lett. **25**, 1326 (1989).
6.120 Keys, R.W., Loni, A., De La Rue, R.M., Electron. Lett. **26**, 624 (1990).
6.121 Bortz, M.L., Fejer, M.M., Opt. Lett. **17**, 704 (1992).
6.122 Cao, X., Srivastava, R., Ramaswamy, R.V., Natour, J., IEEE Photon. Technol. Lett. **3**, 25 (1991).
6.123 Laurell, F., Roelofs, M.G., Hsiung H., Appl. Phys. Lett. **60**, 301 (1992).
6.124 Hsu, W.-Y., Willand, C.S., Gopalan, V., Gupta, M.C., Appl. Phys. Lett. **61**, 2263 (1992).
6.125 Bortz, M.L., Eyres, L.A., Fejer, M.M., Appl. Phys. Lett. **62**, 2012 (1993).
6.126 Veng, T., Skettrup, T., Pedersen, K., Appl. Phys. Lett. **69**, 2333 (1996).
6.127 Rams, J., Agullo-Rueda, F., Cabrera, J.M., Appl. Phys. Lett. **71**, 3356 (1997).
6.128 Ohnishi, N., Jpn. J. Appl. Phys. **16**, 1069 (1977).
6.129 Nakamura, K., Ando, H., Shimizu, H., Appl. Phys. Lett. **50**, 1413 (1987).
6.130 Miyazawa, S., J. Appl. Phys. **50**, 4599 (1979).

6.131 Webjörn, J., Laurell, F., Arvidsson, G., J. Lightwave Technol. **7**, 1597 (1989).
6.132 Fujimura, M., Suhara, T., Nishihara, H., Electron. Lett. **27**, 1207 (1991).
6.133 Haycock, P.W., Townsend, P.D., Appl. Phys. Lett. **48**, 698 (1986).
6.134 Keys, R.W., Loni, A., De La Rue, R.M., Ironside, C.N., Marsh, J.H., Luff, B.J., Townsend, P.D., Electron. Lett. **26**, 188 (1990).
6.135 Yamada, M., Kishima, K., Electron. Lett. **27**, 828 (1991).
6.136 Ito, H., Takyu, C., Inaba, H., Electron. Lett. **27**, 1221 (1991).
6.137 Nutt, A.C.G., Gopalan, V., Gupta, M.C., Appl. Phys. Lett. **60**, 2828 (1992).
6.138 Makio, S., Nitanada, F., Ito, K., Sato, M., Appl. Phys. Lett. **61**, 3077 (1992).
6.139 Yamada, M., Nada, N., Saitoh, M., Watanbe, K., Appl. Phys. Lett. **62**, 435 (1993).
6.140 Nassau, K., Levinstein, H.J., Loiacono, G.M., J. Phys. Chem. Solids **27**, 983 (1966).
6.141 Webjörn, J., J. Lightwave Technol. **11**, 589 (1993).
6.142 Ohnishi, N., Takashi, I., J. Appl. Phys. **46**, 1063 (1974).
6.143 Laurell, F., Webjörn, J., Arvidsson G., Holmberg, J., IEEE J. Lightwave Technol. **10**, 1606 (1992).
6.144 Houé, M., Townsend, P.D., J. Phys. D: Appl. Phys. **28**, 1747 (1995).
6.145 Thaniyavarn, S., Findakly, T., Booher, D., Moen, J., Appl. Phys. Lett. **46**, 933 (1985).
6.146 Peuzin, J.C., Appl. Phys. Lett. **48**, 1104 (1986).
6.147 Huang, L., Jaeger, N.A.F., Appl. Phys. Lett. **65**, 1763 (1994).
6.148 Helmfrid, S., Arvidsson, G., Webjörn, J., J. Opt. Soc. Am. B **10**, 222 (1992).
6.149 Webjörn, J., Siala, S., Nam, D.W., Waarts, R.G., Lang, R.J., IEEE J. Quantum Electron. **33**, 1673 (1997).
6.150 Fujimura, M., Kintaka, K., Suhara, T., Nishihara, H., J. Lightwave Technol. **11**, 1360 (1993).
6.151 Webjörn, J., Pruneri, V., Russell, P.St.J., Barr, J.R.M., Hanna, D.C., Electron. Lett. **30**, 894 (1994).
6.152 Li, W.H., Tavlykaev, R., Ramaswamy, R.V., Samson, S., Appl. Phys. Lett. **68**, 1470 (1996).
6.153 Sonoda, S., Tsuruma, I., Hatori, M., Appl. Phys. Lett. **70**, 3078 (1997).
6.154 Mizuuchi, K., Yamamoto, K., Kato, M., Electron. Lett. **33**, 806 (1997).
6.155 Sonoda, S., Tsuruma, I., Hatori, M., Appl. Phys. Lett. **71**, 3048 (1997).
6.156 Ashkin, A., Boyd, G.D., Dziedzic, J.M., Smith, R.G., Ballman, A.A., Levinstein, J.J., Nassau, K., Appl. Phys. Lett. **9**, 72 (1966).
6.157 Fujiwara, T., Cao, X., Srivastava, R., Ramaswamy, R.V., Appl. Phys. Lett. **61**, 743 (1992).
6.158 Bryan, D.A., Gerson, R., Tomaschke, H.E., Appl. Phys. Lett. **44**, 847 (1984).
6.159 Volk, T.R., Pryalkin, V.I., Rubinina, N.M., Opt. Lett. **15**, 996 (1990).
6.160 Volk, T., Rubinina, N., Woehlecke, M., J. Opt. Soc. Am. B **11**, 1681 (1994).
6.161 Young, W.M., Feigelson, R.S., Fejer, M.M., Digonnet, M.J.F., Shaw, H.J., Opt. Lett. **16**, 995 (1991).
6.162 Schirmer, O.F., Thiemann, O., Woehlecke M., J. Phys. Chem. Solids **52**, 185 (1991).
6.163 Malovichko, G.I., Grachev, V.G., Schirmer, O.F., Sol. State Commun. **89**, 195 (1994).
6.164 Steinberg, S., Göring, R., Hennig, T., Rasch, A., Opt. Lett. **20**, 683 (1995).
6.165 Taya, M., Bashaw, M.C., Fejer, M.M., Opt. Lett. **21**, 857 (1996).

6.166 Eger, D., Arbore, M.A., Fejer, M.M., Bortz, M.L., J. Appl. Phys. **82**, 998 (1997).

6.167 Laurell, F., Arvidsson, G., J. Opt. Soc. Am. B **5**, 292 (1988).

6.168 Fejer, M.M., Digonnet, M.J.F., Byer, R.L., Opt. Lett. **11**, 230 (1986).

6.169 Uesugi, N., Kimura, T., Appl. Phys. Lett. **29**, 572 (1976).

6.170 Regener, R., Sohler, W., J. Opt. Soc. Am. B **5**, 267 (1988).

6.171 Sohler, W., Hampel, B., Regener, R., Ricken, R., Suche, H., Volk, R., J. Lightwave Technol. **LT-4**, 772 (1986).

6.172 Lim, E.J., Matsumoto, S., Fejer, M.M., Appl. Phys. Lett. **57**, 2294 (1990).

6.173 Ishigame, Y., Suhara, T., Nishihara, H., Opt. Lett. **16**, 375 (1991).

6.174 Cao, X., Rose, B., Ramaswamy, R.V., Srivastava, R., Opt. Lett. **17**, 795 (1992).

6.175 Bortz, M.L., Field, S.J., Fejer, M.M., Nam, D.W., Waarts, R.G., Welch, D.F., IEEE Trans. Quantum Electron. **TQE-30**, 2953 (1994).

6.176 Mizuuchi, K., Ohta, H., Yamamoto, K., Kato, M., Opt. Lett **22**, 1217 (1997).

6.177 Kitaoka, Y., Yokoyama, T., Mizuuchi, K., Yamamoto, K., Kato, M., Electron. Lett. **33**, 1638 (1997).

6.178 Kintaka, K., Fujimura, M., Suhara, T., Nishihara, H., J. Lightwave Technol. **14**, 462 (1996).

6.179 Hammer, J.M., Philips, W., Appl. Phys. Lett. **24**, 545 (1974).

6.180 Tangonan, G.L., Barnoski, M.K., Lotspeich, J.F., Lee, A., Appl. Phys. Lett. **30**, 238 (1977).

6.181 Eknoyan, O., Burns, W.K., Moeller, R.P., Frigo, N.J., Appl. Opt. **27**, 114 (1988).

6.182 Eknoyan, O., Yoon, D.W., Taylor, H.F., Appl. Phys. Lett. **51**, 384 (1987).

6.183 Jackel, J.L., Appl. Opt. **19**, 1996 (1980).

6.184 Tada, K., Murain, T., Nakabayashi, T., Iwashima, T., Ishikawa, T., Jpn. J. Appl. Phys. **26**, 503 (1987).

6.185 Findakly, T., Suchoski, P., Leonberger, F., Opt. Lett. **13**, 797 (1988).

6.186 Mizuuchi, K., Yamamoto, K., Taniuchi, T., Appl. Phys. Lett. **58**, 2732 (1991).

6.187 Davis, G.M., Lindop, N.A., J. Appl. Phys. **77**, 6121 (1995).

6.188 Kan, D., Yip, G.L., Appl. Opt. **35**, 5348 (1996).

6.189 Mizuuchi, K., Yamamoto, K., J. Appl. Phys. **72**, 5061 (1992).

6.190 Mizuuchi, K., Yamamoto, K., Appl. Opt. **33**, 1812 (1994).

6.191 Yuhara, T., Tada, K., Li, Y.-S., J. Appl. Phys. **71**, 3966 (1992).

6.192 Matthews, P.J., Mickelson, A.R., Novak, S.W., J. Appl. Phys. **72**, 2562 (1992).

6.193 Åhlfeldt, H., Webjörn, J., Laurell, F., Arvidsson, G., J. Appl. Phys. **75**, 717 (1994).

6.194 Ganshin, Y.A., Korkishko, Yu. N., Phys. Stat. Sol. (a) **119**, 11 (1990).

6.195 Fedorov, V.A., Korkishko, Yu. N., Ferroelectrics **160**, 185 (1994).

6.196 El Hadi, K., Baldi, P., Nouh, S., De Micheli, M.P., Leycuras, A., Fedorov, V.A., Korkishko, Y.N., Opt. Lett. **20**, 1698 (1995).

6.197 Matthews, P.J., Mickelson, A.R., J. Appl. Phys. **71**, 5310 (1992).

6.198 Howerton, M.M., Burns, W.K., Photorefractive effects in proton-exchanged LiTaO$_3$ waveguides, in *Integrated Photonics Research*, Monterey, CA, 9–11 April, 1991 (IEEE, OSA, 1991).

6.199 Nakamura, K., Shimizu, H., Appl. Phys. Lett. **56**, 1535 (1990).

6.200 Mizuuchi, K., Yamamoto, K., Taniuchi, T., Appl. Phys. Lett. **59**, 1538 (1991).

6.201 Mizuuchi, K., Yamamoto, K., Sato, H., J. Appl. Phys. **75**, 1311 (1994).
6.202 Nakamura, K., Hosoya, M., Tourlog, A., J. Appl. Phys. **73**, 1390 (1993).
6.203 Mizuuchi, K., Yamamoto, K., Sato, H., Appl. Phys. Lett. **62**, 1860 (1993).
6.204 Yi, S.-Y., Shin, S.-Y., Jin, Y.-S., Son, Y.-S., Appl. Phys. Lett. **68**, 2493 (1996).
6.205 Matsumoto, S., Lim, E.J., Hertz, H.M., Fejer, M.M., Electron. Lett. **27**, 2040 (1991).
6.206 Hsu, W.-Y., Gupta, M.C., Appl. Phys. Lett. **60**, 1 (1992).
6.207 Mizuuchi, K., Yamamoto, K., Appl. Phys. Lett. **66**, 2943 (1995).
6.208 Mizuuchi, K., Yamamoto, K., Opt. Lett. **21**, 107 (1996).
6.209 Li, Y.-S., Tada, K., Murai, T., Yuhara, T., Jpn. J. Appl. Phys. **28**, L263 (1989).
6.210 Åhlfeldt, H., Laurell, F., Arvidsson, G., Electron. Lett. **29**, 819 (1993).
6.211 Åhlfeldt, H., J. Appl. Phys. **76**, 3254 (1994).
6.212 Yamamoto, K., Mizuuchi, K., Kitaoka, Y., Kato, M., Appl. Phys. Lett. **62**, 2599 (1993).
6.213 Zumsteg, F.C., Bierlein, J.D., Gier, T.E., J. Appl. Phys. **47**, 4980 (1976).
6.214 Scheidt, M., Beier, B., Knappe, R., Boller, K.-J., Wallenstein, R., J. Opt. Soc. Am. B **12**, 2087 (1995).
6.215 Bierlein, J.D., Vanherzeele, H., J. Opt. Soc. Am. B **6**, 622 (1989).
6.216 Vanherzeele, H., Bierlein, J.D., Opt. Lett. **17**, 982 (1992).
6.217 Boulanger, B., Fève, J.P., Marnier, G., Ménaert, B., Cabriol, X., Villeval, P., Bonnin, C., J. Opt. Soc. Am. B **11**, 750 (1994).
6.218 Morris, P.A., Feretti, A., Bierlein, J.D., Loiacono, G.M., J. Crystal Growth **109**, 367 (1991).
6.219 Risk, W.P., Appl. Phys. Lett. **58**, 19 (1991).
6.220 Roelofs, M.G., Morris, P.A., Bierlein, J.D., J. Appl. Phys. **70**, 720 (1991).
6.221 Daneshvar, K., Giess, E.A., Bacon, A.M., Dawes, D.G., Gea, L.A., Boatner, L.A., Appl. Phys. Lett. **71**, 756 (1997).
6.222 Tuschel, D.D., Paz-Pujalt, G.R., Risk, W.P., Appl. Phys. Lett. **66**, 1035 (1995).
6.223 Loiacono, G.M., Loiacono, D.N., McGee, T., Babb, M., J. Appl. Phys. **72**, 2705 (1992).
6.224 Roelofs, M.G., J. Appl. Phys. **65**, 4976 (1989).
6.225 Jongerius, M.J., Bolt, R.J., Sweep, N.A., J. Appl. Phys. **75**, 3316 (1994).
6.226 Kato, K., IEEE J. Quantum Electron. **24**, 3 (1988).
6.227 Risk, W.P., Payne, R.N., Lenth, W., Harder, C., Meier, H., Appl. Phys. Lett. **55**, 1179 (1989).
6.228 Bierlein, J.D., Laubacher D.B., Brown, J.B., Van der Poel, C.J., Appl. Phys. Lett. **56**, 1725 (1990).
6.229 Van der Poel, C.J., Bierlein, J.D., Brown, J.B., Colak, S., Appl. Phys. Lett. **57**, 2074 (1990).
6.230 Laurell, F., Roelofs, M.G., Bindloss, W., Hsiung, H., Suna, A., Bierlein, J.D., J. Appl. Phys. **71**, 4664 (1992).
6.231 Eger, D., Oron, M., Katz, M., J. Appl. Phys. **74**, 4298 (1993).
6.232 Roelofs, M.G., Suna, A., Bindloss, W., Bierlein, J.D., J. Appl. Phys. **76**, 4999 (1994).
6.233 Chen, Q., Risk, W.P., Electron. Lett. **32**, 107 (1996).
6.234 Gupta, M.C., Risk, W.P., Nutt, A.C.G., Lau, S.D., Appl. Phys. Lett. **63**, 1167 (1993).

6.235 Chen, Q., Risk, W.P., Electron. Lett. **30**, 1516 (1994).
6.236 Karlsson, H., Laurell, F., Appl. Phys. Lett. **71**, 3474 (1997).
6.237 Shinozaki, K., Fukunaga, T., Watanabe, K., Kamijoh, T., Appl. Phys. Lett. **59**, 510 (1991).
6.238 Risk, W.P., Lau, S.D., Opt. Lett. **18**, 272 (1993).
6.239 Risk, W.P., Lau, S.D., McCord, M.A., IEEE Photon. Technol. Lett. **6**, 406 (1994).
6.240 Eger, D., Oron, M., Katz, M., Zussman, A., J. Appl. Phys. **77**, 2205 (1995).
6.241 Eger, D., Oron, M., Katz, M., Zussman, A., Appl. Phys. Lett. **64**, 3208 (1994).
6.242 Eger, D., Oron, M., Katz, M., Reizman, A., Rosenman, G., Skliar, A., Electron. Lett. **33**, 1548 (1997).
6.243 Doumuki, T., Tamada, H., Saitoh, M., Appl. Phys. Lett. **65**, 2519 (1994).
6.244 Wiesendanger, E., Ferroelectrics **1**, 141 (1970).
6.245 Wiesendanger, E., Czech. J. Phys. B **23**, 91 (1973).
6.246 Wiesendanger, E., Ferroelectrics **6**, 263 (1974).
6.247 Uematsu, Y., Jpn. J. Appl. Phys. **12**, 1257 (1973).
6.248 Uematsu, Y., Jpn. J. Appl. Phys. **13**, 1362 (1974).
6.249 Günter, P., Opt. Commun. **11**, 285 (1974).
6.250 Baumert, J.-C., Hoffnagle, J., Günter, P., Proc. SPIE **492**, 374 (1984).
6.251 Zgonik, M., Schlesser, R., Biaggio, I., Voit, E., Tscherry, J., Günter, P., J. Appl. Phys. **74**, 1287 (1993).
6.252 Uematsu, Y., Fukuda, T., Jpn. J. Appl. Phys. **4**, 507 (1971).
6.253 Uematsu, Y., Fukuda, T., Jpn. J. Appl. Phys. **12**, 841 (1973).
6.254 Günter, P., Appl. Phys. Lett. **34**, 650 (1979).
6.255 Günter, P., Asbeck, P.M., Kurtz S.K., Appl. Phys. Lett. **35**, 461 (1979).
6.256 Zysset, B., Biaggio, I., Günter, P., J. Opt. Soc. Am. B **9**, 380 (1992).
6.257 Biaggio, I., Kerkoc, P., Wu, L.-S., Günter, P., Zysset, B., J. Opt. Soc. Am. B **9**, 507 (1992).
6.258 Baumert, J.-C., Walther, C., Buchmann, P., Kaufmann, H., Melchior, H., Günter, P., Appl. Phys. Lett. **46**, 1018 (1985).
6.259 Bremer, T., Heiland, W., Hellermann, B., Hertel, P., Krätzig, E., Kollewe, D., Ferroelectrics Lett. **9**, 11 (1988).
6.260 Zhang, L., Chandler, P.J., Townsend, P.D., Ferroelectrics Lett. **11**, 89 (1990).
6.261 Moretti, P., Thevenard, P., Wirl, K., Hertel, P., Hesse, H., Krätzig, E., Godefroy, G., Ferroelectrics **128**, 13 (1992).
6.262 Strohkendl, F.P., Günter, P., Buchal, Ch., Irmscher, R., J. Appl. Phys. **69**, 84 (1991).
6.263 Fluck, D., Günter, P., Irmscher, R., Buchal, Ch., Appl. Phys. Lett. **59**, 3213 (1991).
6.264 Fluck, D., Günter, P., Fleuster, M., Buchal, Ch., J. Appl. Phys. **72**, 1671 (1992).
6.265 Fluck, D., Binder, B., Küpfer, M., Looser, H., Buchal, Ch., Günter, P., Opt. Commun. **90**, 304 (1992).
6.266 Fluck, D., Pliska, T., Günter, P., Bauer, St., Beckers L., Buchal, Ch., Appl. Phys. Lett. **69**, 4133 (1996).
6.267 Schwyn Thöny, S., Günter, P., Lehmann, H.W., Appl. Phys. Lett. **61**, 373 (1992).
6.268 Chow, A.F., Lichtenwalner, D.J., Woolcott Jr, R.R., Graettinger, T.M., Auciello, O., Kingon, A.I., Appl. Phys. Lett. **65**, 1073 (1994).

524   T. Pliska, D. Fluck, and P. Günter

6.269 Zaldo, C., Gill, D.S., Eason, R.W., Mendiola, J., Chandler, P.J., Appl. Phys. Lett. **65**, 502 (1994).

6.270 Beckers, L., Zander, W., Schubert, J., Leinenbach, P., Buchal, Ch., Fluck, D., Günter, P., Mater. Res. Soc. Symp. Proc. **441**, 549 (1997).

6.271 Beckers, L., Schubert, J., Zander, W., Ziesmann, J., Eckau, A., Leinenbach, P., Buchal, Ch., J. Appl. Phys. **83**, 3305 (1998).

6.272 Irmscher, R., Fluck, D., Buchal, Ch., Stritzker, B., Günter, P., Mater. Res. Soc. Symp. Proc. **201**, 399 (1991).

6.273 Biersack, J.P., Haggmark, L.G., Nucl. Instrum. Meth. **174**, 257 (1980).

6.274 Fluck, D., Jundt, D.H., Günter, P., Fleuster, M., Buchal, Ch., J. Appl. Phys. **74**, 6023 (1993).

6.275 Solcia, C., Fluck, D., Pliska, T., Günter, P., Bauer, St., Fleuster, M., Beckers, L., Buchal, Ch., Opt. Commun. **120**, 39 (1995).

6.276 Pliska, T., Solcia, C., Fluck, D., Günter, P., Beckers, L., Buchal, Ch., J. Appl. Phys. **81**, 1099 (1997).

6.277 Fluck, D., Irmscher, R., Buchal, Ch., Günter, P., Ferroelectrics **128**, 79 (1992).

6.278 Zhang, L., Chandler, P.J., Townsend, P.D., Nucl. Instrum. Meth. **B59/60**, 1147 (1991).

6.279 Mazzoldi, P., Arnold, G.W., *Ion Beam Modification of Insulators* (Elsevier, Amsterdam 1987).

6.280 Pliska, T., Fluck, D., Günter, P., Gini, E., Melchior, H., Beckers, L., Buchal, Ch., Appl. Phys. Lett. **72**, 2364 (1998).

6.281 Glavas, E., Zhang, L., Chandler, P.J., Townsend, P.D., Nucl. Instrum. Meth. **B32**, 45 (1988).

6.282 Wong, J.Y.C., Zhang, L., Kakarantzas, G., Townsend, P.D., Chandler, P.J., Boatner, L.A., J. Appl. Phys. **71**, 49 (1992).

6.283 Zhang, L., Chandler, P.J., Townsend, P.D., Appl. Phys. Lett. **53**, 544 (1988).

6.284 Pliska, T., Jundt, D.H., Fluck, D., Günter, P., Rytz, D., Fleuster, M., Buchal, Ch., J. Appl. Phys. **77**, 6114 (1995).

6.285 Fluck, D., Pliska, T., Küpfer, M., Günter, P., Appl. Phys. Lett. **67**, 748 (1995).

6.286 Marcatili, E.A.J., Bell Syst. Tech. J. **48**, 2071 (1969).

6.287 Goell, J.E., Bell Syst. Tech. J. **48**, 2133 (1969).

6.288 Ramaswamy, V., Bell Syst. Tech. J. **53**, 697 (1974).

6.289 Kaminow, I.P., Ramaswamy, V., Schmidt, R.V., Turner, E.H., Appl. Phys. Lett. **24**, 622 (1974).

6.290 Tamir, T. (ed.), *Integrated Optics* (Springer-Verlag, Berlin, Heidelberg, New York, 1975)

6.291 Tamir, T. (ed.), *Guided-Wave Optoelectronics* (Springer-Verlag, Berlin, 1990)

6.292 Pliska, T., Fluck, D., Günter, P., Beckers, L., Buchal, Ch., J. Opt. Soc. Am. B **15**, 628 (1998).

6.293 Jackel, J.L., Veselka, J.J., Appl. Opt. **23**, 197 (1984).

6.294 Haegele, K.H., Ulrich, R., Opt. Lett. **4**, 60 (1979).

6.295 Regener, R., Sohler, W., Appl. Phys. B **36**, 143 (1985).

6.296 Brülisauer, S., Fluck, D., Solcia, C., Pliska, T., Günter, P., Opt. Lett. **20**, 1773 (1995).

6.297 Pliska, T., Fluck, D., Günter, P., Beckers, L., Buchal, C., J. Appl. Phys. **84**, 1186 (1998).

6.298 Günter, P., Huignard, J.P. (eds.), *Photorefractive Materials and Their Applications I* (Springer-Verlag, Berlin, 1988)

6.299 Robertson, E.E., Eason, R.W., Yokoo, Y., Chandler, P.J., Appl. Phys. Lett. **70**, 2094 (1997).

6.300 Pliska, T., Mayer, F., Fluck, D., Günter, P., Rytz, D., J. Opt. Soc. Am. B **12**, 1878 (1995).

6.301 Fluck, D., Zha, M., Günter, P., Fleuster, M., Buchal, Ch., Ferroelectrics **151**, 205 (1994).

6.302 Bosshard, C., Adv. Mater. **8**, 385 (1996).

6.303 Stegeman, G.I., Hagan D.J., Torner, L., Optical and Quantum Electronics **28**, 1691 (1996).

6.304 Schiek, R., Sundheimer, M.L., Kim, D.Y., Beak, Y., Stegeman, G.I., Seibert, H., Sohler, W., Opt. Lett. **19**, 1949 (1994).

6.305 Schiek, R., Baek, Y., Stegeman, G., J. Opt. Soc. Am. B **15**, 2255 (1998).

6.306 Baek, Y., Schiek, R., Stegeman, G.I., Krijnen, G., Baumann, I., Sohler, W., Appl. Phys. Lett. **68**, 2055 (1996).

6.307 Baek, Y., Schiek, R., Stegeman, G.I., Opt. Lett. **20**, 2168 (1995).

6.308 Schiek, R., Baek, Y., Krijnen, G., Stegeman, G.I., Baumann, I., Sohler, W., Opt. Lett. **21**, 940 (1996).

6.309 Schiek, R., Baek, Y., Stegeman, G.I., Phys. Rev. E **53**, 1138 (1996).

6.310 Sundheimer, M.L., Bosshard, Ch., Van Stryland, E.W., Stegeman, G.I., Bierlein, J.D., Opt. Lett. **18**, 1397 (1993).

6.311 Asobe, M., Yokohama, I., Itoh, H., Kaino, T., Opt. Lett. **22**, 274 (1997).

6.312 Vidakovic, P., Lovering, D.J., Levenson, J.A., Webjoern, J., Russell, P.St.J., Opt. Lett. **22**, 277 (1997).

6.313 Torruellas, W.E., Assanto, G., Lawrence, B., Fuerst, R.A., Stegeman, G.I., Appl. Phys. Lett. **68**, 1449 (1996).

6.314 Bosshard, C., Spreiter, R., Zgonik, M., Günter, P., Phys. Rev. Lett. **74**, 2816 (1995).

6.315 Sohler, W., Suche, H., Appl. Phys. Lett. **33**, 518 (1978).

6.316 Arvidsson, G., Laurell, F., Thin Solid Films **136**, 29 (1986).

6.317 Lim, E.J., Fejer, M.M., Byer, R.L., Kozlovsky, W.J., Electron. Lett. **25**, 731 (1989).

6.318 Webjörn, J., Laurell, F., Arvidsson, G., IEEE Photon. Technol. Lett. **1**, 316 (1989).

6.319 Yamamoto, K., Mizuuchi, K., Takeshige, K., J. Appl. Phys. **70**, 1947 (1991).

6.320 Armani, F., Delacourt, D., Lallier, E., Papuchon, M., He, Q., De Micheli, M., Ostrowsky, D.B., Electron. Lett. **28**, 139 (1992).

6.321 Fujimura, M., Suhara, T., Nishihara, H., Electron. Lett. **28**, 721 (1992).

6.322 Cao, X., Srivastava, R., Ramaswamy, R.V., Opt. Lett. **17**, 592 (1992).

6.323 Cao, X., Srivastava, R., Ramaswamy, R.V., Appl. Phys. Lett. **60**, 3280 (1992).

6.324 Fujimura, M., Kintaka, K., Suhara, T., Nishihara, H., Electron. Lett. **28**, 1868 (1992).

6.325 Kintaka, K., Fujimura, M., Suhara, T., Nishihara, H., Electron. Lett. **32**, 2237 (1996).

6.326 Amin, J., Pruneri, V., Webjörn, J., Russell, P.St.J., Hanna, D.C., Wilkinson, J.S., Opt. Commun. **135**, 41 (1997).

6.327 Åhlfeldt, H., Webjörn, J., Arvidsson, G., IEEE Photon. Technol. Lett. **3**, 638 (1991).

6.328 Mizuuchi, K., Yamamoto, K., Appl. Phys. Lett. **60**, 1283 (1992).

6.329 Yamamoto, K., Mizuuchi, K., IEEE Photon. Technol. Lett. **4**, 435 (1992).

6.330 Yamamoto, K., Mizuuchi, K., Taniuchi, T., IEEE J. Quantum Electron. **28**, 1909 (1992).

6.331 Gupta, M.C., Kozlovsky, W., Nutt, A.C.G., Appl. Phys. Lett. **64**, 3210 (1994).

6.332 Yamamoto, K., Mizuuchi, K., Kitaoka, Y., Kato, M., Opt. Lett. **20**, 273 (1995).

6.333 Mizuuchi, K., Kitaoka, Y., Yamamoto, K., Electron. Lett. **31**, 727 (1995).

6.334 Leo, G., Drenten, R.R., Jongerius, M.J., IEEE J. Quantum Electron. **28**, 534 (1992).

6.335 Laurell, F., Electron. Lett. **29**, 1629 (1993).

6.336 Risk, W.P., Lau, S.D., Fontana, R., Lane, L., Nadler, Ch., Appl. Phys. Lett. **63**, 1301 (1993).

6.337 Risk, W.P., Kozlovsky, W.J., Lau, S.D., Bona, G.L., Jaeckel, H., Webb, D.J., Appl. Phys. Lett. **63**, 3134 (1993).

6.338 Yamamoto, Y., Yamaguchi, S., Suzuki, K., Yamada, N., Appl. Phys. Lett. **65**, 938 (1994).

6.339 Risk, W.P., Lau, S.D., Appl. Phys. Lett. **69**, 3999 (1996).

6.340 D. Fluck, M. Fleuster, Ch. Buchal, P. Günter, Conference on Lasers and Electro-Optics, Baltimore (Optical Society of America, 1993) p. 268.

Subject Index

Springer Series in Optical Sciences

Editorial Board: A. L. Schawlow† A. E. Siegman T. Tamir